Frontiers in Fusion Research II

Mitsuru Kikuchi • Masafumi Azumi

Frontiers in Fusion Research II

Introduction to Modern Tokamak Physics

Springer

Mitsuru Kikuchi
Japan Atomic Energy Agency
Naka, Ibaraki, Japan

Masafumi Azumi
AZM Techno Science Soft
Setagaya, Tokyo, Japan

ISBN 978-3-319-18904-8 ISBN 978-3-319-18905-5 (eBook)
DOI 10.1007/978-3-319-18905-5

Library of Congress Control Number: 2011921304

Springer Cham Heidelberg New York Dordrecht London

Printed on acid-free paper

Springer International Publishing AG Switzerland is part of Springer Science+Business Media (www.springer.com)

Preface

The purpose of this monograph is to provide an introduction to modern tokamak physics, as an extension of our review paper "Steady State Tokamak Research—Core Physics" published in Reviews of Modern Physics (2012). This monograph is for graduate students and young researchers in the field of controlled fusion research, while the previous book **Frontier in Fusion Research I—Physics and Fusion** (2011) was intended for more general physics students who may be interested in fusion plasma physics. This monograph is an introduction of modern tokamak physics, and we show some derivations of key equations but some are left for your investigation of the original papers. We introduce Tokamak fusion reactor concepts such as the SSTR (Steady State Tokamak Reactor) (see Fig. 1) in Chap. 1 and some technological background in Chap. 9. Other chapters are devoted to the description of the physics of the Tokamak confinement.

Chapter 2 introduces the magnetic field structure in tokamak, tokamak equilibria, and 1.5D transport equations. Chapter 3 introduces the tokamak operating regimes and the long pulse operation. Chapter 4 introduces the transport process due to Coulomb collision. Chapter 5 gives an introduction to key collective motions in tokamak. Chapter 6 gives fundamentals of ballooning transform and the WKBJ theory of 2D eigenmode problem. Chapter 7 gives an introduction of turbulent transport in tokamak and briefly the plasma confinement scaling. Chapter 8 includes ideal, resistive, and kinetic MHD stabilities. In this book, the divertor and the plasma–wall interaction are omitted except a small introduction in Chap. 3.

Mito, Japan
Tokyo, Japan
March 2015

Mitsuru Kikuchi
Masafumi Azumi

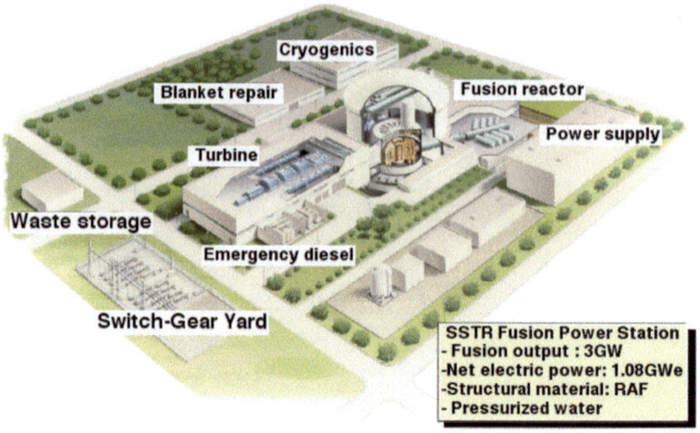

Fig. 1 Bird's eye view of the steady state tokamak reactor. Modified from Kikuchi [433]

Acknowledgements

This book is dedicated to the past Director Hiroshi Kishimoto for his strong leadership in advanced tokamak research in JT-60. The authors acknowledge fruitful discussions with the following peoples in past 30 years.

Prof. R. Conn, Dr. Y. Seki, Prof. F. Najmabadi, Prof. M. Abdou, Dr. K. Tobita for tokamak fusion design. Prof. S. Hamada, Prof. A. Boozer, Dr. S. Tokuda, Prof. T. Fujita, Dr. T. Takizuka, Dr. T. Ozeki for tokamak equilibrium and Current Hole physics. Dr. R. Stambaugh, Dr. T. Taylor, Dr. S. Ishida, Dr. Y. Kamada, Dr. R. Hawryluk, Prof. R. Goldston, Dr. F. Romaneli, Dr. X. Litaudon, Prof. X. Duan, Prof. J. Li, Prof. B. Wan, Dr. M. Kwon, Dr. J. Kwak, Dr. G. Sips, Prof. T. Mutoh, Prof. H. Zushi for the advanced tokamak and long pulse operation. Prof. R. Hazeltine, Prof. R. Sagdeev, Dr. R. Bickerton, Prof. J. Callen, Prof. K.C. Shaing, Prof. P. Helander, Dr. M. Zarnstorff, Prof. N. Fisch, Dr. J. Cordey, Dr. K. Tani, Dr. Y. Sun, Dr. T. Oikawa, Dr. T. Suzuki, Dr. M. Honda for neoclassical theory and experiments. Acad. B. Kadomtsev, Dr. B. Taylor, Prof. L. Chen, Prof. W. Horton, Prof. J. Weiland, Prof. T.S. Hahm, Dr. X. Garbet, Dr. Y. Idomura, Dr. S. Brunner, Prof. Z. Lin for low frequency collective motion and gyrokinetic theory. Prof. P. Diamond, Prof. A. Hasegawa, Prof. L. Chen, Prof. T.S. Hahm, Prof. H. Sugama, Dr. Y. Idomura, Prof. Z. Lin, Dr. F. Jenko, Dr. K. Burrell, Dr. T. Hoang, Dr. F. Ryter, Prof. K. Ida, Prof. A. Fujisawa, Dr. J. Rice, Prof. J.Q. Dong, Prof. G. Tynan, Dr. B. Scott, Dr. C. Angioni, Prof. Y. Kishimoto, Prof. T. Watanabe, Dr. K. Miki, Dr. Y. Koide, Dr. G. Conway, Dr. C. Hidalgo, Prof. G.S. Xu, Prof. T. Ido, Prof. Y. Hamada, Dr. Y. Sakamoto, Dr. H. Urano,Dr. N. Miyato, Dr. K. Nagashima, Dr. MJ Pueschel, Dr. W. Gutterfelder, for turbulent transports. Prof. J. Wesson, Prof. J. Freidberg, Prof. Liu Chen, Dr. F. Zonca, Prof. R. Dewar, Prof. C.Z. Cheng, Dr. J. Manickam, Prof. J. Callen, Dr. M. Chu, Dr. S. Tokuda, Prof. H. Zohm, Prof. R. White, Prof. H. Wilson, Dr. T. Evans, Prof. A. Fasoli, Dr. T. Ozeki, Dr. K.Wong, Dr. M. Okabayashi, Dr. F. Waelbroeck, Dr. J. van Dam, Prof. H. Berk, Prof. Z. Yoshida, Prof. H. Park, Dr. R. Nazikian, Dr. K. Shinohara, Dr. M. Takechi, Prof. S. Sabbagh, Dr. G. Matsunaga, Dr. G. Kurita, Dr. T. Hender, Dr. N. Aiba, Dr. A. Isayama, Dr. N. Oyama, Dr. N. Hayashi, Dr. Y.M. Jeon, Dr. P. Snyder, Dr. Y. Ishii, Dr. M. Lesur, Dr. Y. Kusama, Prof. M. Furukawa, Prof. X.G. Wang, Prof. Z.X. Wang, Prof. P. Zhu, Prof. M. Hirota, Dr. J. Shiraishi for ideal, resistive, and kinetic MHD. Dr. Y. Shimomura, Prof. R. Goldston, Dr. N. Asakura, Prof. T. Tanabe, Dr. G. Janeschitz, Dr. A. Loarte, Dr. H. Takenaga, Dr. B. Lipschultz, Dr. M. Shimada, Dr. T. Takizuka, Dr. K. Shimizu, Dr. N. Ueda, Prof. S. Takamura, Prof. N. Yoshida, Prof. T. Watanabe, Dr. G. Federici, Prof. H. Zushi for SOL/divertor and plasma material interactions. Dr. T. Ando, Dr. H. Tsuji, Dr. K. Okuno, Dr. Y. Takahashi, Dr. K. Yoshida for superconducting magnet system. Prof. A. Kohyama, Dr. A. Hishinuma, Prof. Matsui, Dr. S. Jitsukawa, Prof. K. Abe, Prof. T. Muroga, Prof. A. Kimura, Dr. N. Baluk, Dr. S. Zinkle, Dr. H. Tanigawa, Dr. A.M. Sukegawa for fusion material. Dr. H. Kawamura, Prof. S. Tanaka, Prof. T. Tanabe, Prof. T. Terai, Dr. M. Nakamichi, Dr. J.H. Kim, Dr. K. Konno, Dr. P. Batistoni, Dr. C. Wong, Prof. Y. Wu, Dr. M. Enoeda, Dr. T. Hayashi, Dr. W. Shu, Dr. T. Hoshino for blanket, neutronics, and tritium technologies. Dr. Y. Okumura, Dr. T. Inoue, Dr. A. Kojima, Dr. U. Fantz for negative ion based neutral beam injection system and Dr. K. Sakamoto, Dr. A. Kasugai, Dr. Y. Ikeda, Dr. T. Kobayashi for Gyrotron and ECRF system, Dr. P. Barabaschi for JT-60SA, Dr. Q. Li for HL-2M.

Contents

Acronyms

AE	Alfven Eigenmode
AMMI	Axisymmetric multi-magnetic island
AT	Advanced tokamak
BAAE	Beta induced Alfven acoustic eigenmode
BAE	Beta induced Alfven eigenmode
BCS	Bardeen-Cooper-Schrieffer
BGK	Bernstein-Green-Kruskal
BLM	Barrier localized mode
CGL	Chew-Goldberger-Low
CH	Current hole
CLW	Clearance level waste
CRM	Cyclotron resonance maser
CXRS	Charge exchange recombination spectroscopy
DBTT	Ductile to brittle transition temperature
DIA	Direct interaction approximation
DKE	Drift kinetic equation
DTM	Double tearing mode
EAE	Ellipticity induced Alfven eigenmode
ECCD	Electron cyclotron current drive
EGAM	Energetic Particle driven GAM
ELM	Edge localized mode
EPM	Energetic Particle Mode
ETB	Edge transport barrier
ETG	Electron temperature gradient
EWM	Energetic-particle-driven wall mode
FCT	Flux conserving tokamak
FENDL	Fusion evaluation nuclear data library
FLR	Finite Larmor radius
GAE	Global Alfven eigenmode
GAM	Geodesic acoustic mode
GKE	Gyro kinetic equation

GGAM	Global geodesic acoustic mode
GL	Ginzburg-Landau
GS	Grad-Shafranov
HM	Hasegawa-Mima
IAEA	International Atomic Energy Agency
ICTP	International Center for Theoretical Physics
ITB	Internal transport barrier
ITG	Ion temperature gradient
ITPA	International Tokamak Physics Activity
KAW	Kinetic Alfven wave
KBM	Kinetic ballooning mode
LCO	Limit cycle oscillation
LHCD	Lower hybrid current drive
MIG	Magnetron injection gun
MP	Magnetic perturbation
MSE	Motional stark emission
MSI	Multi-step ionization
NAE	Noncircular triangularity induced Alfven eigenmode
NBCD	Neutral beam current drive
N-NBI	Negative ion based Neutral Beam Injector
NS	Negative shear
NTM	Neoclassical tearing mode
NTV	Neoclassical Toroidal Viscosity
ODE	Ordinary differential equation
ODS	Oxide dispersion strengthened
PCI	Phase contrast imaging
PFC	Plasma-facing component
RAF	Reduced activation Ferrite/Martensitic steel
RMP	Resonant magnetic perturbation
RSAE	Reversed shear Alfven eigenmode
RWM	Resistive wall mode
SAR	Shear Alfven resonance
SOC	Self-organized criticality
SOL	Scrape off layer
SSTR	Steady state tokamak reactor
TAE	Toroidicity induced Alfven eigenmode
TBR	Tritium breeding ratio
TEM	Trapped electron mode
TEP	Turbulent equipartition
WKBJ	Wentzel-Kramers-Brillouin-Jeffries
WS	Weak shear

Chapter 1
Tokamak Fusion Reactor

Abstract The goal of the fusion research is the realization of the fusion power plant, which produces the fusion energy in a useful form (the electricity and/or the hydrogen). Tokamak is the leading concept of the magnetic confinement fusion. The concepts of fusion power stations based on tokamak concept is introduced in this chapter, especially with continuous operation, called the Steady State Tokamak Reactor (SSTR). After the introduction of fusion reaction in Sect. 1.1, tokamak fusion reactor designs based on pulsed and steady state operations are introduced in Sect. 1.2. Basic design features of the SSTR are introduced in Sects. 1.3 and 1.4. Power balance of the tokamak fusion reactor is introduced in Sect. 1.5. Salon 1-1 and 1-2 gives some historical activities such as International school of plasma physics and IAEA fusion energy conferences.

Further Reading:

Books: For quantum mechanics, you may read Schiff [669], Landau [494]. Theory of Atomic Collision by Mott-Massey [563] is quite comprehensive text on Nuclear Physics. Kikuchi [443] Chaps. 1 and 2 are useful for fusion and nuclear physics. Asimov (1991) [30] is well written book on nuclear physics without equations. There are no up-to date good textbook for fusion reactor design but you may find some interests in the text by Gross (1984) [268].

Review Papers: Kikuchi-Conn et al. [432] and Kikuchi [437] are basic papers on tokamak reactor design. IAEA booklet [359] is a good brief summary of 50 years of fusion research.

Design Reports and Home Pages: For fusion reactor design report, you may read SSTR design report at http://jolissrch-inter.tokai-sc.jaea.go.jp.pdfdata/JAERIM-91-081.pdf. You may visit ARIES Program Home Page (http://aries.ucsd.edu/ARIES/). The FIRE place Home Page (http://fire.pppl.gov/) has many informations on fusion.

M. Kikuchi, M. Azumi, *Frontiers in Fusion Research II*,
DOI 10.1007/978-3-319-18905-5_1

1.1 D-T Fusion Reaction

The fusion reaction for current fusion research is the reaction of deuterium and tritium. When they get close enough to each other for the nuclear force to operate beyond the Coulomb barrier, the distance is less than 3 *Fermi* (1 *Fermi* $= 10^{-15}$ m). The barrier height for charged particles, $U_{max} \sim 0.48$ MeV. Fusion will occur if the relative energy of deuterium and tritium is larger than 0.48 MeV, but it is difficult to raise the temperature to this level. However, the tunnel effect enables fusion can occur at low energy (several 10 keV) by penetrating the Coulomb barrier. Scattering and penetration of the particle beam can be investigated by solving the Schrödinger equation $[-(\hbar^2/2m)\partial^2/\partial \boldsymbol{x}^2 + V(\boldsymbol{x})]\psi = i\hbar\partial\psi/\partial t$ under the Coulomb field [669]. This Schrödinger equation was first derived by Austrian physicist Erwin Schrödinger (1887–1961) who was awarded the 1933 Nobel Prize in physics.

The **deuterium** consists of one proton and one neutron. Deuterium was discovered in 1932 by American chemist H.C. Urey (1893–1981), who showed that 1 of every 7000 hydrogen atoms is deuterium. Urey was awarded the 1934 Nobel Prize in Chemistry. Heavy water (chemical symbol D_2O, chemically bonded two deuterium and one oxygen) exists at about 158 ppm in seawater and 140 ppm in the freshwater. Deuterium chemistry is introduced in Sect. 9.5.

Hydrogen with mass number 3 is called **tritium**. The word tritium comes from the Greek word meaning third. The nucleus of tritium consists of one proton and two neutrons. Tritium rarely exists in nature, and is made only in the atmosphere by cosmic rays. This element decays into helium-3 by emitting a high-energy electron beam. This is called beta decay and has a half-life of 12.26 years. Tritium was first produced in the laboratory by Australian physicist M. Oliphant (1901–2000) in 1934 by colliding deuterium. Tritium as fuel of DT fusion is generated by the nuclear reaction of neutron with lithium. The DT fusion produces neutrons and these neutrons can be used to generate tritium by the reaction with lithium. Therefore, high temperature plasma is covered with a device called a **blanket** containing lithium to generate tritium. Lithium has two isotopes (^6L_i and ^7L_i) and the abundance of ^6L_i and ^7L_i in natural L_i is 7.4 and 92.4 %, respectively. Tritium is created by neutron absorption reactions, ^6L_i reaction $^6L_i + n \rightarrow ^3 T +^4 H_e + 4.78$ MeV is an exothermic reaction, while the ^7L_i reaction $^7L_i + n \rightarrow ^3 T +^4 H_e + n - 2.47$ MeV is an endothermic reaction.

The tokamak system has excellent plasma confinement capability with its symmetry in the toroidal direction and will demonstrate fusion energy production of ~500 MW by the ITER project (https://www.iter.org/).

1.2 Tokamak Operation

Tokamak has geometrical symmetry in the toroidal direction and this symmetry provides robustness in maintaining a nested flux surface against various parametrical changes leading this configuration to be a front-runner in fusion research [443].

Tokamak achieved equivalent break-even conditions in large tokamaks such as JT-60U [458] and JET [424] or produced significant fusion power (>10 MW) in TFTR [301] and JET, while other magnetic confinement fusion experiments remained much lower. Geometrical symmetry provides good confinement of energetic charged particles as well as thermal plasmas. This is a reason why the tokamak concept was selected for ITER.

However, this configuration requires a net toroidal plasma current, which is driven mainly by inductive means. This method is quite effective since the electrical conductivity of 10 keV plasma is 20 times higher than that of Cu at the room temperature. But the induction of the toroidal electric field is limited to a finite pulse length (300–500 s in ITER at the nominal plasma current) due to the current limits in the transformer. The tokamak reactor design based on inductive operation was first made in UWMAK studies in 1976 by R. Conn [137], whose cutaway view is shown in Fig. 1.1. This reactor will operate with a pulse length of 1 h but will provide average electric output of ~2 GWe.

This means that a tokamak fusion power station may be pulsed as shown in Fig. 1.2 subject to cyclic loads and it requires large energy reservoir to deliver the electric power continuously. Since present power sources such as oil/coal/natural

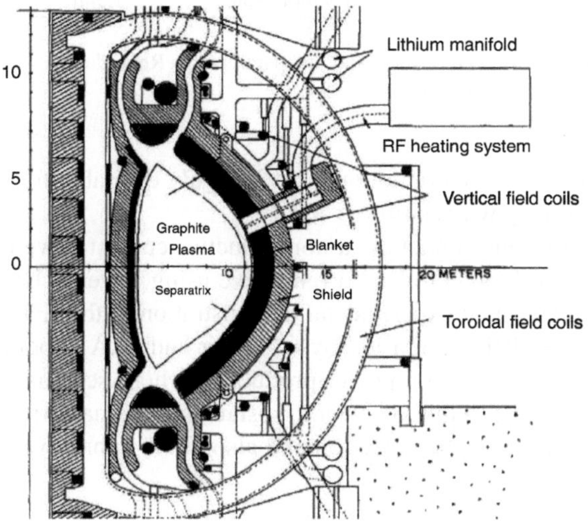

UWMAK III parameter list

Parameter	Value
Plasma current	15.8MA
Toroidal field	4.05T
Thermal power	5GW
Electric power(av)	1.985GW
Burn time	1800s
Coolant	He & Li
Plant efficiency	41.9%
Major radius	8.1m
Minor radius	2.7m

Fig. 1.1 Poloidal cross section of UWMAK-III tokamak fusion reactor and major parameters of UWMAK III pulsed tokamak reactor. Reproduced with permission from Conn [137]. Copyright IAEA Vienna

Fig. 1.2 Schematics of the inductive operation of tokamak

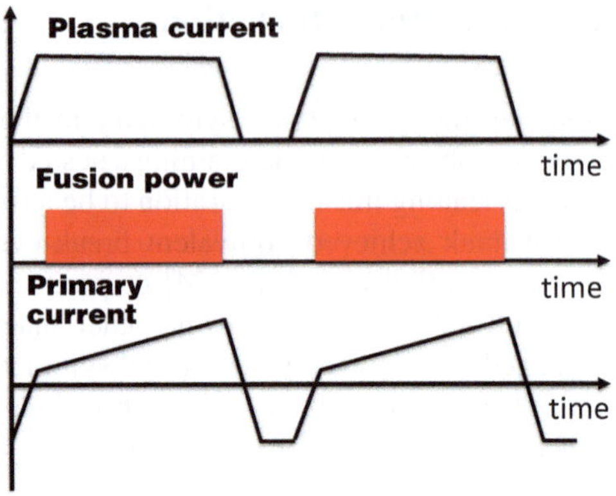

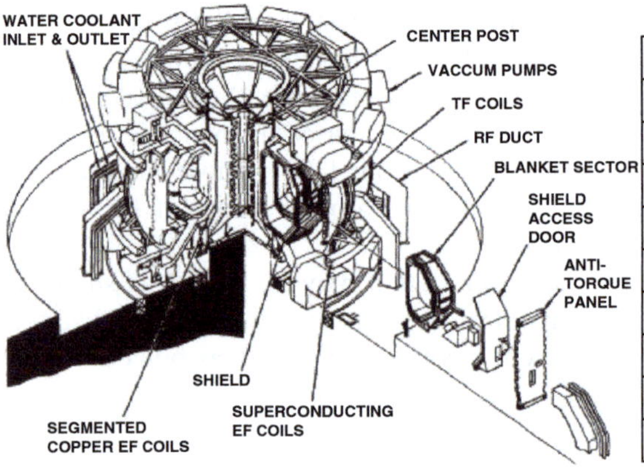

STARFIRE parameter list

Parameter	Value
Plasma current	10MA
Toroidal field	5.8T
Thermal power	4GW
Net electric power	1.2GWe
Burn time	continuous
Coolant	H2O
Plant efficiency	30%
Major radius	7m
Minor radius	1.94m

Fig. 1.3 Birds eye view of STARFIRE fusion reactor and its parameters. Reproduced with permission from Abdou [1]. Copyright IAEA Vienna

gas fired plants and fission plants operate continuously, it is highly desirable for a tokamak reactor to be a steady-state power station.

To achieve continuous operation in a tokamak, a non-inductive current drive is essential. After the theoretical development of a current drive using lower hybrid wave by N.J. Fisch [204] and subsequent experimental demonstration in the JFT-2 by T. Yamamoto [841], the STARFIRE design led by C. Baker and M.A. Abdou [1, 40] was made to realize continuous fusion power production whose schematic view is shown in Fig. 1.3. However, the required power to sustain the plasma current becomes huge if we include effect of trapped electron and also lower hybrid wave absorption due to α particles.

1.3 Steady State Tokamak Reactors

Nature blesses human being by providing a self-generation mechanism of the plasma current called the bootstrap current [65] (see Chap. 4, especially Sect. 4.4.3). The bootstrap current is driven by the pressure gradient in a rare collision regime (called the collisionless plasma), which is a kind of thermo EMF that drives the plasma current in a toroidal direction.

Observation of a large bootstrap current fraction up to 80 % in JT-60 high β_p discharges by Kikuchi [431] immediately stimulated the concept development of the Steady State Tokamak Reactor (SSTR) by Kikuchi [430] and H. Kishimoto (see Fig. 3.1) decided the development of the conceptual design of SSTR [236, 433, 675] with the strong involvement of Japanese industries aiming to show scientific and technical capability of sustained electric power generation by tokamak concept with minimum extrapolation from the knowledge in those days. The bird's eye view of the SSTR plant is shown in Fig. 1 of preface. The side view of main building and a part of the auxiliary building is shown in Fig. 1.4 showing tokamak, four set of steam generator (SG) for the power generation, divertor maintenance rooms, replaceable blanket cask for the vertical maintenance, vacuum pumping system, and polar crane and part of the negative ion based neutral beam injection (N-NBI) system with inverter and convertor power supplies.

Figure 1.5 shows plan view of the main reactor building where tokamak reactor locates at the center and two sets of neutral beam injection lines, two sets of remote maintenance system, four sets of steam generator (SG) for power generation, ECH

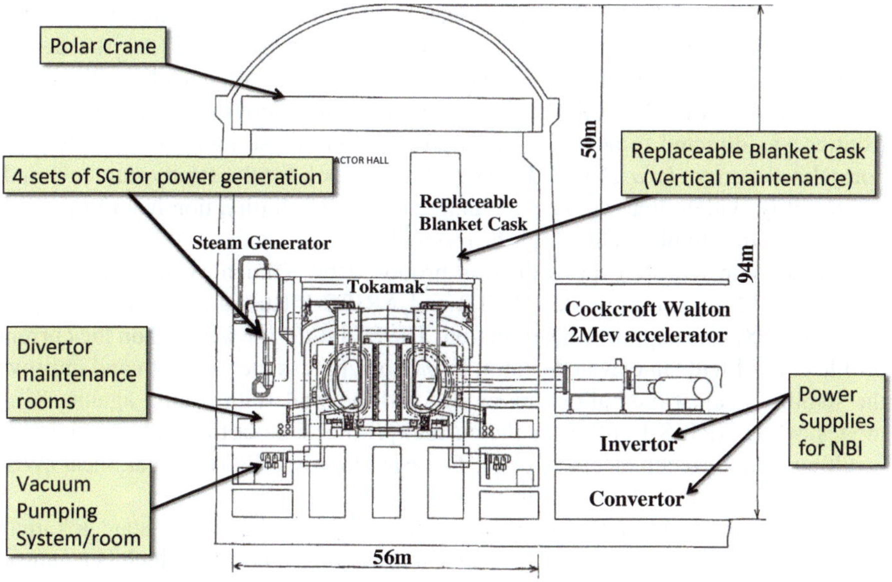

Fig. 1.4 Side view of the SSTR building. SG: steam generator. Modified from Kikuchi [432]

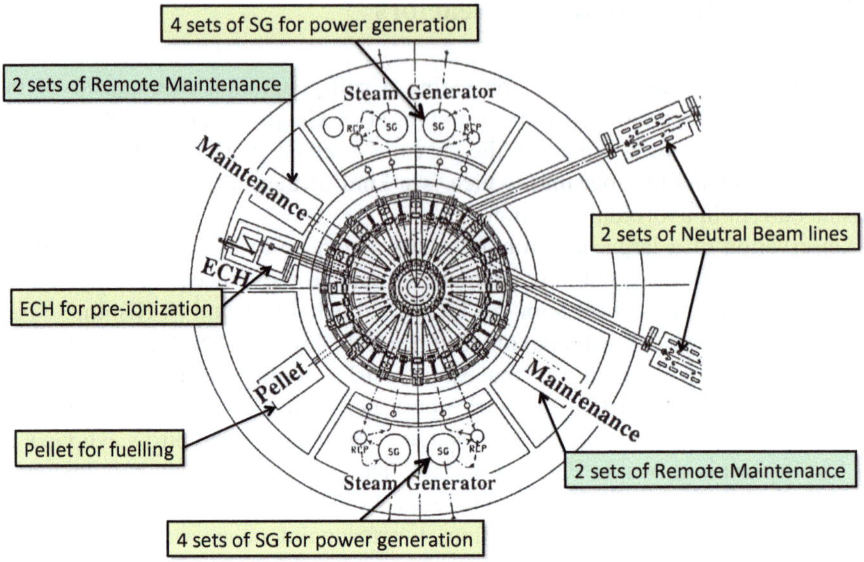

Fig. 1.5 Plan view of the SSTR building. Modified from Kikuchi [432]

for pre-ionization and neoclassical tearing mode (NTM) control, and pellet fueling system are arranged around the tokamak device.

R. Conn et al. [139] also developed another independent power reactor study with more aggressive technical provisions, ARIES-I. When SSTR and ARIES-I concepts are proposed in 1990, world fusion research was directed to increase the plasma current to improve the energy confinement, typically represented by the design change from a plasma current $I_p = 6$ MA in INTOR (International tokamak reactor) to $I_p = 21$ MA in ITER-CDA (CDA: conceptual design activity). Also the research frontier was directed to achieve a high toroidal beta close to 10 % with the high normalized-current I_p/a_pB_t in DIII-D [201]. If we look for the steady state tokamak reactor, it is evident that we have to change our research direction from low q and high β_t research to high q and high β_p research.

"Change" in research direction was proposed by Kikuchi, Conn, Najmabadi, and Seki [432] as a joint work of SSTR and ARIES-I. Russian design given by Y. Sokolov [708] also supports this direction. The SSTR design is based on the current profile control with an elevated central q (q_0) to suppress the sawtooth and improve the stability against the ballooning modes and the wall stabilization against Kink modes, while ARIES-I optimizes the current profile without the wall stabilization, resulting in modest enhancement of q_0. These are jointly called the weak shear (WS) scenario. Here, (magnetic) shear is given by $s = Rdq/dr/q$.

While weak shear (WS) operation is proposed for the steady state tokamak reactor by Kikuchi [432], an active use of hollow current profile (which is a natural profile for the bootstrap current) for the steady state tokamak reactor is proposed by Ozeki [596], and later by Kessel [427] is called the negative-shear (NS) scenario. Tokamak reactor concepts based on NS scenarios are developed such as ARIES-RS

[571] in the United States, and CREST [589] in Japan. These operation scenarios are discussed in Chap. 3.

Salon 1-1: Pioneer and Alliances of Fusion Reactor Design

(a) R. Conn is the pioneer of fusion reactor study and formed a team to study series of fusion reactor studies called "ARIES". He is the President of The Kavli Foundation and Zable Professor and Dean, Emeritus, of the Jacobs School of Engineering at UC San Diego as of 2014 (courtesy of R. Conn).

(b) International school of plasma physics (Piero Caldirola) is a long-standing school activity near Como lake in Italy. Group photo of this school in 1994 to discuss steady state operation of tokamak [54] including R. Bickerton (UKAEA, see Fig. 4.5), D. Robinson (UKAEA), R. Conn (UCSD), J. Wesson (UKAEA) who is famous in his textbook on tokamak, F. Romanelli (ENEA and JET), K. Lackner (Max Planck Institute for plasma physics), H. Zohm (Max Planck Institute for plasma physics), T. Taylor (General Atomics), N. Sauthoff (US ITER), V. Chan (General Atomics), R. Andreani (ENEA), H. Ninomiya (JAEA), D. Stork (UKAEA), S. Jardin (Princeton plasma physics laboratory), M. Okabayashi (Princeton plasma physics laboratory), W. Nevins (Lawrence Livermore National Laboratory), S. Sabbagh (Columbia university), S. Eckstrand (DOE), S. Bernabei (Princeton plasma physics laboratory), M. Kikuchi (author). With kind permission by Dr. E. Sindoni.

1.4 Major Features of SSTR

Among many important parameters, plasma pressure limit is an important constraint for the reactor design. The plasma pressure $\langle p \rangle$ required for a fusion reactor is in a range of 10 atm. This must be confined by the magnetic field (namely, $\boldsymbol{J} \times \boldsymbol{B}$ force). The magnetic field has its pressure $B^2/2\mu_0$ and the ratio of plasma pressure to the magnetic field pressure is called 'beta'. This beta can be defined for the toroidal field and the poloidal field, called the toroidal beta ($\beta_t \equiv \langle p \rangle / (B_t^2/2\mu_0)$) and the poloidal beta ($\beta_p \equiv \langle p \rangle / (B_p^2/2\mu_0)$), respectively.

As we increase the beta, plasma becomes unstable to kink modes or ballooning modes at some value, called the beta limit. The parametric dependences of this beta limit has been extensively investigated during 1980 decade. Swiss physicist F. Troyon [772] identified essential parametric dependences called Troyon scaling theoretically as follows,

$$\beta_t = \beta_N \frac{I_p[MA]}{a_p[m]B_t[T]}, \tag{1.1}$$

where I_p is plasma current, a_p is plasma horizontal minor radius, B_t is toroidal magnetic field, β_N is a constant named the 'normalized beta'. The proportionality of β_t to I_p/a_pB_t is confirmed in DIII-D [201] and also $\langle p \rangle \sim I_pB_t/a_p$ in JT-60U [439] as seen in Fig. 1.6. MHD stability is described in detail in Chap. 8.

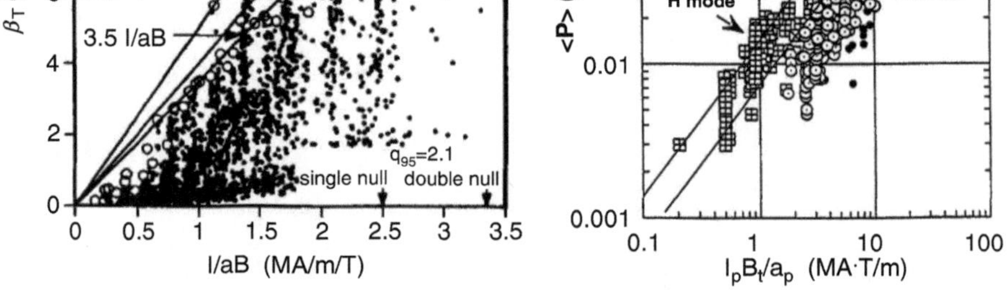

Fig. 1.6 (**a**) Experimental confirmation of Troyon scaling in DIII-D. Reproduced with permission from Ferron [201]. Copyright 1999 AIP Publishing LLC. (**b**) Experimental average plasma pressure $\langle p \rangle$ in proportion to I_pB_t/a_p. Reproduced with permission from Kikuchi [439]. Copyright IAEA Vienna

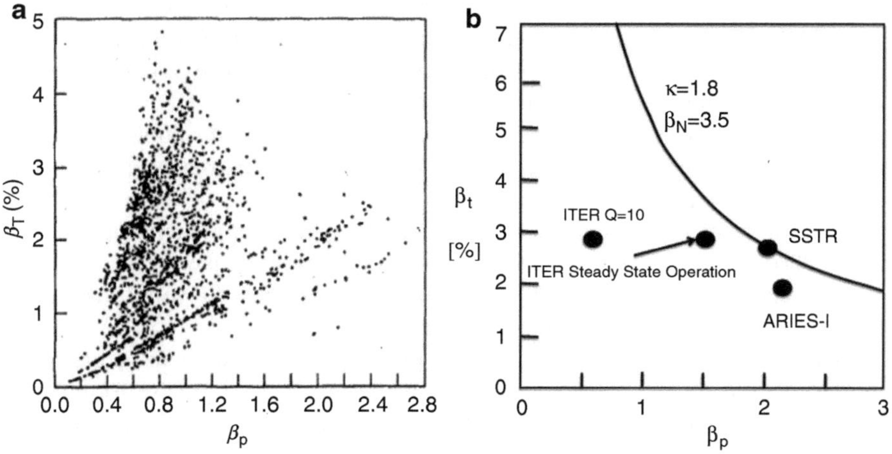

Fig. 1.7 (**a**) Experimental demonstration of $\beta_t = C/\beta_p$ in the Doublet III. Reproduced with permission from Stambaugh [714]. Copyright IAEA Vienna. (**b**) (β_t, β_p) diagram

Combination of this Troyon scaling with the definition of poloidal beta $\beta_p = 4\int pdV/(\mu_0 I_p^2 R_p)$ gives relation between β_t and β_p as,

$$\beta_t \beta_p = \kappa \frac{\beta_N^2}{4}, \tag{1.2}$$

where κ is the vertical plasma elongation. This scaling is confirmed in DIII-D [714] as shown in Fig. 1.7a.

The major feature of SSTR [430] is the maximum utilization of the bootstrap current for the efficient steady state operation. The bootstrap current fraction (f_{boot}) is known to be proportional to the poloidal beta (a ratio of plasma pressure to the magnetic pressure of poloidal magnetic field produced by the plasma current).

While the theory of the bootstrap current is given later in Chap. 4, a simple expression of the bootstrap current fraction is given by Cordey [149] as follows,

$$f_{boot} = c_1 (a/R)^{0.5} \beta_p, \tag{1.3}$$

$$\text{where } \beta_p = 4\int PdV/(\mu_0 I_p^2 R_p) \sim \langle P\rangle/(B_p^2/2\mu_0) \tag{1.4}$$

Here, a is the horizontal plasma minor radius and R is the plasma major radius, and $(a/R)^{0.5}$ is proportional to the trapped particle fraction. Cordey [149] gives $c_1 = 0.67$. Therefore, the reactor should operate in the high $\beta_p \sim 2-3$ regime if the major fraction of the plasma current is driven by the bootstrap current.

Figure 1.7b shows the (β_t, β_p) diagram in which the solid curve corresponds to $\beta_N = 3.5$ and $\kappa = 1.8$, in which the steady-state fusion power concepts SSTR and

ARIES-1 ($\beta_p = 2 - 2.1$) as well as the current ITER steady state design ($\beta_p \sim 1.5$) adopt high β_p operation to increase bootstrap current fraction. Since $\beta_p \sim 1/\beta_t$ for fixed β_N and κ, the steady state tokamak reactor should sacrifice β_t to achieve high β_p and hence a high bootstrap current fraction.

If we want to have 75 % of the plasma current carried by the bootstrap current, we have to operate at $(a/R)^{0.5}\beta_p = 1.12$. Required β_p increases weakly with the aspect ratio, $A = R/a$.

The rest of the plasma current has to be driven by the non-inductive means such as NBCD (neutral beam current drive) and/or RFCD (radio frequency current drive).

If we choose an appropriate value for the normalized beta β_N with some margin from the beta limit, β_t value will be given by Eq. (1.2).

Once the toroidal beta β_t is fixed, the fusion power density can be calculated by the relation $P_f \approx c_2\beta_t^2 B_t^4 \sim \langle n_i T_i\rangle^2$, while much more precise formula can be available in the literatures. Here, the toroidal magnetic field B_t has following geometrical constraint,

$$B_t = B_{max}(R_p - a_p - \Delta)/R_p, \tag{1.5}$$

where Δ is a distance between the inner plasma surface and the major radius of the maximum magnetic field on the toroidal field conductor and B_{max} is the maximum magnetic field specified by the superconductor ($B_{max} \approx 12T$ for Nb_3Sn). The distance Δ is in the range of 1.3–1.5 m. Then, the plasma current I_p is determined from the Troyon scaling.

The efficiency of the current drive is measured by the current drive efficiency.

$$\eta_{CD} \equiv \frac{n_e I_{pCD} R_p}{P_{abs}}, \tag{1.6}$$

which is typically 0.2–0.5× A/m^{2}W depending on the driver choice. For a given η_{CD}, we can calculate power required for the current drive to supplement bootstrap current. For a set of machine parameters, the required confinement enhancement factor over the L-mode energy confinement scaling law can be obtained. Physics basis for a tokamak fusion power plant are summarized by Jardin [395] based on five different operation modes studied in ARIES program [140]. Aspect ratio (ratio of major radius and horizontal minor radius) is one of key parameters in the tokamak reactor design. Wong [831] showed dependence on the aspect ratio.

1.5 Reactor Power Balance

According to the basic procedures in the last section, we may arrive at a set of machine parameters. In case of the SSTR [432], the power balance in the fusion reactor can be illustrated in Fig. 1.8.

Reactor power balance is an important aspect in the steady state tokamak reactor. The general power flow diagram is shown in Fig. 1.9. Here, P_f is fusion power from

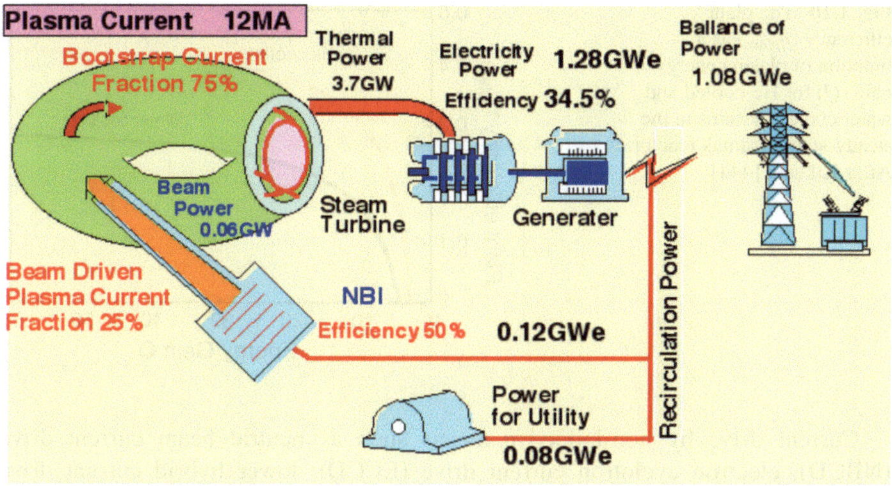

Fig. 1.8 Power flow diagram of SSTR fusion reactor. "Utility" means auxiliary equipments in Fig. 1.9. After Kikuchi [443]

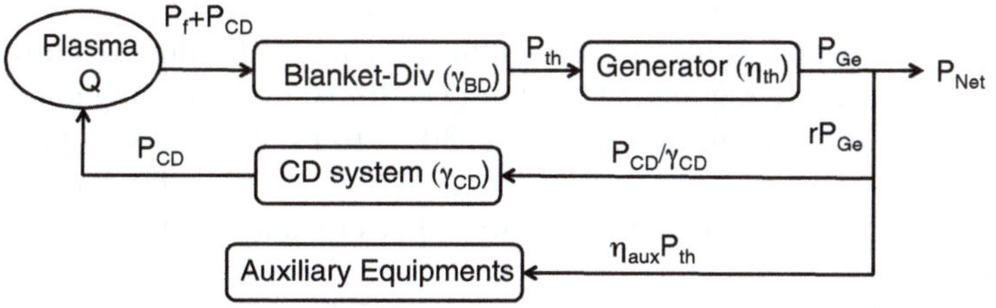

Fig. 1.9 Energy flow diagram of steady state tokamak reactor. The plant efficiency $\eta_{Plant}(= P_{Net}/P_{th})$, the ratio of net electric power output to thermal power output, is given by $\eta_{Plant} = \eta_{th} - \eta_{aux} - 1/(\gamma_{CD}\gamma_{BD}(1+Q))$. After Kikuchi [444]

plasma, P_{CD} is heating and current drive (CD) power, Q is energy gain of confined plasma $Q = P_f/P_{CD}$, γ_{BD} is energy multiplication factor in the blanket-divertor system, P_{Ge} is gross electric power, P_{Net} is net electric power to the grid, rP_{Ge} is re-circulating power (r is re-circulating power fraction), γ_{CD} is overall system efficiency of CD system, $\eta_{aux}P_{th}$ is power required for auxiliary equipments. Here, thermal conversion efficiency η_{th} is 0.345 for water cooling in fission light water reactor, while it is 0.49 for advanced high temperature He cooling system. η_{aux} is reduction of plant efficiency due to auxiliary equipments in a range of 0.03–0.06 depending on coolant.

The (η_{plant}, Q) diagram is shown in Fig. 1.10 for a pressurized-water cooling and a high temperature helium gas cooling. It must be noted that η_{plant} depends weakly on Q around $Q = 30$–50, $\eta_{plant} \sim 0.3$ for pressurized water and $\eta_{plant} \sim 0.4$ for high temperature helium.

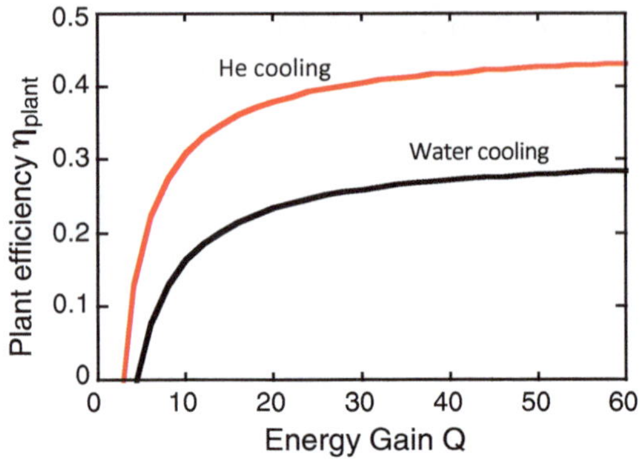

Fig. 1.10 The plant efficiency (η_{plant}) as a function of plasma energy gain (Q) for He-cooled and water cooled systems in the steady-state tokamak reactor. After Kikuchi [444]

Current drive by non-inductive means such as neutral beam current drive (NBCD), electron cyclotron current drive (ECCD), lower hybrid current drive (LHCD), and fast wave current drive (FWCD) are developed as reviewed by Fisch [204]. Efficiency of the non-inductive CD is expressed by the current drive efficiency η_{CD} defined by $\eta_{CD} = I_p^{CD} R_p \langle n_e \rangle / P_{CD}$ and has certain limit $\eta_{CD} \sim 5 \times 10^{19}$ A/Wm^2 for N-NBI (negative-ion based NBI) at $\langle T_e \rangle = 17$ keV and $E_{beam} = 2$ MeV, which is much less than the efficiency of inductive current drive. It is difficult to achieve required Q level of $Q = 30$–50 by only using non-inductive current drive by external means (e.g. 3 GW/30 = 100 MW can drive only 4 MA for $R = 7$ m, $\langle n_e \rangle = 10^{20}$ m^{-3}, $\eta_{CD} = 3 \times 10^{19}$ A/W/m^2). This is a fundamental reason why we have to utilize bootstrap current to realize efficient steady state operation of tokamak reactor.

EU approach to the DEMO design is started from safety assessment called SEAFP project reported by Raeder [620] and the recent approach emphasize comparative studies among several options called PPCS (power plant conceptual study) [531]. In 2012, EFDA (European Fusion Development Agreement) published a roadmap which outlines how to supply fusion electricity to the grid by 2050 [186]. EUROfusion formed in 2014 for coordination of European fusion research as successor of EFDA. Strategic plan has been developed in Korea for the fusion energy development beyond ITER by Kwon [484]. China is timely to design a next step device called CFETR (2012) led by Y.X. Wan, Jiangang Li, and Liu Yong by addressing fusion fuel cycle demonstration to supplement ITER towards DEMO [801].

Salon 1-2: IAEA Fusion Energy Conferences [359]

After the world-war II, peaceful use of the atomic energy is pursued. The United Nations (UN) organized the first conference on the “Peaceful Uses of Atomic Energy” in Geneva. The president of the first conference is Homi

(continued)

Bhabha. In the opening address, he mentioned that 'It is well known that atomic energy can be obtained by fusion processes as in the H-bomb and there is no basic scientific knowledge in our possession today to show that it is impossible for us to obtain this energy from the fusion process in a controlled manner. The technological problems are formidable, but one should remember that it is not yet fifteen years since atomic energy was released in an atomic pile for the first time by Fermi. I venture to predict that a method will be found for liberating fusion energy in a controlled manner within the next two decades. When that happens the energy problem of the world will truly have been solved forever for the fuel will be as plentiful as the heavy hydrogen in the oceans.'

The second UN Conference on Peaceful Uses of Atomic Energy is held at Geneva Palais des Nations in 1–13 September 1958, in which about 5000 delegates, observers and guests discussed over 2150 papers in which 105 covers fusion topics (US 65, USSR 14, UK 9, France 4, Sweden 4, Japan 2, Denmark 1, Germany 1, Poland 1, Romania 1, Switzerland 1), "Monster Conference" Time Magazine 15 September 1958. In spite of smaller number of papers, fusion and its declassification were the focus of this conference with many fusion exhibitions of actual magnetic confinement fusion machines and press conferences.

Academician L.A. Artsimovich in his overview paper (1958) [27] from the Soviet Union says "For the first time, these results will be discussed on the international scale, and this is probably the most important step that has been made towards the solution of this problem". He also mentioned in his paper that "The solution of the problem of thermonuclear fusion will require a maximum concentration of intellectual effort and the mobilization of very appreciable material facilities and complex apparatus".

Edward Teller [760] made his famous comparison with flying "I think we are at a stage similar to the stage at which flying was about one hundred years ago. At that time the question was to understand a very difficult subject, namely the subject of hydrodynamics and in particular the subject of turbulent hydrodynamics. We are now similarly trying to understand the subject of magnetohydrodynamics...".

IAEA booklet [359] says that at least when the conference ended, the situation had become clear: No matter how long it might take, the 'unwritten chapter' on plasma physics had to be written.

Since then, there are number of international conference on plasma physics and controlled fusion research held under the auspice of IAEA, at Saltzburg (1961), Culham (1965), Novosibirsk (1968), Madison (1971), Tokyo (1974), Berchtesgaden (1976), Innsbruck (1978), Brussels (1980), Baltimore (1982), London (1984), Kyoto (1986), Nice (1988), Washington (1990), Würzburg (1992), Seville (1994), Montreal (1996), Yokohama (1998), Sorrento (2000),

(continued)

Lyon (2002), Vilamoura (2004), Chengdu (2006), Geneva (2008), Daejeon (2010), San Diego (2012), St Petersburg (2014) and Kyoto (2016). The conferences are called "Fusion Energy Conference" after Yokohama.

The 22nd Fusion Energy Conference was an anniversary conference to celebrate 50 years of magnetic confinement fusion research. Figure 1.11 shows an opening ceremony on October 13, 2008 when Prof. Tran gives opening address. V.P. Smirnov [698] gave a lecture "Tokamak foundation in USSR/Russia 1950–1990". Jean Jacquinot [390] and Dale Meade [548] gave lectures "Fifty years in fusion and the way forward" and "50 years of fusion research", respectively. Kaname Ikeda [371] gave a plenary "ITER on the road to fusion energy".

Fig. 1.11 Opening ceremony of 22nd Fusion Energy Conference (2008) at Palais des Nations, Geneva, Switzerland cerebrating 50 years of fusion research. From left, M.Q. Tran (President of local organizing committee), M. Kikuchi (Chair of international program committee), Y. Sokolov (IAEA Deputy Director General), J. Beagle (Deputy Director-General of the United Nations Office in Geneva), Z. Stancic (EU DG Research), G.S. Lee (Chair of International Fusion Research Council, right end on the desk), far right end Dr. G. Mank (Head of physics section as conference secretary)

Exercise 1.1. Derive Schrödinger equation $[-(\hbar^2/2m)\partial^2/\partial \boldsymbol{x}^2 + V(\boldsymbol{x})]\psi = i\hbar\partial\psi/\partial t$ from the energy conservation law $\boldsymbol{p}^2/2m + V(\boldsymbol{x}) = E$, the Einstein's relation $E = \hbar\omega$ and the de Broglie relation $\boldsymbol{p} = \hbar\boldsymbol{k}$.

Answer. Substituting the Einstein and de Broglie relations into energy conservation relation, we have $\hbar^2\boldsymbol{k}^2/2m + V(\boldsymbol{x}) = \hbar\omega$. The wave function with the angular frequency ω and the wave number $\boldsymbol{k}$ is given by $\psi = exp(i\boldsymbol{k}\cdot\boldsymbol{x} - i\omega t)$ and the relations $\partial\psi/\partial\boldsymbol{x} = i\boldsymbol{k}\psi$ and $\partial\psi/\partial t = -i\omega\psi$ holds. This means $i\boldsymbol{k} \to \partial/\partial\boldsymbol{x}$ and $-i\omega \to \partial/\partial t$. Substituting these relations into $\hbar^2\boldsymbol{k}^2/2m + V(\boldsymbol{x}) = \hbar\omega$, we have relation of differential operator, $-(\hbar^2/2m)\partial^2/\partial\boldsymbol{x}^2 + V(\boldsymbol{x}) = i\hbar\partial/\partial t$. Operating the wave function from right, we have Schrödinger equation.

Exercise 1.2. Derive Einstein's relation $E = mc^2$ from the variational principle with the action integral $S = -m_0c^2\int_1^2 d\tau$, where $\tau = t(1-(d\boldsymbol{x}/cdt)^2)^{1/2}$ is the proper time.

Answer. Substituting the expression of τ into the action integral, we have $S = \int Ldt = -m_0c^2\int\sqrt{1-(d\boldsymbol{x}/cdt)^2}dt$, i.e. $L = -m_0c^2\sqrt{1-(d\boldsymbol{x}/cdt)^2}$. Defining the canonical momentum $\boldsymbol{p} \equiv \partial L/\partial\dot{\boldsymbol{x}} = m_0\boldsymbol{v}/\sqrt{1-v^2/c^2}$, we have expression for the Hamiltonian $H = \boldsymbol{p}\cdot\dot{\boldsymbol{x}} - L$, which is an expression for the energy E.

$$E = \frac{m_0c^2}{\sqrt{1-v^2/c^2}}$$

See [443] for detail.

Exercise 1.3. Derive Yukawa's meson equation from the relativistic momentum balance equation, $-\boldsymbol{p}^2 - m_0^2c^2 + (E/c)^2 = 0$.

$$-\left[\hbar^2\frac{\partial^2}{\partial\boldsymbol{x}^2} - m_0^2c^2 - \frac{\hbar^2}{c^2}\frac{\partial^2}{\partial t^2}\right]U = 0$$

Answer. See [443] Sect. 2.2.

Exercise 1.4. Explain why ^4He has large binding energy, $E_b \sim 7\,\text{MeV}$.

Answer. See [443] Sect. 2.5.

Exercise 1.5. Calculate the gravitational energy of the Sun $E_g = GM^2/R_{Sun}$, using $G = 6.67\times10^{-11}\,\text{m}^3/\text{kg s}^2$, $M = 2\times10^{30}\,\text{kg}$, $R_{Sun} = 70\times10^4\,\text{km}$. Using the luminosity of the Sun $L = 3.86\times10^{26}\,\text{W}$, calculate decay time of the gravitational energy by radiation if there is no fusion reaction in the Sun.

Answer. Substituting parameters, we have $E_g = 3.8\times10^{41}\,\text{J}$. The time constant τ is $\tau = 3.8\times10^{41}\,\text{J}/3.86\times10^{26}\,\text{W} \sim 10^{15}\,\text{s} \sim= 32\,\text{M years}$.

Exercise 1.5. While abundance of ^7Li which absorb energy by nuclear reaction (endothermic) is much larger than that of ^6Li which can produce energy by nuclear reaction (exothermic), why blanket can produce additional energy?

Answer. It is because nuclear cross section for ^6Li is much larger than that for ^7Li. See Sect. 9.3.

Exercise 1.6. Compare Gamov's tunnel factor $P(E/E_c) = \sqrt{E_c/E}e^{-\sqrt{E_c/E}}$ with exact Mott-Massey formula $P(E/E_c) = \sqrt{E_c/E}/(e^{\sqrt{E_c/E}} - 1)$ as a function of E/E_c.

Answer. Plot by yourself. They are close if $E/E_c < 1$ and different if $E/E_c > 1$.

Exercise 1.7. Plot $D - T$ fusion cross section given by the following formula from Li [503]. Here E_l is the deuterium energy in keV in the laboratory frame. A_d and A_t are mass number of deuterium and tritium, respectively.

$$\sigma_r(E_l)(barn) = -\frac{16389C_3(1 + A_d/A_t)^2 A_d E_l[e^{31.40\sqrt{A_d/E_l}} - 1]}{(C_1 + C_2E_l)^2 + (C_3 - 2\pi/e^{31.40\sqrt{A_d/E_l}})^2}$$

$$C_1 = -0.5405, \quad C_2 = 0.005546, \quad C_3 = -0.3909$$

Answer. See Li [503] Fig. 2.

Chapter 2
Plasma Equilibrium in Tokamak

Abstract Plasma equilibrium is the state of force balance and is fundamental to understand charged particle orbit and the neoclassical transport, collective phenomena and MHD stability in the toroidal plasma. After introduction of topology of the magnetic confinement bottle in Sect. 2.1, the Hamilton structure of the magnetic field and the flux coordinates in general toroidal equilibrium in Sect. 2.2. Then so-called Grad-Shafranov equation for the axisymmetric Tokamak equilibrium, Current Hole equilibrium, and the modified Grad-Shafranov equation including the anisotropic pressure and the toroidal flow, general tensor equilibrium relations are introduced in Sect. 2.3. A set of the evolution equation for the plasma equilibrium called 1.5 D transport equations is introduced in Sect. 2.4 including energy and particle conservation equations, magnetic diffusion equation, and the equilibrium equation, and the FCT equilibrium. Salon includes Activity of International Center for Theoretical Physics (ICTP) related to plasma physics.

Further Reading:
Books: Freidberg [211], Hazeltine-Meiss [308], White [818] and Kikuchi [443] are useful for basic understanding of the magnetic field structure in tokamak. Lichtenberg [508] is quite comprehensive in Hamilton structure of magnetic field. Jardin (2010) [394] provides methods to calculate time evolution of 1.5D tokamak equilibrium and transport. Shafranov [677] and Miyamoto [552] are still good introductory textbooks. Landau-Lifshitz [495] is useful to understand fluid equations. For Hamilton mechanics, you may read Arnolds [26].

Review Papers: Boozer review [73] is an useful article. Some part of Hinton-Hazeltine review [321] provides good description of 1.5D transport equations. For topology of magnetic fields, you may read Woltjer [828] and Moffatt [557].

2.1 Topology of Magnetic Confinement

Topologies of 2 dimensional manifold are identified as sphere, torus, two hole torus, etc. as shown in Fig. 2.1a. The "Sun" is a natural fusion reactor. Dense and hot

M. Kikuchi, M. Azumi, *Frontiers in Fusion Research II*,
DOI 10.1007/978-3-319-18905-5_2

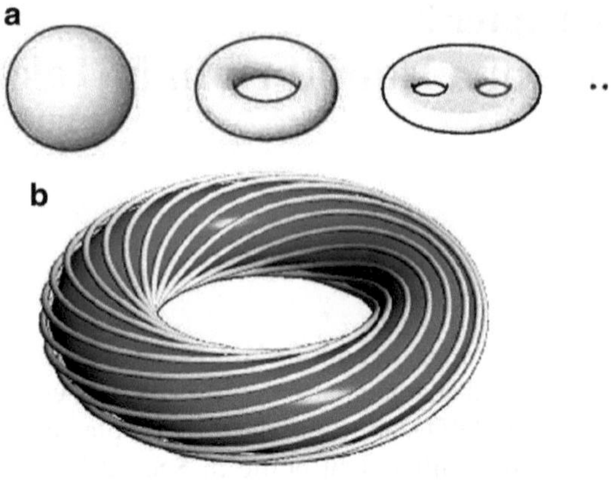

Fig. 2.1 (**a**) Topologies of 2 dimensional closed surface (sphere, torus (doughnut), two-hole torus (pretzel), etc.), (**b**) Torus—topology which can be covered with non-zero vector field

plasmas are confined with the gravitational field. Characteristic of this force is that it is a central force and the force acts in the direction of field line. For this reason, the confinement bottle has a topology of Sphere.

In the man-made fusion reactor, a high temperature plasma is confined by trapping charged particles with the Lorentz force in the magnetic field to sustain reaction with a small dimension of 100 millionth of the Sun. Characteristic of this force is that the force acts in the direction perpendicular to the field line. For this reason, the confinement bottle has a topology of Torus. The force balance is,

$$\boldsymbol{j} \times \boldsymbol{B} = \nabla p \tag{2.1}$$

French mathematician Henri Poincare proved a theorem "Closed surface that can be covered with the vector field without fixed point (null point) is restricted to a torus", which is called "Poincare theorem" (see Kikuchi [443] for the proof of this theorem). Meaning of Poincare theorem is important for the fusion research. Consider the boundary surface of the magnetically confined plasma, the plasma will leak from the zero point of magnetic field vector. The surface must be covered by non-zero magnetic field to confine the hot plasma. This is why we use toroidal geometry for magnetic confinement. In the magnetic confinement system, the toroidal plasma equilibrium is formed by the infinite set of nested toroidal surface, each of which is characterized by a constant plasma pressure. Each toroidal surface is densely covered by the helically wounded magnetic field as shown in Fig. 2.1b.

Topology of magnetic flux tube in the ideal MHD fluid is also of particular importance as manifested by the Woltjer's helicity conservation (1958) [828] and Moffatt's invariant in the linked flux tube (1990) [557]. J.B. Taylor's famous "Taylor relaxation state" [756] comes from the Woltjer's helicity conservation law.

2.2 Magnetic Field Structure of Tokamak

2.2.1 Hamilton Structure of the Magnetic Field B

Since $\boldsymbol{B}$ is incompressible flow, $\boldsymbol{B}$ has vector potential. Consider the toroidal geometry characterized by arbitrary poloidal and toroidal angles θ and ζ. The vector potential $\boldsymbol{A}$ is given by,

$$\boldsymbol{A} = \phi\nabla\theta - \psi\nabla\zeta + \nabla G, \tag{2.2}$$

where G is the "gauge term" (see Exercise 2.1 for its proof). Then, $\boldsymbol{B}$ is given by,

$$\boldsymbol{B} = \nabla\phi \times \nabla\theta - \nabla\psi \times \nabla\zeta \tag{2.3}$$

This is called the "symplectic form" of the magnetic field and ϕ and ψ are called Clebsch potentials. Magnetic field line trajectory along the toroidal direction ζ is,

$$\frac{d\theta}{d\zeta} = \frac{\partial\psi}{\partial\phi}, \quad \frac{d\phi}{d\zeta} = -\frac{\partial\psi}{\partial\theta} \tag{2.4}$$

This is the Hamilton equation by regarding ζ as "time", ψ as "Hamiltonian", θ as "canonical coordinate", ϕ as "canonical momentum" conjugate to θ (see Exercises 2.2 and 2.3 for its proof).

Variational principle of field line is given by the analogy to Hamilton action integral (see Appendix B.3), $S = \int L dt = \int [\boldsymbol{p} \cdot d\mathbf{x}/dt - H]dt$. Substituting relationship $\boldsymbol{p} \to \phi$, $d\boldsymbol{x}/dt \to d\theta/d\zeta$, $H \to \psi$, $t \to \zeta$, we have $S = \int [\phi d\theta/d\zeta - \psi] d\zeta = \int \boldsymbol{A} \cdot d\boldsymbol{x}$. Thus the Hamilton equation of the magnetic field line trajectory (2.4) is given by a following variational principle (see Exercise 2.4 for its direct proof).

$$\delta \int \boldsymbol{A} \cdot d\boldsymbol{x} = 0 \tag{2.5}$$

In case we need to obtain approximate magnetic field structure $\boldsymbol{B}_1$, the incompressibility is not guaranteed if we make an approximation directly to $\boldsymbol{B}$. If we make an approximation to the vector potential $\boldsymbol{A}$, the incompressibility of the magnetic field is always guaranteed. It may be important to note that R. White showed that ideal MHD perturbation $\boldsymbol{B}_1 = \nabla \times (\boldsymbol{\xi} \times \boldsymbol{B})$ do not conserve magnetic topology while it satisfy the incompressibility condition [820].

The curvature of the magnetic field $\boldsymbol{\kappa}$ is defined as $\boldsymbol{\kappa} \equiv \boldsymbol{b} \cdot \nabla\boldsymbol{b}$ has following relations and is perpendicular to $\boldsymbol{B}$ ($\boldsymbol{b} \cdot \boldsymbol{\kappa} = 0$) (see Exercise 2.5 for its proof).

$$\boldsymbol{\kappa} \equiv \boldsymbol{b} \cdot \nabla\boldsymbol{b} = -\boldsymbol{b} \times (\nabla \times \boldsymbol{b}) \tag{2.6}$$

$$\nabla \times \boldsymbol{b} = \boldsymbol{b} \times \boldsymbol{\kappa} + (\boldsymbol{b} \cdot \nabla \times \boldsymbol{b})\boldsymbol{b} \quad (\boldsymbol{b}\times (2.6)) \tag{2.7}$$

2.2.2 Magnetic Field in Flux Coordinates

The magnetic field in the symplectic form $\boldsymbol{B} = \nabla\phi \times \nabla\theta - \nabla\psi \times \nabla\zeta$ (2.2) has two cross products of gradient vectors to describe incompressible flow. But the magnetic field in plasma equilibrium $\boldsymbol{J} \times \boldsymbol{B} = \nabla P$ is more constrained to ensure the magnetic field on the flux surface ($\boldsymbol{B} \cdot \nabla\psi = 0$) given by $\boldsymbol{B} = \nabla\psi \times \nabla\alpha$ where $\alpha = q\theta_m - \zeta$. This form is called the "Clebsch form".

Let u as the flux surface label, and θ and ζ as poloidal and toroidal angle variables, respectively. Since the magnetic field lies on the flux surface in plasma equilibrium, it is given by the linear combination of two tangent vectors: $\boldsymbol{B} = c_1\partial\boldsymbol{x}/\partial\theta + c_2\partial\boldsymbol{x}/\partial\zeta$. Using the dual relation $\partial\boldsymbol{x}/\partial\theta = J\nabla\zeta \times \nabla u$ and $\partial\boldsymbol{x}/\partial\zeta = J\nabla u \times \nabla\theta$ (A.67), we may write the magnetic field $\boldsymbol{B} = b_1\nabla\zeta \times \nabla u + b_2\nabla u \times \nabla\theta$. Substituting this into $\nabla \cdot \boldsymbol{B} = 0$, we have $\partial b_1/\partial\theta + \partial b_2/\partial\zeta = 0$. This leads to the existence of the stream function h to satisfy $b_1 = -\partial h/\partial\zeta, b_2 = \partial h/\partial\theta$. Thus, the magnetic field is given in the Clebsch form: $\boldsymbol{B} = \nabla u \times \nabla h$. The periodicity requirement for b_1 and b_2 in (θ, ζ) leads to the expression $h = h_1(u)\theta + h_2(u)\zeta + \tilde{h}(u, \theta, \zeta)$, where $\tilde{h}$ is periodic function in (θ, ζ). If we make a coordinate transformation $\theta_m \equiv \theta + \tilde{h}/h_1$, the stream function is give by $h = h_1(u)\theta_m + h_2(u)\zeta$.

Considering the geometry in Fig. 2.2, we define the toroidal flux ϕ and the poloidal flux ψ enclosed by a flux surface as follows,

$$\phi = \frac{1}{2\pi}\int \boldsymbol{B} \cdot d\boldsymbol{a}_\zeta \,, \; \psi = \frac{-1}{2\pi}\int \boldsymbol{B} \cdot d\boldsymbol{a}_\theta \tag{2.8}$$

These fluxes are related to the stream function as $d\phi/du = h_1(u)$ and $d\psi/du = -h_2(u)$. If we take $u = \phi$ or $u = \psi$, $\boldsymbol{B}$ and stream function are given as follows.

$$\boldsymbol{B} = \nabla\phi \times \nabla h \,, \quad h = \theta_m - \zeta/q(\phi) \tag{2.9}$$

$$\boldsymbol{B} = \nabla\psi \times \nabla\alpha \,, \quad \alpha = q(\psi)\theta_m - \zeta \tag{2.10}$$

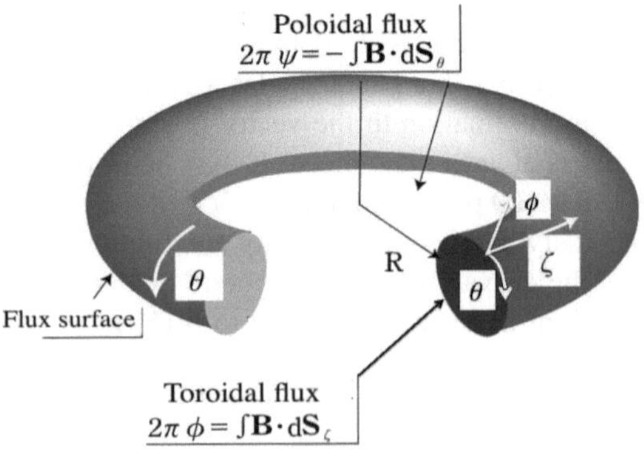

Fig. 2.2 Toroidal coordinates (ϕ, θ, ζ) in a plasma equilibrium and toroidal and poloidal fluxes

Here $q(\phi) \equiv d\phi/d\psi = q(\psi)$. The coordinates (ϕ, θ_m, ζ) (or (ψ, θ_m, ζ)) are called flux coordinates. In this coordinates, the magnetic field line becomes straight and is also called "straight field line coordinates". This coordinate has a simple meaning that we can construct a coordinates in which the magnetic field becomes a straight field in (θ_m, ζ) plane on the flux surface by an appropriate choice of θ to θ_m. The toroidal angle ζ in (2.9) or (2.10) is arbitrary.

For the flux coordinates, we may replace ϕ or ψ to any kind of flux label such as volume inside the flux surface, V or normalized radius ρ defined in Sect. 2.4.

Specific choices of ζ produce important coordinates such as the Hamada coordinates [282] and the Boozer coordinates [71]. In the Hamada coordinates, the toroidal angle ζ is chosen so that $\boldsymbol{j}$ as well as $\boldsymbol{B}$ becomes straight on the flux surface. You may read previous publication Kikuchi [443] for the detail.

The normal vector $\boldsymbol{n}$ is defined as a unit vector normal to the flux surface: $\boldsymbol{n} = \nabla\psi/|\nabla\psi|$. The bi-normal vector $\boldsymbol{s}$ is the defined as $\boldsymbol{s} = \boldsymbol{b} \times \boldsymbol{n}$. The set of unit vector $(\boldsymbol{b}, \boldsymbol{n}, \boldsymbol{s})$ forms the right-handed orthogonal set on the flux surface.

$$\boldsymbol{b} = \boldsymbol{B}/B\,, \boldsymbol{n} = \nabla\psi/|\nabla\psi|\,, \boldsymbol{s} = \boldsymbol{b} \times \boldsymbol{n} \tag{2.11}$$

In the general flux coordinates, the differential operator parallel to $\boldsymbol{B}$ is given by,

$$\boldsymbol{B} \cdot \nabla = J^{-1}\left(\frac{\partial}{\partial\theta_m} + q(\psi)\frac{\partial}{\partial\zeta}\right), \tag{2.12}$$

where J is the Jacobian:

$$J \equiv \frac{\partial \boldsymbol{x}}{\partial\psi} \cdot \frac{\partial \boldsymbol{x}}{\partial\theta_m} \times \frac{\partial \boldsymbol{x}}{\partial\zeta} = \frac{1}{\nabla\psi \times \nabla\theta_m \cdot \nabla\zeta} = \sqrt{g} = \frac{1}{\boldsymbol{B} \cdot \nabla\zeta} \tag{2.13}$$

Choice of ζ as a symmetric coordinate gives,

$$\boldsymbol{B} \cdot \nabla = J^{-1}\partial/\partial\theta_m \tag{2.14}$$

The flux surface average of a function f is defined as the volume average between the flux surface ψ and $\psi + d\psi$.

$$< f >= \frac{\int f dV}{\int dV} = \frac{\oint d\theta f/\boldsymbol{B} \cdot \nabla\theta}{\oint d\theta/\boldsymbol{B} \cdot \nabla\theta} = \frac{\oint f dl_p/B_p}{\oint dl_p/B_p} \tag{2.15}$$

The flux surface average is annihilator of this differential operator $\boldsymbol{B} \cdot \nabla$.

$$\langle \boldsymbol{B} \cdot \nabla f \rangle = 0 \tag{2.16}$$

In Chap. 4, you will see following type of differential equation along the magnetic field called the "magnetic differential equation" by Newcomb.

$$\boldsymbol{B} \cdot \nabla f = S \tag{2.17}$$

Because the flux surface average $\langle\rangle$ is annihilator of $\boldsymbol{B} \cdot \nabla$, we have a constraint for the source function S, which is called the 'solvability condition' of f.

$$\langle S \rangle = 0 \text{ , or } \int_0^{2\pi} \frac{S d\theta}{\boldsymbol{B} \cdot \nabla\theta} = 0 \tag{2.18}$$

Fig. 2.3 (**a**) Abdus Salam is the 1979 Nobel Laureate in physics and also a founder of ICTP (International center for theoretical physics). (**b**) Trieste from the hill. (**c**) Plasma school held at ICTP in 2012 organized by ICTP/IAEA. Headmasters of this school are S. Mahajan and Z. Yoshida

Salon 2-1: Abdus Salam and ICTP
Abdus Salam (1926–1996) (Fig. 2.3a) is a Pakistani theoretical physicist who received 1979 Nobel Prize in physics with Steven Weinberg and Sheldon Lee Glashow for their contribution to electroweak unification. In 1964, he established the Abdus Salam International Centre for Theoretical Physics (ICTP) at Trieste in Italy. ICTP has been a driving force to advance scientific expertise in the developing world for more than 50 years. There are many leading plasma physicists assigned at ICTP in the past such as M. Rosenbluth. In 2012, the author joined Joint ICTP-IAEA College on Plasma Physics (Fig. 2.3c) to give a lecture on Hamilton mechanics of magnetic field lines.

2.3 Tokamak Equilibrium

2.3.1 *Grad-Shafranov Equilibrium*

Tokamak is one of axisymmetric toroidal equilibrium configurations. Other configurations are such as the Spherical Torus (ST) and the Reversed Field Pinch (RFP). The tokamak is characterized by its high toroidal magnetic field (B_t) for the plasma stability. Tokamak has geometrical symmetry in toroidal direction ensuring the existence of flux surface under wide operating conditions. The force balance equation in axisymmetric torus leads to so-called Grad-Shafranov (GS) equation [677], which is nothing but a ζ component of the Ampere's law (B.13) with $\partial/\partial t = 0$ and the constraints to J_ζ from the equilibrium condition.

We will derive the Grad-Shafranov equation. Taking the ζ component of the static Ampere's law, $\nabla^2 \boldsymbol{A} = -\mu_0 \boldsymbol{J}$, and using Eq. (A.54) and the axisymmetry ($\partial/\partial\zeta = 0$), we have following equation for $\psi = RA_\zeta$.

$$\left[\frac{\partial^2}{\partial R^2} - \frac{1}{R}\frac{\partial}{\partial R} + \frac{\partial^2}{\partial Z^2}\right]\psi = -\mu_0 R J_\zeta \tag{2.19}$$

Now the important point is what kind of constraints are imposed for the toroidal plasma current density J_ζ from the equilibrium force balance equation, $\boldsymbol{J}\times\boldsymbol{B} = \nabla p$.

From $\boldsymbol{B} = \nabla\times\boldsymbol{A}$ and $\mu_0\boldsymbol{J} = \nabla\times\boldsymbol{B}$ and the axisymmetry ($\partial/\partial\zeta = 0$), we have

$$B_R = -\frac{1}{R}\frac{\partial\psi}{\partial Z}, \; B_Z = \frac{1}{R}\frac{\partial\psi}{\partial R} \qquad \text{in vector form: } \boldsymbol{B}_p = \nabla\zeta\times\nabla\psi \tag{2.20}$$

$$\mu_0 J_R = -\frac{1}{R}\frac{\partial F}{\partial Z}, \; \mu_0 J_Z = \frac{1}{R}\frac{\partial F}{\partial R} \qquad \text{in vector form: } \mu_0\boldsymbol{J}_p = \nabla\zeta\times\nabla F, \tag{2.21}$$

where $F = RB_\zeta$. We obtain following relations by substitution of above equations (2.20) and (2.21) into $\boldsymbol{B} \cdot \nabla p = 0$ and $\boldsymbol{J} \cdot \nabla p = 0$, which results from the equilibrium force balance equation $\boldsymbol{J} \times \boldsymbol{B} = \nabla p$.

$$\frac{\partial \psi}{\partial R}\frac{\partial p}{\partial Z} - \frac{\partial \psi}{\partial Z}\frac{\partial p}{\partial R} = \frac{\partial(\psi, p)}{\partial(R, Z)} = 0 \qquad \text{i.e, } p = p(\psi) \tag{2.22}$$

$$\frac{\partial F}{\partial R}\frac{\partial p}{\partial Z} - \frac{\partial F}{\partial Z}\frac{\partial p}{\partial R} = \frac{\partial(F, p)}{\partial(R, Z)} = 0 \qquad \text{i.e., } F = F(p) = F(\psi) \tag{2.23}$$

Thus, $\boldsymbol{B}$ in tokamak is given by $\boldsymbol{B} = \nabla\zeta \times \nabla\psi + F(\psi)\nabla\zeta$. Substitution of $\boldsymbol{B} = \boldsymbol{B}_p + F\nabla\zeta$ and $\boldsymbol{J} = \boldsymbol{J}_p + RJ_\zeta\nabla\zeta$ into $\boldsymbol{J} \times \boldsymbol{B} = \nabla p$, we have,

$$\boldsymbol{J}_p \times F\nabla\zeta + RJ_\zeta\nabla\zeta \times \boldsymbol{B}_p = \nabla p \tag{2.24}$$

Substituting $\boldsymbol{B}_p = \nabla\zeta \times \nabla\psi$ and $\mu_0\boldsymbol{J}_p = \nabla\zeta \times \nabla F$ and noting $p = p(\psi)$ and $F = F(\psi)$, we obtain following expression for the toroidal current density J_ζ.

$$\mu_0 RJ_\zeta = \mu_0 R^2\frac{dp}{d\psi} + F\frac{dF}{d\psi} \tag{2.25}$$

Combination of (2.19) and (2.25) gives so-called Grad-Shafranov equation.

$$\left[\frac{\partial^2}{\partial R^2} - \frac{1}{R}\frac{\partial}{\partial R} + \frac{\partial^2}{\partial Z^2}\right]\psi = -\mu_0 R^2\frac{dp}{d\psi} - F\frac{dF}{d\psi} \tag{2.26}$$

Grad [261] further studied the adiabatic evolution of this Grad-Shafranov equation showing that adiabatic equilibrium evolution is characterized as a generalized differential equation (GDE) such that this equation is the elliptic partial differential equation (PDE) seeing from left and is an ordinary differential equation (ODE) seen from right.

Typical equilibrium solution of this GS equation is shown in Fig. 2.4, which has nested flux surfaces. Tokamak system usually operate with the divertor, where the plasma is bounded by the separatrix. Region outside of the separatrix is called SOL (scrape-off-layer), which is an open field line geometry to the diverter plate. The magnetic field is helically wounded on the flux surface around the torus and the safety factor is defined as $q = \Delta\zeta/2\pi$, where $\Delta\zeta$ is the change in the toroidal angle when the magnetic field line returned to the same poloidal angle. The safety factor is given in a different form as $q = d\phi/d\psi$, where $\phi = (2\pi)^{-1}\int B_\zeta dS_\zeta$ is the normalized toroidal flux inside the flux surface.

Fig. 2.4 Solution of Grad-Shafranov equation showing an equi-contour plot of the poloidal flux function with a divertor. Photos of Prof. Halold Grad and Academician Vitaly D. Shafranov (courtesy of Kurchatov institute)

2.3.2 Current Hole Equilibrium

In 2001, tokamak equilibria with almost zero current region in the plasma center is discovered during the advanced tokamak research (see Chap. 3) in JT-60U by Fujita [227], and in JET by Hawkes [300], called the Current Hole (CH), which is shown in Fig 2.5a. This state exists stably without clear evidence of MHD instabilities [227]. This CH is explained by an equilibrium bifurcation from nested to non-nested (or multi-island) equilibrium (Fig. 2.5b) [742].

The dynamics of the CH operation is that it is created by strong heating during current ramp, leading to build up of large bootstrap current [227] or non-inductive CD [300] associated with negative ohmic current with negative E_ζ. This negative E_ζ penetrates to the plasma center and central J_ζ goes to very low but can not be negative due to loss of equilibrium with nested flux surface but Axisymmetric Multi-Magnetic Island (AMMI) equilibrium with slightly negative J_ζ can exist for CH operation. The reason for the current clamp [228] to small negative value is also attributed to the loss of MHD equilibrium by the high $\epsilon\beta_p$ at the current hole region. Number of investigations of AMMI equilibrium has been published since then [125, 634, 804]. Rodrigues [634] in particular showed excellent agreement with JT-60U experimental data as shown in Fig. 2.6.

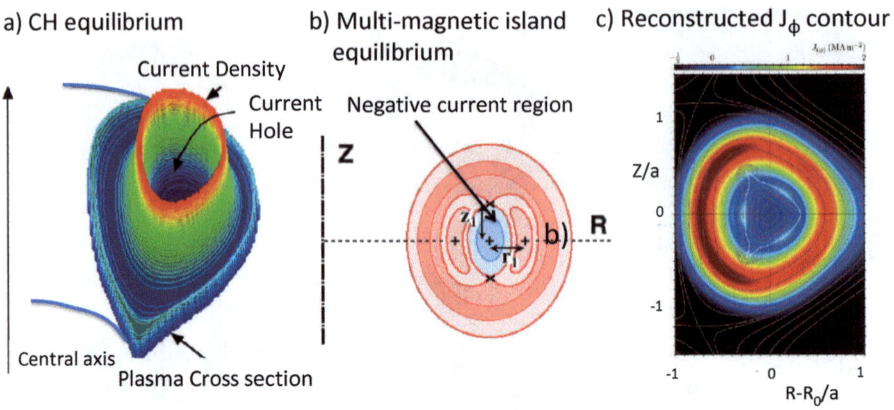

Fig. 2.5 (**a**) Current Hole equilibrium, (**b**) schematics of AMMI equilibrium. After Takizuka [742], and (**c**) reconstructed tri-magnetic Island equilibrium. Reproduced with permission from Rodrigues [634]. Copyright American Physical Society

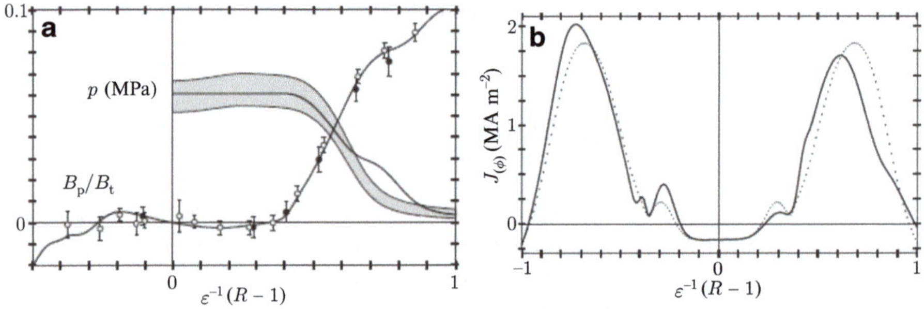

Fig. 2.6 (**a**) Comparison of poloidal field in JT-60U CH and numerical profile of B_p/B_t. (**b**) Reconstructed J_ζ profile showing slightly negative region in the plasma center. Reproduced with permission from Rodrigues [634]. Copyright American Physical Society

2.3.3 *Anisotropic Pressure Equilibrium*

In high temperature tokamak plasma, plasma becomes highly collisionless and the conservation of the magnetic moment produces pressure anisotropy (see Chap. 4). Effect of pressure anisotropy on the toroidal equilibrium is investigated by [260, 711, 738]. Such anisotropy may be important to investigate kinetic MHD stability with energetic particles [119]. Basic equations for the tokamak may be given as,

$$\boldsymbol{J} \times \boldsymbol{B} = \nabla \cdot \boldsymbol{P}, \tag{2.27}$$

where the pressure tensor $\boldsymbol{P}$ is given by the following Chew-Goldberger-Low (CGL) form in the magnetized plasma [121],

$$\boldsymbol{P} = p_\perp \boldsymbol{I} - (p_\perp - p_\parallel)\boldsymbol{bb} \tag{2.28}$$

Noting $\boldsymbol{b}\cdot\nabla\cdot(p_\perp \boldsymbol{I}) = b_i\partial_j(p_\perp\delta_{ji}) = b_i\partial_i p_\perp = \boldsymbol{b}\cdot\nabla p_\perp$, and $\boldsymbol{b}\cdot\nabla\cdot[(p_\parallel - p_\perp)\boldsymbol{bb}] = b_i\partial_j[(p_\parallel - p_\perp)b_j b_i] = b_i(p_\parallel - p_\perp)b_j\partial_j b_i + b_i b_i \partial_j[(p_\parallel - p_\perp)b_j] = (p_\parallel - p_\perp)\boldsymbol{b}\cdot(\boldsymbol{b}\cdot\nabla\boldsymbol{b}) + \nabla\cdot[(p_\parallel - p_\perp)\boldsymbol{B}/B]$. As we see in (2.6), the curvature is perpendicular to the magnetic field: $\boldsymbol{b}\cdot(\boldsymbol{b}\cdot\nabla\boldsymbol{b}) = \boldsymbol{b}\cdot\boldsymbol{\kappa} = 0$ and using $\nabla\cdot\boldsymbol{B} = 0$, we have $\boldsymbol{b}\cdot\nabla\cdot[(p_\parallel - p_\perp)\boldsymbol{bb}] = \boldsymbol{B}\cdot\nabla[(p_\parallel - p_\perp)/B] = \boldsymbol{b}\cdot\nabla(p_\parallel - p_\perp) - [(p_\parallel - p_\perp)/B]\boldsymbol{b}\cdot\nabla B$. The inner product $\boldsymbol{b}\cdot$(2.27) reads $B\partial p_\parallel/\partial s + (p_\perp - p_\parallel)\partial B/\partial s = 0$ where $\partial/\partial s = \boldsymbol{b}\cdot\nabla$ as given by J.B. Taylor [754]. It is convenient to express $p_\perp$ and $p_\parallel$ in terms of (ψ, B) and define σ, τ and $\boldsymbol{K}$ as follows.

$$\sigma \equiv 1 - \frac{\mu_0}{B}\frac{\partial p_\parallel}{\partial B}\Big|_\psi,\ \tau \equiv 1 + \frac{\mu_0}{B}\frac{\partial p_\perp}{\partial B}\Big|_\psi,\ \boldsymbol{K} \equiv \frac{1}{\mu_0}\nabla\times(\sigma\boldsymbol{B}) \tag{2.29}$$

Then the tensor force balance equation (2.27) is reduced to,

$$\boldsymbol{K}\times\boldsymbol{B} = \frac{\partial p_\parallel}{\partial\psi}\Big|_B \nabla\psi \tag{2.30}$$

$$\frac{\mu_0}{B}\frac{\partial p_\parallel}{\partial B}\Big|_\psi = 1 - \sigma = \frac{\mu_0(p_\parallel - p_\perp)}{B^2} \tag{2.31}$$

Using the expression of the magnetic field in tokamak $\boldsymbol{B} = \nabla\zeta\times\nabla\psi + F\nabla\zeta$, $\nabla\zeta$ component of (2.30) reads $\boldsymbol{K}\cdot\nabla\psi = 0$.

Using (2.29), we find $I = \sigma F$ is a flux function. Taking the $\nabla\psi$ component of Eq. (2.30) and using the relations between $\boldsymbol{J}, \boldsymbol{B}, \boldsymbol{K}$, we obtain the anisotropic Grad-Shafranov equation derived by H. Grad [260].

$$L\psi + \nabla\psi\cdot\nabla ln\sigma = -\frac{\mu_0 R^2}{\sigma}\frac{\partial p_\parallel}{\partial\psi}\Big|_B - \frac{1}{\sigma^2}I\frac{dI}{d\psi} \tag{2.32}$$

$$L \equiv \left[\frac{\partial^2}{\partial R^2} - \frac{1}{R}\frac{\partial}{\partial R} + \frac{\partial^2}{\partial Z^2}\right]$$

Important observation here is that the current density vector $\boldsymbol{J}$ does not lie on the flux surface. Instead, the vector $\boldsymbol{K}$ lies on the flux surface.

2.3.4 Equilibrium with Flow

Due to the toroidal symmetry, the tokamak plasma can rotate easily in the toroidal direction up to an order of sound velocity but is more difficult to rotate poloidally (see Sect. 4.5.1). Toroidal flow can be driven by the external torque, symmetry breaking in neoclassical flow (see Chap. 4) or in plasma turbulence (see Chap. 7). Even small toroidal rotation can affect stability of the resistive wall mode (see Chap. 8).

Effect of flow on the toroidal equilibrium is investigated by Hameiri [286], Takeda and Tokuda [738]. Toroidal flow is important to stabilize resistive wall mode and also modifies energy principle as discussed in Chap. 8. Basic equations for the tokamak may be given as,

$$\rho \boldsymbol{u} \cdot \nabla \boldsymbol{u} + \nabla P = \boldsymbol{J} \times \boldsymbol{B} \tag{2.33}$$

$$\nabla \times (\boldsymbol{u} \times \boldsymbol{B}) = 0 \tag{2.34}$$

$$\boldsymbol{B} = \nabla \zeta \times \nabla \psi + F \nabla \zeta \tag{2.35}$$

Taking $\nabla\times$ (2.35) and using L defined in (2.32), we have

$$\mu_0 \boldsymbol{J} = (L\psi)\nabla \zeta + \nabla F \times \nabla \zeta \tag{2.36}$$

From Eq. (2.34), we have $\boldsymbol{u} \times \boldsymbol{B} = -\nabla \Phi$. This means $\boldsymbol{B} \cdot \nabla \Phi = 0$ and the Φ is a flux function. Therefore, we may write $\nabla \Phi = \Omega(\psi)\nabla\psi$. For purely toroidal rotation, we have $\boldsymbol{u} = R^2 \Omega \nabla \zeta$ by using (2.35). Substituting $\boldsymbol{u} = R^2\Omega\nabla\zeta$ into Eq. (2.33), we have

$$\nabla P = \boldsymbol{J} \times \boldsymbol{B} + \rho R \Omega^2 \nabla R \tag{2.37}$$

The second term of the RHS is the centrifugal force due to the plasma toroidal rotation. Due to this term, plasma pressure can not be a flux function. We use a curvilinear coordinates (R, ζ, ψ) instead of cylindrical coordinates (R, ζ, Z). We use the orthogonal relation (A.66): $\nabla u^i \cdot (\partial \boldsymbol{x}/\partial u^j) = \delta_{ij}$. Taking the toroidal component (2.37)$\cdot\nabla\zeta$, we have $(\boldsymbol{J} \times \boldsymbol{B}) \cdot \nabla \zeta = 0$. Substituting (2.36) and (2.35), we have $(\nabla\psi \times \nabla F) \cdot \nabla \zeta = 0$, i.e. $F = F(\psi)$.

$$\boldsymbol{J} \times \boldsymbol{B} = -\frac{FF'(\psi) + L\psi}{\mu_0 R^2} \nabla \psi \tag{2.38}$$

Thus, $\boldsymbol{J} \times \boldsymbol{B}$ force is still perpendicular to the flux surface ψ, while ∇P force is not.

Taking the radial component (2.37)$\cdot(\partial \boldsymbol{x}/\partial R)$ and using the orthogonal relations $\nabla R \cdot \partial \boldsymbol{x}/\partial R = 1$ and $\nabla \psi \cdot \partial \boldsymbol{x}/\partial R = 0$, we have

$$\partial P(\psi, R)/\partial R = \rho R \Omega(\psi)^2 \tag{2.39}$$

Taking (2.37)$\cdot\partial \boldsymbol{x}/\partial\psi$ and noting orthogonal relations $\nabla\psi\cdot\partial\boldsymbol{x}/\partial\psi = 1$ and $\nabla R\cdot\partial\boldsymbol{x}/\partial\psi = 0$, we have following modified Grad-Shafravov equation,

$$L\psi = -\mu_0 R^2 \frac{\partial P(\psi, R)}{\partial\psi} - F\frac{dF(\psi)}{d\psi} \tag{2.40}$$

2.3.5 General Tensor Equilibrium

In case plasma equilibrium has both the plasma flow and the pressure anisotropy, we have to include both effect. Moreover, if we apply non-axisymmetric field to tokamak, we have 3 dimensional plasma equilibrium. The equilibrium in this case is governed by most general tensor equilibrium relation as follows,

$$\rho(\boldsymbol{u}\cdot\nabla)\boldsymbol{u} = \boldsymbol{J}\times\boldsymbol{B} - \nabla\cdot\boldsymbol{P}, \tag{2.41}$$

where the pressure tensor $\boldsymbol{P}$ is given by the CGL form (2.28). The $\boldsymbol{J}\times\boldsymbol{B}$ force can be expressed by the Maxwell tensor in (B.19) and the convective derivative $(\boldsymbol{u}\cdot\nabla)\boldsymbol{u}$ can be given by the Reynolds stress in the Note. Then the equilibrium is expressed as,

$$\frac{\partial\Pi_{ij}}{\partial x_i} = 0\,,\ \text{or}\ \nabla\cdot\boldsymbol{\Pi} = 0 \tag{2.42}$$

$$\Pi_{ij} = (p_\perp + \frac{B^2}{2\mu_0})\delta_{ij} - \left(1 + \frac{\mu_0(p_\perp - p_\parallel)}{B^2}\right)\frac{B_iB_j}{\mu_0} + \rho u_i u_j, \tag{2.43}$$

where Π_{ij} may be called the 'generalized pressure tensor' or the "momentum flux density" (see Column 2-1). The generalized pressure tensor Π_{ij} is a symmetric tensor.

$$\Pi_{ij} = \Pi_{ji} \tag{2.44}$$

Using the tensor identity, $\partial(x_k\Pi_{ij})/\partial x_i = \Pi_{ij}\delta_{ki} + x_k\partial\Pi_{ij}/\partial x_i$ and noting second term of RHS (right hand side) vanishes at equilibrium by the relation (2.42), we have following integral relation.

$$\oint x_k\Pi_{ij}dS_j = \int \Pi_{ij}\delta_{kj}dV, \tag{2.45}$$

where we used Gauss's theorem to change volume integral to surface integral for the LHS (left hand side). Using explicit expression for Π_{ij}, we have following general 'Virial theorem' by S. Chandrasekhar [103] and S. Yoshikawa [848] .

$$\oint x_k\left((p+\frac{B^2}{2\mu_0})\delta_{ij}-\frac{B_iB_j}{\mu_0}+\rho u_iu_j\right)dS_j$$
$$=\int\left((p+\frac{B^2}{2\mu_0})\delta_{ij}-\frac{B_iB_j}{\mu_0}+\rho u_iu_j\right)\delta_{kj}dV \quad (2.46)$$

Taking the trace ($k=j=i=1,3$), we have following form by V. Shafranov [677].

$$\oint\left[\left(p+\frac{B^2}{2\mu_0}\right)\boldsymbol{x}\cdot d\boldsymbol{S}-\frac{(\boldsymbol{B}\cdot\boldsymbol{x})(\boldsymbol{B}\cdot d\boldsymbol{S})}{\mu_0}+\rho(\boldsymbol{u}\cdot\boldsymbol{x})(\boldsymbol{u}\cdot d\boldsymbol{S})\right]$$
$$=\int\left(3p+\frac{B^2}{2\mu_0}+\rho u^2\right)dV \quad (2.47)$$

Column 2-1: Momentum Flux Density in Ideal Neutral Fluid
In ideal neutral fluid, the continuity equation and the Euler equation are:

$$\frac{\partial\rho}{\partial t}+\frac{\partial\rho u_i}{\partial x_i}=0$$
$$\rho\left(\frac{\partial u_i}{\partial t}+u_j\frac{\partial u_i}{\partial x_j}\right)=-\frac{\partial p}{\partial x_i} \quad (2.48)$$

This Euler equation can be converted to following conservative form.

$$\frac{\partial\rho u_i}{\partial t}=-\frac{\partial\Pi_{ik}}{\partial x_k}, \quad (2.49)$$
$$\text{where}\quad \Pi_{ik}=p\delta_{ik}+\rho u_iu_k$$

Here, Π_{ki} is called the "momentum flux density" and is a symmetric tensor (see Landau-Lifschitz [495]). Here ρu_iu_j is called the "Reynolds stress". Integrating in volume and using the Green's theorem, we have:

$$\partial/\partial t\int\rho u_idV=-\oint\Pi_{ik}dS_k \quad (2.50)$$

The $\Pi_{ik}dS_k$ is the i-th component of the momentum flowing out from the volume through the surface dS.

Column 2-2: Helicity and Kelvin's Circulation Theorem
The Euler equation (2.48) has eight global conserved quantities (the kinetic energy $K=\int\frac{1}{2}\rho u^2dV$, the helicity $H=\int\boldsymbol{u}\cdot\boldsymbol{\omega}dV$, the momentum $\boldsymbol{P}=\int\rho\boldsymbol{u}dV$, and the angular momentum $\boldsymbol{M}=\int\boldsymbol{x}\times\boldsymbol{u}dV$), and infinite number of

(continued)

Fig. 2.7 William Thomson (Lord Kelvin) (1824–1907). England Physicist. He is famous for introduction of absolute temperature, Joule-Thomson effect, Kelvin's circulation theorem, etc.

local conserved quantities. The circulation Γ is defined as,

$$\Gamma = \oint \boldsymbol{u} \cdot d\boldsymbol{l} \equiv \oint \boldsymbol{u} \cdot \delta \boldsymbol{x} \tag{2.51}$$

The last equation is just to separate differential operator and integration. The time derivative of the circulation is given by,

$$\frac{d\Gamma}{dt} = \frac{d}{dt}\oint \boldsymbol{u} \cdot \delta\boldsymbol{x} = \oint \frac{d\boldsymbol{u}}{dt} \cdot \delta\boldsymbol{x} + \oint \boldsymbol{u} \cdot \frac{d\delta\boldsymbol{x}}{dt}$$

The second term becomes total differential $\delta u^2/2$ and its closed integral is zero. In case of isentropic system where $\rho = \rho(p)$, Euler equation can be expressed as $d\boldsymbol{u}/dt = -\nabla\Phi$, where $d\Phi = dp/\rho$. Applying the Stokes theorem, we have conservation of the circulation using $\nabla \times \nabla\Phi \equiv 0$, called Kelvin's circulation theorem.

$$\frac{d}{dt}\oint \boldsymbol{u} \cdot d\boldsymbol{l} = 0 \tag{2.52}$$

2.4 1.5D Transport Equations in Tokamak

Basic equations for the time evolution of the tokamak equilibrium and transport are given in this section based on the development of TOPICS code at JAEA. These equation are used in the time dependent current diffusion analysis to identify bootstrap current (Fig. 4.4), for example.

In this section, the moving coordinates (ρ, θ, ζ) is employed, where ρ is a flux surface label, θ is the poloidal angle, ζ is the toroidal angle (ignorable coordinate). For moving coordinates, try two exercises 2.10 and 2.11 and also read [321] for the details.

2.4.1 Integral Relations

A volume integral of a function f inside the flux label ρ is given by,

$$\int f d\tau = \int_0^\rho d\rho \left[\int fJ d\theta d\zeta \right], \tag{2.53}$$

where $d\tau = J d\rho d\theta d\zeta$ is a volume element and $J = 1/\nabla\rho \cdot (\nabla\theta \times \nabla\zeta)$ is the Jacobian. Let V is volume inside ρ, we have

$$\frac{\partial V}{\partial \rho} = \int J d\theta d\zeta \tag{2.54}$$

The surface average of f is

$$\int fJ d\theta d\zeta = \frac{\partial V}{\partial \rho} < f > \tag{2.55}$$

For the surface integral of a vector $\boldsymbol{f}$ on the flux surface ρ, we have

$$\oint \boldsymbol{f} \cdot d\boldsymbol{S}_\rho = \oint \boldsymbol{f} \cdot \nabla\rho J d\theta d\zeta = \frac{\partial V}{\partial \rho} < \boldsymbol{f} \cdot \nabla\rho > \tag{2.56}$$

The moving coordinate $\rho = \rho(\psi, t)$ moves with a speed of $\boldsymbol{u}_\rho$, i.e., $d\rho/dt = \partial\rho/\partial t + \boldsymbol{u}_\rho \cdot \nabla\rho = 0$. Using this equation, the temporal evolution of the surface quantity $\partial f/\partial t$ in the real coordinate is expressed in the moving coordinate ρ as.

$$\left.\frac{\partial f}{\partial t}\right|_{\boldsymbol{x}} = \left.\frac{\partial f}{\partial t}\right|_\rho + \left.\frac{\partial \rho}{\partial t}\frac{\partial f}{\partial \rho}\right|_{\boldsymbol{x}} = \left.\frac{\partial f}{\partial t}\right|_\rho - \boldsymbol{u}_\rho \cdot \nabla\rho \left.\frac{\partial f}{\partial \rho}\right|_{\boldsymbol{x}} \tag{2.57}$$

In the moving coordinate ρ, we have following integral relation.

$$\frac{\partial}{\partial t}\int^\rho f d\tau = \int^\rho \frac{\partial f}{\partial t} d\tau + \oint f\boldsymbol{u}_\rho \cdot d\boldsymbol{S}_\rho, \tag{2.58}$$

where the volume integral is taking inside the ρ surface.

In the following, we derive the temporal evolution of plasma parameters and magnetic fluxes in the moving coordinate ρ.

2.4.2 *Energy and Particle Conservation Equations*

We start from the derivation of a reduced set of the transport equations in the direction of ρ for the particle and the energy in the time-varying 2 dimensional equilibrium. The derivation from the kinetic equation can be seen in the beginning of Chap. 4.

2.4.2.1 Particle Conservation Equation

The particle conservation equation can be expressed in a following form.

$$\frac{\partial n_a}{\partial t} + \nabla \cdot (n_a \boldsymbol{u}_a) = S_a, \tag{2.59}$$

where 'a' represents species a and S_a is a particle source density.

Taking the volume integration inside the flux surface ρ and considering ρ is moving ($d\rho/dt = \partial\rho/\partial t + \boldsymbol{u}_\rho \cdot \nabla\rho = 0$), we obtain

$$\begin{aligned} \frac{\partial}{\partial t}\int^{\rho} n_a d\tau\bigg|_{\rho} &= \int^{\rho} \frac{\partial n_a}{\partial t} d\tau + \oint^{\rho} n_a \boldsymbol{u}_\rho \cdot d\boldsymbol{S}_\rho \\ &= -\oint n_a(\boldsymbol{u}_a - \boldsymbol{u}_\rho)\cdot d\boldsymbol{S}_\rho + \int S_a d\tau \end{aligned} \tag{2.60}$$

Taking the ρ derivative of the above equation and using Eqs. (2.55), (2.56), we have

$$\frac{\partial}{\partial t}\left[\langle n_a\rangle \frac{\partial V}{\partial\rho}\right] + \frac{\partial}{\partial\rho}\left[\langle n_a(\boldsymbol{u}_a - \boldsymbol{u}_\rho)\cdot\nabla\rho\rangle \frac{\partial V}{\partial\rho}\right] = \langle S_a\rangle \frac{\partial V}{\partial\rho} \tag{2.61}$$

We simplify the symbol $n_a = \langle n_a\rangle$. We also use the following diffusion approximation, which is not always true.

$$\langle n_a(\boldsymbol{u}_a - \boldsymbol{u}_\rho)\cdot\nabla\rho\rangle = \langle D_a \nabla n_a \cdot \nabla\rho\rangle = D_a \langle|\nabla\rho|^2\rangle \frac{\partial n_a}{\partial\rho}, \tag{2.62}$$

We have following particle diffusion equation.

$$\frac{1}{V'}\frac{\partial}{\partial t}\left[\langle n_a\rangle V'\right] + \frac{1}{V'}\frac{\partial}{\partial\rho}\left[V' D_a \langle|\nabla\rho|^2\rangle \frac{\partial n_a}{\partial\rho}\right] = \langle S_a\rangle \quad , \quad V' \equiv \frac{\partial V}{\partial\rho} \tag{2.63}$$

2.4.2.2 Energy Conservation Equation

We derive a reduced equation for the following simple energy balance equation.

$$\frac{3}{2}\frac{\partial P_a}{\partial t} + \frac{3}{2}\nabla\cdot(P_a \boldsymbol{u}_a) + P_a \nabla\cdot\boldsymbol{u}_a = -\nabla\cdot\boldsymbol{q}_a + Q_a \tag{2.64}$$

Taking the volume integration inside the flux surface ρ, we obtain

$$
\begin{aligned}
&\frac{3}{2}\frac{\partial}{\partial t}\int^{\rho} P_a d\tau\bigg|_{\rho} \\
&\quad = -\frac{3}{2}\int^{\rho} P_a(\boldsymbol{u}_a - \boldsymbol{u}_\rho)\cdot d\boldsymbol{S}_\rho - \int p_a \nabla\cdot \boldsymbol{u}_a d\tau - \int^{\rho} \boldsymbol{q}_a \cdot d\boldsymbol{S}_\rho + \int^{\rho} Q_a d\tau \\
&\quad = -\int^{\rho}\left[\frac{5}{2}P_a(\boldsymbol{u}_a - \boldsymbol{u}_\rho) + \boldsymbol{q}_a\right]\cdot d\boldsymbol{S}_\rho \\
&\qquad + \int^{\rho} (\boldsymbol{u}_a - \boldsymbol{u}_\rho)\cdot\nabla P_a d\tau - \int P_a \nabla\cdot\boldsymbol{u}_\rho d\tau + \int Q_a d\tau
\end{aligned}
\tag{2.65}
$$

Taking the ρ derivative of the above equation and using Eqs. (2.55), (2.56), we have

$$
\begin{aligned}
&\frac{3}{2}\frac{\partial}{\partial t}\left[P_a\frac{\partial V}{\partial \rho}\right] + \frac{\partial}{\partial \rho}\left[\frac{\partial V}{\partial \rho}\left[\frac{5}{2}\langle P_a(\boldsymbol{u}_a - \boldsymbol{u}_\rho)\cdot\nabla\rho\rangle + \langle \boldsymbol{q}\cdot\nabla\rho\rangle\right]\right] \\
&\qquad -\big\langle(\boldsymbol{u}_a - \boldsymbol{u}_\rho)\cdot\nabla P_a\big\rangle\frac{\partial V}{\partial \rho} + \big\langle P_a\nabla\cdot\boldsymbol{u}_\rho\big\rangle\frac{\partial V}{\partial \rho} = \langle Q_a\rangle\frac{\partial V}{\partial \rho}
\end{aligned}
\tag{2.66}
$$

To derive a final form, we use following approximation.

$$
\big\langle P_a(\boldsymbol{u}_a - \boldsymbol{u}_\rho)\cdot\nabla\rho\big\rangle = T_a\big\langle n_a(\boldsymbol{u}_a - \boldsymbol{u}_\rho)\cdot\nabla\rho\big\rangle = D_a\big\langle|\nabla\rho|^2\big\rangle T_a\frac{\partial n_a}{\partial \rho} \tag{2.67}
$$

$$
\langle \boldsymbol{q}_a\cdot\nabla\rho\rangle = \langle \kappa\nabla T_a\cdot\nabla\rho\rangle = \kappa\big\langle|\nabla\rho|^2\big\rangle\frac{\partial T_a}{\partial \rho} \tag{2.68}
$$

$$
\big\langle(\boldsymbol{u}_a - \boldsymbol{u}_\rho)\cdot\nabla P_a\big\rangle = \big\langle(\boldsymbol{u}_a - \boldsymbol{u}_\rho)\cdot\nabla\rho\big\rangle\frac{\partial P_a}{\partial \rho} = D_a\big\langle|\nabla\rho|^2\big\rangle\frac{1}{n_a}\frac{\partial n_a}{\partial \rho}\frac{\partial P_a}{\partial \rho} \tag{2.69}
$$

$$
\big\langle P_a\nabla\cdot\boldsymbol{u}_\rho\big\rangle\frac{\partial V}{\partial \rho} = P_a\big\langle\nabla\cdot\boldsymbol{u}_\rho\big\rangle\frac{\partial V}{\partial \rho} = P_a\frac{\partial}{\partial \rho}\int\boldsymbol{u}_\rho\cdot d\boldsymbol{S}_\rho = P_a\frac{\partial}{\partial t}\left(\frac{\partial V}{\partial \rho}\right) \tag{2.70}
$$

The final form of the energy conservation equation is given by,

$$
\begin{aligned}
&\frac{1}{(V')^{5/3}}\frac{\partial}{\partial t}\left[P_a(V')^{5/3}\right] + \frac{1}{V'}\frac{\partial}{\partial \rho}\left[V'\big\langle|\nabla\rho|^2\big\rangle\left(\frac{5}{2}D_a T_a\frac{\partial n_a}{\partial \rho} + \kappa\frac{\partial T_a}{\partial \rho}\right)\right] \\
&\qquad - D_a\big\langle|\nabla\rho|^2\big\rangle\frac{1}{n_a}\frac{\partial n_a}{\partial \rho}\frac{\partial P_a}{\partial \rho} = \langle Q_a\rangle
\end{aligned}
\tag{2.71}
$$

2.4.3 Magnetic Diffusion Equation

We derive equations of magnetic flux from following Faraday's law.

$$\frac{\partial \boldsymbol{B}}{\partial t} = -\nabla \times \boldsymbol{E} \tag{2.72}$$

$$\boldsymbol{B} = F\nabla\zeta + \nabla\zeta \times \nabla\psi \tag{2.73}$$

$$\boldsymbol{E}^* = \boldsymbol{E} + \boldsymbol{u} \times \boldsymbol{B} \tag{2.74}$$

Taking $\nabla\times$ (2.73) and using $\Delta^*\psi = R^2\nabla\cdot(R^{-2}\nabla\psi)$, we obtain

$$\mu_0\boldsymbol{J} = \nabla\times\boldsymbol{B} = \nabla\times(F\nabla\zeta + \nabla\zeta\times\nabla\psi) = \nabla F\times\nabla\zeta + \Delta^*\psi\nabla\zeta \tag{2.75}$$

2.4.3.1 Poloidal Flux Diffusion Equation

Taking $\nabla\zeta\times$ (2.72) and using the vector formula $\boldsymbol{a}\times(\boldsymbol{b}\times\boldsymbol{c}) = \boldsymbol{b}(\boldsymbol{a}\cdot\boldsymbol{c}) - \boldsymbol{c}(\boldsymbol{a}\cdot\boldsymbol{b})$,

$$\frac{\partial}{\partial t}\frac{\nabla\psi}{R^2} = \nabla\zeta\times(\nabla\times\boldsymbol{E}) \tag{2.76}$$

Using the formula $\nabla(\boldsymbol{a}\cdot\boldsymbol{b}) = (\boldsymbol{b}\cdot\nabla)\boldsymbol{a} + \boldsymbol{b}\times(\nabla\times\boldsymbol{a}) + (\boldsymbol{a}\cdot\nabla)\boldsymbol{b} + \boldsymbol{a}\times(\nabla\times\boldsymbol{b})$,

$$\begin{aligned}\nabla\zeta\times(\nabla\times\boldsymbol{E}) &= \nabla(\boldsymbol{E}\cdot\nabla\zeta) - (\nabla\zeta\cdot\nabla)\boldsymbol{E} - (\boldsymbol{E}\cdot\nabla)\nabla\zeta \\ &= \nabla\left(\frac{E_\zeta}{R}\right) + \frac{-E_R\boldsymbol{e}_\zeta + E_\zeta\boldsymbol{e}_R}{R^2} + \frac{E_R\boldsymbol{e}_\zeta + E_\zeta\boldsymbol{e}_R}{R^2} \\ &= \nabla\left(\frac{E_\zeta}{R}\right) + \frac{2E_\zeta}{R^2}\boldsymbol{e}_R = \frac{1}{R^2}\nabla(RE_\zeta),\end{aligned} \tag{2.77}$$

where $\boldsymbol{e}_R$ and $\boldsymbol{e}_\zeta$ are unit vector in R and ζ directions, respectively. Substituting this into (2.76) gives $R^{-2}\nabla[\partial\psi/\partial t - RE_\zeta] = 0$ and integration in space gives,

$$\frac{\partial\psi}{\partial t} = RE_\zeta + c(t) \tag{2.78}$$

Taking $\nabla\zeta\cdot$(2.73), we have $(E^*_\zeta - E_\zeta)/R = \nabla\zeta\cdot\boldsymbol{u}\times(\nabla\zeta\times\nabla\psi) = \boldsymbol{u}\cdot\nabla\psi/R^2$. Therefore, we have $RE_\zeta = -\boldsymbol{u}\cdot\nabla\psi + RE^*_\zeta$. Substituting this equation to (2.74),

$$\frac{\partial\psi}{\partial t} + \boldsymbol{u}\cdot\nabla\psi = RE^*_\zeta + c(t) \tag{2.79}$$

Substitution of (2.57) to (2.79) gives following relations.

$$\left.\frac{\partial\psi}{\partial t}\right|_{\rho} + (\boldsymbol{u} - \boldsymbol{u}_p)\cdot\nabla\rho\frac{\partial\psi}{\partial\rho} = RE_{\zeta}^{*} + c(t) \tag{2.80}$$

Finally, by surface-averaging, the following equation is obtained.

$$\left.\frac{\partial\psi}{\partial t}\right|_{\rho} + U\frac{\partial\psi}{\partial\rho} =< RE_{\zeta}^{*} > +c(t) \tag{2.81}$$

$$U \equiv< (\boldsymbol{u} - \boldsymbol{u}_p)\cdot\nabla\rho > \tag{2.82}$$

For the next subsection, another expression of ψ is

$$\left.\frac{\partial\psi}{\partial t}\right|_{\rho} + \frac{(< \boldsymbol{u} - \boldsymbol{u}_p)\cdot\nabla\rho/R^2 >}{< 1/R^2 >}\frac{\partial\psi}{\partial\rho} = \frac{< E_{\zeta}^{*}/R >}{< 1/R^2 >} + c(t) \tag{2.83}$$

2.4.3.2 Toroidal Flux Diffusion Equation

Taking $\nabla\zeta\cdot$(2.72) and using the vector formula $\nabla\cdot(\boldsymbol{a}\times\boldsymbol{b}) = \boldsymbol{b}\cdot(\nabla\times\boldsymbol{a}) - \boldsymbol{a}\cdot(\nabla\times\boldsymbol{b})$,

$$\frac{\partial}{\partial t}\frac{F}{R^2} = -\nabla\zeta\cdot\nabla\times\boldsymbol{E} = \nabla\cdot(\nabla\zeta\times\boldsymbol{E}) \tag{2.84}$$

Taking $\nabla\zeta\times$ (2.74) and using the vector formula $\boldsymbol{a}\times(\boldsymbol{b}\times\boldsymbol{c}) = \boldsymbol{b}(\boldsymbol{a}\cdot\boldsymbol{c}) - \boldsymbol{c}(\boldsymbol{a}\cdot\boldsymbol{b})$,

$$\nabla\zeta\times\boldsymbol{E} = (\boldsymbol{u}\cdot\nabla\zeta)\boldsymbol{B} - \frac{F}{R^2}\boldsymbol{u} + \nabla\zeta\times\boldsymbol{E}^{*} \tag{2.85}$$

Substituting this into (2.84), we obtain

$$\frac{\partial}{\partial t}\frac{F}{R^2} = \nabla\cdot\left[(\boldsymbol{u}\cdot\nabla\zeta)\boldsymbol{B} - \frac{F}{R^2}\boldsymbol{u} + \nabla\zeta\times\boldsymbol{E}^{*}\right] \tag{2.86}$$

Taking this volume integration inside the ρ surface and using Gauss's theorem and noting $\boldsymbol{B}\cdot d\boldsymbol{S}_{\rho} = 0$ since ρ is also a flux label, we obtain

$$\int^{\rho}\frac{\partial}{\partial t}\frac{F}{R^2}d\tau = \oint\left[-\frac{F}{R^2}\boldsymbol{u} + \nabla\zeta\times\boldsymbol{E}^{*}\right]\cdot d\boldsymbol{S}_{\rho} \tag{2.87}$$

Substituting $f = F/R^2$ in the integral relation (2.58) in the moving coordinates,

$$\frac{\partial}{\partial t}\int^{\rho}\frac{F}{R^2}d\tau = \oint\left[-\frac{F}{R^2}(\boldsymbol{u} - \boldsymbol{u}_{\rho}) + \nabla\zeta\times\boldsymbol{E}^{*}\right]\cdot d\boldsymbol{S}_{\rho} \tag{2.88}$$

The toroidal flux function $\chi(\rho)$ is introduced as

$$\chi(\rho) = 2\pi \int \boldsymbol{B} \cdot d\boldsymbol{S}_\zeta = 2\pi \int^V \frac{F(\psi)}{R^2} J dV d\theta = \int^V \frac{F(\psi)}{R^2} d\tau \tag{2.89}$$

$$\frac{\partial \chi}{\partial V} = \frac{\partial}{\partial V} \int^V \frac{F}{R^2} d\tau = F < 1/R^2 > \tag{2.90}$$

By using these relations between F and χ, the equation of toroidal flux evolution is

$$\left.\frac{\partial \chi}{\partial t}\right|_\rho = -\frac{\partial \chi}{\partial \rho} \frac{< (\boldsymbol{u} - \boldsymbol{u}_\rho) \cdot \nabla \rho / R^2 >}{< 1/R^2 >} - < \zeta \times \nabla \rho \cdot \boldsymbol{E}^* > \frac{\partial V}{\partial \rho} \tag{2.91}$$

Using Eqs. (2.81), (2.83),

$$\frac{< (\boldsymbol{u} - \boldsymbol{u}_p) \cdot \nabla \rho / R^2 >}{< 1/R^2 >} = < (\boldsymbol{u} - \boldsymbol{u}_p) \cdot \nabla \rho > + \frac{1}{\partial \psi / \partial \rho} \left[\frac{< E_\zeta^* / R >}{< 1/R^2 >} - < R E_\zeta^* > \right] \tag{2.92}$$

we have

$$\left.\frac{\partial \chi}{\partial t}\right|_\rho + U \frac{\partial \chi}{\partial \rho} = -\frac{\partial V}{\partial \psi} < \boldsymbol{B} \cdot \boldsymbol{E}^* > + \frac{\partial \chi}{\partial \psi} < R E_\zeta^* > \tag{2.93}$$

2.4.3.3 Magnetic Pitch Diffusion Equation

Let define a quantity proportional to pitch of the magnetic field.

$$\nu \equiv \frac{\partial \psi}{\partial \chi} = \frac{1}{4\pi^2 q} \tag{2.94}$$

Using $\partial \psi / \partial \rho = \nu \partial \chi / \partial \rho$, the time derivative of ν is given as,

$$\left.\frac{\partial \nu}{\partial t}\right|_\rho = \frac{\partial}{\partial t} \left[\frac{\partial \psi / \partial \rho}{\partial \chi / \partial \rho} \right] = \frac{\partial \rho}{\partial \chi} \left[\frac{\partial}{\partial \rho} \left(\frac{\partial \psi}{\partial t} \right)_\rho - \nu \frac{\partial}{\partial \rho} \left(\frac{\partial \chi}{\partial t} \right)_\rho \right] \tag{2.95}$$

Using (2.82) and (2.94) and noting $\partial^2 \psi / \partial \rho^2 = (\partial \nu / \partial \rho)(\partial \chi / \partial \rho) + \nu (\partial^2 \chi / \partial \rho^2)$, we have,

$$\left.\frac{\partial \nu}{\partial t}\right|_\rho = \left.\frac{\partial \chi}{\partial t}\right|_\rho \frac{\partial \nu}{\partial \chi} + \frac{\partial}{\partial \chi} \left[\frac{\partial V}{\partial \chi} < \boldsymbol{B} \cdot \boldsymbol{E}^* > \right] \tag{2.96}$$

Since $\frac{\partial \nu}{\partial t}|_\rho = \frac{\partial \nu}{\partial t}|_\chi + \frac{\partial \nu}{\partial \chi}\frac{\partial \chi}{\partial t}|_\rho$ for $\nu = \nu(\chi(\rho, t), t)$, we have

$$\frac{\partial \nu}{\partial t}\bigg|_\chi = \frac{\partial}{\partial \chi}\left[\frac{\partial V}{\partial \chi} < \boldsymbol{B} \cdot \boldsymbol{E}^* >\right] \tag{2.97}$$

Note that the equation does not have the velocity U of the radial coordinate ρ, appeared in Eqs. (2.81) and (2.93). This helps the definition of the radial coordinate ρ. In tokamaks, the toroidal magnetic file is almost constant, which means the toroidal magnetic flux can be used as the radial coordinate. We define the radial coordinate ρ by

$$\chi(\rho) = \chi_a(\rho/\rho_a)^2 \tag{2.98}$$

Then, the equation of ν evolution is

$$\frac{\partial \nu}{\partial t}\bigg|_\rho = (\frac{\rho_a^2}{2\chi_a})^2 \frac{1}{\rho}\frac{\partial}{\partial \rho}\left[\frac{V'}{\rho} < \boldsymbol{B} \cdot \boldsymbol{E}^* >\right] \tag{2.99}$$

Additional equations for magnetic fluxes (χ, ψ) are

$$\frac{\partial \chi}{\partial t}\bigg|_\rho = 0 \ , \ \ \psi = \frac{2\chi_a}{\rho_a^2}\int_0^\rho \nu\rho d\rho \tag{2.100}$$

2.4.3.4 Generalized Ohm's Law

In order to close the diffusion equations of magnetic fluxes (2.103), we use the generalized Ohm's law to express $< \boldsymbol{B} \cdot \boldsymbol{E}^* >$ in terms of the resistivity and the current. The generalized Ohm's law is :

$$\boldsymbol{E}^* = \eta_\parallel \boldsymbol{J}_\parallel + \eta_\perp \boldsymbol{J}_\perp = \eta_\parallel \boldsymbol{J} + (\eta_\perp - \eta_\parallel)\boldsymbol{J}_\perp, \tag{2.101}$$

where $\eta_\parallel$ and $\eta_\perp$ are plasma resistivity parallel and perpendicular to the magnetic field, respectively and $\boldsymbol{J}_\parallel, \boldsymbol{J}_\perp$ are parallel current, and perpendicular current to the magnetic field, respectively. The effect of non-inductive current (just to change $\boldsymbol{J}$ to $\boldsymbol{J} - \boldsymbol{J}^{ni}$) is taken into account at the end, where the non-inductive current consists of the bootstrap current, beam driven current, RF driven current etc.

For the pitch time evolution, we derive following expression for $< \boldsymbol{B} \cdot \boldsymbol{E}^* >$ by using the expressions for $\boldsymbol{B}$ (2.68), $\boldsymbol{J}$ (2.70) and $< \nabla\cdot(\nabla\psi/R^2) >= \partial[< |\nabla V/R|^2 > (\partial\psi/\partial V)]/\partial V$.

$$
\begin{aligned}
<\boldsymbol{B}\cdot\boldsymbol{E}^*> &= \eta_{\parallel} <\boldsymbol{B}\cdot\boldsymbol{J}> \\
&= \frac{\eta_{\parallel}}{\mu_0} < (F\nabla\zeta)\cdot(\Delta^*\psi\nabla\zeta) + (\nabla\zeta\times\nabla\psi)\cdot(\nabla F\times\nabla\zeta) > \\
&= \frac{\eta_{\parallel}}{\mu_0}\left\langle F\nabla\cdot\left[\frac{\nabla\psi}{R^2}\right] - \frac{\partial F}{\partial\psi}\frac{|\nabla\psi|^2}{R^2}\right\rangle \\
&= \frac{\eta_{\parallel}}{\mu_0}\left[F\frac{\partial}{\partial V}\left[<|\nabla V/R|^2>\frac{\partial\psi}{\partial V}\right] - <|\nabla V/R|^2>\frac{\partial\psi}{\partial V}\frac{\partial F}{\partial V}\right] \quad (2.102)
\end{aligned}
$$

Then the explicit form of $<\boldsymbol{B}\cdot\boldsymbol{E}^*>$ in the ρ coordinate is given by

$$
<\boldsymbol{B}\cdot\boldsymbol{E}^*> = \frac{\eta_{\parallel}}{\mu_0}\Big(\frac{2\chi_a\rho/\rho_a^2}{V'<1/R^2>}\Big)^2\frac{1}{V'}\frac{\partial}{\partial\rho}\Big(\nu<1/R^2><|\nabla V/R|^2>\Big) \quad (2.103)
$$

2.4.4 Equilibrium Equation

The radial coordinate ρ moves in time due to the change of MHD equilibrium, and the surface averaged equilibrium quantities do change also. On the other hand, the surface averaged transport equations described above give the constraints on MHD equilibrium; that is, the toroidal flux function at the plasma surface $\rho = \rho_a$ has to keep the prescribed value of χ_a, and the number of plasma particle, internal energy and magnetic pitch have to be conserved. This means that, for the MHD equilibrium consistent with surfed averaged plasma transport equations, $(\nu(\rho), \Theta(\rho))$ have to be conserved in solving the equilibrium equation, where

$$
\Theta(\rho) = P(\rho)(V')^{5/3}\ , \quad P(\rho) = \sum_k p^k(\rho), \quad F(\rho) = \frac{2\chi_a\rho/\rho_a^2}{V'<1/R^2>}
$$

Under these constraints, 2D dimensional MHD equilibrium equation is expressed as

$$
L\psi(R,Z) = -\frac{R^2}{2\chi_a\nu}\frac{d}{d\rho}\Big[\Theta(V')^{-5/3}\Big] - \frac{\chi_a}{\rho_a^2\nu}\frac{d}{d\rho}\Big[\frac{\rho}{V'<1/R^2>}\Big]^2 \quad (2.104)
$$

This equation is the generalized differential equation, introduced by Grad, that is, the lhs is the partial differential equation of (R,Z),while the rhs is the ordinary differential equation of ρ. Note that $(\nu(\rho), \Theta(\rho))$ are the function of ρ. Then for the calculation of the r.h.s of Eq. (2.104), the relation between ρ and the equilibrium quantity, say the plasma volume profile V, is required. This relation is calculated by solving the following the surfaced averaged equilibrium equation

$$\Big[\nu^2 < |\frac{\nabla V}{R}|^2 > + \frac{1}{< 1/R^2 >} + \frac{5}{3}\Theta(\frac{d\chi}{dV})^{-1/3}\Big]\frac{d^2\chi}{dV^2}$$

$$+ \frac{d}{dV}(\frac{1}{< 1/R^2 >}) + \frac{d\Theta}{dV}(\frac{d\chi}{dV})^{-1/3}\Big]\frac{d\chi}{dV} = 0 \tag{2.105}$$

Here, the toroidal flux function $\chi(V)$ is used as the variable instead of ρ, and is solved under the boundary condition of $\chi(V = 0) = 0$ and $\chi(V_a) = \chi_a$. The solution of this equation gives the relation between the radial coordinate ρ and the surface average equilibrium quantities.

2.4.5 Summary of 1.5D Transport Equations

The 1.5D transport equations are summarized in the following

particle transport equation

$$\frac{1}{V'}\frac{\partial}{\partial t}\left[\langle n_a\rangle V'\right] + \frac{1}{V'}\frac{\partial}{\partial \rho}\left[V' D_a \langle|\nabla\rho|^2\rangle \frac{\partial n_a}{\partial \rho}\right] = \langle S_a\rangle$$

energy transport equation

$$\frac{1}{(V')^{5/3}}\frac{\partial}{\partial t}\left[P_a (V')^{5/3}\right] + \frac{1}{V'}\frac{\partial}{\partial \rho}\left[V'\langle|\nabla\rho|^2\rangle\left(\frac{5}{2}D_a T_a\frac{\partial n_a}{\partial \rho} + \kappa\frac{\partial T_a}{\partial \rho}\right)\right]$$

$$-D_a\langle|\nabla\rho|^2\rangle\frac{1}{n_a}\frac{\partial n_a}{\partial \rho}\frac{\partial P_a}{\partial \rho} = \langle Q_a\rangle$$

magnetic pitch equation

$$\left.\frac{\partial \nu}{\partial t}\right|_\chi = \frac{\partial}{\partial \chi}\left[\frac{\partial V}{\partial \chi}\frac{\eta_\parallel}{\mu_0}\left(\frac{2\chi_a\rho/\rho_a^2}{V' < 1/R^2 >}\right)^2\frac{1}{V'}\frac{\partial}{\partial \rho}\left(\nu < 1/R^2 >< |\nabla V/R|^2 >\right)\right]$$

magnetic flux equation

$$\chi(\rho) = \chi_a(\rho/\rho_a)^2 \ , \ \psi(\rho, t) = (2\chi_a/\rho_a^2)\int_0^\rho \nu(\rho, t)\rho d\rho$$

2D mhd equilibrium equation

$$L\psi(R, Z) = -\frac{R^2}{2\chi_a\nu}\frac{d}{d\rho}\Big[\Theta(V')^{-5/3}\Big] - \frac{\chi_a}{\rho_a^2\nu}\frac{d}{d\rho}\Big[\frac{\rho}{V' < 1/R^2 >}\Big]^2$$

$$\Theta(\rho) = (V')^{5/3}\sum_a P_a(\rho)$$

1D surface averaged equilibrium equation

$$\left[\nu^2 < |\frac{\nabla V}{R}|^2 > + \frac{1}{< 1/R^2 >} + \frac{5}{3}\Theta(\frac{d\chi}{dV})^{-1/3}\right]\frac{d^2\chi}{dV^2}$$
$$+ \frac{d}{dV}(\frac{1}{< 1/R^2 >}) + \frac{d\Theta}{dV}(\frac{d\chi}{dV})^{-1/3}\Big]\frac{d\chi}{dV} = 0$$

2.4.6 Flux Conserving Tokamak Equilibrium

It is well known that tokamak equilibrium is not necessarily a nested equilibrium with single axis. As poloidal beta value (β_p (1.4)) increases, there is a possible equilibrium with the internal separatrix shown by Artsimovich [28] and Callen [91] as shown in Fig. 2.8. The internal separatrix is associated with appearance of region of reversed current (Fig. 2.8c).

The appearance of this internal separatrix can be suppressed if the plasma is rapidly heated up and the dissipation of magnetic flux is negligible. One of this process is the adiabatic compression of a tokamak and another one is the so-called FCT (Flux Conserving Tokamak) equilibria shown by Clarke [130] and Dory [174] as shown in Fig. 2.9. When the magnetic flux conserves, the magnetic topology does not change, which means that FCT equilibria do not produce any internal separatrix though they produces large Shafranov shift (up to a half of the minor plasma radius) and the current density profile is more shifted to outboard surface area. This

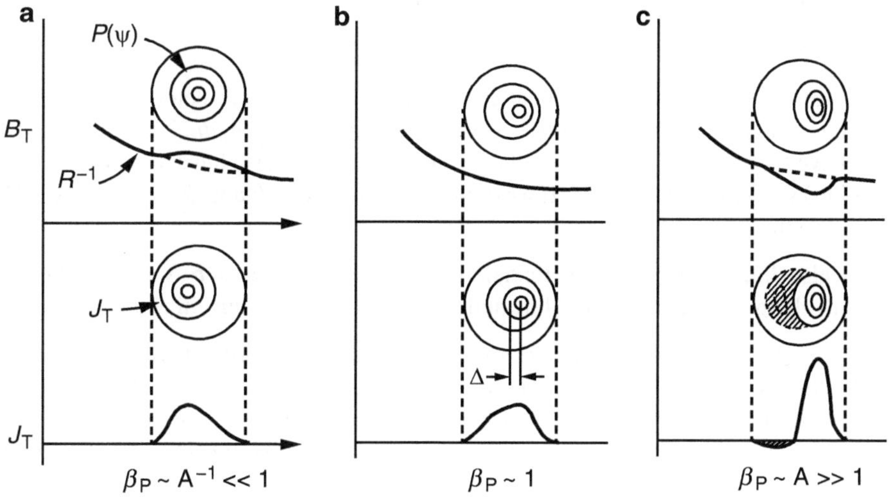

Fig. 2.8 Tokamak equilibria with β_p to see appearance of internal separatrix. Reproduced with permission from Callen [91]. Copyright AIP Publishing LLC

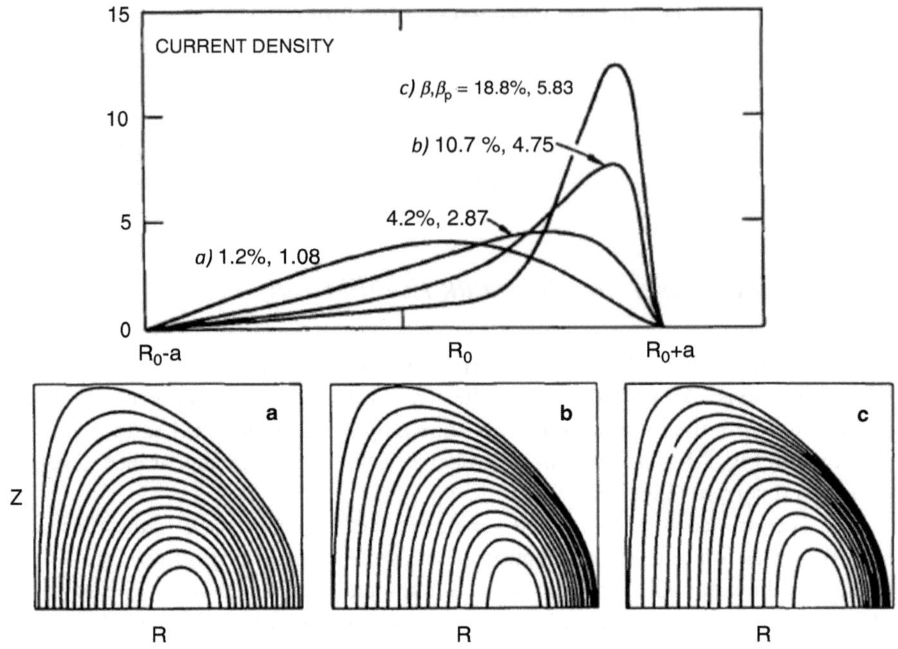

Fig. 2.9 Flux Conserving Tokamak Equilibria and their current profiles. Reproduced with permission from Dory [174]. Copyright IAEA Vienna

evolution can be simulated by a set of 1.5D transport equations in Sect. 2.4.5. These FCT equilibria implies there is no practical equilibrium beta limit.

Exercise 2.1. Derive Eq. (2.2): $\boldsymbol{A} = \phi\nabla\theta - \psi\nabla\zeta + \nabla G$.

Answer. In the general curvilinear coordinates (u, θ, ζ), any vector field $\boldsymbol{A}$ can be expressed as $\boldsymbol{A} = A_u\nabla u + A_\theta\nabla\theta + A_\zeta\nabla\zeta$. Let define $G = \int A_u du$, i.e. $dG/du = A_u$. We have $\nabla G = \partial G/\partial u\nabla u + \partial G/\partial\theta\nabla\theta + \partial G/\partial\zeta\nabla\zeta$. Then, we have $\boldsymbol{A} = \nabla G + (A_\theta - \partial G/\partial\theta)\nabla\theta + (A_\zeta - \partial G/\partial\zeta)\nabla\zeta$. If we define $\phi = A_\theta - \partial G/\partial\theta$ and $\psi = -A_\zeta + \partial G/\partial\zeta$, we have the expression.

Exercise 2.2. Show the magnetic field line equation: $du^i/ds = \boldsymbol{b}\cdot\nabla u^i$. Here s is the length along the magnetic field line and $\{u^i\}(i = 1, 3)$ is the curvilinear coordinates.

Answer. Let $\boldsymbol{x}$ is the position vector in the cartesian coordinates. The magnetic field line is given as $d\boldsymbol{x}/ds = \boldsymbol{b}$ where $\boldsymbol{b} \equiv \boldsymbol{B}/B$ is the unit vector along the magnetic field. Taking the inner product $(d\boldsymbol{x}/ds = \boldsymbol{b}) \cdot \nabla u^i$ and substituting the relation $d\boldsymbol{x}/ds = \sum(\partial\boldsymbol{x}/\partial u^i)du^i/ds$, we have the expression using the orthogonal relation (A.66).

Exercise 2.3. Derive Eq. (2.4): $d\theta/d\zeta = \partial\psi/\partial\phi$, $d\phi/d\zeta = -\partial\psi/\partial\theta$.

Answer. The magnetic field line is given as $d\phi/ds = \boldsymbol{b}\cdot\nabla\phi$, $d\theta/ds = \boldsymbol{b}\cdot\nabla\theta$, $d\zeta/ds = \boldsymbol{b}\cdot\nabla\zeta$ in the curvilinear coordinates (ϕ, θ, ζ) using Exercise 2.2. Since $d\phi/d\zeta = (d\phi/ds)/(d\zeta/ds)$, we have $d\phi/d\zeta = \boldsymbol{b}\cdot\nabla\phi/\boldsymbol{b}\cdot\nabla\zeta = \boldsymbol{B}\cdot\nabla\phi/\boldsymbol{B}\cdot\nabla\zeta$. Substituting the symplectic form $\boldsymbol{B} = \nabla\phi\times\nabla\theta - \nabla\psi\times\nabla\zeta$ and expanding $\nabla\psi = (\partial\psi/\partial\phi)\nabla\phi + (\partial\psi/\partial\theta)\nabla\theta + (\partial\psi/\partial\zeta)\nabla\zeta$, we have $d\phi/d\zeta = -\partial\psi/\partial\theta$. Similarly, we have $d\theta/d\zeta = \boldsymbol{B}\cdot\nabla\theta/\boldsymbol{B}\cdot\nabla\zeta$. Substituting the symplectic form $\boldsymbol{B}$, we have $d\theta/d\zeta = \partial\psi/\partial\phi$.

Exercise 2.4. Show that magnetic field line Hamilton equation can be given by the variational principle $\delta S = 0$, where $S = \int \boldsymbol{A}\cdot d\boldsymbol{x}$.

Answer. Substituting (2.2) into this action integral, we have $S = \int(\phi(d\theta/d\zeta)d\zeta - \psi d\zeta + dG)$. The gauge term $\int dG = G(2) - G(1)$ is the difference of start and end point. Since the variational principle is formulated with fixed value at the start and end points, $\delta\int dG = 0$. The variation δS is given as $\delta S = \int(\delta\phi d\theta/d\zeta + \phi d\delta\theta/d\zeta - \delta\psi)d\zeta$. Expanding $\delta\psi = (\partial\psi/\partial\phi)\delta\phi + (\partial\psi/\partial\theta)\delta\theta + (\partial\psi/\partial\zeta)\delta\zeta$ and making partial integration for $d\delta\theta/d\zeta$, we have following relation: $\delta S(\theta,\phi) = \int\left[\left(\frac{d\theta}{d\zeta} - \frac{\partial\psi}{\partial\phi}\right)\delta\phi - \left(\frac{d\phi}{d\zeta} + \frac{\partial\psi}{\partial\theta}\right)\delta\theta + \frac{d(\phi\delta\theta)}{d\zeta}\right]d\zeta$. Since the third term is total derivative like the gauge term, the variation is zero. Thus we obtain Hamilton form of the magnetic field line.

Exercise 2.5. Prove the magnetic curvature relation $\boldsymbol{\kappa} \equiv \boldsymbol{b}\cdot\nabla\boldsymbol{b} = -\boldsymbol{b}\times(\nabla\times\boldsymbol{b})$.

Answer. Substituting the vector formula (A.12): $\nabla(\boldsymbol{a}\cdot\boldsymbol{b}) = (\boldsymbol{b}\cdot\nabla)\boldsymbol{a} + \boldsymbol{b}\times(\nabla\times\boldsymbol{a}) + (\boldsymbol{a}\cdot\nabla)\boldsymbol{b} + \boldsymbol{a}\times(\nabla\times\boldsymbol{b})$ into the relation $\nabla(\boldsymbol{b}\cdot\boldsymbol{b}) = 0$, we obtain this relation.

Exercise 2.6. Derive so-called axisymmetric relation : $\boldsymbol{b}\times\nabla\psi = F(\psi)\boldsymbol{b} - R^2 B\nabla\zeta$.

Answer. Use the magnetic field expression in tokamak $\boldsymbol{B} = \nabla\times\nabla\psi + F\nabla\zeta$ and calculate $\nabla\psi\times\boldsymbol{B}$ and use $|\nabla\psi|^2 = R^2(B^2 - B_\zeta^2)$, we have this axisymmetric relation.

Exercise 2.7. Derive Grad-Shafranov equation in the axisymmetric equilibrium using orthogonal relation in (R, ζ, ψ) coordinates.

Answer. Use $(\boldsymbol{J}\times\boldsymbol{B})\cdot\nabla\zeta = \nabla P\cdot\nabla\zeta = 0$. Substitute expressions of $\boldsymbol{B} = \nabla\zeta\times\nabla\psi + F\nabla\zeta$ and $\mu_0\boldsymbol{J} = (L\psi)\nabla\zeta + \nabla F\times\nabla\zeta$ into this equation, we have $(\nabla\psi\times\nabla F)\cdot\nabla\zeta = 0$, i.e. $F = F(\psi)$. Also we have $\boldsymbol{J}\times\boldsymbol{B} = -((FF' + L(\psi))/\mu_0 R^2)\nabla\psi$. Taking $(\boldsymbol{J}\times\boldsymbol{B} - \nabla P = 0)\cdot(\partial\boldsymbol{x}/\partial R)$, we have $\partial P(\psi, R)/\partial R = 0$, i.e. $P = P(\psi)$. Taking $(\boldsymbol{J}\times\boldsymbol{B} - \nabla P = 0)\cdot(\partial\boldsymbol{x}/\partial\psi)$, we have the Grad-Shafranov equation, $L\psi = -\mu_0 R^2(dP(\psi)/d\psi) - FdF(\psi)/d\psi$. Here, we use $\nabla u^i\cdot\partial\boldsymbol{x}/\partial u^j = \delta_{ij}$.

Exercise 2.8. Show that there exits a selection of coordinates (θ, ζ) in which both $\boldsymbol{B}$ and $\boldsymbol{J}$ can be expressed by the straight field lines.

Answer. Since $\boldsymbol{B}$ and $\boldsymbol{J}$ are integrable (stay on the flux surface), there exist stream functions $b = b_1(\phi)\theta + b_2(\phi)\zeta + \tilde{b}(\phi,\theta,\zeta)$ and $j = j_1(\phi)\theta + j_2(\phi)\zeta + \tilde{j}(\phi,\theta,\zeta)$ so that $\boldsymbol{B} = \nabla\phi\times\nabla b$ and $\boldsymbol{J} = \nabla\phi\times\nabla j$. Here $\tilde{j}$ and $\tilde{b}$ are periodic function of θ

and ζ. Consider the coordinate transformation to remove these periodic functions by $\theta_h = \theta + \theta_1$ and $\zeta_h = \zeta + \zeta_1$. We obtain $\theta_1 = (\tilde{b}j_2 - \tilde{j}b_2)/(b_1j_2 - b_2j_1)$ and $\zeta_1 = (-\tilde{b}j_1 + \tilde{j}b_1)/(b_1j_2 - b_2j_1)$. The magnetic field and current density are given as $\boldsymbol{B} = \nabla\phi \times \nabla(b_1(\phi)\theta_h + b_2(\phi)\zeta_h)$ and $\boldsymbol{J} = \nabla\phi \times \nabla(j_1(\phi)\theta_h + j_2(\phi)\zeta_h)$. This means both are straight lines in (θ_h, ζ_h) coordinates, which is named Hamada coordinates.

Exercise 2.9. Show that there exits a coordinates (θ, ζ) in which $\boldsymbol{B}$ can be expressed as $\boldsymbol{B} = g(\phi)\nabla\zeta_b + f(\phi)\nabla\theta_b + \beta_*\nabla\phi$ also Clebsch form $\boldsymbol{B} = \nabla\phi \times \nabla(\theta_m - \zeta/q)$.

Answer. We start from flux coordinates (ϕ, θ_m, ζ) in which $\boldsymbol{B} = \nabla\phi\times\nabla(\theta_m - \zeta/q)$. Using the answer of the previous exercise, let express $\mu_0\boldsymbol{J} = \nabla(f'(\phi)\theta + g'(\phi)\zeta - \nu(\phi, \theta, \zeta)) \times \nabla\phi$, where $\mu_0 j_1 = -f'(\phi)$, $\mu_0 j_2 = -g'(\phi)$, and $\mu_0\tilde{j} = \nu$. Using the vector formula (A.8): $\nabla \times (f\boldsymbol{a}) = \nabla f \times \boldsymbol{a} + f\nabla \times \boldsymbol{a}$ and noting $\nabla \times \nabla f = 0$, we have $\nabla(f'(\phi)\theta) \times \nabla\phi = \nabla\theta \times \nabla f(\phi) = \nabla \times (f\nabla\theta)$. Noting $\nabla \times (-\nu\nabla\phi) = -\nabla\nu \times \nabla\phi$, we obtain $\mu_0\boldsymbol{J} = \nabla \times [f(\phi)\nabla\theta + g(\phi)\nabla\zeta) - \nu(\phi, \theta, \zeta)\nabla\phi]$. Since we have gauge arbitrariness, we have $\boldsymbol{B} = f(\phi)\nabla\theta + g(\phi)\nabla\zeta) - \nu(\phi, \theta, \zeta)\nabla\phi + \nabla G$. Boozer [70] found a coordinate transform $(\theta_b, \zeta_b) = (\theta_m + \eta, \zeta + q(\phi)\eta)$ to eliminate the gauge term ∇G. For $\eta = G/(gq + f)$, we have $\boldsymbol{B} = g(\phi)\nabla\zeta_b + f(\phi)\nabla\theta_b + \beta_*\nabla\phi$ where $\beta_* = -\nu + \eta(qg'(\phi) + f'(\phi))$.

Exercise 2.10. Derive the time derivative relation $(\partial\psi/\partial t)_\phi = -(\partial\phi/\partial t)_\psi/q$.

Answer. Substituting the relation $(\partial\phi/\partial t)_\phi = (\partial\phi/\partial t)_\psi + (\partial\phi/\partial\psi)(\partial\psi/\partial t)_\phi$ into the identity $(\partial\phi/\partial t)_\phi \equiv 0$ and using $(\partial\phi/\partial\psi) = q$, we have this relation.

Exercise 2.11. Derive the relation $2\pi(\partial\phi/\partial t)_\psi = -\langle E_\parallel B\rangle(\partial V/\partial\psi)$.

Answer. Substituting $\rho = \psi$ and $f = \boldsymbol{B} \cdot \nabla\zeta$ into (2.58), we have $2\pi(\partial\phi/\partial t)_\psi = \int(\partial\boldsymbol{B}/\partial t) \cdot \nabla\zeta d\tau + \oint(\boldsymbol{B} \cdot \nabla\zeta)(\boldsymbol{u}_\psi \cdot \nabla\psi)dS_\psi/|\nabla\psi|$. Here, $d\boldsymbol{S}_\psi = (dS_\psi/|\nabla\psi|)\nabla\psi$ is used. By definition, we have $\boldsymbol{u}_\psi \cdot \nabla\psi \equiv -(\partial\psi/\partial t)_x = -RE_\zeta$. Taking the (2.72) $\cdot\nabla\zeta$, we have $(\partial\boldsymbol{B}/\partial t) \cdot \nabla\zeta = -\nabla \cdot (\boldsymbol{E} \times \nabla\zeta)$. Integration over the volume enclosed by the surface ψ, we have $\int(\partial\boldsymbol{B}/\partial t) \cdot \nabla\zeta d\tau = -\oint\boldsymbol{E} \cdot \boldsymbol{B}_p dS_\psi/|\nabla\psi|$. Combining these equations, we have $2\pi(\partial\phi/\partial t)_\psi = -\oint\boldsymbol{E} \cdot \boldsymbol{B}dS_\psi/|\nabla\psi|$. Using the definition of the surface average, we have $2\pi(\partial\phi/\partial t)_\psi = -\langle E_\parallel B\rangle(\partial V/\partial\psi)$.

Chapter 3
Advanced Tokamak Regime

Abstract For the efficient steady state operation, an operation at a high bootstrap current fraction is essential. The bootstrap current is hollow in the radial direction, which tends to be unstable to the MHD modes. But there is certain operating regime, which is stable to MHD modes and has excellent energy and particle confinement property. This regime is called advanced tokamak regime. After some introduction of advanced tokamak research in Sect. 3.1, operation regimes in (q, l_i) and $(n_e, I_p/\pi a^2)$ are shown in Sect. 3.2. Characteristics and ideal MHD stability of the weak shear operation scenario are introduced in Sect. 3.3. Characteristics and ideal MHD stability of the negative shear and current hole operations are introduced in Sects. 3.4 and 3.5, respectively. We also briefly describe the status and issues of the long pulse operation as an important research direction in Sect. 3.6. New advanced tokamak devices are introduced in Sect. 3.7 and representative data of tokamak and helical device LHD are shown in Sect. 3.8. Activity of IAE Large Tokamak agreement is introduced in a Salon.

Further Reading:
Books: For basic understanding of MHD modes, you may read Freidberg (1987) [211] and Bateman (1978)[45].

Review Papers: You may read review papers related advanced tokamak research such as Kikuchi-Azumi [444], Kikuchi [442], Kishimoto [458], Taylor [759], Goldston [252], Kikuchi [437]. For current hole operation, you may read Fujita [229]. Many of DIII-D research papers are relevant for advanced tokamak research.

As for the International Tokamak Physics Activity (ITPA), you may read Special issue in Nuclear Fusion volume 39, Number 12, 1999 (http://iopscience.iop.org/0029-5515/39/12) and Special issue in Nuclear Fusion volume 47, Number 6, 2007 (http://iopscience.iop.org/0029-5515/47/6). Many of DIII-D research papers are relevant for advanced tokamak research.

M. Kikuchi, M. Azumi, *Frontiers in Fusion Research II*,
DOI 10.1007/978-3-319-18905-5_3

3.1 Introduction

For the advanced tokamak operation, the control of the current profile is important since the profile of the bootstrap current ($J_{bs} \sim \sqrt{r/R}(dP/dr)/B_p$) is hollow. All advanced tokamak operations suppress sawtooth instability by tayloring $q(0) > 1$ and a hollow q profile called negative shear (NS) operation is actively exploited. There is certain stable operating regime in hollow current profile. The current profile is difficult to change at high plasma temperature and current profile control before intensive auxiliary heating is important from the operational point of view.

Kishimoto [458] shows the schematics of the steady state operation regime at high bootstrap current fraction and the access to this regime from the low bootstrap current fraction plasma (Fig. 3.1a). Optimization of q profile at high bootstrap current fraction and approach from Ohmic phase (OH) are key research element.

In 1993, the first comprehensive review on the prospects of the steady state tokamak reactor was given by Kikuchi [437], which covered the physics requirements of the high bootstrap current fraction, the confinement enhancement factors, the non-inductive current drive, the MHD stability including the disruption probability, the power and particle control and the need for new research directions was stressed, in addition to some engineering features of the magnet, neutral beam, coolant and material selection. The steady state tokamak research is called by the name of the **Advanced Tokamak Research**. Research directions have shifted to the demonstration of a high bootstrap high normalized-beta regime since the appearance of the SSTR concept.

First systematic experimental studies addressing the steady state tokamak regimes are given from JT-60U by Kikuchi [439] and from DIII-D by R. Stambaugh [715]. Reviews of the advanced tokamak research are given by R. Goldston [252], by T. Taylor [759] and the advanced tokamak research on JT-60 is summarized by H. Kishimoto [458].

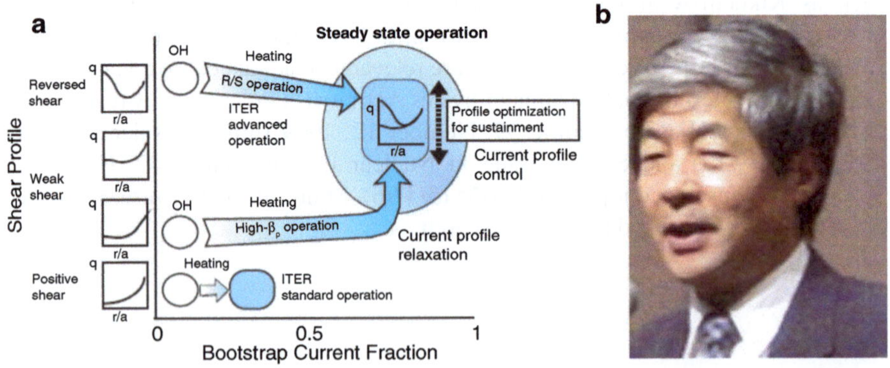

Fig. 3.1 (**a**) Approach to advanced tokamak operation in (bootstrap current fraction, q profile) space. Reproduced with permission from Kishimoto [458]. Copyright IAEA Vienna. (**b**) H. Kishimoto (1942–2005), who took a leadership in JT-60 project and ITER negotiation

3.2 Current Profile and Density Limit

In the standard tokamak operation, the current profile is peaked at the plasma center and the safety factor profile is monotonically increasing in the minor radius direction. If the safety factor at the plasma center $q(0) \leq 1$, sawtooth instability may be excited as observed by Von Goeler [789] which has many puzzling phenomena [604, 605, 850]. At the low order rational surface such as $q(r_s) = m/n = 2/1$, so-called tearing mode [232] may be unstable to form a magnetic island and the radial width of the magnetic island evolves nonlinearly according to the so-called Rutherford equation. If the island size becomes sufficiently large or magnetic islands of different modes overlaps, the magnetic field line becomes stochastic and the plasma may terminate. The current profile is characterized by the 'internal inductance', l_i as a ratio of volume-averaged poloidal field energy density and that at the surface defined as $l_i = \frac{\int B_p^2 dV/V_p}{(\mu_0 I_p / \int dl)^2}$, where dl is a differential poloidal length and dV is differential volume.

The (q, l_i) diagram (Fig. 3.2a) is one of important operational diagrams of a tokamak [118]. The lower l_i boundary is limited by the surface kink modes or locked mode while the upper l_i boundary is limited by the tearing modes. Between upper and lower boundaries, we have stable operation regime, in which we have sawtooth boundary shown by the dotted line in Fig. 3.2a. During the tokamak start up phase, the appearance of the hollow current profile is expected due to the skin effect, which may result in the current profile re-distribution or plasma termination (called major disruption) by the double tearing mode [355, 718]. To eliminate hollow current profile during the current ramp up phase, plasma minor radius expansion is effective especially by using so-called 'constant q_a operation' [429]. Weak and negative shear regimes are below the sawtooth boundary.

Another important operation boundary is the density limit discovered by Greenwald [264] as $n_G = I_p/\pi a^2$ (Fig. 3.2b). Here n_G, I_p, a are density limit in $10^{20}\,\mathrm{m}^{-3}$, plasma current in MA, horizontal minor radius in m, respectively. This scaling

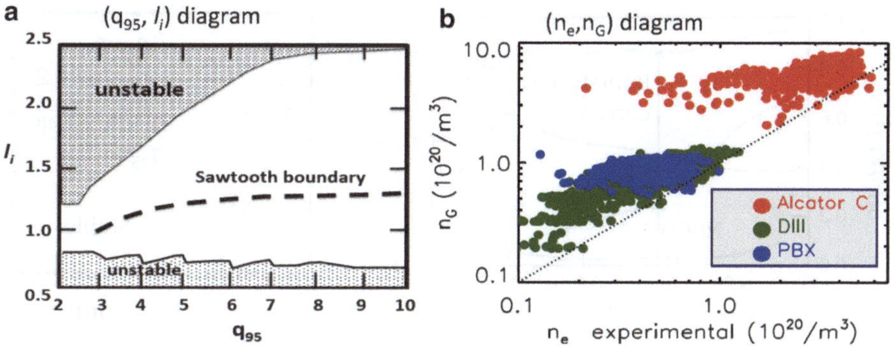

Fig. 3.2 (**a**) Schematic operating diagram in (q_{95}, l_i) plane. (**b**) Density operation regime. Here $n_G = I_p/\pi a^2$. Reproduced with permission from Greenwald [265]. Copyright IOP Publishing

correctly incorporates effect of plasma elongation through the increase in plasma current. When the density profile is peaked, the density limit is increased implying that the density limit is the edge phenomena and correlated with the edge density.

3.3 Weak Shear Operation

The radial profile of the bootstrap current is hollow. To obtain a monotonic q profile, an active central current drive has to be applied to compensate this hollow bootstrap current profile. Central active current drive can be made either by NBCD (neutral beam current drive) and ECCD (electron cyclotron current drive).

Figure 3.3 shows WS profile and stability calculation using central NBCD for the SSTR. By keeping $q(0) > 1$, the sawtooth instability is suppressed. The ballooning mode becomes stable when $q(0)$ becomes higher than 2. Ideal MHD stability calculation using the ERATO-J code shows that $n = 1, 2, 3$ and ∞ ideal MHD modes also become stable when $q(0) > 2$ if the wall stabilization is effective at $r_{wall}/a = 1.2$. While WS scenario with $q(0) > 2$ is effective in enhancing beta limit with wall stabilization, ideal MHD stability limit without wall stabilization is reduces with $q(0)$ ($\beta_N \sim 1/q(0)$) shown by Ramos [621]. So, it is important to have a wall stabilization. The effectiveness of this wall stabilization is discussed in Sect. 8.8.

To keep ideal MHD stability without the wall stabilization, Manickham [537] gives this scenario at $q(0) \sim 1$ under vanishing edge bootstrap current as shown in Fig. 3.4. With wall stabilization, the stability limit is determined by the $n = \infty$ ballooning mode as $\beta_N \sim 4$, while it is determined by the $n = 1$ kink-ballooning mode as $\beta_N \sim 3.5$ without wall stabilization.

Further profile optimization without the wall stabilization has been made by Lin-Liu [514] with a finite edge bootstrap current. He showed stable high $\beta_N = 4.0$ and high bootstrap current fraction $f_{bs} \sim 70\,\%$ equilibrium with relatively high

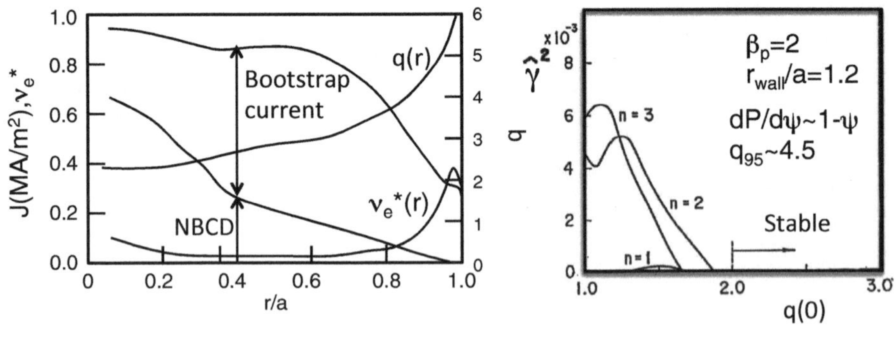

Fig. 3.3 WS profiles using central NBCD and $n = 1, 2, 3$ ideal MHD stability as a function of central safety factor q_0 for the SSTR. Modified from Kikuchi [437]

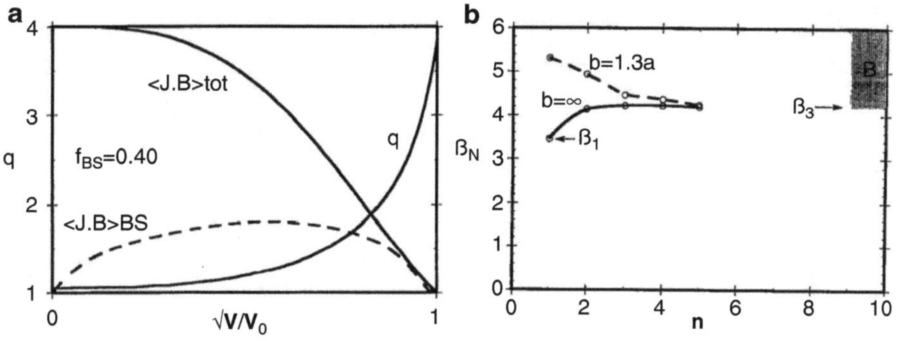

Fig. 3.4 (**a**) Current and q profiles of WS operation with a bootstrap current fraction of 40 % as a function of normalized radius $\rho = \sqrt{V/V_0}$. (**b**) Maximum stable β_N in WS operation as a function of the toroidal mode number n with (*solid line*, $b = 1.3a$) and without (*dashed line*, $b = \infty$) wall stabilization. Reproduced with permission from Manickam [537]. Copyright AIP Publisher LLC

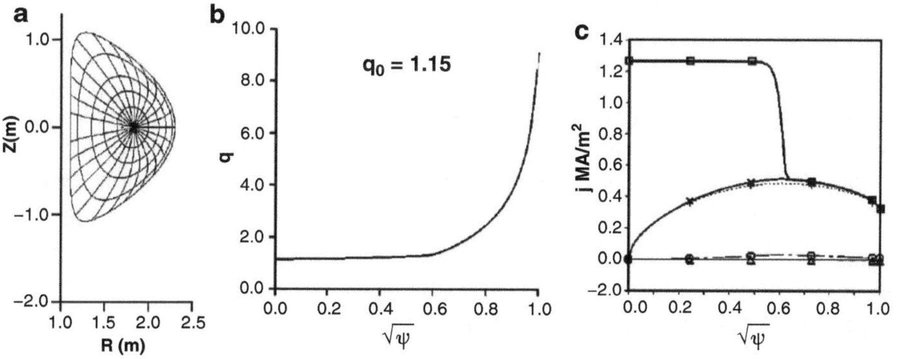

Fig. 3.5 WS optimized scenario with $f_{bs} = 70\,\%$ and $\beta_N = 4$ stable to ideal MHD modes without wall stabilization. (**a**) Cross section of the plasma, (**b**) q profile, (**c**) toroidal plasma current density. Pressure gradient is monotonically decreasing to half of the central value at the edge. Reproduced with permission from LinLiu [514]. Copyright AIP Publisher LLC

$l_i \sim 0.92$ with the aspect ratio $A = R_p/a_p = 2.83$ (Fig. 3.5). He calculated only $n = 1$ and $n = \infty$ while stability against some infernal modes (see Sect. 8.4.3) in the intermediate n has to be checked. All these scenarios are based on the smooth pressure profiles without the local transport barrier. Formation of the edge and the internal transport barriers bring other issues associated with the excitation of the localized bootstrap current to be discussed in Chap. 8 as well.

This WS regime has been investigated in many tokamaks and called in different names such as the 'high β_p regime' [376] and the 'high β_p H-mode regime' [463] in JT-60, the 'supershot regime' in TFTR [719], the 'improved H-mode' in ASDEX Upgrade [696], 'optimized shear regime' in JET [258], 'hybrid regime' in DIII-D [527]. In these regimes, improved core confinement is observed with sawtooth

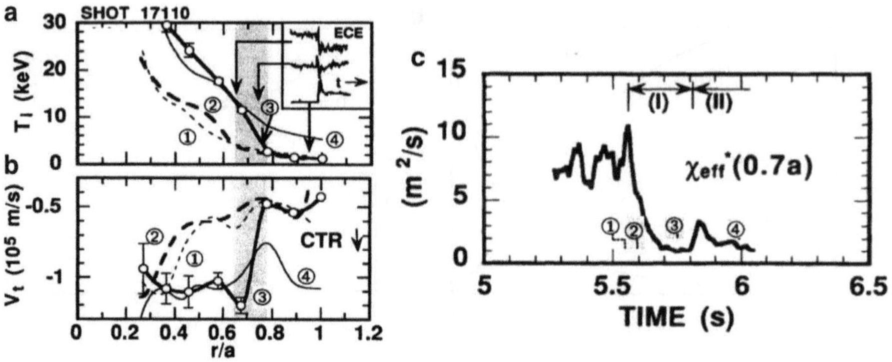

Fig. 3.6 Enhanced confinement regime with the ITB in WS plasma. (**a**) The ion temperature T_i in excess of 30 keV is achieved with ITB near $r/a \sim 0.7$, which is relaxed to excite H-mode. (**b**) Strong sheared toroidal rotation (V_t) is formed in the ITB. (**c**) Time evolution of the effective thermal diffusivity $\chi^*_{eff}(0.7a)$. The formation of the ITB reduces $\chi^*_{eff}(0.7a)$. Reproduced with permission from Koide [463]. Copyright American Physical Society

suppression. Figure 3.6 shows typical time evolution of enhanced confinement regime in WS plasma (called high β_p H-mode). The improved core confinement is associated with the **internal transport barrier** (ITB) named by Koide in 1994 [463]. It is also interesting that ITB is associated with large flow shear in the toroidal rotation.

3.4 Negative Shear Operation

Ozeki in 1992 [596] showed that a hollow current profile with a reduced pressure gradient near q_{min} can be stable to ideal MHD modes, called NS operation. Since bootstrap current profile is hollow, it is much easier to use hollow current profile to minimize the active current drive. Since magnetic shear plays an important role in stabilizing pressure driven ideal MHD modes, reduced pressure gradient near q_{min} (low shear region) is important for NS scenario. A low pressure gradient near q_{min} means that the bootstrap current is low. So, it becomes essential to drive plasma current non inductively near the pitch minimum location. They proposed to use off-axis NBCD to realize a negative shear profile, as shown in Fig. 3.7.

Ideal MHD stability of NS scenario has been studied by Manickam [537], which shows that the beta limit is quite high $\beta_N = 5$ with the wall stabilization but becomes fairly lower $\beta_N = 2$ without the wall stabilization determined by $n = 1$ modes, as shown in Fig. 3.8. The reason for this is quite simple that wall stabilization is easier if the current is closer to the wall, but such a surface current is unstable if the wall is not effective. This results implies that wall stabilization should be securely maintained in reactor by a combination of rotational stabilization and the feedback stabilization.

Turnbull [779] made a systematic parametric scans on shaping, pressure peaking and wall stabilization on ideal $n = 1$ mode for NS plasma as shown in Fig. 3.9.

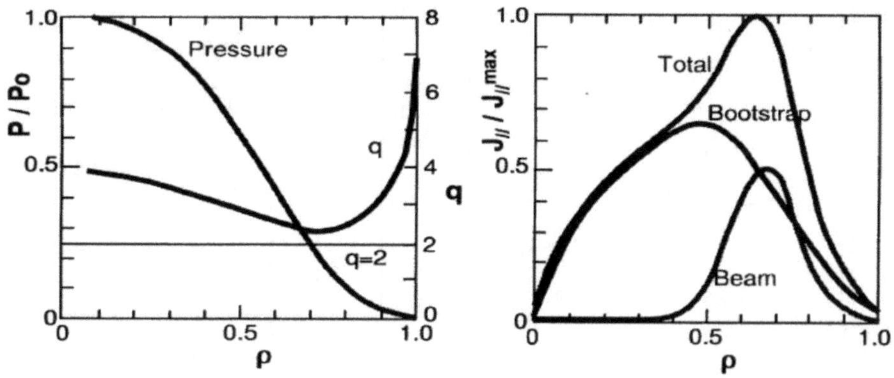

Fig. 3.7 NS scenario for the SSTR, where off-axis NBCD is used to form NS. Reproduced with permission from Ozeki [596]. Copyright IAEA Vienna

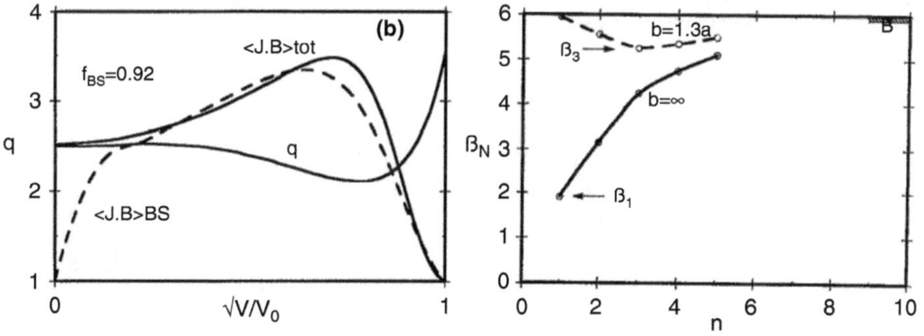

Fig. 3.8 NS plasma profile and Ideal MHD stability limit with ($b = 1.3a$) and without ($b = \infty$) wall stabilization. Reproduced with permission from Manickam [537]. Copyright AIP Publisher LLC

The β_N limit increases with reduced pressure peaking factor for both circular and D-shaped plasmas. But the improvement in D-shaped plasma is much higher than that in the circular plasma. Effect of wall stabilization to increase the β_N limit is also much stronger for D-shaped plasma. This NS regime has been investigated in many tokamaks such as enhanced reversed shear (ERS) regime in TFTR [502], negative central shear (NCS) regime in DIII-D [720], reversed shear regime in JT-60 [223, 224] and stationary magnetic shear reversal regime in Tore Supra [517]. Figure 3.10 shows NS plasma profiles in TFTR and D-III-D. In these regimes, very high confinement is observed with strong ITB. Key issue is control of transport to match optimum pressure profile stable to high β_N operation. Access to stable high β_N in current ramp up phase is also an important issue, especially when q_{min} passes low m/n rational value to avoid kink-ballooning and double tearing modes (see Sect. 8.5.3).

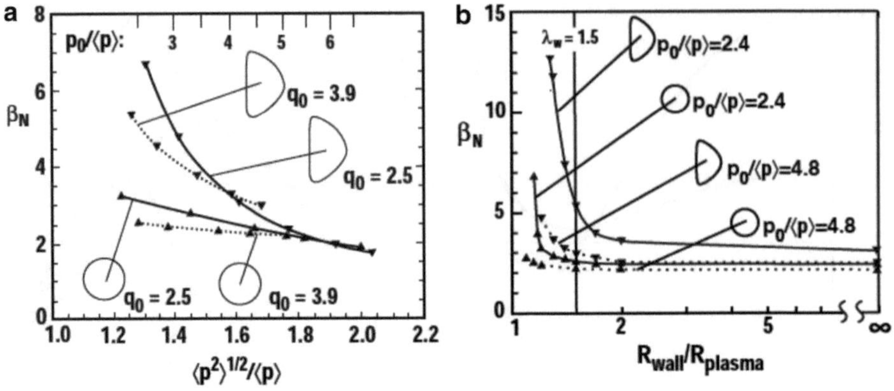

Fig. 3.9 (**a**) β_N limit as a function of pressure peaking factor $\langle p^2\rangle^{1/2}/\langle p\rangle$ for circular and D-shaped plasmas and $q_0 = 3.9, 2.5$. (**b**) β_N limit as a function of wall location R_{wall}/R_{plasma} for circular and D-shaped plasmas and $p_0/\langle p\rangle = 2.4, 4.8$. Reproduced with permission from Turnbull [779]. Copyright AIP Publisher LLC

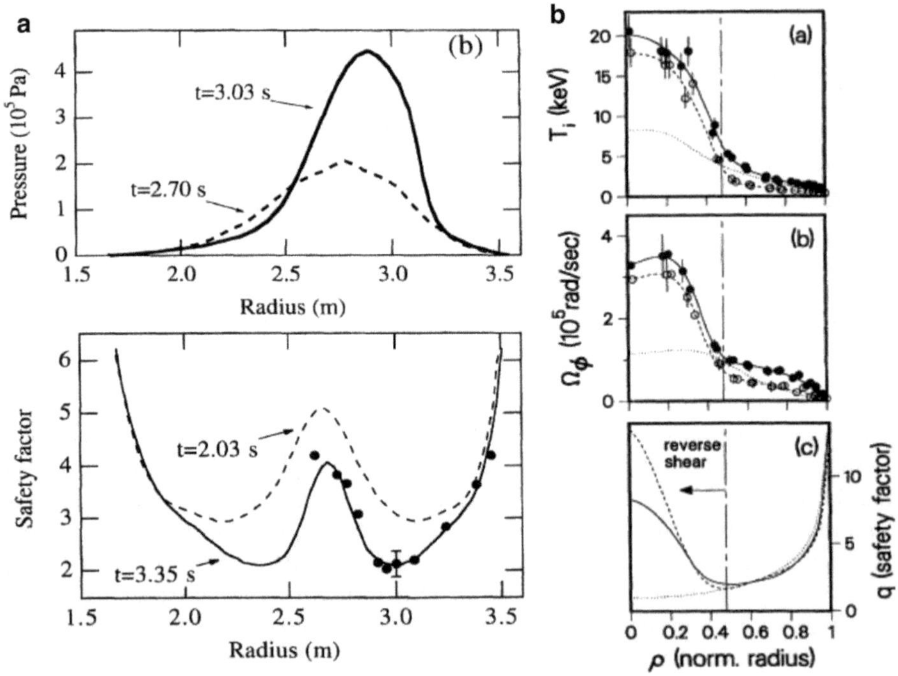

Fig. 3.10 (**a**) NS regime (called ERS) in TFTR. (**b**) NS regime (called NCS) in DIII-D. Reproduced with permission from Levinton [502] and Strait [720]. Copyright American Physical Society

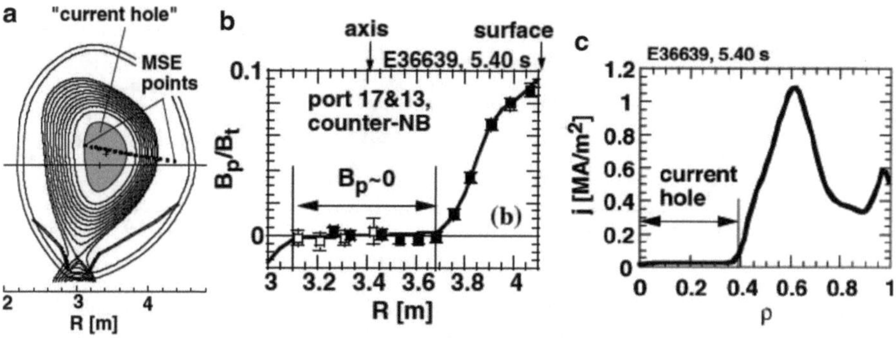

Fig. 3.11 (**a**) Current hole equilibrium in JT-60U, (**b**) poloidal field measurement by MSE (Motional Stark Emission) spectroscopy, (**c**) deduced toroidal current density profile. Reproduced with permission from Fujita [227]. Copyright American Physical Society

3.5 Current Hole Operation

As extreme situation in NS configuration, equilibrium with zero plasma current in the central regime called 'Current Hole (CH)' (Fig. 3.11) was formed in JT-60U [227] and JET [300]. Current Hole is an interesting structure formation in tokamak magnetic configuration, which is robust against perturbation.

Key mechanism for the CH formation involves the bootstrap current. During the current ramp up phase, strong auxiliary heating powers such as NBI and ECRF are applied to form a hollow current profile through skin effect. The early current profile shows positive current density in the plasma center. But a formation of hollow q profile leads to a formation of ITB (internal transport barrier) and this ITB excites large bootstrap current at $\rho \sim 0.4$–0.8. When the local bootstrap current density exceed total current density, the negative toroidal electric field ($E_\zeta < 0$) is induced. This negative electric field diffuses into the central region and reduces the central current density down to zero. When the current density goes to negative, nested flux surfaces in the central region are changed to the multi magnetic island (MMI) equilibrium discussed in Sect. 2.3.2.

High central ion and electron temperatures are supported at the ITB region ($\rho \sim 0.5$–0.7) as shown in Fig. 3.12. Temperature and density profiles are flat in the CH region consistent with the absence of confining poloidal magnetic field. Such flat temperature regions are observed in high performance NS discharges [225, 378], implying these are also in the CH regime.

Time evolution of the CH is quite interesting that ITB is formed near the plasma center and moves outward. Figure 3.13 shows time evolution of ion temperature profile in the CH discharge. The radius of q_{min} is also moving outward and stagnates at $\rho \sim 0.8$, while radius of ion ITB foot stays slightly inside of q_{min} radius. Although the core confinement inside the CH is poor (almost no confinement), an ITB is formed just outside of the CH showing good confinement close to ion neoclassical transport [305, 743]. As a whole global energy confinement is quite good up to

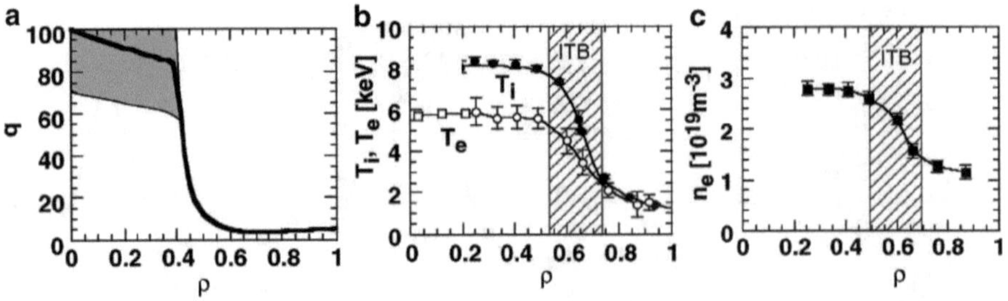

Fig. 3.12 (**a**) Safety factor (q) profile (shaded area gives uncertainty). (**b**) Ion (T_i) and electron (T_e) temperature profiles, (**c**) electron density profile. Reproduced with permission from Fujita [227]. Copyright American Physical Society

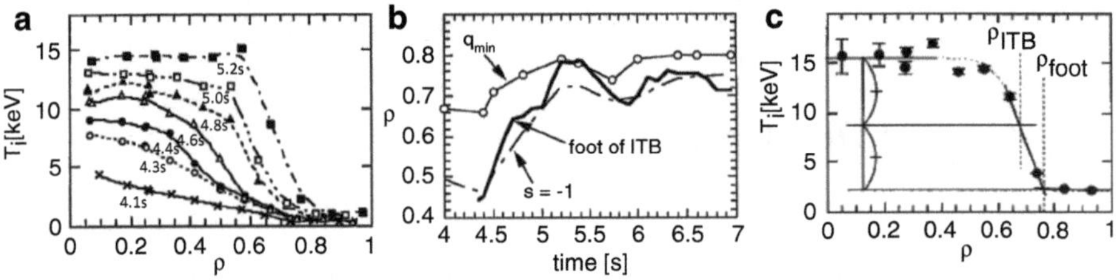

Fig. 3.13 (**a**) Time evolution of the ion temperature profile, (**b**) time evolution of q_{min} radius, radius of the ion ITB foot and radius of magnetic shear $s = -1$. (**c**) Definition of ITB foot. Reproduced with permission from Fujita [225]. Copyright IAEA Vienna

$HH_{y2} = 1.45$ in JT-60U [227]. Here HH_{y2} is defined as an enhancement factor of plasma energy confinement time over the ITER IPB98(y,2) H-mode scaling [387] and $HH_{y2} = 1$ corresponds to ITER standard operation.

One of important Current Hole characteristics is current clamp to $J_\zeta \sim 0$ shown by Fujita [228]. Huysmans solved following time dependent reduced MHD equations (see Sect. 8.5.3) for the electrostatic potential ϕ, the poloidal flux ψ, and the pressure p in the quasi cylindrical geometry [356]. It is found that central current density is clamped to nearly zero due to $n = 0$ low m resistive instabilities in case negative current density is expected without taking into account of these modes as shown in Fig. 3.14. The growth rate is largest fo $m = 1$ and scales as $\gamma \sim \eta^{1/3}$ (same as internal kink mode), while $\gamma \sim \eta^{3/5}$ for $m = 2$ mode (same as tearing mode).

This CH regime can be stably sustained for several seconds. This regime is interesting from the control viewpoint that it has low l_i and is easier to get positionally stable elongated plasma and also is easier to get high bootstrap current fraction. On the other hand, CH regime is subject to higher ripple loss of α particles in future reactors and sets tight constraints on maximum toroidal field ripple as well as low no-wall beta limit similar to NS regime.

While there is no confinement inside the CH, it is possible to sustain high plasma pressure by dP/dr outside the CH. In fact, it was found that the stability limit is not so different between a strong reversed shear case with $q(0)/q_{min} = 20$ (model equilibrium of CH since there is no equilibrium solution for $q(0) = \infty$) and a moderate reversed shear case with $q(0)/q_{min} = 2$ [599]. Standard ideal MHD stability calculation for CH by Ozeki [600] showed beta limit could be $\beta_N \sim 5$.

One of interesting observation in the ideal MHD potential energy of CH is that plasma behaves as if it were bordered by an extra internal vacuum region as shown by Chu [126].

3.6 Long Pulse Operation

Even if steady state tokamak operation becomes possible utilizing the bootstrap current mechanism, demonstration of long pulse operation is essential for future advanced tokamak research.

The power system should operate reliably so that off-normal event should be less than once a year. This requires very reliable operation with NTM suppression by ECCD, RWM suppression by rotation, ELM suppression, control of the divertor detachment and particle exhaust. In case of off-normal event, effect of plasma shut down should be significantly reduced on current and thermal quenches in tokamak.

Sputtering of the first wall by fast neutral hydrogen atoms is another important issue. Evaluation of energy spectrum of neutral flux to the 1st wall is important to evaluate possible 1st wall damage due to energetic neutral.

Figure 3.14 shows neutral calculation using DEGAS [315] for the JT-60U equilibrium shown in Fig. 3.14a. Figure 3.14b shows the estimated average energy and neutral flux to the 1st wall and Fig. 3.14c shows the energy distribution of neutrals under the saturated wall condition (E44020) [441]. Average energy of neutrals at the 1st wall is 100 eV and flux of a few $\times 10^{19}$ particles/s/m^2 are estimated. Figure 3.14c shows the distribution of particle fluxes for energy range of 1–1000 eV. It was found that there is non-negligible population of high-energy

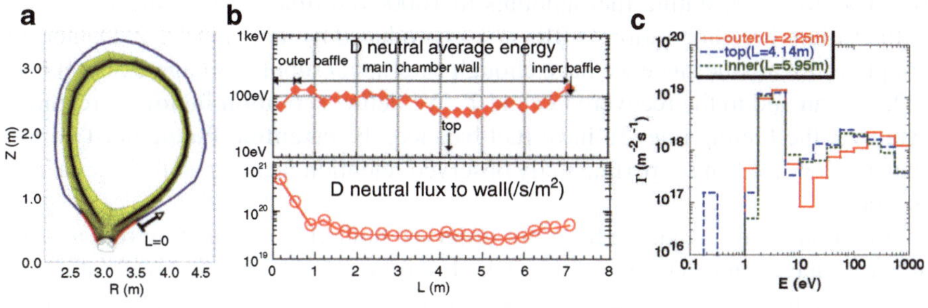

Fig. 3.14 (**a**) Equilibrium configuration, (**b**) average neutral energy along poloidal circumference. (**c**) Neutral energy distribution at three locations. Reproduced with permission from Kikuchi [441]. Copyright Elsevier

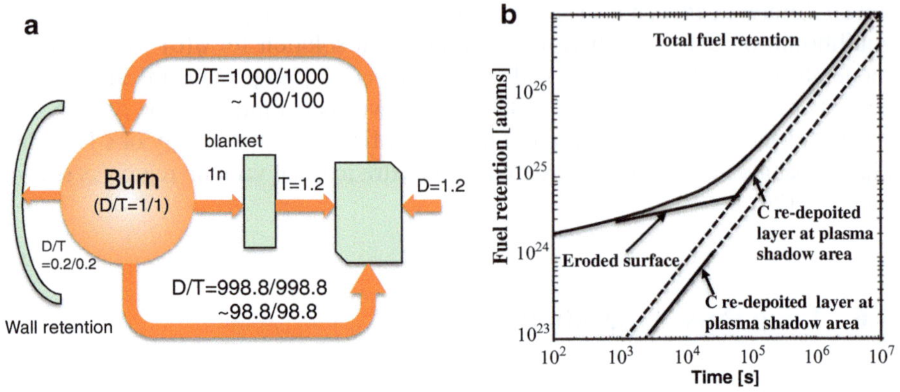

Fig. 3.15 (**a**) Example of marginal situation in the fuel balance in tokamak reactor. (**b**) Change-over of hydrogen retention implied from JT-60U experiments. Modified from Yoshida [847]

(a few 100 eV–1 keV) neutrals. Existence of such energetic neutrals as well as the existence of the second SOL observed by Asakura [29] is of concern for future long pulse operation of fusion reactor [441] and left for future studies.

For long pulse operation, the fuel circulation is also an important subject since the tritium retention in the first wall has big effect for the tritium self-sufficiency. Figure 3.15a is an example of fuel circulation in a tokamak reactor. In case of $3GW$ fusion power station, DT burn rate is 2.0×10^{21}/s (see for example, SSTR [236]). For a particle confinement time $\tau_p = 0.5$ s, the particle out flux from the separatrix is $\Gamma_p = 2 \times 10^{23}\,\mathrm{s}^{-1}$. If we assume total DT fuel circulation is this particle out flux from the separatrix, namely $\Gamma_c = \Gamma_p$, Γ_c is 100 times the DT burn rate.

If we include strong DT gas puff required to reduce plasma temperature in front of the divertor plate estimated in SSTR design [432], total fuel circulation may go up to 1000 times the fuel burn up rate, $\Gamma_c \sim 2 \times 10^{24}\,\mathrm{s}^{-1}$.

Thus, the burning efficiency (ratio of fuel burn rate to fuel circulation rate) in magnetic confinement is fairly low at 0.1–1 %. For a burning efficiency of 0.1–1 %, total circulating fuel amounts to 1000–100 times of the burn rate. Even if the tritium breeding ration (TBR) is 1.2, produced tritium is just 1.2. Deuterium supply is not a issue since we have enough external D supply. Adding external D of 1.2, we can add to the recovered D and T. We define **retention factor** by retention rate over the fueling rate. To have fuel balance, the **retention factor** must be less than 0.02–0.2 %. The experimentally observed retention factor is up to 50 % in Tore Supra.

Key issue here is that we have not observed reduction of retention rate to zero, while we have some evidence of transient desorption [575]. The local wall saturation time estimated by Takenaga [741] is less than 1 s at the divertor plates and the divertor dome and 10 s at the lower half of the baffle plates and 100 s at the main chamber wall. The formation of the redeposited layers on plasma facing surfaces significantly increases the retention rate as indicated from JT-60U experiment (see

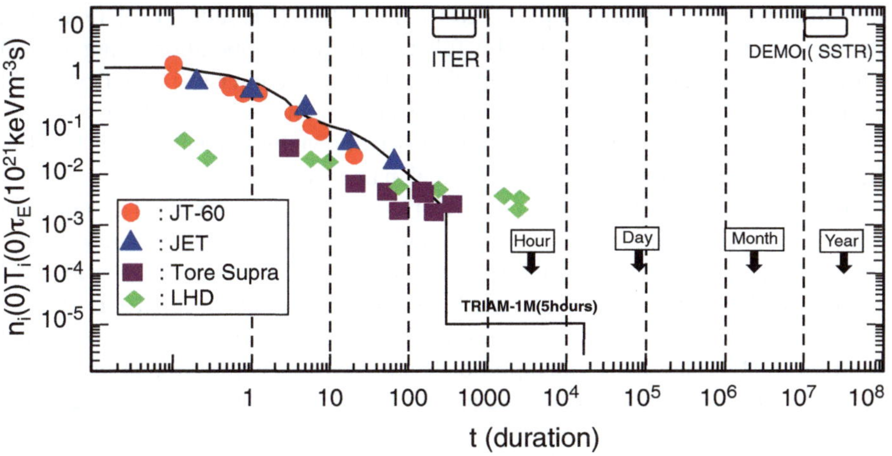

Fig. 3.16 Fusion product $(n_i(0)T_i(0)\tau_E, t)$ from tokamak and helical devices. See individual data in the Appendix of the chapter

Fig. 3.15b) by Yoshida [847], and Tanabe [747]. They show that the retention rate changes in the time scale of **One Day** for carbon. While *W* divertor is actively pursued, the formation of *W* nano-structure [410] implies *W* may not be free from enhanced retention in addition to the DBTT and re-crystallization issues for the *W*.

Unfortunately, there is no tokamak operating in this time frame nor there is no technical proposal to continuously recover retention tritium from the wall, while recovery of tritium from breeding blanket is designed and validated as discussed in Chap. 9. It should also be remembered that even if there is no other loss, the tritium decays with a half life of 12.323 year (decay of 5 %/year).

Figure 3.16 shows present status and future operation points of ITER and DEMO on fusion triple product as a function of sustained duration. DEMO such as SSTR has to operate almost year-long. The ITER under construction is expected to achieve DEMO relevant fusion triple product. While ITER equips with full superconducting magnets, the pulse length is limited less than 1 h. Large tokamaks such as JT-60U and JET uses normal conductor for the magnet and are limited to short pulse length less than a few tens seconds. Longest pulse length in tokamaks and magnetic confinement system is the record data from TRIAM-1M. S. Itoh and TRIAM-1M team demonstrated the longest operation up to 2 h in TRIAM-1M by Itoh [386], Itoh [388] and subsequent extension to 3 h by Zushi [872] and 5 h Zushi [873] is still a world record of plasma discharge duration in fusion experiments. TRIAM-1M also reported high ion temperature ($T_{(0)} \sim 3.5$ keV) by Zushi [869], enhanced current drive (ECD) by Zushi [870, 871] and existence of multiple recycling time constant by Sakamoto [650].

Both Tore Supra and LHD operated up to 300 s. Tore Supra equips with superconducting toroidal magnet and its initial operation is reported in 1988 by Aymar [32]. Saint-Laurent reported demonstration of high power long-pulse full non inductive current drive up to 120 s (2 min) [647]. Houtte [787] reported a demonstration of 6 min and 1 GJ by full non inductive operation using LHCD. Further progress has been reported by Colas [135], Chatelier [107], Joffrin [404], Giruzzi [248], Saoutic [657]. Sudden occurrence of the operational limit through MARFE phenomena from re-deposition layer is shown by Ekedahl [183]. The Tore Supra will be modified to a divertor tokamak named WEST [84, 262].

Both EAST [504, 505, 800] and KSTAR [3, 483, 485, 497, 498, 843] are equipped with full superconducting coils. It is a merit of medium size superconducting tokamaks the power required for long pulse is much smaller than ITER or JT-60SA [379, 414]. Li (2013) showed that the EAST tokamak established LHCD H-mode operation with reduced ELM activities and Wan showed the extended pulse length to 400 s [802]. Long pulse operation of heating system is progressing in KSTAR [482]. Some of the representative data from JET, JT-60U, Tore supra, LHD are in Table 3.1.

As seen from Fig. 3.15b, shortest time scale by re-deposition for carbon wall is order of **DAY** near the divertor and may be much longer for other areas of the first wall. The problem of carbon wall is that it forms hydro-carbon. Thus the large tritium inventory is expected even for ITER. Thus, the tungsten is adopted for the divertor material. Shu [694] showed that the retention rate for the tungsten is $\leq 10^{-5}$. The deuterium induced blistering by Shu [695] is also an issue for the use of tungsten in the divertor. To clarify issues and countermeasures for DEMO relevant long pulse operation, it may be important to explore **Month-long** operation at least in the medium tokamaks such as EAST [504] in China, and KSTAR [485] in Korea as well as small superconducting tokamak, such as SST-1 [667] in India. Key is to see whether retention can be stopped even with accumulation of re-deposition layers and find countermeasures if not as well as how small ELM is necessary for such a long pulse operation.

3.7 New Tokamaks for Advanced Tokamak Research

During the 1990 decade, large tokamaks such as JT-60U, JET, TFTR, DIII-D played important role in advancing tokamak physics. In the 2000 decades, joint effort among various sized tokamak played key role to extrapolate to ITER. This activities is called ITPA (International tokamak physics activity) whose initial legal framework has been provided by the IEA Large Tokamak Agreement (see Salon).

Tokamak research toward fusion energy development still needs a lot of efforts to understand physics and technologies for fusion energy. New machines such as JT-60SA (Ishida [379], Kamada [414], Barabaschi [43]) located in Japan (Fig. 3.17a) and HL-2M (Li [506]) in China (Fig. 3.17b) are under construction as of 2015.

Major scientific challenges are the power and particle control for long pulse operation, efficient continuous operation with high bootstrap current fraction and control and sustainment of high pressure plasma to increase fusion power density.

Salon: IEA Large Tokamak Agreement and ITPA

IEA (International Energy Agency:http://www.iea.org/) is an autonomous organization to coordinate actions to ensure reliable, affordable and clean energy for member countries and beyond. IEA also provides legal framework on cooperation in Fusion research. IEA implementing agreement (IA) on cooperation on large tokamak facilities was one of major IAs under IEA. This IA is initiated on February 19, 1986 as the cooperation among the three large tokamak facilities (JET, JT-60 and TFTR). This implementing agreement works very well to enhance cooperation and competition among three large tokamaks. Especially in case of JT-60, personal exchanges under this IA are extremely helpful to learn from US and EU physicist and engineers. One of the author (MK) stayed for 6 months in Princeton to join TFTR experiments learned a lot from leading scientists. There are many visitors from JET and TFTR to JT-60. Learning from two fusion institutions led to the early modification of the original JT-60 to JT-60U by which we made a significant advances in fusion plasma performance to have many world records [458].

After the shut down of TFTR in 1997, this IA is renewed as IEA Large Tokamak Agreement. Under this IA, various Tasks (transport physics, confinement database and modeling, MHD, disruption and Control, Edge and pedestal physics, SOL and divertor physics, energetic particles and steady state operation, tritium and remote-handling technologies, and others), Workshops (examples are Plasma Shaping, Real Time Control of ITB, Electron Transport, Management of Waste from Fusion Facilities, High Beta Steady State Tokamak, Physics of Current Hole, Long Pulse, Burning Plasma Physics and Simulation, Implementation of the ITPA Coordinated Research Recommendations, etc.), Personal Assignments (short and long term visit) has been implemented and reported to the FPCC (Fusion Power Coordinating Committee).

ITPA (International Tokamak Physics Activity) is an valuable activities to resolve critical physics issues relevant for ITER. This IAE-IA in cooperation with other Fusion IAs (poloidal divertor IA and TEXTOR IA) provided a legal framework for implementation of ITPA (International Tokamak Physics Activity). The first planning meeting for this ITPA/IEA Coordination was held at MIT (W52) in November 2002, the second (W54) at Naka in November 2003 and the third meeting (W58) took place near Oxford in December 2004. In case of W58 workshop, the organizing committee invited leaders of medium-sized and large-sized tokamaks in EU, Japan, US, Russian Federation and China (JET, JT-60U, DIII-D, ASDEX-U, C-MOD, NSTX, FTU, MAST, JFT-2M, TRIAM-1M, TEXTOR, TCV, Tore Supra, T-10, T-11M, Globus-M,Tuman-3M, FT-2, HL2A, HT-7) to consider the implementation of ITPA joint experiments in their research programs.

Figure 3.18a shows a group photo of this workshop at the Eynshaw Hall, which was previously a manor house in the eighteenth century and Fig. 3.18b at NAKA.

Most of the scientific career of one of the authors (MK) is spent for the research and management of this large tokamak experiment.

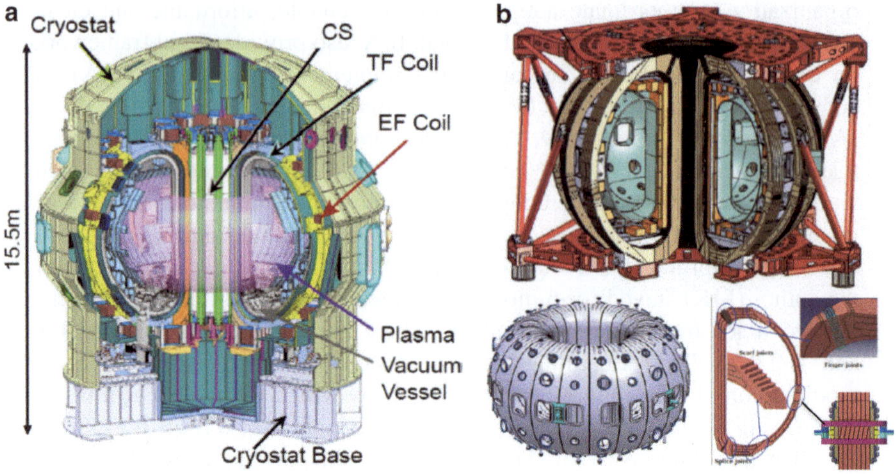

Fig. 3.17 (**a**) Birds eye views of JT-60SA. $R_p \sim 3\,\mathrm{m}, V_p \sim 130\,\mathrm{m}^3, I_p = 5.5\,\mathrm{MA}, B_t = 2.25\,\mathrm{T}$. This machine is constructed under the EU-Japan broader approach agreement. After Kamada [414] with the permission of IAEA. (**b**) Birds eye views of HL-2M. $R_p \sim 1.78\,\mathrm{m}, a_p \sim 0.65\,\mathrm{m}, I_p = 2.5(3)\,\mathrm{MA}, B_t = 2.2(3)\,\mathrm{T}$. This machine is capable of flexible shaping and may reach an equivalent break even plasma regime. Courtesy of X. Duan (SWIP, China)

3.8 Appendix: Tokamak/Helical Representative Data

Representative data from JT-60U, JET, Tore Supra, LHD are shown. Data provided by JT-60U (Drs Y. Kamada, S. Ide), JET (Dr. G. Sips), Tore Supra (Drs X. Litaudon, G.T. Hoang), and LHD (Prof. T. Mutoh; *; [566], private communication) (Table 3.1).

Fig. 3.18 (**a**) IEA Large Tokamak workshop W58 "Implementation of the ITPA coordinated research recommendations" in December 2004 held at Eynshaw Hall (eighteenth century manor house transformed into a conference center), Witney near Oxford, UK. (**b**) IEA Large Tokamak Workshop W65 "Implementation of the ITPA coordinated research recommendations" in December 2006 held at Naka, Japan

Table 3.1 Representative short and long pulse data from JT-60U, JET, Tore Supra (TS) and LHD [$n_i(0)\tau_E T_i(0)$ in 10^{21} keV s m^{-3}, $n_i(0)$ in 10^{19} m^{-3}]

Device	Type	Shot no.	Dur. (s)	$n_i(0)$	$T_i(0)$ (keV)	τ_E (s)	$n_i(0)\tau_E T_i(0)$	Ref.
JT-60U	Record T_i	26939	0.1	4.53	45	0.75	1.53	[377]
JT-60U	Record $Q_{DT}^{equ.}$	31872	0.1	4.8	16.8	1.07	0.86	[226, 378]
JT-60U	RS	40259	0.55	4.4	18.3	0.72	0.58	[413]
JT-60U	High β_p H	21140	0.62	5.6	30	0.31	0.51	[413]
JT-60U	High β_p H	21143	0.87	5.0	29	0.28	0.40	[413]
JT-60U	RS	34292	0.8	3.6	12.9	0.89	0.42	[413]
JT-60U	High β_p H	21282	1.55	4.6	32	0.30	0.44	[413]
JT-60U	High β_p H	30006	4.5	4.5	11	0.32	0.16	[413]
JT-60U	RS full CD	43046	7.4	2.3	8.7	0.45	0.089	[652]
JT-60U	High β_p H	29941	8.5	4.1	8.8	0.18	0.065	[413]
JT-60U	High β_p H	48158	25	1.35	7.1	0.22	0.021	[594]
JET	Record DT	42976	0.2	3.4	28	0.92	0.87	[423]
JET	α heating	42847	1	3.3	17.7	1.0	0.6	[423]
JET	DT H-mode	42982	5	6.6	8	0.45	0.23	[423]
JET	20 s H-mode	62065	20	3	8	0.2	0.05	[403]
JET	60 s pulse	56552	60	1.6	5	0.32	0.025	[403]
TS	Improved L	23419	3	7.5	3.1	0.13	0.03	[332]
TS	IC+LH	33898	21	5.5	3	0.043	0.007	[335]
TS	IC+LH	34181	65	4.5	2.7	0.045	0.0055	[335]
TS	LH	19980	87	2.3	1.5	0.06	0.002	[786]
TS	LH	46596	150	3.6	2.5	0.06	0.005	[179]
TS	IC+LH	47979	150	4.2	2.6	0.06	0.006	[179]
TS	LH	31917	173	2.7	1.5	0.06	0.002	[389]
TS	LH	32299	365	2.8	1.6	0.06	0.003	[787]
LHD	IC+ECH	124617	233	1.5	2.5	0.13	0.005	*
LHD	IC+ECH	124612	1713	1.2	2.1	0.15	0.0038	*
LHD	IC+ECH	124530	2223	1.2	1.4	0.17	0.0027	*
LHD	IC+ECH	124576	2673	1.2	1.5	0.16	0.0028	*
LHD	IC+ECH	124579	2859	1.2	2.0	0.15	0.0035	*

Chapter 4
Collisional Transport in Tokamak

Abstract Collisional parallel transport physics plays an essential role in current and rotation drives for the steady state tokamak reactor and is discussed here. The moment equations is derived from Vlasov Fokker-Planck equation and the flux-surface averaged parallel momentum and heat flow balance equation is derived in Sect. 4.1. The characteristics of Coulomb collision operator is discussed in Sect. 4.2. The drift kinetic equation is derived and the parallel friction and viscosity are derived from the drift kinetic equation in Sect. 4.3. In Sect. 4.4, we describe the generalized Ohm's laws including the electrical conductivity, the bootstrap current, neutral beam current drive, and EC current drive. Neoclassical rotation in tokamak including offset toroidal rotation due to symmetry breaking is discussed in Sect. 4.5. Fluid viscosity, gyro-viscosity, guiding center velocity are introduced as columns.

Further Reading:
Books: Braginskii [76] is well written classical text on the derivation of moment equations. Helander-Sigmar [317] is a well-written textbook on Coulomb collision, solution of the drift kinetic equation. Hazeltine-Waelbloeck [309] is also useful for introduction to the basic kinetic equation. For the neoclassical cross field transports based on moment equation approach skipped in this monograph, see Chap. 8 of Kikuchi [443]. For the neutral beam current drive, you may read Chap. 5 of Kikuchi-Lackner-Tran [445].

Review Papers: You may read Hirshman and Sigmar [330] and Hinton-Hazeltine [321] for neoclassical transport. For the basic current drive physics, you may read Fisch [204]. As for the bootstrap current, you may read Kikuchi-Azumi [440] and Kikuchi-Azumi [444].

M. Kikuchi, M. Azumi, *Frontiers in Fusion Research II*,
DOI 10.1007/978-3-319-18905-5_4

4.1 Moment Equations in Tokamak

4.1.1 Moment Equations

The kinetic equation to describe the time evolution of the distribution function f_a of a species a under Coulomb collision in a multiple ion species plasma is the Vlasov Fokker-Planck equation given by,

$$\frac{\partial f_a}{\partial t} + \boldsymbol{v} \cdot \frac{\partial f_a}{\partial \boldsymbol{x}} + \frac{e_a}{m_a}(\boldsymbol{E} + \boldsymbol{v} \times \boldsymbol{B}) \cdot \frac{\partial f_a}{\partial \boldsymbol{v}} = C_a(f_a) + S_a \tag{4.1}$$

where S_a is an external particle source term and C_a is the Coulomb collision operator which is given by a summation of collisions with each particle species including self collision: $C_a = \sum_b C_{ab}$. We describe derivation of moment equation for n_a, u_{ai}, p_a, q_{ai} given in Braginskii [76] and Hirshman-Sigmar [330].

For an arbitrary physical quantity $A = A(\boldsymbol{x}, \boldsymbol{v}, t)$, we can define the velocity average $\langle A \rangle_v \equiv \int A f_a d\boldsymbol{v} / \int f_a d\boldsymbol{v} = \int A f_a d\boldsymbol{v} / n_a$. We have general moment relation by multiplying A to the Vlasov equation:

$$\frac{\partial n_a \langle A \rangle_v}{\partial t} - n_a \left\langle \frac{\partial A}{\partial t} \right\rangle_v + \frac{\partial n_a \langle v_i A \rangle_v}{\partial x_i} - n_a \left\langle \frac{\partial v_i A}{\partial x_i} \right\rangle_v - \frac{n_a}{m_a} \left\langle F_i \frac{\partial A}{\partial v_i} \right\rangle = \int A(C_a + S_a) d\boldsymbol{v}$$

Here we note $\int A \partial f_a / \partial t d\boldsymbol{v} = \partial (n \langle A \rangle_v)/\partial t - n_a \partial \langle A \rangle_v / \partial t$, $\int A v_j \partial f_a / \partial x_j d\boldsymbol{v} = \partial (n \langle v_j A \rangle_v)/\partial x_j - n_a \partial \langle v_j A \rangle_v / \partial x_j$, and $\int A F_j \partial f_a / \partial v_j d\boldsymbol{v} = -n_a \langle \partial A F_j / \partial v_j \rangle_v$ after the partial integration. Here $F_j = e_a (\boldsymbol{E} + \boldsymbol{v} \times \boldsymbol{B})_j$ satisfies $\partial F_j / \partial v_j = 0$ and $\int A F_j \partial f_a / \partial v_j d\boldsymbol{v} = -n_a \langle F_j \partial A / \partial v_j \rangle_v$. In case of velocity moments ($A = A(\boldsymbol{v})$), we have $\partial A / \partial t = 0$ and $\partial v_j A / \partial x_j = 0$ and we have following general velocity moment relation [637]:

$$\frac{\partial n_a \langle A \rangle_v}{\partial t} + \frac{\partial n_a \langle v_j A \rangle_v}{\partial x_j} - \frac{n_a}{m_a} \left\langle F_j \frac{\partial A}{\partial v_j} \right\rangle_v = \int A(C_a + S_a) d\boldsymbol{v} \tag{4.2}$$

Using the general moment relation (4.2), we can obtain a set of moment equation using velocity average $\langle A \rangle \equiv \int A f_a d\boldsymbol{v} / \int f_a d\boldsymbol{v}$, where we drop subscript "$_v$" for simplicity. We use the Einstein's summation convention rule hereafter.

The velocity moment (1) of Vlasov equation (4.1) gives the particle conservation equation.

$$\frac{\partial n_a}{\partial t} + \nabla \cdot (n_a \boldsymbol{u}_a) = S_{na} \tag{4.3}$$

This is obtained by substituting $A = 1$ into the general velocity moment relation (4.2) noting $\partial 1 / \partial v_j = 0$ as $\partial n_a / \partial t + \partial n_a u_{ai} / \partial x_i = S_{na}$. Here, $n_a = \int f_a d\boldsymbol{v}$ is density, $u_{ai} = \int v_i f_a d\boldsymbol{v} / n_a = \langle v_i \rangle$ is the flow velocity, and $S_{na} = \int S_a d\boldsymbol{v}$ is the particle source. We note a relation $\int C_a d\boldsymbol{v} = 0$.

The velocity moment ($m_a\boldsymbol{v}$) of (4.1) gives the momentum balance equation.

$$m_a n_a \frac{d\boldsymbol{u}_a}{dt} = e_a n_a(\boldsymbol{E} + \boldsymbol{u} \times \boldsymbol{B}) - \nabla p_a - \nabla \cdot \boldsymbol{\pi}_a + \boldsymbol{F}_{a1} + \boldsymbol{S}_{ua} \tag{4.4}$$

$$p_a = \int \frac{1}{3} m_a n_a(\boldsymbol{v} - \boldsymbol{u}_a)^2 f_a d\boldsymbol{v} \ , \ \boldsymbol{\pi}_a = \int m_a n_a((\boldsymbol{v} - \boldsymbol{u}_a)(\boldsymbol{v} - \boldsymbol{u}_a)$$

$$- \frac{1}{3}(\boldsymbol{v} - \boldsymbol{u}_a)^2 \boldsymbol{I}) f_a d\boldsymbol{v}$$

$$\boldsymbol{F}_{a1} \equiv \int m_a(\boldsymbol{v} - \boldsymbol{u}_a) C_a(f_a) d\boldsymbol{v} \ , \ \boldsymbol{S}_{ua} = \int m_a(\boldsymbol{v} - \boldsymbol{u}_a) S_a d\boldsymbol{v}$$

To obtain this, firstly we substitute $A = m_a v_i$ into (4.2). A moment relation: $m_a \partial n_a \langle v_i \rangle / \partial t + m_a \partial n_a \langle v_i v_j \rangle / \partial x_j - e_a n_a(\boldsymbol{E} + (\boldsymbol{u} \times \boldsymbol{B}))_i = \int m_a v_i (C_a + S_a) d\boldsymbol{v}$ is obtained noting $\partial v_i / \partial v_j = \delta_{ij}$. Using the random velocity variable $v'_j = v_j - u_{aj}$, we decompose $\langle v_i v_j \rangle = u_{ai} u_{aj} + \langle v'_i v'_j \rangle = u_{ai} u_{aj} + \langle v'^2 \rangle \delta_{ij}/3 + (\langle v'_i v'_j - v'^2 \delta_{ij}/3 \rangle)$. If we define the pressure by $p_a = m_a n_a \langle v'^2 \rangle / 3$ and the viscous stress tensor by $\pi_{aij} = m_a n_a \langle v'_i v'_j - v'^2 \delta_{ij}/3 \rangle$, we have the expression for the stress tensor $P_{aij} \equiv m_a n_a \langle v_i v_j \rangle = m_a n_a u_{ai} u_{aj} + p_a \delta_{ij} + \pi_{aij}$. Substituting $\partial n_a / \partial t = S_{na} - \partial n_a u_{ai} / \partial x_i$ into the above moment relation, we obtain $m_a n_a du_{ai}/dt = e_a n_a (E_i + (\boldsymbol{u} \times \boldsymbol{B})_i) - \partial p_a / \partial x_i - \partial \pi_{aij} / \partial x_j + F_{a1i} + S_{uai}$, where $F_{a1i} \equiv \int m_a (v_i - u_{ai}) C_a(f_a) d\boldsymbol{v}$ is the friction force and $S_{uai} = \int m_a (v_i - u_{ai}) S_a d\boldsymbol{v}$ is the momentum source.

The viscous stress tensor $\boldsymbol{\pi}_a$ in a magnetized plasma is different from that in the neutral fluid shown in Column 4–1 and $\boldsymbol{\pi}_a$ in tokamak is dominated by the parallel viscous stress tensor $\boldsymbol{\pi}_{a\|}$ originating from the difference in $\langle v_\|^2 \rangle$ and $\langle v_\perp^2 \rangle$ due to the conservation of the magnetic moment under spatially varying magnetic field. There is also a gyro viscous stress tensor $\boldsymbol{\pi}_{ag}$ originating from the gyro motion as given in the Column 4–2. This gyro viscosity is sometimes called as magneto viscosity. This gyro viscosity plays some role in the drift wave dynamics.

The velocity moment ($\frac{1}{2} m_a v^2$) of (4.1) gives the energy balance equation.

$$\frac{3}{2} \frac{\partial p_a}{\partial t} + \nabla \cdot (\boldsymbol{q}_a + \frac{5}{2} p_a \boldsymbol{u}_a) = -\boldsymbol{\pi}_a : \nabla \boldsymbol{u}_a + \boldsymbol{u}_a \cdot \nabla p_a + Q_a + S_{pa} \tag{4.5}$$

$$\boldsymbol{q}_a = \int \frac{1}{2} m_a (\boldsymbol{v} - \boldsymbol{u}_a)^2 (\boldsymbol{v} - \boldsymbol{u}_a) f_a d\boldsymbol{v} \ , \ Q_a = \int \frac{1}{2} m_a (\boldsymbol{v} - \boldsymbol{u}_a)^2 C_a d\boldsymbol{v} \ ,$$

$$S_{Ea} = \int \frac{1}{2} m_a v^2 S_a d\boldsymbol{v}$$

To obtain this, firstly we substitute $A = \frac{1}{2} m_a v^2$ into (4.2). A moment relation: $\frac{1}{2} m_a \partial n_a \langle v^2 \rangle / \partial t + \frac{1}{2} m_a \partial n_a \langle v^2 v_i \rangle / \partial x_i - e_a n_a \boldsymbol{E} \cdot \boldsymbol{u} = \int \frac{1}{2} m_a v^2 (C_a + S_a) d\boldsymbol{v}$ is obtained noting $(\boldsymbol{v} \times \boldsymbol{B})_j v_j = 0$. We decompose $\langle v^2 v_i \rangle = \langle v'^2 v'_i \rangle + \langle v'^2 \rangle u_{ai} + 2 u_{aj} \langle v'_j v'_i \rangle + u_a^2 u_{ai}$. If we define the total energy flux by $Q_{ai} = \frac{1}{2} m_a n_a \langle v^2 v_i \rangle$ and the conduction heat flux by $q_{ai} = \frac{1}{2} m_a n_a \langle v'^2 v'_i \rangle$ and use the definitions of p_a and π_{aij}, we have

$Q_{ai} = q_{ai} + \frac{5}{2}p_a u_{ai} + u_{aj}\pi_{aji} + \frac{1}{2}m_a n_a u_a^2 u_{ai}$. Using this decomposition and noting $\frac{1}{2}m_a\langle v^2\rangle = \frac{1}{2}m_a\langle v'^2\rangle + \frac{1}{2}m_a u_a^2$ and $\int \frac{1}{2}m_a v^2 C_a d\boldsymbol{v} = \int \frac{1}{2}m_a v'^2 C_a d\boldsymbol{v} + \boldsymbol{u}_a \cdot \int m_a \boldsymbol{v}' C_a d\boldsymbol{v}$ (since $\int C_a d\boldsymbol{v} = 0$), the moment relation is converted to:

$$\frac{\partial}{\partial t}(\frac{3}{2}p_a + \frac{1}{2}n_a m_a u_a^2) + \frac{\partial}{\partial x_i}(q_{ai} + \frac{5}{2}p_a u_{ai} + u_{aj}\pi_{aji} + \frac{1}{2}m_a n_a u_a^2 u_{ai})$$
$$= (e_a n_a E_i' + F_{a1i})u_{ai} + Q_a + S_{Ea}$$

where $Q_a = \int \frac{1}{2}m_a v'^2 C_a d\boldsymbol{v}$ is the collisional heat generation rate, $S_{Ea} = \int \frac{1}{2}m_a v^2 S_a d\boldsymbol{v}$ is the energy source rate. If we subtract (4.5)$\cdot\boldsymbol{u}_a$ and (4.3)$\times m_a u_a^2/2$ from above, we have: $\frac{3}{2}\partial p_a/\partial t + \partial(q_{ai} + \frac{5}{2}p_a u_{ai})/\partial x_i = -\pi_{aij}\partial u_{ai}/\partial x_j + u_{ai}\partial p_a/\partial x_i + Q_a + S_{pa}$, where $S_{pa} = S_{Ea} - \boldsymbol{u}_a \cdot \boldsymbol{S}_{ua} - (m_a u_a^2/2)S_{na}$ and using the definition of the double dot product, $\pi_{aij}\partial u_{ai}/\partial x_j = \boldsymbol{\pi}_a : \nabla \boldsymbol{u}_a$.

The velocity moment $(\frac{1}{2}m_a v^2 \boldsymbol{v})$ of (4.1) gives energy flow equation.

$$\frac{\partial \boldsymbol{Q}_a}{\partial t} + \nabla \cdot \boldsymbol{R}_a = \frac{e_a}{m_a}\left((\frac{3}{2}p_a + m_a n_a u_a^2)\boldsymbol{E} + \boldsymbol{E}\cdot\boldsymbol{P}_a + \boldsymbol{Q}_a \times \boldsymbol{B} + \boldsymbol{G}_a\right) + \boldsymbol{S}_{Qa} \tag{4.6}$$

$$\boldsymbol{Q}_a = \boldsymbol{q}_a + \frac{5}{2}p_a\boldsymbol{u}_a + \boldsymbol{u}_a\cdot\boldsymbol{\pi}_a + \frac{1}{2}m_a n_a u_a^2 \boldsymbol{u}_a\,,\ \boldsymbol{P}_a \equiv m_a n_a\langle \boldsymbol{v}\boldsymbol{v}\rangle = m_a n_a \boldsymbol{u}_a\boldsymbol{u}_a + p_a\boldsymbol{I} + \boldsymbol{\pi}_a$$

$$\boldsymbol{R}_a = \int \frac{m_a v^2}{2}\boldsymbol{v}\boldsymbol{v} f_a d\boldsymbol{v}\,,\ \boldsymbol{G}_a = \int \frac{m_a v^2}{2}\boldsymbol{v} C_a(f_a) d\boldsymbol{v}\,,\ \boldsymbol{S}_{Qa} = \int \frac{m_a v^2}{2}\boldsymbol{v} S_a d\boldsymbol{v}$$

To obtain this, we substitute $A = \frac{1}{2}m_a v^2 v_i$ into (4.2) to obtain a moment relation:

$$\frac{1}{2}m_a\frac{\partial}{\partial t}n_a\langle v^2 v_i\rangle + \frac{1}{2}m_a\frac{\partial}{\partial x_j}n_a\langle v^2 v_i v_j\rangle = \frac{e_a}{m_a}E_j n_a\langle \frac{1}{2}m_a v^2\delta_{ij} + m_a v_j v_i\rangle$$
$$+ \frac{e_a}{m_a}n_a\langle(\boldsymbol{v}\times\boldsymbol{B})_j(\frac{1}{2}m_a v^2\delta_{ij} + m_a v_j v_i)\rangle + \int v^2 v_i(C_a + S_a)d\boldsymbol{v}$$

Here, we note $\frac{\partial A}{\partial v_j} = \frac{1}{2}m_a v^2\delta_{ij} + m_a v_i v_j$. As for the electric field term, we decompose $n_a\langle \frac{1}{2}m_a v^2\delta_{ij} + m_a v_j v_i\rangle = (\frac{3}{2}p_a + \frac{1}{2}m_a n_a u_a^2)\delta_{ij} + P_{aji}$. As for the Lorentz force term, we note $(\boldsymbol{v}\times\boldsymbol{B})_j v_j = (\boldsymbol{v}\times\boldsymbol{B})\cdot\boldsymbol{v} = 0$ and $\langle \frac{1}{2}m_a v^2 \boldsymbol{v}\rangle = \boldsymbol{Q}_a$. Therefore, we have: $\partial Q_{ai}/\partial t + \partial R_{aji}/\partial x_j = \frac{e_a}{m_a}[(\frac{3}{2}p_a + m_a n_a u_a^2)E_i + E_j P_{aji} + (\boldsymbol{Q}_a\times\boldsymbol{B})_i + G_{ai}] + S_{Qai}$. Here, $Q_{ai} = \frac{1}{2}m_a n_a\langle v^2 v_i\rangle$ is the total energy flux and $R_{aij} = \frac{1}{2}m_a n_a\langle v^2 v_i v_j\rangle$ is the energy weighted total stress tensor, $G_{ai} = \int \frac{1}{2}m_a v^2 v_i C_a d\boldsymbol{v}$ is the collisional heat generation rate. If we neglect the inertial term $m_a n_a u_a^2$ in the right hand side of (4.6), we have the same result with Hinton-Hazeltine [321] and Hirshman-Sigmar [330].

For strongly magnetized plasma such as tokamak plasmas, the pressure tensor $\boldsymbol{p}_a \equiv m_a n_a\langle \boldsymbol{v}'\boldsymbol{v}'\rangle$ and the energy weighted stress tensor $\boldsymbol{r}_a \equiv \frac{1}{2}m_a n_a\langle v'^2\boldsymbol{v}'\boldsymbol{v}'\rangle$ have anisotropy given in the Chew-Goldberg-Low (CGL) form [121].

$$\boldsymbol{p}_a = p_a\boldsymbol{I} + (p_{\parallel a} - p_{\perp a})(\boldsymbol{bb} - \frac{1}{3}\boldsymbol{I}) + O(\epsilon^2) = p_a\boldsymbol{I} + \boldsymbol{\pi}_a \tag{4.7}$$

$$\boldsymbol{r}_a = r_a\boldsymbol{I} + (r_{\parallel a} - r_{\perp a})(\boldsymbol{bb} - \frac{1}{3}\boldsymbol{I}) + O(\epsilon^2) \tag{4.8}$$

$$p_{\parallel a} = \int m_a(v_\parallel - u_{\parallel a})^2 f_a d\boldsymbol{v}\ ,\ p_{\perp a} = \int \frac{1}{2}m_a v_\perp^2 f_a d\boldsymbol{v}$$

$$r_{\parallel a} = \int \frac{1}{2}m_a(\boldsymbol{v} - \boldsymbol{u}_a)^2(v_\parallel - u_{\parallel a})^2 f_a d\boldsymbol{v}\ ,\ r_{\perp a} = \int \frac{1}{4}m_a(\boldsymbol{v} - \boldsymbol{u}_a)^2 v_\perp^2 f_a d\boldsymbol{v},$$

where small perpendicular flow term is neglected for $p_{\perp a}$ and $r_{\perp a}$. Here $r_a \equiv Tr\frac{1}{3}\boldsymbol{r}_a = \frac{1}{6}m_a n_a\langle v'^4\rangle$ and $r_a = \frac{5T_a}{2m_a}p_a$ for a Maxwellian plasma. We decompose $r_{aij} \equiv \frac{1}{2}m_a n_a\langle v'^2 v_i' v_j'\rangle = \frac{1}{6}m_a n_a\langle v'^4\rangle\delta_{ij} + \frac{1}{2}m_a n_a\langle v'^2(v_i'v_j' - \frac{1}{3}v'^2\delta_{ij})\rangle$. If we define the heat viscous stress tensor $\Theta_{aij} \equiv m_a n_a\langle(\frac{m_a v'^2}{2T_a} - \frac{5}{2})(v_i'v_j' - \frac{1}{3}v'^2\delta_{ij})\rangle$ and substitute $\frac{1}{6}m_a n_a\langle v'^4\rangle = \frac{5T_a}{2m_a}p_a$, we have $r_{aij} = \frac{5T_a}{2m_a}p_{aij} + \frac{T_a}{m_a}\Theta_{aij}$. If we define $x_a^2 = m_a v'^2/2T_a$, the heat viscous stress tensor $\boldsymbol{\Theta}_a = \frac{m_a}{T_a}\boldsymbol{r}_a - \frac{5}{2}\boldsymbol{p}_a$ is given by:

$$\boldsymbol{\Theta}_a = \int m_a\left(\boldsymbol{v}'\boldsymbol{v}' - \frac{v'^2}{3}\boldsymbol{I}\right)(x_a^2 - \frac{5}{2})f_a d\boldsymbol{v} = (\Theta_{\parallel a} - \Theta_{\perp a})(\boldsymbol{bb} - \frac{1}{3}\boldsymbol{I})$$

$$\Theta_{\parallel a} = \int m_a v_\parallel^2(x_a^2 - \frac{5}{2})f_a d\boldsymbol{v}\ ,\ \Theta_{\perp a} = \int \frac{1}{2}m_a v_\perp^2(x_a^2 - \frac{5}{2})f_a d\boldsymbol{v} \tag{4.9}$$

The collisional heat generation rate $\boldsymbol{G}_a$ in (4.6) can be decomposed similarly.

$$\boldsymbol{G}_a = \frac{T_a}{m_a}\left(\frac{5}{2}\boldsymbol{F}_{a1} + \boldsymbol{F}_{a2}\right)$$

$$\boldsymbol{F}_{a2} = \int m_a\boldsymbol{v}(x_a^2 - \frac{5}{2})C_a(f_a)d\boldsymbol{v} \tag{4.10}$$

where $\boldsymbol{F}_{a1}$ is friction force defined in (4.4) and $\boldsymbol{F}_{a2}$ is called the heat friction force.

From now on, we assume flow u_a is much smaller than the thermal speed $v_{Ta} = \sqrt{2T_a/m_a}$, namely $\epsilon_u = u_a/v_{Ta} \ll 1$. In this case, we have $\boldsymbol{Q}_a = \boldsymbol{q}_a + \frac{5}{2}p_a\boldsymbol{u}_a + O(\epsilon_u^2)$ and $\boldsymbol{R}_a = \boldsymbol{r}_a + O(\epsilon_u^2)$, where $\boldsymbol{r}_a \equiv \frac{1}{2}m_a n_a\langle v'^2\boldsymbol{v}'\boldsymbol{v}'\rangle$ is the energy weighted stress tensor. Using $\epsilon_u = u_a/v_{Ta} \ll 1$ approximation for the first term of (4.6) and using (4.4) and (4.5) to eliminate $\frac{\partial \boldsymbol{u}_a}{\partial t}$ and $\frac{\partial p_a}{\partial t}$, we have $\frac{\partial Q_{ai}}{\partial t} \approx \frac{\partial q_{ai}}{\partial t} + \frac{5e_a}{2m_a}p_a(\boldsymbol{E} + \boldsymbol{u}\times\boldsymbol{B})_i - \frac{5T_a}{2m_a}\frac{\partial p_{aji}}{\partial x_j} + \frac{5T_a}{2m_a}F_{a1i}$, where we neglected terms $O(\epsilon_u^2)$ as well as a second order term in flow $-\frac{5}{3}u_{ai}\frac{\partial q_{aj}}{\partial x_j}$. Similar algebra gives $\frac{\partial R_{aji}}{\partial x_j} \approx \frac{\partial r_{aji}}{\partial x_j} = \frac{\partial}{\partial x_j}(\frac{5T_a}{2m_a}p_{aji} + \frac{T_a}{m_a}\Theta_{aji}) \approx \frac{5T_a}{2m_a}\frac{\partial p_{aji}}{\partial x_j} + \frac{5p_a}{2m_a}\frac{\partial T_a}{\partial x_i} + \frac{T_a}{m_a}\frac{\partial \Theta_{aji}}{\partial x_j}$, where we used $p_a \gg \pi_{aji}, \Theta_{aji}$. The right hand side of Eq. (4.6) is given as $\frac{e_a}{m_a}[\frac{5}{2}p_a(\boldsymbol{E} + \boldsymbol{u}\times\boldsymbol{B})_i + (\boldsymbol{q}_a\times\boldsymbol{B})_i + \frac{T_a}{m_a}(\frac{5}{2}F_{a1i} + F_{a2i})]$ under this approximation. Thus we obtain following equation for the heat flow $\boldsymbol{q}_a$.

$$\frac{m_a}{T_a}\frac{\partial \boldsymbol{q}_a}{\partial t} = \frac{e_a}{T_a}\boldsymbol{q}_a \times \boldsymbol{B} - \frac{5}{2}n_a \nabla T_a - \nabla \cdot \boldsymbol{\Theta}_a + \boldsymbol{F}_{a2}$$

$$\boldsymbol{\Theta}_a = \int m_a \left(\boldsymbol{v}'\boldsymbol{v}' - \frac{v'^2}{3}\boldsymbol{I}\right)(x_a^2 - \frac{5}{2})f_a d\boldsymbol{v}\ , \boldsymbol{F}_{a2} = \int m_a \boldsymbol{v}(x_a^2 - \frac{5}{2})C_a(f_a)d\boldsymbol{v} \tag{4.11}$$

If we add external momentum and heat flow sources $\boldsymbol{M}_a$ and $\boldsymbol{Q}_a$ in the right hand side of (4.4) and (4.11), we have following set of flow equations.

$$m_a n_a \frac{d\boldsymbol{u}_a}{dt} = e_a n_a (\boldsymbol{E} + \boldsymbol{u}_a \times \boldsymbol{B}) - \nabla p_a - \nabla \cdot \boldsymbol{\pi}_a + \boldsymbol{F}_{a1} + \boldsymbol{M}_a \tag{4.12}$$

$$\frac{m_a}{T_a}\frac{\partial \boldsymbol{q}_a}{\partial t} = \frac{e_a}{T_a}\boldsymbol{q}_a \times \boldsymbol{B} - \frac{5}{2}n_a \nabla T_a - \nabla \cdot \boldsymbol{\Theta}_a + \boldsymbol{F}_{a2} + \boldsymbol{Q}_a \tag{4.13}$$

Total viscous stress tensor $\boldsymbol{\Pi}_a$ is defined and related to viscous stress tensor $\boldsymbol{\pi}_a$ as,

$$\Pi_{aij} \equiv < m_a(v_i v_j - \frac{1}{3}v^2\delta_{ij})f_a >_v = \pi_{aij} + m_a n_a u_{ai} u_{aj} \tag{4.14}$$

If the flow is subsonic, there is only $O(\epsilon_u^2 v_{Ta})$ difference between $\boldsymbol{\Pi}_a$ and $\boldsymbol{\pi}_a$ and we will make no distinction between them unless otherwise specified. Hirshman and Sigmar in Fig. 4.1 developed a moment approach to collisional transport.

Fig. 4.1 (**a**) Steven P. Hirshman and (**b**) Dieter Sigmar, who developed moment approach and analytical friction and viscosity coefficients for the neoclassical transport. Courtesy of S. P. Hirshman and MIT plasma science and fusion center

Column 4-1: Fluid Viscosity; Landau-Lifschitz [495], Tani [751]
In ideal neutral fluid, the continuity equation and the Euler equation are:

$$\frac{\partial \rho}{\partial t} + \frac{\partial \rho u_i}{\partial x_i} = 0 \text{ , and } \rho\left(\frac{\partial u_i}{\partial t} + u_j \frac{\partial u_i}{\partial x_j}\right) = -\frac{\partial p}{\partial x_i} \tag{4.15}$$

This Euler equation can be converted to following conservative form.

$$\frac{\partial \rho u_i}{\partial t} = -\frac{\partial \Pi_{ik}}{\partial x_k} \text{ , where } \Pi_{ik} = p\delta_{ik} + \rho u_i u_k \tag{4.16}$$

Here, Π_{ki} is called the momentum flux density and is a symmetric tensor (Landau-Lifschitz [495]). Integrating in volume and using the Gauss's theorem, we have: $\partial/\partial t \int \rho u_i dV = -\oint \Pi_{ik} dS_k$. The $\Pi_{ik} dS_k$ is the i-th component of the momentum flowing out from the volume through the surface dS_k.

In the viscous fluids, there is another source of momentum flux through the surface element through the collisions between gas molecules and interaction in liquid molecules, called the viscous momentum flux density ($\pi = -\tau$), where τ is called the viscous stress. If π is proportional to the velocity gradient ($\pi_{xy} = -\mu du_x/dy$), this fluid is called the "Newtonian fluid" originating from Isaac Newton (1687). This viscous momentum flux becomes tensor in general geometry (viscous stress tensor). We note that velocity differential can be decomposed as the sum of the rotation and the rate-of-strain as follows.

$$du_i = \frac{\partial u_i}{\partial x_j} dx_j = \frac{1}{2}(\omega_{ij} + s_{ij}) dx_j \tag{4.17}$$

$$\omega_{ij} = \frac{1}{2}\left(\frac{\partial u_i}{\partial x_j} - \frac{\partial u_j}{\partial x_i}\right) \text{ , } s_{ij} = \frac{1}{2}\left(\frac{\partial u_i}{\partial x_j} + \frac{\partial u_j}{\partial x_i}\right) \text{ ,} \tag{4.18}$$

where ω_{ij} is called the vorticity tensor and s_{ij} is called the rate-of-strain tensor [428, 751]. The vorticity tensor ω_{ij} and the vorticity $\boldsymbol{\omega} = \nabla \times \boldsymbol{u}$ are related as $\omega_i = \epsilon_{ijk}\omega_{jk}$, $\omega_{ij} = \epsilon_{ijk}\omega_k/2$, where $\epsilon_{ijk} = (\boldsymbol{e}_i \times \boldsymbol{e}_j) \cdot \boldsymbol{e}_k$ is the Eddington ϵ (or Levi-Civita symbol), ω_i is i-th component of the vorticity.

The viscous momentum flux is proportional to the rate-of-strain tensor since the rotation do not contribute to the viscous momentum flux. The viscous momentum flux in the Newtonian incompressible flow is given as $\pi_{ij} = -2\mu s_{ij}$. In case of incompressible fluid, this is traceless. In the general compressible fluid, Trace$\{s_{ij}\} = \nabla \cdot \boldsymbol{u}$ and we define the traceless rate-of-strain tensor S_{ij} as follows (factor 2 is to match Braginskii definition).

(continued)

$$S_{ij} = \frac{\partial u_i}{\partial x_j} + \frac{\partial u_j}{\partial x_i} - \frac{2}{3}\nabla \cdot \boldsymbol{u}\delta_{ij} \tag{4.19}$$

The momentum flux density is given by adding a term proportional to $\nabla \cdot \boldsymbol{u}$ as: $\Pi_{ij} = p\delta_{ij} - \mu S_{ij} + \mu_v \nabla \cdot \boldsymbol{u}\delta_{ij} + \rho u_i u_j$, where μ_v is called the bulk viscosity.

Column 4-2: Gyro Viscosity; Kaufman [419], Braginskii [76]

In the magnetized plasma, the physical mechanism to produce viscous force is quite different. The lowest order velocity distribution function f_0, which is a solution of the drift kinetic equation $v_\| \nabla_\| f_0 + \boldsymbol{v}_d \cdot \nabla f_0 = C(f_0)$, depends on guiding center position $\boldsymbol{X}$ rather than the particle position $\boldsymbol{r} = \boldsymbol{X} + \boldsymbol{\rho}$. Therefore, the distribution function at $\boldsymbol{r}$ is $f(\boldsymbol{r}) \sim f_0(\boldsymbol{X}) \sim f_0(\boldsymbol{r}) - \boldsymbol{\rho} \cdot \nabla f_0(\boldsymbol{r})$. The term $-\boldsymbol{\rho} \cdot \nabla f_0(\boldsymbol{r})$ produces the diamagnetic flow, diamagnetic heat flow, and the momentum flux. The gyro viscosity $\boldsymbol{\pi}_g$ is given by:

$$\boldsymbol{\pi}_g = -\int m_a(\boldsymbol{v}'\boldsymbol{v}' - (v'^2/3)\boldsymbol{I})\boldsymbol{\rho} \cdot \nabla f_0(\boldsymbol{r})d^3\boldsymbol{v} \tag{4.20}$$

The relation between viscous stress tensor π_{ij} and the rate of strain tensor S_{ij} is simple in the absence of the magnetic field similar to the case of a neutral fluid ($\pi_{ij} \sim -\mu S_{ij}$) as given by Chapman-Cowling [105] and refined by Braginskii [76] ($\mu_i = 1.36 n_i T_i \tau_{ii}$ for ion and $\mu_e = 0.733 n_e T_e \tau_{ei}$ for electron). Here, S_{ij} is the trace-less rate of strain tensor defined in (4.19).

The presence of the magnetic field (we assume $\boldsymbol{B} = B\boldsymbol{e}_z$) leads to a complicated relation between π_{ij} and S_{ij}. This gyro viscous tensor is a traceless symmetric tensor having five independent components, π_{zz}, π_{xy}, π_{xz}, π_{yz}, $\pi_{xx} - \pi_{yy}$, which are to be expressed in terms of the invariant set of rate of strain components against rotation around the z axis, S_{zz}, S_{xy}, S_{xz}, S_{yz}, and $S_{xx} - S_{yy}$.

The viscous stress component $\pi_{zz} = -\mu S_{zz}$ is same as in the absence of the magnetic field. The traceless condition gives $\pi_{xx} + \pi_{yy} = -\pi_{zz}$. In the strongly magnetized plasma ($\Omega_a \tau_a \gg 1$), part of the viscous stress goes to zero as $\Omega_a \tau_a \to \infty$ but there are some remaining term $\pi_{ij} \sim p_a/\Omega_a$ as seen from Chapman-Cowling formula [105, p. 378].

Physical account is given by Kaufman [419]. The viscous stress tensor from du_x/dy associated with dE_y/dy is obtained by solving the equation of motion as $\langle \dot{x}^2 \rangle = \frac{1}{2}\rho_a^2(\Omega_a^2 - 2\Omega_a \frac{\partial u_x}{\partial y})$ and $\langle \dot{y}^2 \rangle = \frac{1}{2}\rho_a^2(\Omega_a^2 - \Omega_a \frac{\partial u_x}{\partial y})$. This gives rise to the stress tensor component, $\pi_{xx} - \pi_{yy} = -(p_{a\perp}/\Omega_a)\frac{\partial u_x}{\partial y}$. In terms of the rate of strain tensor S_{ij}, we have $\pi_{xx} - \pi_{yy} = -S_{xy} p_{a\perp}/\Omega_a$.

(continued)

The viscous stress tensor from $\partial u_x/\partial z$ associated with dE_y/dz is obtained noting $\frac{dz}{dt} = v_\parallel$. When $dE_y/dz \neq 0$, we have an acceleration $\frac{du_x}{dt} = \frac{\partial u_x}{\partial z} v_\parallel$. This gives rise to an additional force drift $u_y = \frac{du_x}{dt}/\Omega_a = \frac{\partial u_x}{\partial z} v_\parallel/\Omega_a$ in the y-direction. This gives rise to the stress tenser component, $\pi_{yz1} = m_a n_a \langle u_y v_\parallel \rangle = (p_{a\parallel}/\Omega_a)\frac{\partial u_x}{\partial z}$. Similarly, the viscous stress tensor from $\partial u_z/\partial x$ is obtained as $\pi_{yz2} = (p_a/\Omega_a)\frac{du_z}{dx}$. Summation gives rise to $\pi_{yz} = \pi_{yz1} + \pi_{yz2} = S_{xz}(p_{\parallel a} + p_{\perp a})/\Omega_a + \omega_{xz}(p_{\parallel a} - p_{\perp a})/\Omega_a$, where ω_{ij} is the vorticity tensor.

Kaufman [419] noted that the stress tensor can be driven by the vorticity for the anisotropic pressure plasma.

4.1.2 Flux-Surface Averaged Moment Equations

We define the flux surface average operator $\langle\rangle$ by,

$$\langle A \rangle \equiv \frac{\int_0^{2\pi} \sqrt{g} A d\theta}{\int_0^{2\pi} \sqrt{g} d\theta} = \frac{\int_0^{2\pi} d\theta A/\boldsymbol{B}\cdot\nabla\theta}{\int_0^{2\pi} d\theta/\boldsymbol{B}\cdot\nabla\theta} \tag{4.21}$$

This flux surface average is an annihilator of $\boldsymbol{B}\cdot\nabla = (1/\sqrt{g})\partial/\partial\theta$ as $\langle \boldsymbol{B}\cdot\nabla A\rangle = 0$.

The flux surface averaged parallel momentum and heat flux balance equations for thermal electron(e), ion(i), and impurity(I) in transport time scale ($\partial/\partial t \sim 0$) are obtained by taking the flux surface average of $\boldsymbol{B}$·(4.12) and $\boldsymbol{B}$·(4.13) as follows,

$$\langle \boldsymbol{B}\cdot\nabla\cdot\boldsymbol{\Pi}_a\rangle = \langle \boldsymbol{B}\cdot\boldsymbol{F}_{a1}\rangle + e_a n_a \langle \boldsymbol{B}\cdot\boldsymbol{E}\rangle + \langle \boldsymbol{B}\cdot\boldsymbol{M}_a\rangle \tag{4.22}$$

$$\langle \boldsymbol{B}\cdot\nabla\cdot\boldsymbol{\Theta}_a\rangle = \langle \boldsymbol{B}\cdot\boldsymbol{F}_{a2}\rangle + \langle \boldsymbol{B}\cdot\boldsymbol{Q}_a\rangle \tag{4.23}$$

The first order flows of species a are decomposed to parallel ($u_{\parallel a}\boldsymbol{b}$ and $q_{\parallel a}\boldsymbol{b}$) and perpendicular ($\boldsymbol{u}_{\perp a}^{(1)}$ and $\boldsymbol{q}_{\perp a}^{(1)}$) components to the magnetic field as,

$$\boldsymbol{u}_a^{(1)} = u_{\parallel a}\boldsymbol{b} + \boldsymbol{u}_{\perp a}^{(1)}, \boldsymbol{q}_a^{(1)} = q_{\parallel a}\boldsymbol{b} + \boldsymbol{q}_{\perp a}^{(1)} \tag{4.24}$$

The perpendicular flows are given by taking $\boldsymbol{b}\times$ (4.12) and (4.13) by setting $\partial/\partial t = 0$ and neglecting higher order terms such as $\boldsymbol{\pi}_a, \boldsymbol{\Theta}_a, \boldsymbol{F}_{a1}, \boldsymbol{F}_{a2}, \boldsymbol{M}_a, \boldsymbol{Q}_a$ as:

$$\boldsymbol{u}_{\perp a}^{(1)} = \frac{BV_{1a}}{F}\frac{\nabla\psi\times\boldsymbol{b}}{B}, \boldsymbol{q}_{\perp a}^{(1)} = \frac{5P_a}{2}\frac{BV_{2a}}{F}\frac{\nabla\psi\times\boldsymbol{b}}{B} \tag{4.25}$$

Here $BV_{1a} = -F[d\Phi/d\psi + (1/e_a n_a)dP_a/d\psi]$ and $BV_{2a} = -(F/e_a)dT_a/d\psi$ are thermodynamic forces and $F(\psi) = RB_\zeta$. Considering the poloidal flow in unit flux tube, we have following relation.

$$u_{a\theta}(\psi) = \frac{\boldsymbol{u}_a^{(1)} \cdot \nabla\theta}{\boldsymbol{B} \cdot \nabla\theta}, q_{a\theta}(\psi) = \frac{\boldsymbol{q}_a^{(1)} \cdot \nabla\theta}{\boldsymbol{B} \cdot \nabla\theta} \tag{4.26}$$

Here we note that $u_{a\theta}(\psi)$ and $q_{a\theta}(\psi)$ do not have a dimension of flow.

From the scalar product (4.24)$\cdot\nabla\theta$, we obtain following relations:

$$B^2 u_{a\theta}(\psi) = Bu_{\parallel a} - BV_{1a} , B^2 \frac{2q_{a\theta}(\psi)}{5P_a} = B\frac{2q_{\parallel a}}{5P_a} - BV_{2a} \tag{4.27}$$

Substituting axisymmetric relation $\boldsymbol{b} \times \nabla\psi = F\boldsymbol{b} - R^2 B \nabla\zeta$ (see Exercise ??) into (4.25), first order flow relation (4.24) is transformed by using (4.27) as follows.

$$\boldsymbol{u}_a^{(1)} = u_{a\theta}(\psi)\boldsymbol{B} + \frac{BV_{1a}}{F} R^2 \nabla\zeta , \boldsymbol{q}_a^{(1)} = q_{a\theta}(\psi)\boldsymbol{B} + \frac{5P_a}{2}\frac{BV_{2a}}{F} R^2 \nabla\zeta \tag{4.28}$$

Taking the toroidal (ζ) component of (4.28) and using the flux surface average of (4.27), we obtain following equations for the toroidal flows,

$$u_{a\zeta}^{(1)} = \frac{B_\zeta}{\langle B^2 \rangle}\langle Bu_{\parallel a}\rangle + \left[1 - \frac{B_\zeta^2}{\langle B^2\rangle}\right]\frac{BV_{1a}}{B_\zeta} \tag{4.29}$$

Here, second terms of the right-hand-side are called Pfirsch-Schlüter terms. Flux surface averaged parallel viscous forces $\langle \boldsymbol{B} \cdot \nabla \cdot \boldsymbol{\Pi}_a \rangle$, $\langle \boldsymbol{B} \cdot \nabla \cdot \boldsymbol{\Theta}_a \rangle$ and friction forces $\langle \boldsymbol{B} \cdot \boldsymbol{F}_{a1} \rangle$, $\langle \boldsymbol{B} \cdot \boldsymbol{F}_{a2} \rangle$ are related to first order flows. Friction forces are given as:

$$\begin{bmatrix} \langle \boldsymbol{B} \cdot \boldsymbol{F}_{a1} \rangle \\ \langle \boldsymbol{B} \cdot \boldsymbol{F}_{a2} \rangle \end{bmatrix} = \frac{m_a n_a}{\tau_{aa}} \sum_b \begin{bmatrix} \hat{l}_{11}^{ab} & -\hat{l}_{12}^{ab} \\ -\hat{l}_{21}^{ab} & \hat{l}_{22}^{ab} \end{bmatrix} \begin{bmatrix} \langle Bu_{\parallel a} \rangle \\ \dfrac{2\langle Bq_{\parallel a}\rangle}{5P_a} \end{bmatrix} \tag{4.30}$$

Here $\hat{l}_{ij}^{ab} = (\tau_{aa}/m_a n_a) l_{ij}^{ab}$ is the normalized friction coefficient. The l_{ij}^{ab} for thermal species and those between fast ion and thermal species are given in Sect. 4.3.1.

The parallel viscous force is related to the poloidal flow since viscous force operates when the particle moves poloidally in response to the variation of the toroidal magnetic field [327, 329].

$$\begin{bmatrix} \langle \boldsymbol{B} \cdot \nabla \cdot \boldsymbol{\Pi}_a \rangle \\ \langle \boldsymbol{B} \cdot \nabla \cdot \boldsymbol{\Theta}_a \rangle \end{bmatrix} = \frac{m_a n_a \langle B^2 \rangle}{\tau_{aa}} \begin{bmatrix} \hat{\mu}_{a1} & \hat{\mu}_{a2} \\ \hat{\mu}_{a2} & \hat{\mu}_{a3} \end{bmatrix} \begin{bmatrix} u_{a\theta} \\ \frac{2q_{a\theta}}{5P_a} \end{bmatrix} \tag{4.31}$$

Here $\hat{\mu}_{ai} = (\tau_{aa}/m_a n_a)\mu_{ai}$ is the normalized viscosity coefficient. The same $\hat{\mu}_{a2}$ appears in two places in off-diagonal part of the viscosity matrix, which is the special example of the Onsager symmetry and comes from the self-adjointness of the Coulomb collision operator. It is also important to note that the viscous force is related only to its own flow fields, which comes from the dominance of the test particle portion of the Coulomb collision operator [330].

While the variational method is effective to obtain transport coefficients numerically [321], a novel method is developed by Tsang-Callen [774] using model Coulomb collision operator. Basic idea is to divide velocity space into Pfirsch-Schlüter, plateau, banana, and boundary layer to obtain transport coefficients so that transport coefficients are given by a velocity space integral. This method is further improved by Hirshman [327], Shaing [679], Houlberg [349] using approximate Coulomb collision operator which conserves energy and momentum [325]. Substituting these formulas for viscosity and friction coefficients into (4.22) and (4.23) and using (4.27), following balance equations of friction and viscous forces are obtained.

$$\begin{bmatrix} \hat{\mu}_{a1} & \hat{\mu}_{a2} \\ \hat{\mu}_{a2} & \hat{\mu}_{a3} \end{bmatrix} \begin{bmatrix} \langle Bu_{\|b}\rangle - BV_{1a} \\ \langle B\frac{2q_{\|b}}{5P_b}\rangle - BV_{2a} \end{bmatrix} = \sum_b \begin{bmatrix} \hat{l}_{11}^{ab} & -\hat{l}_{12}^{ab} \\ -\hat{l}_{21}^{ab} & \hat{l}_{22}^{ab} \end{bmatrix} \begin{bmatrix} \langle Bu_{\|b}\rangle \\ \langle B\frac{2q_{\|b}}{5P_b}\rangle \end{bmatrix} + \frac{e_a\tau_{aa}}{m_a n_a} \begin{bmatrix} \langle B(M_{\|a} + n_a E_{\|})\rangle \\ \langle BQ_{\|a}\rangle \end{bmatrix} \tag{4.32}$$

Here, $M_{\|a}, Q_{\|a}$ are parallel momentum source and parallel heat source, respectively. If we write Eq. (4.32) for electron, ion, impurity and fast ion (only momentum balance is considered for fast ion since heat flow by fast ion is comparatively small), we obtain following system of linear equations.

$$\hat{\boldsymbol{M}}\left(\boldsymbol{U}_{\|} - \boldsymbol{V}_{\perp}\right) = \hat{\boldsymbol{L}}\boldsymbol{U}_{\|} + \hat{\boldsymbol{E}} + \hat{\boldsymbol{S}}_{\|} \tag{4.33}$$

where

$\hat{\boldsymbol{M}}$: normalized viscosity matrix, $\hat{\boldsymbol{L}}$: normalized friction matrix
$\boldsymbol{U}_{\|}$: parallel flow vector, $\boldsymbol{V}_{\perp}$: thermodynamic force vector
$\hat{\boldsymbol{E}}$: electric field acceleration vector, $\hat{\boldsymbol{S}}_{\|}$: parallel source vector
Using (4.33), we obtain following expressions for parallel flow.

$$\langle \boldsymbol{B}\cdot\boldsymbol{u}_a\rangle = \sum_{b=1}^{4} \frac{e_b\tau_{bb}\hat{c}_{ab}}{m_b}\langle \boldsymbol{B}\cdot\boldsymbol{E}\rangle + \sum_{b=1}^{7}\left[\hat{\alpha}_{ab}V_{\perp b} + \frac{\tau_{bb}\hat{c}_{ab}}{m_b n_b}\hat{S}_{\|b}\right] \tag{4.34}$$

$$\hat{\alpha} = (\hat{\boldsymbol{M}} - \hat{\boldsymbol{L}})^{-1}\hat{\boldsymbol{M}}\,, \hat{c} = (\hat{\boldsymbol{M}} - \hat{\boldsymbol{L}})^{-1} \tag{4.35}$$

Before we go into the details of kinetic theory to calculate the parallel friction and viscous coefficients in the following sections, we show the relation of the moment variables and expansion of velocity distribution function [449].

The moment equations (4.12) and (4.13) has 18 variables $(n_a, T_a, \boldsymbol{u}_a, \boldsymbol{q}_a, \boldsymbol{\Pi}_a, \boldsymbol{\Theta}_a)$. These variables are related to the structure of the velocity distribution function in the velocity coordinates (v, θ, φ). Using the orthogonality of Legendre polynomials in pitch angle θ and the Sonnine polynomials in v, we have following expansion.

$$f_a = f_{aM} + f_{a1}^{(1)} + f_{a1}^{(2)} \tag{4.36}$$

$$f_{a1}^{(1)} = \frac{2\boldsymbol{v}}{v_{Ta}^2} \cdot \left[\boldsymbol{u}_{a0} L_0^{(3/2)} + \boldsymbol{u}_{a1} L_1^{(3/2)} + \boldsymbol{u}_{a2} L_2^{(3/2)} + -- \right] f_{aM}$$

$$f_{a1}^{(2)} = 2\frac{\boldsymbol{vv} - (v^2/3)\boldsymbol{I}}{m_a n_a v_{Ta}^4} : \left[\boldsymbol{\Pi}_a L_0^{(5/2)} + \frac{2}{7}(\boldsymbol{\Theta}_a + \boldsymbol{\Pi}_a) L_1^{(5/2)} + -- \right] f_{aM}$$

Here, the upper subscript of $f_{a1}^{(1)}$ and $f_{a1}^{(2)}$ implies Legendre harmonics, $P_1(cos\theta)$ and $P_2(cos\theta)$. The gyro phase average of the distribution function is obtained noting that $v_\parallel = vcos\theta$ and $2v_\parallel^2 - v_\perp^2 = (3cos^2\theta - 1)v^2 = 2P_2(cos\theta)v^2$ and following gyro-phase average relation:

$$\langle \boldsymbol{vv} \rangle = \frac{v_\perp^2}{2}(\boldsymbol{I} - \boldsymbol{bb}) + v_\parallel^2 \boldsymbol{bb} \tag{4.37}$$

The gyro phase average of the distribution function is then given by:

$$f_{a1}^{(1)} = \frac{2v_\parallel}{v_{Ta}^2} \cdot \left[u_{a\parallel} - \frac{2q_{a\parallel}}{5p_a}\left(\frac{5}{2} - x_a^2\right) + u_{a2\parallel} L_2^{(3/2)} + -- \right] f_{aM} \tag{4.38}$$

$$f_{a1}^{(2)} = \frac{2v_\parallel^2 - v_\perp^2}{m_a n_a v_{Ta}^4} \left(\boldsymbol{bb} - \frac{1}{3}\boldsymbol{I}\right) : \left[\boldsymbol{\Pi}_a + (\boldsymbol{\Theta}_a + \boldsymbol{\Pi}_a)\left(1 - \frac{2x_a^2}{7}\right) + -- \right] f_{aM} \tag{4.39}$$

4.2 Coulomb Collision

4.2.1 Collision Operator

Coulomb collision operator given by L.D. Landau [492] from Boltzman form is,

$$C_{ab} = -\frac{e_a^2 e_b^2 ln\Lambda}{8\pi\varepsilon_0^2 m_a} \sum_{k,i} \frac{\partial}{\partial v_k} \int U_{ki} \left[\frac{f_a(\boldsymbol{v})}{m_b} \frac{\partial f_b(\boldsymbol{v}_b)}{\partial v_{bi}} - \frac{f_b(\boldsymbol{v}_b)}{m_a} \frac{\partial f_a(\boldsymbol{v})}{\partial v_i} \right] d\boldsymbol{v}_b, \tag{4.40}$$

where U_{ki} is given as a function of the relative velocity $\boldsymbol{u} = \boldsymbol{v}_a - \boldsymbol{v}_b$ by,

$$U_{ki} \equiv \frac{u^2 \delta_{ki} - u_k u_i}{u^3}$$

Since the Boltzmann's collision integral is derived for short range force, Rosenbluth derived the Coulomb collision term which eventually agrees with Landau form using Markov process formulation [638] written in a Fokker-Planck form as,

$$C_{ab}(f_a, f_b) = -\frac{\partial \langle \Delta v_k/\Delta t \rangle f_a}{\partial v_k} + \frac{\partial^2 \langle \Delta v_k \Delta v_i / 2\Delta t \rangle f_a}{\partial v_k \partial v_i}, \tag{4.41}$$

where $\langle \Delta v_k/\Delta t \rangle$ and $\langle \Delta v_k \Delta v_i/2\Delta t \rangle$ are the average change of Δv_k and $\Delta v_k \Delta v_i$ per unit time due to the collision and are called the dynamical friction coefficient and the diffusion tensor, respectively.

$$C_{ab} = \frac{e_a^2 e_b^2 ln\Lambda}{4\pi m_a^2 \varepsilon_0^2} \sum_{k,i} \left[-\frac{\partial}{\partial v_k} \left(\frac{\partial h_{ab}}{\partial v_k} f_a \right) + \frac{1}{2} \frac{\partial^2}{\partial v_k \partial v_i} \left(\frac{\partial^2 g_{ab}}{\partial v_k \partial v_i} f_a \right) \right], \tag{4.42}$$

where h_{ab} and g_{ab} are called Rosenbluth potentials:

$$h_{ab}(\boldsymbol{v}) = \left(1 + \frac{m_a}{m_b} \right) \int \frac{f_b(\boldsymbol{v}_b)}{u} d\boldsymbol{v}_b \tag{4.43}$$

$$g_{ab}(\boldsymbol{v}) = \int u f_b(\boldsymbol{v}_b) d\boldsymbol{v}_b \tag{4.44}$$

If the background species b has the isotropic velocity distribution function, the Rosenbluth potentials h_{ab}, g_{ab} only depends on the speed v in the spherical velocity coordinates (v, θ, φ). If we write this Fokker-Planck equation as $C_{ab}(f_a) = \partial/\partial v_k [A_k^{ab} f_a + \partial(D_{ki}^{ab} f_a)/\partial v_i]$, A_k^{ab} and D_{ki}^{ab} are called the dynamical friction coefficient and the diffusion tensor, respectively. If we define $J_k^{ab} = A_k^{ab} f_a + \partial(D_{ki}^{ab} f_a)/\partial v_i$, $\boldsymbol{J}^{ab} = \{J_k^{ab}\}$ is a particle density flux in the a species velocity space as discussed by Sivukhin [697]. Taking the first velocity coordinate as the relative velocity direction, we find following non-zero components.

$$A_1^{ab} = v\nu_s^{ab}, D_{11}^{ab} = (v^2/2)\nu_\parallel^{ab}, D_{22}^{ab} = D_{33}^{ab} = (v^2/2)\nu_D^{ab}, \tag{4.45}$$

where the slowing down frequency $\nu_s^{ab}(v)$, the deflection frequency $\nu_D^{ab}(v)$, and the parallel velocity diffusion frequency $\nu_\parallel^{ab}$ are defined as:

$$\nu_s^{ab}(v) = -\frac{e_a^2 e_b^2 ln\Lambda}{4\pi m_a^2 \varepsilon_0^2} \frac{h'_{ab}(v)}{v} \tag{4.46}$$

$$\nu_D^{ab}(v) = \frac{e_a^2 e_b^2 ln\Lambda}{4\pi m_a^2 \varepsilon_0^2} \frac{g'_{ab}(v)}{v^3} \tag{4.47}$$

$$\nu_\parallel^{ab}(v) = \frac{e_a^2 e_b^2 ln\Lambda}{4\pi m_a^2 \varepsilon_0^2} \frac{g''_{ab}(v)}{v^2} \tag{4.48}$$

Using these quantities, the Fokker-Planck collision operator can be written as,

$$C_{ab}(f_a, f_b) = \nu_D^{ab} \mathscr{L}(f_a) + \frac{1}{v^2} \frac{\partial}{\partial v} \left[v^3 \left(\frac{\nu_s^{ab}}{1 + m_b/m_a} f_a + \frac{1}{2} \nu_\parallel^{ab} v \frac{\partial f_a}{\partial v} \right) \right] \tag{4.49}$$

Here $\mathscr{L}(f_a)$ is the Lorentz scattering operator defined by,

$$\mathscr{L}(f_a) = \frac{1}{2}\left[\frac{1}{sin\theta}\frac{\partial}{\partial\theta}\left(sin\theta\frac{\partial f_a}{\partial\theta}\right) + \frac{1}{sin^2\theta}\frac{\partial^2 f_a}{\partial\varphi^2}\right] \tag{4.50}$$

Usually, the azimuthal dependence is absent and the Lorentz scattering operator can be given as follows by using $\xi = cos\theta$.

$$\mathscr{L}(f_a) = \frac{1}{2}\frac{1}{sin\theta}\frac{\partial}{\partial\theta}\left(sin\theta\frac{\partial f_a}{\partial\theta}\right) = \frac{1}{2}\frac{\partial}{\partial\xi}\left[(1-\xi^2)\frac{\partial f_a}{\partial\xi}\right] \tag{4.51}$$

In case background species b is the Maxwellian velocity distribution with the temperature T_b, the velocity distribution function $f_b(\boldsymbol{v})$ is given as,

$$f_b(\boldsymbol{v}) = \frac{n_b}{\pi^{3/2}v_{Tb}^3}e^{-x_b^2} \text{ where, } v_{Tb} = \left(\frac{2T_b}{m_b}\right)^{1/2}, x_b = \frac{v}{v_{Tb}}, \tag{4.52}$$

the Rosenbluth potentials h_{ab} and g_{ab} are just functions of speed v.

$$h_{ab}(\boldsymbol{v}) = h_{ab}(v) = \left(1+\frac{m_a}{m_b}\right)\frac{n_b}{v_{Tb}}\frac{\Phi_{err}(x_b)}{x_b} \tag{4.53}$$

$$g_{ab}(\boldsymbol{v}) = g_{ab}(v) = \frac{n_b v_{Tb}}{2x_b}\left[(1+2x_b^2)\Phi_{err}(x_b) + x_b\Phi'_{err}(x_b)\right], \tag{4.54}$$

where $\Phi_{err}(x)$ is the error function.

$$\Phi_{err}(x) = \frac{2}{\sqrt{\pi}}\int_0^x e^{-y^2}dy\,,\ \Phi'_{err}(x) = \frac{2}{\sqrt{\pi}}e^{-x^2} \tag{4.55}$$

Substitution of (4.53), (4.54) into Eqs. (4.46)–(4.48) gives following formula for three frequencies.

$$\nu_D^{ab} = \gamma_{ab}\frac{\Phi_{err}(x_b) - G(x_b)}{2x_a^3} \tag{4.56}$$

$$\nu_s^{ab} = \gamma_{ab}\frac{2T_a}{T_b}\left(1+\frac{m_b}{m_a}\right)\frac{G(x_b)}{x_a} \tag{4.57}$$

$$\nu_\parallel^{ab} = 2\gamma_{ab}\frac{G(x_b)}{x_a^3} \tag{4.58}$$

$$G(x) = \frac{\Phi_{err}(x) - x\Phi'_{err}(x)}{2x^2}\,,\ \gamma_{ab} = \frac{e_a^2 e_b^2 n_b ln\Lambda}{4\pi\varepsilon_0^2 m_a^2 v_{Ta}^3}, \tag{4.59}$$

where $G(x)$ is called the Chandrasekar function Chandrasekar [102]. The γ_{ab} has a dimension of frequency and is used as the collision time by Trubnikov [773] and early publications.

$$G(x) \sim \frac{2x}{3\sqrt{\pi}} \text{ for } x \ll 1,\, G(x) \sim \frac{1}{2x^2} \text{ for } x \gg 1 \tag{4.60}$$

Since the ion and impurity thermal speed is much smaller than the electron thermal speed, $v_{TI} < v_{Ti} \ll v_{Te}$, the ion and impurity velocity distribution functions $f_i(\boldsymbol{v})$ seen from the electron can be approximated by the delta function, $f_i(\boldsymbol{v}) = n_i\delta(\boldsymbol{v} - \boldsymbol{u}_i)$, and $f_I(\boldsymbol{v}) = n_I\delta(\boldsymbol{v} - \boldsymbol{u}_I)$. Using this approximation and $x_e = v/v_{Te}$, the electron-ion/impurity collision operator $C_{ej}(f_e)$ $(j = i, I)$ becomes,

$$C_{ej}(f_e) = \gamma_{ej}(x_e)\left(\mathscr{L}(f_e) + \frac{m_e\boldsymbol{v}\cdot\boldsymbol{u}_j}{T_e}f_{eM}\right),\; j = i, I \tag{4.61}$$

where, $\gamma_{ej}(x_e) \equiv \gamma_{ej}/x_e^3$ i.e. $\gamma_{ej}(1) = \gamma_{ej}$. Since any ion behaves as if it has ∞ mass for the collision with the electron, the electron-ion collision operator do not have any ion mass dependence. If all ion flows are negligible $\boldsymbol{u}_j = 0$ $(j = i, I)$, the total electron-ion collision operator can be written as,

$$\sum_{j=i,I} C_{eb}(f_e) = \gamma_{ej}(x_e)Z_{eff}\mathscr{L}(f_e)\,,\, Z_{eff} = \frac{1}{n_e}\sum_{j=i,I} n_jZ_j^2 \tag{4.62}$$

Here, Z_{eff} is called the "effective charge".

For the ion-electron collision, the ion are collided by the light and fast moving electrons and its motion is similar to the Brownian motion. Using this approximation, the ion-electron collision operator becomes,

$$C_{ie}(f_i) = \frac{\boldsymbol{R}_{ei}}{n_im_i}\cdot\frac{\partial f_i}{\partial \boldsymbol{v}} + \frac{m_e}{m_i}\frac{n_e}{n_i\tau_{ei}}\frac{\partial}{\partial \boldsymbol{v}}\cdot\left[(\boldsymbol{v} - \boldsymbol{u}_i)f_i + \frac{T_e}{m_i}\frac{\partial f_i}{\partial \boldsymbol{v}}\right], \tag{4.63}$$

where $\boldsymbol{R}_{ei}$ is the collisional friction on the electron from the ion.

4.2.2 Linearized Collision Operator

While the full Coulomb collision operator is bi-linear in f_a and f_b and is nonlinear for the self collision. If the system is close to thermodynamic equilibrium (namely close to Maxwellian), the distribution function can be divided into $f_a = f_{aM} + f_{a1}$, where $f_{a1} \ll f_{aM}$. The linearized collision operator C_{a1}, C_{ab1} are defined by,

$$C_{a1}(f_{a1}) \equiv \sum_b C_{ab1}(f_{a1}, f_{b1}) \equiv \sum_b C_{ab}(f_{a1}, f_{bM}) + C_{ab}(f_{aM}, f_{b1}) \tag{4.64}$$

This linearized collision operator for different species, C_{ab1} are used to calculate the collisional friction $\boldsymbol{R}_{ab}$ and the energy exchange rate Q_{ab} between a and b.

$$\boldsymbol{R}_{ab} = \int m_a \boldsymbol{v} C_{ab1} d\boldsymbol{v} \tag{4.65}$$

$$Q_{ab} = \int \frac{1}{2} m_a v^2 C_{ab1} d\boldsymbol{v} \tag{4.66}$$

These quantities can be expressed by using the momentum conservation law and the energy conservation law as,

$$\boldsymbol{R}_{ab} = -\int \boldsymbol{v}[m_a \nu_s^{ab} f_{a1} + m_b \nu_s^{ba} f_{b1}] d\boldsymbol{v} \tag{4.67}$$

$$Q_{ab} = -\int \frac{1}{2} v^2 [m_a \nu_E^{ab} f_{a1} + m_b \nu_E^{ba} f_{b1}] d\boldsymbol{v}, \tag{4.68}$$

where ν_E^{ab} is the energy exchange frequency as defined by,

$$\nu_E^{ab} = 2\nu_s^{ab} - 2\nu_D^{ab} - \nu_\parallel^{ab} \tag{4.69}$$

The approximate linearized collision operator given by Hirshman [325] is:

$$C_{a1}(f_{a1}) = \sum_b \left[\nu_D^{ab} \mathscr{L}(f_{a1}) + [\nu_D^{ab} - \nu_s^{ab}] \frac{v_\parallel u_{a1}(v)}{v^2} f_{aM} + \frac{2 v_\parallel r_{ba}}{v_{Ta}^2} \nu_s^{ab} f_{aM} \right] \tag{4.70}$$

$$\text{where, } \boldsymbol{r}_{ba} = \frac{3}{2} \int m_b \boldsymbol{v} \nu_s^{ba} f_{b1} d\boldsymbol{v} \Big/ \int m_a x_a^2 \nu_s^{ab} f_{aM} d\boldsymbol{v} \tag{4.71}$$

$$n_{a1}(v) = \int \frac{d\Omega}{4\pi} \frac{f_{a1}}{f_{aM}}, \quad \boldsymbol{u}_{a1}(v) = \int \frac{3 d\Omega}{4\pi} \boldsymbol{v} \frac{f_{a1}}{f_{aM}} \tag{4.72}$$

Here, $\mathscr{L}$ is the Lorentz collision operator given by (4.50) and ν_D^{ab} and ν_s^{ab} are given in (4.56) and (4.57), respectively.

This linearized Coulomb collision operator satisfies self-adjointness for identical particles and H-theorem and conserves particle, momentum and energy. But it does not satisfy self-adjointness for collision between different species with different temperatures. The linearized collision operator with accurate energy diffusion relevant for gyrokinetic simulation is given by Sugama [722].

4.2.3 Collisionality Regimes in Tokamaks

We use a standard definition of the collision frequency of the species a with the species b, ν_{ab}; and the collision time, τ_{ab} as,

$$\nu_{ab}(s^{-1}) = \frac{1}{\tau_{ab}} = \frac{e_a^2 e_b^2 n_b ln\Lambda}{6\pi\sqrt{2\pi}\varepsilon_0^2 m_a^{1/2} T_a^{3/2}} = \frac{e_a^2 e_b^2 n_b ln\Lambda}{3\pi^{3/2}\varepsilon_0^2 m_a^2 v_{Ta}^3}, \quad (4.73)$$

where $ln\Lambda$ is Coulomb logarithm.

If we go back to the original papers and early textbooks, you may find different definitions of collision frequencies. Key differences are not only difference in units (SI units in this text and CGS units in many papers) but also some difference in the coefficients. This definition of collision frequency originates from Braginskii [76].

The collisional transport depends on the relative magnitude of the collision frequency ν_{aa} and the transit frequency $\omega_{ta} = v_{Ta}/qR$. We define the Pfirsch-Schlüter regime (or collisional regime) by $\nu_{aa}/\omega_{ta} \gg 1$. In the Pfirsch-Schlüter regime, the smallness parameter is $\epsilon = \omega_{ta}/\nu_{aa}$. The opposite regime $\nu_{aa}/\omega_{ta} \ll 1$ is called banana-plateau regime. If the aspect ratio is large $A = R_p/a_p \gg 1$, the banana-plateau regime is divided into plateau and banana regimes. The plateau regime is defined by $\epsilon^{3/2} \ll \nu_{aa}/\omega_{ta} \ll 1$ and the banana regime is defined by $\nu_{aa}/\omega_{ta} \ll \epsilon^{3/2}$. The collisionality of the plasma is defined by,

$$\nu_{a*} = \frac{\nu_{aa}/\epsilon}{\omega_b} = \frac{\nu_{aa}}{\epsilon^{3/2}\omega_{ta}}, \quad (4.74)$$

where ν_{aa}/ϵ is the effective de-trapping frequency and $\omega_b = \epsilon^{1/2}\omega_{ta}$ is the bounce frequency of the trapped particle.

Since $\nu_{ee}/\nu_{ii} \sim v_{Te}/v_{Ti}$, it is similar for both electron and ion if the temperatures are similar. But the impurity can be more collisional than electron and main ions.

Figure 4.2 shows the collisionality dependence of the normalized neoclassical diffusion coefficient for a very small ϵ (say $\epsilon \sim 10^{-4}$) and $\epsilon = 0.2$. We can see that we have only two regimes except for the plasma center and the concept of "plateau" is quite vague. But the theory is developed by dividing into three regime and then connected by interpolation.

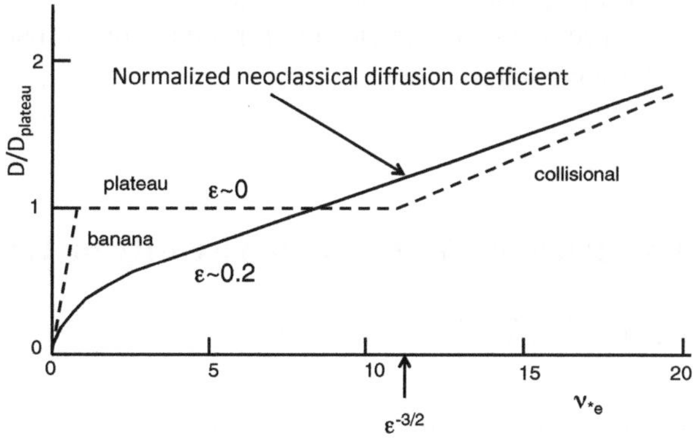

Fig. 4.2 Collisionality dependence of neoclassical diffusion coefficient for $\epsilon = 0.2$ and $\epsilon \sim 0$. Modified from Hinton [321]

Column 4-3: Collision Times and Coulomb Logarithm

Collision time τ_{ab} is defined as collision time of species a with species b. Electron-electron collision time τ_{ee}, electron-ion collision time τ_{ei}, ion-electron collision time τ_{ie}, and ion-ion collision time τ_{ii} are given as follows,

$$\tau_{ab}(s) = \frac{6\pi\sqrt{2\pi m_a}\varepsilon_0^2 T_a^{\frac{3}{2}}}{n_b e_a^2 e_b^2 ln\Lambda} = \frac{4.67\times 10^{17} A_a^{1/2} T_a(keV)^{3/2}}{n_a(m^{-3}) Z_a^2 Z_b^2 ln\Lambda} \tag{4.75}$$

$$\tau_{ee}(s) = \frac{6\pi\sqrt{2\pi m_e}\varepsilon_0^2 T_e^{3/2}}{n_e e^4 ln\Lambda} = \frac{1.09\times 10^{16} T_e(keV)^{3/2}}{n_e(m^{-3}) ln\Lambda} \tag{4.76}$$

$$\tau_{ei}(s) = \frac{6\pi\sqrt{2\pi m_e}\varepsilon_0^2 T_e^{3/2}}{n_i Z_i^2 e^4 ln\Lambda} = \frac{1.09\times 10^{16} T_e(keV)^{3/2}}{n_i(m^{-3}) Z_i^2 ln\Lambda} \tag{4.77}$$

$$\tau_{ie}(s) = \frac{6\pi\sqrt{2\pi m_i}\varepsilon_0^2 T_i^{3/2}}{n_e Z_i^2 e^4 ln\Lambda} = \frac{4.67\times 10^{17} A_i^{1/2} T_i(keV)^{3/2}}{n_i(m^{-3}) Z_i^2 ln\Lambda} \tag{4.78}$$

$$\tau_{ii}(s) = \frac{6\pi\sqrt{2\pi m_i}\varepsilon_0^2 T_i^{3/2}}{n_e Z_i^2 e^4 ln\Lambda} = \frac{4.67\times 10^{17} A_i^{1/2} T_i(keV)^{3/2}}{n_i(m^{-3}) Z_i^4 ln\Lambda} \tag{4.79}$$

The Coulomb logarithm is defined by $ln\Lambda \equiv ln(9N) = ln(b_{max}/b_{min})$, where N is the number of particles in a spree of Debye length λ_D and b_{max} and b_{min} are the maximum and minimum values of the impact parameter of the classical particle moving with the thermal speed. Braginskii [76] gives,

$$ln\Lambda = 18.4 - 1.15 log_{10} n[m^{-3}] + 2.30 log_{10} T_e[eV]\ (T_e > 50eV) \tag{4.80}$$

$$ln\Lambda = 16.5 - 1.15 log_{10} n[m^{-3}] + 3.45 log_{10} T_e[eV]\ (T_e < 50eV) \tag{4.81}$$

Sivukhin [697] give the Coulomb logarithm for different species collision assuming Debye shielding is dominated by electron. Honda [337] gives refined Coulomb logarithm formulae for collision between species with different temperatures.

4.3 Parallel Friction and Viscosity in Tokamak

4.3.1 Drift Kinetic Equation

We start from general kinetic equation in phase space canonical variables $\boldsymbol{z} = (\boldsymbol{x}, \boldsymbol{p})$,

$$\frac{\partial f_a}{\partial t} + \frac{\partial}{\partial \boldsymbol{z}} \cdot [\dot{\boldsymbol{z}} f_a] = C(f_a) \tag{4.82}$$

Since the phase space flow is incompressible from the Hamilton equation i.e. $\partial \dot{z}/\partial z = 0$, above equation is reduced to,

$$\frac{\partial f_a}{\partial t} + \dot{z} \cdot \frac{\partial f_a}{\partial z} = C(f_a) \tag{4.83}$$

Noting $\dot{z}_i \partial f_a/\partial z_i = (\partial z_i/\partial Z_j)\dot{Z}_j(\partial f_a/\partial Z_k)(\partial Z_k/\partial z_i) = \delta_{jk}\dot{Z}_j(\partial f_a/\partial Z_k)$, this form of kinetic equation is independent of our choice of phase space variables including non canonical variables. For the drift kinetic equation, we use guiding center phase space (non-canonical) variables $\boldsymbol{Z} = (\boldsymbol{X}, E, \mu, \varphi)$:

$$\frac{\partial f_a}{\partial t} + \dot{\boldsymbol{X}} \cdot \frac{\partial f_a}{\partial \boldsymbol{X}} + \dot{E}\frac{\partial f_a}{\partial E} + \dot{\mu}\frac{\partial f_a}{\partial \mu} + \dot{\varphi}\frac{\partial f_a}{\partial \varphi} = C(f_a), \tag{4.84}$$

where $E = m_a v^2/2 + e_a \Phi$, μ is the magnetic moment and φ is the gyro phase. We use the guiding center velocity in the Note ($\dot{\boldsymbol{X}} = \boldsymbol{v}_g, \dot{\mu} = 0, \dot{\varphi} = -\Omega_a$) to see,

$$\frac{\partial f_a}{\partial t} + \boldsymbol{v}_g \cdot \frac{\partial f_a}{\partial \boldsymbol{X}} + \dot{E}\frac{\partial f_a}{\partial E} - \Omega_a \frac{\partial f_a}{\partial \varphi} = C(f_a) \tag{4.85}$$

The last term of the left hand side of equation is largest and can be annihilated by taking the gyro phase average. We define the gyro phase average of $A(\boldsymbol{X}, E, \mu, \varphi, t)$:

$$\bar{A}(\boldsymbol{X}, E, \mu, t) = \oint \frac{d\varphi}{2\pi} A(\boldsymbol{x}, E, \mu, \varphi, t) \tag{4.86}$$

Taking the gyro phase average of (4.85), we obtain,

$$\frac{\partial \bar{f}_a}{\partial t} + \boldsymbol{v}_g \cdot \frac{\partial \bar{f}_a}{\partial \boldsymbol{X}} + \dot{E}\frac{\partial \bar{f}_a}{\partial E} = \overline{C(f_a)}, \tag{4.87}$$

where $\dot{E} = \mu \partial B/\partial t + e_a \partial \Phi/\partial t - e_a v_\parallel \partial A_\parallel/\partial t$ (Eq. (196) in [73]). Equation (4.87) is called drift kinetic equation (DKE). Use of gyro averaged Lagrangian significantly simplifies the algebra having self consistent gyro radius ordering. If we use Lagrangian accurate for higher order in gyro radius expansion, we can obtain refined drift kinetic equation. When we take the leading order terms in guiding center equation, we obtain following familiar drift kinetic equation using $E_\parallel^A = -\partial A_\parallel/\partial t$.

$$\frac{\partial \bar{f}_a}{\partial t} + (v_\parallel \boldsymbol{b} + \boldsymbol{v}_d) \cdot \frac{\partial \bar{f}_a}{\partial \boldsymbol{X}} + e_a v_\parallel E_\parallel^A \frac{\partial \bar{f}_a}{\partial E} = \overline{C(f_a)} \tag{4.88}$$

Column 4-4: Guiding Center Velocity

The guiding center drift velocity $\boldsymbol{v}_g$ is given by Alfven [9] as,

$$\boldsymbol{v}_g = v_\parallel \boldsymbol{b} + \frac{\boldsymbol{E} \times \boldsymbol{B}}{B^2} + \frac{m_a v_\perp^2}{2e_a B^2} \boldsymbol{b} \times \nabla B + \frac{m_a v_\parallel^2}{e_a B} \boldsymbol{b} \times \boldsymbol{\kappa} + \frac{m_a v_\parallel}{e_a B} \boldsymbol{b} \times \frac{\partial \boldsymbol{b}}{\partial t}, \tag{4.89}$$

where $\boldsymbol{\kappa}$ is the curvature vector (2.6) and the last $(\partial \boldsymbol{b}/\partial t)$ term is usually negligibly small. For a static field $(\partial \boldsymbol{A}/\partial t = 0)$, the guiding center velocity $\boldsymbol{v}_g$ is given by

$$\boldsymbol{v}_g = v_\parallel \boldsymbol{b} + \boldsymbol{v}_d \,, \boldsymbol{v}_d = \frac{\boldsymbol{b}}{e_a B} \times [e_a \nabla \Phi + \mu \nabla B + m_a v_\parallel^2 \boldsymbol{\kappa}] \tag{4.90}$$

or in a simple form given by Morozov-Solovev [562],

$$\boldsymbol{v}_g = \frac{v_\parallel}{B} \nabla \times (\boldsymbol{A} + \rho_\parallel \boldsymbol{B}) \tag{4.91}$$

$$v_\parallel = \pm\sqrt{2(E - \mu B - e_a \Phi)/m_a} \,, \rho_\parallel = m_a v_\parallel / e_a B, \tag{4.92}$$

where $\boldsymbol{A}^* \equiv \boldsymbol{A} + \rho_\parallel \boldsymbol{B}$ is called the modified vector potential. Using the vector formula (A.8), the charged particle drift (4.91) is expressed as,

$$\boldsymbol{v}_g = v_\parallel \boldsymbol{b} + v_\parallel \nabla \rho_\parallel \times \boldsymbol{b} + \frac{v_\parallel \rho_\parallel}{B} \mu_0 \boldsymbol{j} \tag{4.93}$$

Using (2.35) and (4.91), the drift across the magnetic surface is given by,

$$\boldsymbol{v}_g \cdot \nabla \psi = \boldsymbol{v}_d \cdot \nabla \psi = v_\parallel \boldsymbol{b} \cdot \nabla (\rho_\parallel F) = v_\parallel F(\psi) \nabla_\parallel \rho_\parallel \tag{4.94}$$

The time required for a particle to complete its poloidal orbit is called the bounce time (its inverse is called bounce frequency) and is given by,

$$\tau_b = \frac{1}{\omega_b} = \oint \frac{d\theta}{v_\parallel \boldsymbol{b} \cdot \nabla \theta} \tag{4.95}$$

The toroidal drift velocity $\boldsymbol{v}_d$ is given by,

$$\boldsymbol{v}_d = \langle \boldsymbol{v} \cdot \left(\frac{\partial (\boldsymbol{b}/\Omega_a)}{\partial \boldsymbol{x}} \times \boldsymbol{v} \right) \cdot (\boldsymbol{bb} - \boldsymbol{I}) \rangle = \boldsymbol{b} \times \left[\frac{v_\perp^2}{2\Omega_a} \nabla lnB + \frac{v_\parallel^2}{\Omega_a} \boldsymbol{b} \cdot \nabla \boldsymbol{b} \right] \tag{4.96}$$

4.3.2 Parallel Friction from Drift Kinetic Equation

As we see in Sect. 4.1, moment approach is very effective to analyze neoclassical transport. To close the moment equations, it is essential to find the friction coefficients and the parallel viscosity coefficients by solving DKE in various collisionality regime. If we find these coefficients, most of the neoclassical transport can be understood from the moment equations.

The DKE (4.88) in the transport time scale ($\partial/\partial t \sim 0$) is given as [330],

$$v_\parallel \nabla_\parallel f_a + \boldsymbol{v}_d \cdot \nabla f_a + e_a v_\parallel E_\parallel^A \frac{\partial f_a}{\partial E} = C_a(f_a), \tag{4.97}$$

where $\nabla_\parallel = \boldsymbol{b} \cdot \nabla$, $v_\parallel$ is the parallel velocity, $E_\parallel^A$ is the inductive parallel electric field, $E = m_a v^2/2 + e_a \Phi$ is the total energy. Here, we dropped over-bar of f_a as an gyro phase average, for simplicity. Since $C_a(f_{aM}) = 0$, we expand $f_a = f_{aM} + f_{a1}$. Considering $v_\parallel \sim v_{ta} \gg v_d$, we seek a solution to satisfy following DKE.

$$v_\parallel \nabla_\parallel f_{aM} + e_a v_\parallel E_\parallel^A \frac{\partial f_{aM}}{\partial E} = C_{a1}(f_{a1}) \tag{4.98}$$

Here, the Maxwell distribution function f_{aM} is given by,

$$f_{aM} = \frac{n_a}{(2\pi)^{3/2} v_{Ta}^3} e^{-m_a v^2/2T_a} = \frac{n_a}{(2\pi)^{3/2} v_{Ta}^3} e^{-(E - e_a \Phi)/T_a} \tag{4.99}$$

Using $\nabla_\parallel f_{aM} = (\nabla_\parallel n_a/n_a - 3\nabla_\parallel T_a/2T_a + m_a v^2 \nabla_\parallel T_a/2T_a^2 + e_a \nabla_\parallel \Phi/T_a) f_{aM}$, we obtain following classical Spitzer equation [326].

$$\left[\frac{\nabla_\parallel p_a}{p_a} - \frac{e_a E_\parallel}{T_a} - \left(\frac{5}{2} - \frac{m_a v^2}{2T_a}\right)\frac{\nabla_\parallel T_a}{T_a}\right] v_\parallel f_{aM} = C_{a1}(f_{a1}), \tag{4.100}$$

where $E_\parallel = E_\parallel^A - e_a \nabla_\parallel \Phi$. We note that $f_{aM} \sim e^{-m_a v^2/2T_a}$, and not $f_{aM} \sim e^{-E/T_a}$. We observe that $L_0^{(3/2)} = 1$ and $(5/2 - m_a v^2/2T_a) = 5/2 - x_a^2 = L_1^{(3/2)}(x_a^2)$ where $L_k^{(3/2)}$ is the Sonnine polynomial (generalized Laguerre polynomial) of order (3/2). The definition of the Sonnine polynomials is shown in Appendix A.6.

Multiplying Eq. (4.100) by $m_a v_\parallel L_k^{(3/2)}$ and integrating over velocity space and using the orthogonal relation (A.99), we have,

$$p_a A_{1a} \delta_{0k} - \frac{3}{2} p_a A_{2a} \delta_{1k} = \int m_a v_\parallel L_k^{(3/2)} C_{a1} d\boldsymbol{v}, \tag{4.101}$$

where $A_{1a} = \nabla_\parallel ln p_a - (e_a/T_a) E_\parallel$, and $A_{2a} = \nabla_\parallel ln T_a$ are generalized thermo dynamic forces. We define parallel friction and heat friction forces by,

$$F_{a1\|} = \int m_a v_\| C_{a1}(f_{a1}) d\mathbf{v} \tag{4.102}$$

$$F_{a2\|} = \int m_a v_\| \left[x_a^2 - \frac{5}{2} \right] C_{a1}(f_{a1}) d\mathbf{v} \tag{4.103}$$

Using these quantities, we obtain classical parallel momentum and heat flow balance equations.

$$\nabla_\| p_a - e_a n_a E_\| = F_{a1\|} \, , \, \frac{3}{2} n_a \nabla_\| T_a = F_{a2\|} \tag{4.104}$$

$$\int m_a v_\| L_k^{(3/2)} C_{a1}(f_{a1}) d\mathbf{v} = 0 \, , k \geq 2 \tag{4.105}$$

Now it is clear that it is useful to expand f_{a1} in a Sonnine polynomials.

$$\begin{aligned} f_{a1} &= \sum_j f_{a1j} = \frac{2v_\|}{v_{Ta}^2} \left[u_{\|a0} L_0^{(3/2)} + u_{\|a1} L_1^{(3/2)} + u_{a2\|} L_2^{(3/2)} + -- \right] f_{aM} \\ &= \frac{2v_\|}{v_{Ta}^2} \left[u_{\|a} - \frac{2}{5} \frac{q_{\|a}}{p_a} \left(\frac{5}{2} - x_a^2 \right) + u_{a2\|} L_2^{(3/2)} + -- \right] f_{aM} \end{aligned} \tag{4.106}$$

Using the orthogonality relation of Sonnine polynomials, we have

$$u_{ak\|} = \frac{3 \int L_k^{(3/2)}(x_a^2) v_\| f_{a1} d\mathbf{v}}{2 \int x_a^2 \left[L_k^{(3/2)}(x_a^2) \right]^2 f_{aM} d\mathbf{v}} \text{ ,where} \tag{4.107}$$

$$L_0^{(3/2)}(x_a^2) = 1 \, , L_1^{(3/2)}(x_a^2) = \frac{5}{2} - x_a^2 \, , L_2^{(3/2)}(x_a^2) = \frac{35}{8} - \frac{7}{2} x_a^2 + \frac{1}{2} x_a^4 \tag{4.108}$$

It is straight forward to show

$$u_{a0\|} = u_{a\|} \, , u_{a1\|} = -\frac{2q_{a\|}}{5p_a} \, , u_{a2\|} = \frac{3 \int L_2^{(3/2)}(x_a^2) v_\| f_{a1} d\mathbf{v}}{2 \int x_a^2 \left[L_2^{(3/2)}(x_a^2) \right]^2 f_{aM} d\mathbf{v}} \tag{4.109}$$

where $u_{a\|}$ and $q_{a\|}$ are the parallel flow speed and the parallel heat flux, respectively.

$$u_{a\|} = \frac{1}{n_a} \int v_\| f_{a1} d\mathbf{v} \, , q_{a\|} = \int \left[\frac{1}{2} m_a v^2 - \frac{5}{2} T_a \right] v_\| f_{a1} d\mathbf{v} \tag{4.110}$$

Substituting (4.106) into (4.102) and (4.103), we arrive at following linear friction-flow relation considering that the linearized collision operator $C_{a1}(f_{a1})$ is linearly proportional to f_{a1}.

$$F_{a1\|} = \sum_b \left[l_{11}^{ab} u_{b\|} - l_{12}^{ab} \frac{2q_{b\|}}{5p_b} + l_{13}^{ab} u_{b2\|} \right], \tag{4.111}$$

$$F_{a2\|} = \sum_b \left[-l_{21}^{ab} u_{b\|} + l_{22}^{ab} \frac{2q_{b\|}}{5p_b} - l_{23}^{ab} u_{b2\|} \right], \tag{4.112}$$

where l_{ij}^{ab} is called the friction coefficient defined by,

$$l_{ij}^{ab} = \left[\sum_k \frac{n_a m_a}{\tau_{ak}} M_{ak}^{i-1,j-1} \right] \delta_{ab} + \frac{n_a m_a}{\tau_{ab}} N_{ab}^{i-1,j-1} \tag{4.113}$$

Here, the matrix elements M_{ab}^{ij} and N_{ab}^{ij} are the moments of the linearized collision operator for the Sonnine polynomial $L_j^{(3/2)}$ part of the perturbed distribution function, $f_{a1j} = (2v_\| / v_{Ta}^2) L_j^{(3/2)}(x_a^2) f_{aM}$.

$$M_{ab}^{ij} = \frac{\tau_{ab}}{n_a} \int v_\| L_i^{(3/2)}(x_a^2) C_{ab}(f_{a1j}, f_{bM}) dv \tag{4.114}$$

$$N_{ab}^{ij} = \frac{\tau_{ab}}{n_a} \int v_\| L_i^{(3/2)}(x_a^2) C_{ab}(f_{aM}, f_{b1j}) dv \tag{4.115}$$

Self-adjointness of the C_{ab} results in the following symmetry property.

$$M_{ab}^{ij} = M_{ab}^{ji}, \; (T_a v_{Ta})^{-1} N_{ab}^{ij} = (T_b v_{Tb})^{-1} N_{ba}^{ji}, \tag{4.116}$$

$$\text{hence, } l_{ij}^{ab} = l_{ji}^{ba} \tag{4.117}$$

The momentum conservation law gives rise to the following relation.

$$M_{ab}^{j0} + N_{ab}^{j0} = 0 \tag{4.118}$$

This relation guarantee the Galilean invariance of the friction force (i.e. $F_{aj\|}$ only depends on relative flow). Explicit expressions for M_{ab}^{ji} and N_{ab}^{ij} are as follows, with $x_{ab} \equiv v_{Tb}/v_{Ta}$:

$$M_{ab}^{00} = -N_{ab}^{00} = -(1 + \frac{m_a}{m_b})(1 + x_{ab}^2)^{-3/2} \tag{4.119}$$

$$M_{ab}^{01} = M_{ab}^{10} = -N_{ab}^{10} = -\frac{3}{2}(1 + \frac{m_a}{m_b})(1 + x_{ab}^2)^{-5/2} \tag{4.120}$$

$$M_{ab}^{11} = -(\frac{13}{4} + 4x_{ab}^2 + \frac{15}{2} x_{ab}^4)(1 + x_{ab}^2)^{-5/2} \tag{4.121}$$

$$N_{ab}^{11} = \frac{27}{4} \frac{T_a}{T_b} x_{ab}^2 (1 + x_{ab}^2)^{-5/2} \tag{4.122}$$

$$M_{ab}^{02} = -x_{ba}N_{ba}^{02} = -\frac{15}{8}(1+\frac{m_a}{m_b})(1+x_{ab}^2)^{-7/2} \quad (4.123)$$

$$M_{ab}^{12} = -(\frac{69}{16}+6x_{ab}^2+\frac{63}{4}x_{ab}^4)(1+x_{ab}^2)^{-7/2} \quad (4.124)$$

$$N_{ab}^{12} = \frac{225}{16}\frac{T_a}{T_b}x_{ab}^4(1+x_{ab}^2)^{-7/2} \quad (4.125)$$

While we usually neglect $u_{b2\|}$ and $q_{b2\|}$, Hirshman-Sigmar [330] give renormalized friction coefficient f_{ij}^{ab} to include finite l_{13}^{ab} since their effect is not always negligible. Recently Honda [338] showed it is important for the multiple hydrogenic ions or impurities with similar light masses.

4.3.3 Parallel Viscosity from Drift Kinetic Equation

The gyro-phase averaged perturbed velocity distribution function f_{a1} in the previous section is the $l = 1$ spherical harmonics of the velocity distribution function $f_{a1}^{(1)}$ in Eq. (4.36), which is proportional to Legendre polynomial $P_1(cos\theta) = cos\theta$, where $\xi \equiv cos\theta = v_\|/v$. This gives rise to the first order flows.

Because of the conservation of the magnetic moment μ in the non-uniform magnetic field, the $l = 2$ spherical harmonics of the velocity distribution function $f_{a1}^{(2)}$ in Eq. (4.36) appears, which is proportional to $P_2(\xi) = \frac{3}{2}\xi^2 - \frac{1}{2}$. This gives rise to the pressure anisotropy in tokamak.

The collisional transport theory solves this equation in three collisionality regimes; Pfirsch-Schlüter regime, plateau regime and banana regime.

The parallel viscous force is proportional to the pressure anisotropy as follows,

$$\langle \boldsymbol{B}\cdot\nabla\cdot\boldsymbol{\Pi}_a\rangle = \langle (p_{\perp a}-p_{\| a})\nabla_\| B\rangle \quad (4.126)$$

$$\langle \boldsymbol{B}\cdot\nabla\cdot\boldsymbol{\Theta}_a\rangle = \langle (\Theta_{\perp a}-\Theta_{\| a})\nabla_\| B\rangle, \quad (4.127)$$

where we use following relation for any scalar F noting $\boldsymbol{\kappa}\cdot\boldsymbol{b} = 0$ and $\langle \boldsymbol{B}\cdot\nabla F\rangle = 0$.

$$\boldsymbol{B}\cdot\nabla\cdot\left[F\left(\boldsymbol{bb}-\frac{1}{3}\boldsymbol{I}\right)\right] = B_i\partial_j(Fb_ib_j) - B_i\partial_iF/3 = B_ib_i\partial_j(Fb_j) + B_ib_jF\partial_jb_i - B_i\partial_iF/3$$

$$= B\nabla\cdot\left(\frac{F\boldsymbol{B}}{B}\right) + F\boldsymbol{B}\cdot\boldsymbol{\kappa} - B\nabla_\| F/3 = 2B\nabla_\| F/3 - F\nabla_\| B \quad (4.128)$$

Here, $p_{\| a} - p_{\perp a}$ and $\Theta_{\perp a} - \Theta_{\| a}$ are written using the Legendre polynomial $P_2 = (3\xi^2-1)/2$ and Sonnine polynomial $L_1^{(3/2)} = 5/2 - x_a^2$ as,

$$p_{\|a} - p_{\perp a} = \int m_a v^2 P_2(\xi) f_a d\boldsymbol{v} \tag{4.129}$$

$$\Theta_{\|a} - \Theta_{\perp a} = \frac{m_a}{T_a}(r_{\|a} - r_{\perp a}) - \frac{5}{2}(p_{\|a} - p_{\perp a}) = \int m_a v^2 L_1^{(3/2)} P_2(\xi) f_a d\boldsymbol{v} \tag{4.130}$$

4.3.3.1 Pfirsch-Schlüter Regime

In the collisional (Pfirsch-Schlüter) regime, there are two smallness parameters: one is $\epsilon \equiv \omega_{ta}/\nu_{aa} \ll 1$ and the other is $\delta \equiv \rho_a/a$. In the collisional regime, the dominant term of the DKE is the right hand side of (4.98). Since $C_a(f_{aM}) = 0$, we expand $f_a = f_{aM} + f_{a1} + --$, where $f_{a1} = O(\epsilon)$. Noting $\partial f_{aM}/\partial E = -f_{aM}/T_a$ and the ratio of the drift velocity to the parallel velocity $v_d/v_\| = O(\delta)$, the first order DKE ($O(\epsilon)$) is given by,

$$v_\| \nabla_\| (f_{a1} + f_{aM}) + \boldsymbol{v}_d \cdot \nabla f_{aM} - \frac{e_a v_\| E_\|^A}{T_a} f_{aM} = C_{a1}(f_{a1}) \tag{4.131}$$

Neglecting smaller terms $v_\| \nabla_\| f_{a1}, \boldsymbol{v}_d \cdot \nabla f_{aM}$, we obtain following classical Spitzer equation using $\nabla_\| f_{aM} = (\nabla_\| n_a/n_a - 3\nabla_\| T_a/2T_a + m_a v^2 \nabla_\| T_a/2T_a^2 + e_a \nabla_\| \Phi/T_a) f_{aM}$.

$$\left[\frac{\nabla_\| p_a}{p_a} - \frac{e_a E_\|}{T_a} + \left(\frac{m_a v^2}{2T_a} - \frac{5}{2}\right)\frac{\nabla_\| T_a}{T_a}\right] v_\| f_{aM} = C_{a1}(f_{a1}), \tag{4.132}$$

where $E_\| = E_\|^A - e_a \nabla_\| \Phi$. We note that $f_{aM} \sim e^{-m_a v^2/2T_a}$, and not $f_{aM} \sim e^{-E/T_a}$.

In the toroidal plasma, we expand f_{a1} in power of small gyro radius parameter δ as $f_{a1} = f_{a1}^{(0)} + f_{a1}^{(1)} + -$, where $f_{a1}^{(0)} = O(\delta^0)$ and $f_{a1}^{(1)} = O(\delta)$. This means the first and second order equations are,

$$\delta^0 \epsilon^1 \text{ order} : v_\| \nabla_\| f_{aM} = C_{a1}(f_{a1}^{(0)}) \tag{4.133}$$

$$\delta^1 \epsilon^1 \text{ order} : v_\| \nabla_\| f_{a1}^{(0)} + \boldsymbol{v}_d \cdot \nabla\psi \frac{\partial f_{aM}}{\partial \psi} - \frac{e_a v_\| E_\|^A}{T_a} f_{aM} = C_{a1}(f_{a1}^{(1)}) \tag{4.134}$$

The first equation (4.133) has a form of the Spitzer problem just mentioned above and will give a solution proportional to the Legendre polynomial $P_1(\xi) = v_\|/v$ and the solution $f_{a1}^{(0)}$ does not contribute to the parallel viscosity. The parallel electric field term in the second equation can be neglected since it does not contribute to the parallel viscosity. Noting the relation (4.94): $\boldsymbol{v}_d \cdot \nabla\psi = v_\| \nabla_\| (\rho_\| F)$ and $\nabla_\| f_{aM} = 0$ at the lowest order ($\rho_\| \equiv m_a v_\|/e_a B$, $F \equiv RB_\zeta$), the second equation (4.134) is:

$$C_{a1}(f_{a1}^{(1)}) = v_{\parallel}\nabla_{\parallel}(f_{a1}^{(0)} - F_a) \equiv v_{\parallel}\nabla_{\parallel}g_a \tag{4.135}$$

Here, $C_{a1}(f_{a1}^{(1)})$ is given by (4.64) and/or (4.70). F_a is given by,

$$F_a = -\rho_{\parallel}F\frac{df_{aM}}{d\psi} = -\rho_{\parallel}F\left[\frac{dlnp_a}{d\psi} + \frac{e_a}{T_a}\frac{d\Phi}{d\psi} + \left(\frac{m_a v^2}{2T_a} - \frac{5}{2}\right)\frac{dlnT_a}{d\psi}\right] \tag{4.136}$$

We expand g_a in Sonnine polynomials as follows,

$$g_a = f_{aM}\frac{m_a v_{\parallel}}{T_a}\sum_{j=0}^{\infty} u_{aj}L_j^{3/2}(x^2) \tag{4.137}$$

Using the relations $u_{a0} = u_{a\theta}(\psi)B$ and $u_{a1} = -(2q_{a\theta}(\psi)/5p_a)B$, we have:

$$\begin{aligned} v_{\parallel}\nabla_{\parallel}g_a &= f_{aM}\frac{m_a v_{\parallel}}{T_a}\nabla_{\parallel}(v_{\parallel}B)\left(u_{a\theta}L_0^{(3/2)} - \frac{2q_{a\theta}}{5p_a}L_1^{(3/2)} + --\right) \\ &= f_{aM}\frac{m_a v^2}{T_a}P_2(\xi)\left(u_{a\theta}L_0^{(3/2)} - \frac{2q_{a\theta}}{5p_a}L_1^{(3/2)} + --\right)\nabla_{\parallel}B \end{aligned} \tag{4.138}$$

Here, the gradient $\nabla_{\parallel}$ is taken at constant magnetic moment and we have the relation, $v_{\parallel}\nabla_{\parallel}(v_{\parallel}B) = (v_{\parallel}^2 - v_{\perp}^2/2)\nabla_{\parallel}B = v^2 P_2(\xi)\nabla_{\parallel}B$ using $v_{\parallel}\nabla_{\parallel}v_{\parallel} = -(\mu/m_a)\nabla_{\parallel}B$ while we keep E and μ constant. Here, $P_2(\xi) = 3\xi^2/2 - 1/2$.

Using (4.126) to (4.127), we have following expression for the parallel viscosity.

$$\langle \boldsymbol{B}\cdot\nabla\cdot\boldsymbol{\Pi}_a\rangle = -\langle(\nabla_{\parallel}B)\int m_a v^2 P_2(\xi)f_{a1}^{(1)}d^3v\rangle \tag{4.139}$$

$$\langle \boldsymbol{B}\cdot\nabla\cdot\boldsymbol{\Theta}_a\rangle = -\langle(\nabla_{\parallel}B)\int m_a v^2 L_1^{(3/2)}P_2(\xi)f_{a1}^{(1)}d^3v\rangle \tag{4.140}$$

Hirshman-Sigmar [330] solved the drift kinetic equation in three ways. Firstly, they solved exactly through conversion to the variational quadratic form as in Robinson [632]. Secondly, they solved (4.135) by using approximate Krook collision term. Thirdly, they solved analytically by neglecting field particle perturbation term. Here, we shows Krook collision term case: $C_{a1}(f_{a1}^{(1)}) = -\nu_T^a(v)f_{a1}^{(1)}$. We define following velocity-space average.

$$\{F(v)\} \equiv \int F\frac{m_a v_{\parallel}^2}{n_a T_a}f_{aM}d^3v = \frac{8}{3\sqrt{\pi}}\int_0^{\infty}F(x)e^{-x^2}x^4dx \tag{4.141}$$

By expanding $f_{a1}^{(1)}$ into Sonnine polynomials, we have:

$$\langle \boldsymbol{B}\cdot\nabla\cdot\boldsymbol{\Pi}_a\rangle \approx \frac{6p_a}{5}\langle(\nabla_{\parallel}B)^2\rangle\left\{\frac{x^2}{\nu_T^a}\left(u_{a\theta}L_0^{(3/2)} - \frac{2q_{a\theta}}{5p_a}L_1^{(3/2)}\right)\right\} \tag{4.142}$$

$$\langle \boldsymbol{B} \cdot \nabla \cdot \boldsymbol{\Theta}_a \rangle \approx \frac{6p_a}{5} \langle (\nabla_\| B)^2 \rangle \left\{ \frac{x^2(x^2 - \frac{5}{2})}{\nu_T^a} \left(u_{a\theta} L_0^{(3/2)} - \frac{2q_{a\theta}}{5p_a} L_1^{(3/2)} \right) \right\} \tag{4.143}$$

The normalized parallel viscosity coefficients $\hat{\mu}_{aj}$ defined in (4.31) are expressed in the following formula noting $L_0^{(3/2)}(x) = 1$ and $L_1^{(3/2)}(x) = 5/2 - x^2$.

$$\hat{\mu}_{ak} = \frac{3\langle (\nabla_\| B)^2 \rangle v_{Ta}^2 \tau_{aa}}{5\langle B^2 \rangle} \left\{ \frac{x^2}{\nu_T^a(x)} \left(x^2 - \frac{5}{2} \right)^{k-1} \right\} \tag{4.144}$$

4.3.3.2 Banana Regime

We follows the method given by Helander-Sigmar [317]. More direct derivation of banana regime parallel viscosity is given in Appendix C.1. Similarly, derivation of parallel viscosity for fast ion in Appendix C.2. Linearizing the drift kinetic equation (4.97), we have:

$$v_\| \nabla_\| f_{a1} + \boldsymbol{v}_d \cdot \nabla \psi \frac{\partial f_{aM}}{\partial \psi} + e_a v_\| E_\|^A \frac{\partial f_{aM}}{\partial E} = C_{a1}(f_{a1}) \tag{4.145}$$

$$C_{a1}(f_{a1}) = \sum_b C_{ab}(f_{a1}, f_{bM}) + C_{ab}(f_{aM}, f_{b1})$$

The source of pressure anisotropy is $\boldsymbol{v}_d \cdot \nabla f_{aM} = (\boldsymbol{v}_d \cdot \nabla \psi) \partial f_{aM} / \partial \psi$. Using the guiding center relation (4.94) $\boldsymbol{v}_d \cdot \nabla \psi = v_\| \nabla_\| (F(\psi) \rho_\|)$, the linearized drift kinetic equation in first order in gyro radius expansion can be written as:

$$v_\| \nabla_\| \left(f_{a1} + F \rho_\| \frac{\partial f_{aM}}{\partial \psi} \right) + e_a v_\| E_\|^A \frac{\partial f_{aM}}{\partial E} = C_{a1}(f_{a1}) \tag{4.146}$$

The collision less plasma in tokamak is characterized by the collisionality (4.74) $\nu_a^* \ll 1$. Defining $g_a(E, \mu, \psi) = f_{a1} + F_a$, the relevant DKE is:

$$v_\| \nabla_\| g_a = C_{a1}(f_{a1}) \tag{4.147}$$

Here, we consider the case of $E_\|^A = 0$ and no momentum sources. The solvability condition for g_a (see (2.18)) can be applied for the circulating particles as:

$$\left\langle \frac{B}{v_\|} C_{a1}(f_{a1}) \right\rangle = 0 \tag{4.148}$$

For the trapped particle, $g_a = 0$. For the collision integral in the solvability condition, we use the Coulomb collision operator (4.70) given by Hirshman as follows,

$$C_{a1}(f_{a1}) = \sum_b \left[\nu_D^{ab}\mathscr{L}(f_{a1}) + [\nu_D^{ab} - \nu_s^{ab}]\frac{v_\parallel u_{a1}(v)}{v^2} f_{aM} + \frac{2v_\parallel r_{ba}}{v_{Ta}^2}\nu_s^{ab} f_{aM} \right]$$
$$\equiv \nu_D^a[\mathscr{L}(g_a) + v_\parallel X_a(v,\psi,\theta) f_{aM}] \tag{4.149}$$

Here, we use the fact that F_a do not have pitch angle dependence.

We define a pitch angle variable $\lambda = \mu B_0/(E - e_a\Phi) = hv_\perp^2/v^2$, $h \equiv B_0/B$ and $B_0 = \langle B^2 \rangle^{1/2}$. Then, the Lorentz operator is given by,

$$\mathscr{L} = \frac{2hv_\parallel}{v^2}\frac{\partial}{\partial\lambda}\lambda v_\parallel \frac{\partial}{\partial\lambda} \tag{4.150}$$

Then the solvability condition (4.148) is converted to the following equation for the perturbed distribution function g_a for the passing particle.

$$\frac{\partial}{\partial\lambda}\lambda\langle v_\parallel\rangle\frac{\partial g_a}{\partial\lambda} = -\frac{v^2}{2}\langle X_a/h\rangle f_{aM} \tag{4.151}$$

The solution can be given using the Heaviside function H as:

$$g_a = H(\lambda_c - \lambda)V_\parallel(\lambda, v)\langle X_a/h\rangle f_{aM} \tag{4.152}$$

$$V_\parallel(\lambda, v) = \pm\frac{v}{2}\int_\lambda^{\lambda_c}\frac{d\lambda'}{\langle\sqrt{1-\lambda'/h(\theta)}\rangle} \tag{4.153}$$

To calculate parallel viscosities, we use the relations (4.22) and (4.23).

$$\langle \boldsymbol{B}\cdot\nabla\cdot\boldsymbol{\Pi}_a\rangle = \langle \boldsymbol{B}\cdot\boldsymbol{F}_{a1}\rangle = \left\langle B\int m_a v_\parallel C_{a1}(f_{a1})d\boldsymbol{v}\right\rangle \tag{4.154}$$

$$\langle \boldsymbol{B}\cdot\nabla\cdot\boldsymbol{\Theta}_a\rangle = \langle \boldsymbol{B}\cdot\boldsymbol{F}_{a2}\rangle = \left\langle B\int m_a v_\parallel \left(x_a^2 - \frac{5}{2}\right) C_{a1}(f_{a1})d\boldsymbol{v}\right\rangle \tag{4.155}$$

for $E_\parallel = M_{a\parallel} = Q_{a\parallel} = 0$. Substituting (4.151) and using the self adjointness of the Lorentz operator ($\int \boldsymbol{v}\mathscr{L}(g_a)d\boldsymbol{v} = \int g_a\mathscr{L}(\boldsymbol{v})d\boldsymbol{v} = -\int g_a\boldsymbol{v}d\boldsymbol{v}$) [317], we have:

$$\langle \boldsymbol{B}\cdot\nabla\cdot\boldsymbol{\Pi}_a\rangle = \left\langle B\int m_a v_\parallel \nu_D^a(v_\parallel X_a f_{aM} - g_a)d\boldsymbol{v}\right\rangle$$
$$= B_0\left\langle\int m_a v_\parallel \nu_D^a\langle X_a/h\rangle(v_\parallel - HV_\parallel/h)f_{aM}d\boldsymbol{v}\right\rangle \tag{4.156}$$

$$\langle \boldsymbol{B}\cdot\nabla\cdot\boldsymbol{\Theta}_a\rangle = \left\langle B\int m_a v_\parallel\left(x_a^2 - \frac{5}{2}\right)\nu_D^a(v_\parallel X_a f_{aM} - g_a)d\boldsymbol{v}\right\rangle$$
$$= B_0\left\langle\int m_a v_\parallel \nu_D^a\langle X_a/h\rangle\left(x_a^2 - \frac{5}{2}\right)(v_\parallel - HV_\parallel/h)f_{aM}d\boldsymbol{v}\right\rangle \tag{4.157}$$

Now, we use following relation for an arbitrary function A by Helander [317].

$$\left\langle \int A(v) \frac{m_a v_\parallel}{T_a} (v_\parallel - HV_\parallel / h) f_{aM} dv \right\rangle = f_t n_a \{A\} \tag{4.158}$$

Then, we have:

$$\langle \boldsymbol{B} \cdot \nabla \cdot \boldsymbol{\Pi}_a \rangle = f_t B_0 p_a \{ \nu_D^a \langle X_a / h \rangle \} = f_t B_0 p_a \left\{ \frac{\nu_D^a g_a}{HV_\parallel f_{aM}} \right\} \tag{4.159}$$

$$\langle \boldsymbol{B} \cdot \nabla \cdot \boldsymbol{\Theta}_a \rangle = f_t B_0 p_a \{ \nu_D^a \langle X_a / h \rangle (x_a^2 - \frac{5}{2}) \} = f_t B_0 p_a \left\{ \frac{\nu_D^a g_a (x_a^2 - \frac{5}{2})}{HV_\parallel f_{aM}} \right\} \tag{4.160}$$

Here, we used Eq. (4.152). We expand this g_a in Sonnine polynomials as:

$$g_a = \frac{m_a HV_\parallel}{T_a} \frac{B_0}{f_c} \left[u_{a\theta} - \frac{2 q_{a\theta}}{5 p_a} L_1^{(3/2)}(x^2) + - \right] f_{aM}, \tag{4.161}$$

where f_c is the passing particle fraction given by:

$$f_c = \frac{3}{4} \int_0^{\lambda_c} \frac{\lambda d\lambda}{\left\langle \sqrt{1 - \lambda / h} \right\rangle} \tag{4.162}$$

Substitution of (4.161) into (4.159) and (4.160), we have:

$$\hat{\mu}_{ak} = \frac{f_t}{f_c} \left\{ \nu_D^a \tau_{aa} \left(x^2 - \frac{5}{2} \right)^{k-1} \right\} \tag{4.163}$$

Here $f_t = 1 - f_c$ is the trapped particle fraction.

4.3.3.3 Velocity-Space Partitioned Parallel Viscosity

If we compare (4.144) and (4.163), we come to the concept of velocity-space partitioning of total collision frequency $\nu_{tot}^a(v)$, which becomes $\sim \nu_D^a(v)$ in collision less limit and $\sim 1/\nu_T^a(v)$ in Pfirsch-Schlüter regime. Hirshman-Sigmar [330] gives following form of total collision frequency for the parallel viscosity coefficient.

$$\hat{\mu}_{ak} = \frac{f_t}{f_c} \left\{ \nu_{tot}^a(v) \tau_{aa} \left(x^2 - \frac{5}{2} \right)^{k-1} \right\} \tag{4.164}$$

$$\nu_{tot}^a(v) = \nu_D^a(v) \left[1 + \frac{\bar{\nu}_{*a} \nu_D^a(v) \tau_{aa}}{x_a} \right]^{-1} \left[1 + \frac{5\pi \nu_T^a(v)}{8 x_a \omega_{Ta}} \right]^{-1} \tag{4.165}$$

$$\bar{\nu}_{*a} \equiv \frac{8}{3\pi}\frac{f_t}{f_c}\frac{\omega_{Ta}\langle B^2\rangle}{v_{Ta}^2\langle(\nabla_\parallel B)^2\rangle}\frac{1}{\tau_{aa}} \quad \text{(corrected from [330](4.74))} \tag{4.166}$$

$$\omega_{Ta} = v_{Ta}/L_c^* \tag{4.167}$$

$$L_c^* \sim Rq, \tag{4.168}$$

where ω_{Ta} is the surface averaged transit frequency and exact form of the effective connection length L_c^* is given in [330], $\bar{\nu}_{*a}$ is the re-normalized collisionality related to the collisionality ν_{*a} asymptotically matches to Eq. (4.174) as follows [446].

$$\bar{\nu}_{*a} = (\frac{8}{3\pi} \times 2 \times 1.46)\nu_{*a} = 2.48\nu_{*a} \tag{4.169}$$

$$\nu_{*a} = \frac{1}{1.46}\frac{f_t}{f_c}\frac{\omega_{ta}}{v_{Ta}^2\tau_{aa}}\frac{\langle B^2\rangle}{2\langle(\nabla_\parallel B)^2\rangle} \tag{4.170}$$

This collisionality becomes well known form of $\nu_{*a} \rightarrow 1/\epsilon^{3/2}\omega_{ta}\tau_{aa}$ in the large aspect ratio limit. This form of total collision frequency incorporates the transition from banana to plateau regimes by $2.48\nu_{*a}\nu_D^a\tau_{aa}/x_a = 1$ and the transition from plateau to Pfirsch-Schlüter regimes by $(5\pi/8)\nu_T^a/x_a\omega_{ta} = 1$ but may not be accurate enough in Pfirsch-Schlüter regime since it uses simple Krook collision operator.

4.4 Ohm's Law in Tokamak

4.4.1 Generalized Ohm's Law

Using Eq. (4.34), we can calculate flux surface averaged parallel current density as follows,

$$\langle \boldsymbol{B}\cdot\boldsymbol{J}\rangle = \sum_a e_a n_a \langle \boldsymbol{B}\cdot\boldsymbol{u}_a\rangle = \langle \boldsymbol{B}\cdot\boldsymbol{J}\rangle_{oh} + \langle \boldsymbol{B}\cdot\boldsymbol{J}\rangle_{bs} + \langle \boldsymbol{B}\cdot\boldsymbol{J}\rangle_{ni}, \tag{4.171}$$

where "oh" , "bs" and "ni" are abbreviations of ohmic current, bootstrap current, non-inductive current, respectively.

$$\langle \boldsymbol{B}\cdot\boldsymbol{J}\rangle_{oh} = \sum_{a,b=1}^{4} \frac{n_a e_a e_b \tau_{bb}}{m_b}\hat{c}_{ab}\langle \boldsymbol{B}\cdot\boldsymbol{E}\rangle \tag{4.172}$$

$$\langle \boldsymbol{B}\cdot\boldsymbol{J}\rangle_{bs} = \sum_{a=1}^{4} e_a n_a \sum_{b=1}^{7} \hat{\alpha}_{ab} V_{\perp b} \tag{4.173}$$

$$\langle \boldsymbol{B} \cdot \boldsymbol{J} \rangle_{ni} = \sum_{a=1}^{4} \sum_{b=1}^{7} \frac{n_a e_a \tau_{bb}}{m_b n_b} \hat{c}_{ab} \hat{S}_{\parallel b} \tag{4.174}$$

Equation (4.171) is called Generalized Ohm's law. Using Eq. (4.29), local toroidal current density is given by,

$$J_\zeta = \frac{B_\zeta}{\langle B^2 \rangle} \langle B J_\parallel \rangle + \left[1 - \frac{B_\zeta^2}{\langle B^2 \rangle} \right] R \frac{dP}{d\psi} \tag{4.175}$$

Then, flux surface averaged toroidal current density $J_\zeta(\rho) = \langle J_\zeta / R \rangle / \langle 1/R \rangle$ is given by,

$$J_\zeta(\rho) = J_{\zeta oh} + J_{\zeta bs} + J_{\zeta ni} + J_{\zeta \nabla P} \tag{4.176}$$

$$J_{\zeta s} = \frac{\langle B_\zeta / R \rangle}{\langle B^2 \rangle \langle \frac{1}{R} \rangle} \langle \boldsymbol{B} \cdot \boldsymbol{J} \rangle_s \ (\mathrm{s} \ = \ \mathrm{oh, bs, ni}) \tag{4.177}$$

$$J_{\zeta \nabla P} = -\frac{\langle B_\theta^2 \rangle \frac{dP}{d\psi}}{\langle B^2 \rangle \langle \frac{1}{R} \rangle} \tag{4.178}$$

Here, we used $B^2 = B_\zeta^2 + B_\theta^2$. The new term $J_{\zeta \nabla P}$ comes from Pfirsch-Schlüter term (last term of right hand side of Eq. (4.175)) and is important for low aspect ratio tokamaks where poloidal field becomes comparable to toroidal field.

4.4.2 Electrical Conductivity

The generalized Ohm's law (4.171) includes current induced by the parallel electric field (4.172). Theoretical expression for the electrical conductivity in fully ionized plasma has been obtained by Spitzer-Härm [712], called "Spitzer conductivity". Key observation is importance of electron-electron collision as well as electron-ion collision to almost double the electrical conductivity. In high temperature plasma like Tokamak, important modification to electrical conductivity happens due to parallel viscosity (trapped particle effect). From Eq. (4.172), we obtain following form of electrical conductivity in tokamak geometry, called neoclassical (NC) electrical conductivity.

$$\sigma_\parallel^{NC} = \sum_{a,b=1}^{4} \frac{n_a e_a e_b \tau_{bb}}{m_b} (\hat{\boldsymbol{M}} - \hat{\boldsymbol{L}})_{ab}^{-1} \tag{4.179}$$

Here, $\hat{\boldsymbol{L}}$ represents the collisional friction forces among various species, and $\hat{\boldsymbol{M}}$ represents effect of trapped particle.

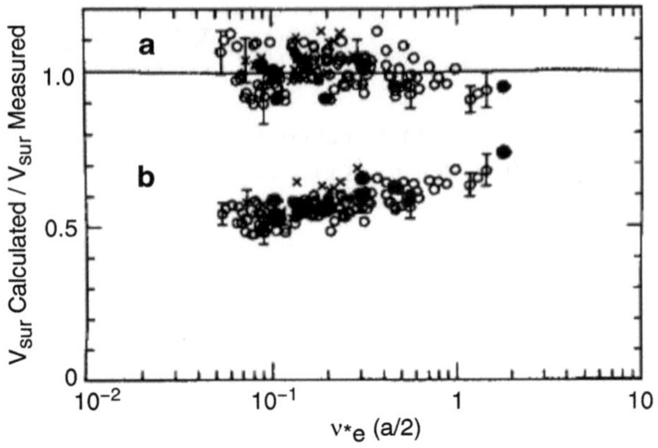

Fig. 4.3 Comparison of experimental and simulated surface voltages with and without trapped particle correction. *Time symbol, open circle, closed circle* correspond to different analysis method and plasma species. Reproduced with permission from Zarnstorff [855]. Copyright AIP Publishing LLC

This summation is dominated by the electron term due to $1/\sqrt{m_b}$ mass dependence of τ_{bb}/m_b. If there is no trapped particles, the viscosity matrix $\hat{\boldsymbol{M}} = 0$ and conductivity σ is given in this case as follows,

$$\sigma_{\|}^{Spitzer} = -\sum_{a,b=1}^{4} \frac{n_a e_a e_b \tau_{bb}}{m_b} \hat{\boldsymbol{L}}_{ab}^{-1} \tag{4.180}$$

Equation (4.180) corresponds to Spitzer conductivity. The electrical conductivity (4.179) is reduced due to viscosity $\hat{\boldsymbol{M}}$, which becomes significant in collisionless regime. In large tokamaks such as JT-60 [458], JET [424], TFTR [301], it is possible to produce collisionless plasma even in ohmically heated plasma.

Experimental resistive loop voltage or surface voltage are consistent with electrical conductivity including trapped particle correction as shown by Kikuchi [431] and Zarnstorff [855]. Typical experimental result from TFTR is shown in Fig. 4.3.

Further verification of electrical conductivity in tokamak was done in TFTR by Batha [47] and in JET by Kelliher [425] showing that measured time variation of local poloidal field by motional stark emission (MSE) spectroscopy is better described by the time dependent simulation using neoclassical conductivity as shown in Fig. 4.4. The trapped particles does not contribute to the current. It creates a frictional force by the relative velocity to circulating electrons. If the plasma is not collisionless enough, the difference in loop voltages calculated with NC and Spitzer conductivity is small. So, small and medium sized tokamak experiments are controversial in identifying viscosity effect (trapped particle correction) to the electrical conductivity.

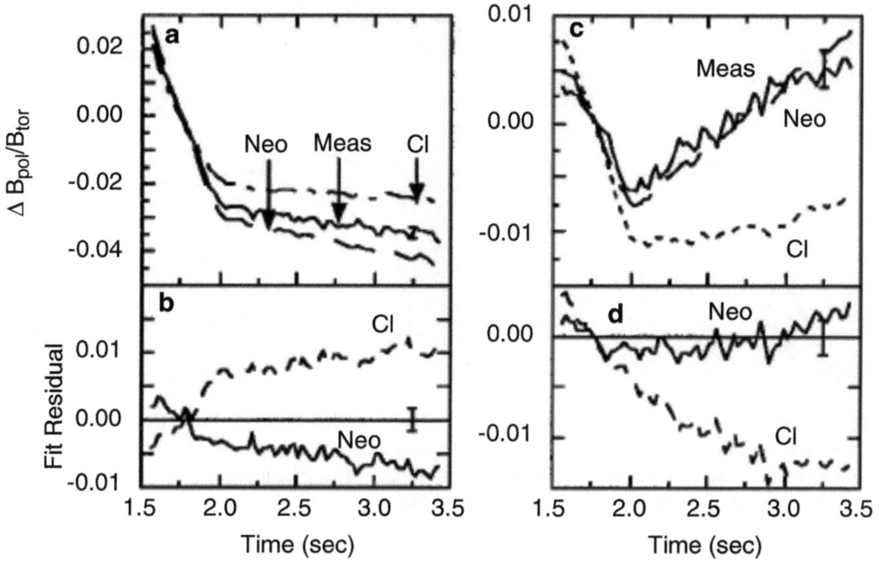

Fig. 4.4 Comparison of experimental and simulated local poloidal field with and without trapped particle correction in TFTR. Better agreement with experiment is obtained using neoclassical resistivity. Reproduced with permission from Batha [47]. Copyright AIP Publishing LLC

4.4.3 Bootstrap Current

The bootstrap current was predicted theoretically by Galeev [237] (see also Sagdeev-Galeev [646]) and its importance for the steady state operation of tokamak was first noted by Bickerton-Connor-Taylor [65] (see Fig. 4.5).

The generalized Ohm's law in (4.171) includes current driven by the thermodynamic forces V_{1a} and V_{2a} as $\langle \boldsymbol{B} \cdot \boldsymbol{J} \rangle_{bs} = \sum_{a=1}^{4} \sum_{b=1}^{7} e_a n_a \hat{\alpha}_{ab} V_{\perp b}$. Here $\hat{\alpha}_{ab}$ is the matrix element of $\hat{\boldsymbol{\alpha}} = (\hat{\boldsymbol{M}} - \hat{\boldsymbol{L}})^{-1}\hat{\boldsymbol{M}}$. Substituting expressions for thermodynamic forces (V_{1a}, V_{2a}) into (4.173), we obtain following form for the bootstrap current.

$$\langle \boldsymbol{B} \cdot \boldsymbol{J} \rangle_{bs} = -\sum_{a=1}^{4} \frac{F n_e}{|Z_a|} \left[L_{31}^a \frac{1}{n_a} \frac{dP_a}{d\psi} + L_{32}^a \frac{dT_a}{d\psi} \right] \tag{4.181}$$

$$L_{31}^a = \sum_{b=1}^{4} \frac{|Z_a|}{Z_a} \frac{Z_b n_b}{n_e} \hat{\alpha}_{ab} \, , \, L_{32}^a = \sum_{b=1}^{3} \frac{|Z_a|}{Z_a} \frac{Z_b n_b}{n_e} \hat{\alpha}_{a,b+4} \tag{4.182}$$

Although V_{1a} includes electrostatic potential term, this term ($J \sim [\sum Z_b n_b \hat{\alpha}_{ab}] F \frac{d\Phi}{d\psi}$) vanishes for the axisymmetric plasma due to charge neutrality. Sauter [663] gives more accurate fitted formula for electrical conductivity and bootstrap current

Fig. 4.5 From left, Dr. Roy Bickerton (by permission of his family), Prof. Roald Sagdeev (by permission of himself), and Prof. Albert A. Galeev (by permission of his family), who made original contribution to the theory of bootstrap current. The prediction of the bootstrap current is said to be an important triumph of neoclassical theory

coefficients using full Fokker-Planck operator since Hirshman-Sigmar [330] and Shaing [679] viscosity uses approximate Coulomb collision operator.

First observation of bootstrap current was made in multipole by Zarnstorff [852] and subsequently in tokamak [853] by driving $\sim 1/3$ of plasma current by bootstrap current. Later, up to 80 % of the plasma current was driven by the bootstrap current in JT-60 by Kikuchi [431]. Time evolution of the surface voltage is measured during the high power perpendicular neutral beam injection with almost no beam driven current and is compared with the numerical simulation using the time dependent 1.5D transport equations in Sect. 2.4 results with and without the bootstrap current. The measured surface voltage is consistent with the existence of the bootstrap current as shown in Fig. 4.6. Time evolution of the internal inductance l_i is also measured through magnetic fitting analysis and agrees with numerical simulation.

Figure 4.7 [440] shows experimental and numerical bootstrap current fractions as a function of poloidal beta. This clearly showed the bootstrap current fraction is proportional to poloidal beta $f_{boot} \sim \beta_p$. This opened good prospect towards efficient steady-state operation of tokamak reactor [430, 432].

In the H-mode [794], steep pressure gradient is formed near the plasma edge, called edge transport barrier (ETB). This ETB induces edge bootstrap current if the edge is deeply collisionless. The excitation of edge bootstrap current sensitively affects time evolution of parallel electric field at ETB, $E_{\parallel} = \langle B_\phi^2 \rangle (\partial\psi/\partial\rho)/F$. In the H-mode, comparison of measured surface voltage with simulation including bootstrap current was done in JET [402] and in more detailed comparison of time evolution of edge parallel electric field with theoretical prediction by Wade [791] consistent with existence of edge bootstrap current.

Direct comparison of measured and numerical local bootstrap current densities is made by using bootstrap current dominated discharges in JT-60U by Sakamoto

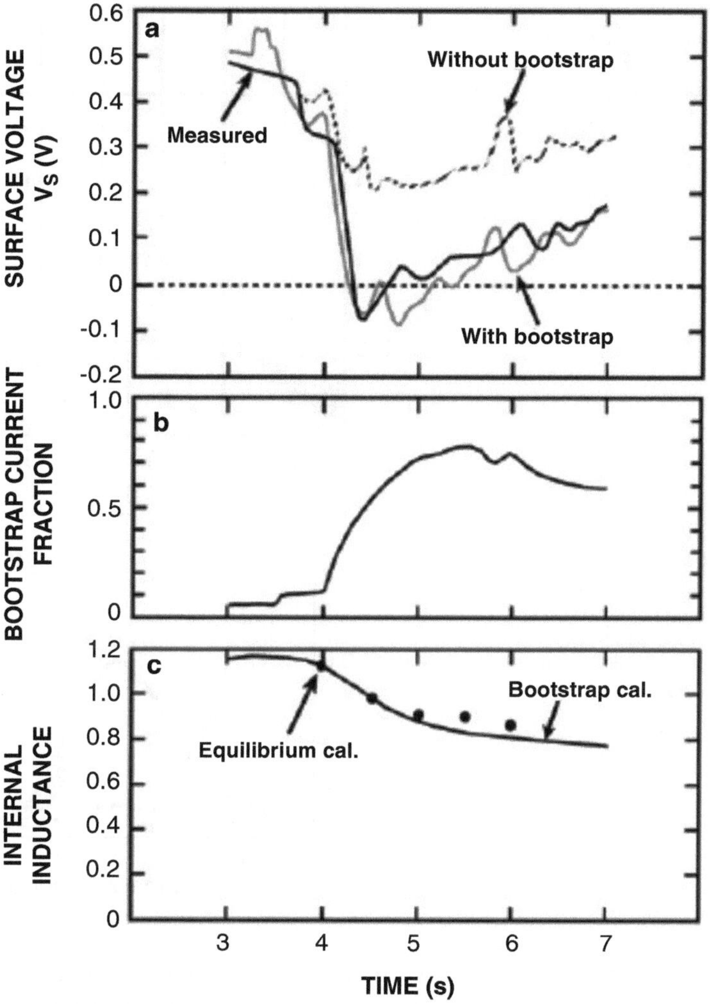

Fig. 4.6 Time evolution of surface voltage (**a**), bootstrap fraction (**b**), and internal inductance (**c**) from measurement and calculation. High power neutral beam injection (NBI) heating power is injected perpendicularly so that NBI do not drive significant beam-driven current. Measured surface loop voltage is consistent with 80 % bootstrap current faction. Reproduced with permission from Kikuchi [440]. Copyright IOP Publishing

[652] as shown in Fig. 4.8. Measured current profile is consistent with prediction of collisional transport theory including the edge region ($0.8 < \rho/a < 1$).

Conventional neoclassical theory predicts zero bootstrap current density at the plasma center and requires seed current to sustain bootstrap current in the plasma center [330]. Shaing [680] showed that the bootstrap current can be sustained without seed current since so-called potato particle can drive viscous force to

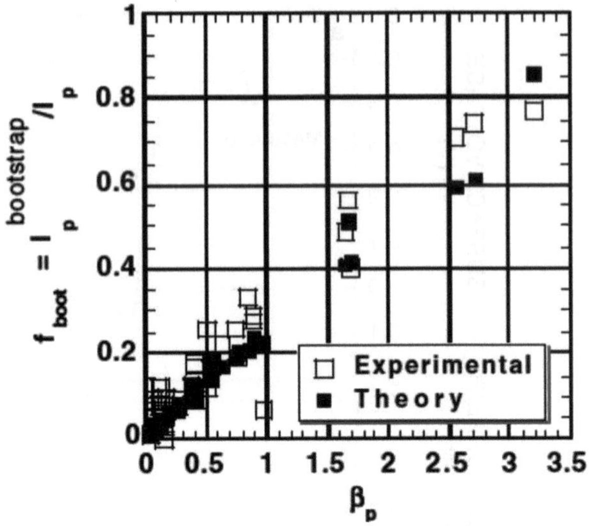

Fig. 4.7 Comparison of experimental and numerical bootstrap current fractions as a function of poloidal beta. Reproduced with permission from Kikuchi [440]. Copyright IOP Publishing

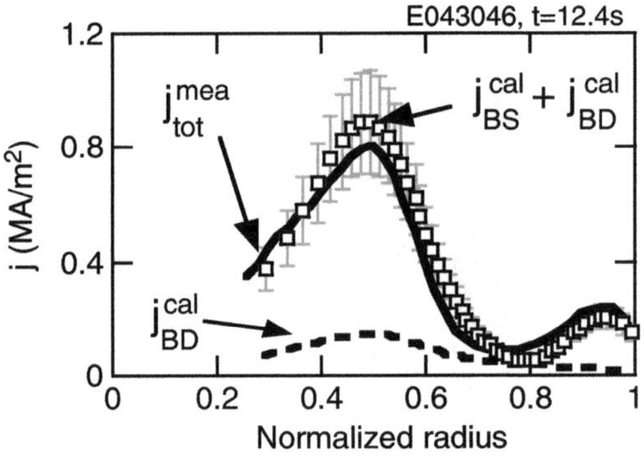

Fig. 4.8 Comparison of measured and calculated current profile in bootstrap current dominated JT-60U discharge. Reproduced with permission from Sakamoto [652]. Copyright IAEA Vienna

passing particle at the plasma center providing finite source current to form poloidal magnetic field. Actually, plasma discharge fully driven by bootstrap current was produced in JT-60U by Takase [735] and more recently for longer time scale of several current diffusion time in TCV by Coda [132].

4.4.4 Neutral Beam Current Drive

Fisch [204] reviewed early experimental results of DITE NBCD [129] ($T_e(0) \sim 0.6\,\text{keV}$, $E_b = 24\,\text{keV}$) noting that trapped electron plays a negligible role. Key parameters of NBCD are beam energy E_b and the electron temperature T_e. Since then, both central electron temperature and the beam energy are extended to a reactor relevant regime, especially in JT-60 ($T_e(0) \sim 15\,\text{keV}$, $E_b = 350\,\text{keV}$) by using N-NBI [585, 586].

When fast neutral beam is injected tangent to the torus, circulating fast ion produces fast ion current (J_{fast}) by multiple circulations around the torus. Collision with bulk electrons produces shielding current (J_{shield}) by the induced drift in the same direction as the fast ion. This shielding is not perfect due to existence of trapped electron and impurities. Sum of fast ion and shielding currents are called beam-driven current $J_{bd}(= J_{fast} + J_{shield})$. Since fast ion velocity distribution function deviates strongly from Maxwellian, J_{fast} has to be obtained from Fokker-Planck equation.

The flux surface averaged fast ion current $\langle \boldsymbol{B} \cdot \boldsymbol{J} \rangle_{fast}$ is obtained from the velocity distribution function of fast ions f as a solution of the Fokker-Planck equation valid for $v_{Ti} \ll v \ll v_{Te}$ as given by Cordey [148].

$$\tau_{se}\frac{\partial f}{\partial t} = \frac{1}{v^2}\frac{\partial}{\partial v}\left[(v_c^3 + v^3)f\right] + \frac{\beta v_c^3}{v^3 \eta \langle v/v_\| \rangle}\frac{\partial}{\partial \eta}\left[\frac{1-\eta^2}{\eta}\left\langle\frac{v_\|}{v}\right\rangle\frac{\partial f}{\partial \eta}\right] + \tau_{se}S(v,\eta) \tag{4.183}$$

$$\tau_{se} = \frac{3(2\pi)^{3/2}\epsilon_0^2 M_f T_e^{3/2}}{e^4 Z_f^2 n_e m_e^{1/2} ln\Lambda} = \frac{0.20 A_f T_e[keV]^{\frac{3}{2}}}{Z_f^2 n_e[10^{20}/m^3] ln\Lambda}(s), \beta = \frac{Z_{eff}}{2A_f\bar{Z}} \tag{4.184}$$

$$Z_{eff} = \sum_{j=i,I}\frac{n_j Z_j^2}{n_e}, \bar{Z} = \sum_{j=i,I}\frac{n_j Z_j^2}{n_e A_j}, v_c = \left(\frac{2E_c}{m_f}\right)^{\frac{1}{2}}, E_c = \left(\frac{9\pi m_p}{16 m_e}\right)^{\frac{1}{3}}\bar{Z}^{\frac{2}{3}}A_f T_e \tag{4.185}$$

$$\langle\frac{v}{v_\|}\rangle = \frac{2}{\pi}K\left[\left(\frac{\eta_t}{\eta}\right)^2\right], \langle\frac{v_\|}{v}\rangle = \frac{2}{\pi}E\left[\left(\frac{\eta_t}{\eta}\right)^2\right], \tag{4.186}$$

where $\eta = (1 - \mu B_{min}/E)^{1/2}$ is $v_\|/v$ at $B = B_{min}$, $S(v,\eta)$ is the bounce averaged fast ion source rate per unit volume, τ_{se} is beam-electron slowing down time, v_c is critical velocity, Z_{eff} is the effective charge, K and E are complete elliptic integrals of first and second kind, respectively. For more explanation, see [445].

In the framework of moment equation, we only use momentum source term from the fast ion $\hat{S}_{\|f}$ for the neutral beam current drive. The flux surface averaged beam driven current may be decomposed as [443], [440],

$$\langle \boldsymbol{B} \cdot \boldsymbol{J} \rangle_{bd} = \langle \boldsymbol{B} \cdot \boldsymbol{J} \rangle_{fast} + \langle \boldsymbol{B} \cdot \boldsymbol{J} \rangle_{shield} \tag{4.187}$$

$$\langle \boldsymbol{B} \cdot \boldsymbol{J} \rangle_{fast} = e_f n_f (\hat{\boldsymbol{M}} - \hat{\boldsymbol{L}})^{-1}_{ff} S_{\|f} \tag{4.188}$$

$$\langle \boldsymbol{B} \cdot \boldsymbol{J} \rangle_{shield} = \sum_{a=e,i,I} e_a n_a (\hat{\boldsymbol{M}} - \hat{\boldsymbol{L}})^{-1}_{af} S_{\|f} \tag{4.189}$$

Here $S_{\|f} = \tau_{ee}\hat{S}_{\|f}/m_e n_e$. While expression for the fast ion current (4.188) is not accurate enough, the ratio of beam driven current J_{bd} over the fast ion current J_{fast}, shielding factor $F = J_{bd}/J_{fast}$ can be accurately as follows,

$$F = \frac{\langle \boldsymbol{B} \cdot \boldsymbol{J} \rangle_{bd}}{\langle \boldsymbol{B} \cdot \boldsymbol{J} \rangle_{fast}} = 1 + \sum_{a=e,i,I} \frac{Z_a n_a}{Z_f n_f} \frac{(\hat{\boldsymbol{M}} - \hat{\boldsymbol{L}})^{-1}_{af}}{(\hat{\boldsymbol{M}} - \hat{\boldsymbol{L}})^{-1}_{ff}} \tag{4.190}$$

Above equation is valid for all aspect ratio, multi-species and all collisionalities. Start-Cordey [716] calculated this shielding factor F assuming zero collisionalities to see parametric dependences on Z_{eff} and ϵ in arbitrary aspect ratio ($0 \leq \epsilon \leq 1$). Calculation by Eq. (4.190) under $\nu_a^* = 0$ agrees well with Start-Cordey calculation as shown in Fig. 4.9.

Lin-Liu [513] found this shielding factor is related to bootstrap coefficient L_{31}^e and gave an analytic expression in the collisionless limit. Finite collisionality has an important effect on shielding factor as calculated recently by Honda [336] using above formula (4.190) as well as Sauter's L_{31}^e formula [663] valid for all collisionality regime as shown in Fig. 4.10. Deviation from collisionless limit starts at $\nu_{*e} \sim 10^{-2}$ depending on the model but also indicating needs for re-assessment of some comparison between measured and calculated NBCD efficiency [731].

Beam ionization by various atomic processes is one of important process to determine the neutral beam driven current profile. Since a historical review of atomic data by Riviere [631], significant efforts are placed to compile ionization cross section and rate coefficients especially under the auspice of IAEA [393]. Those from ground state (Fig. 4.11a) are refined for charge-exchange with bulk ion (p. 78 in [393]), ionization by ion impact (p. 68 in [393]), ionization by impurity (Eq. (15) in [392]) and ionization by electron impact (reaction 2.1.5 of p. 258 in [391]). But, there are number of processes from excited states (Fig. 4.11b). And the process becomes multi-step including excitation from ground state, called multi step ionization (MSI) as noted by Janev [392]. The MSI processes are refined by Suzuki [728] and enhancement of the stopping cross section due to MSI becomes as much as 20–40 % for the beam energy $E_b = 0.1 - 1$ MeV and a good agreement of shine-through rate was obtained between experiments and calculation in JT-60.

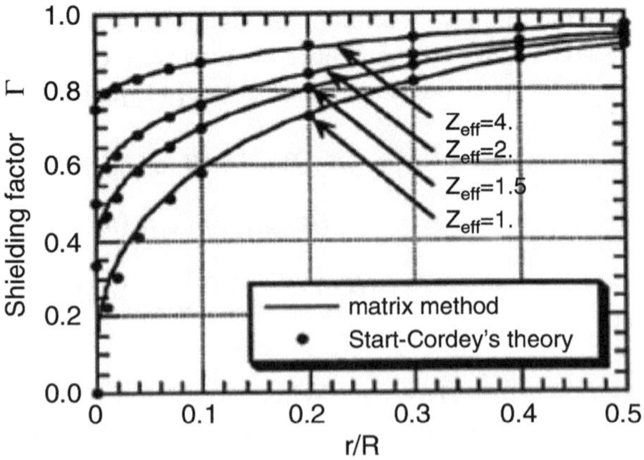

Fig. 4.9 Comparison of shielding factors between Start-Cordey [716] calculation and Eq. (5.39) with $\nu_a^* = 0$. Reproduced with permission from Kikuchi [440]. Copyright IOP Publishing

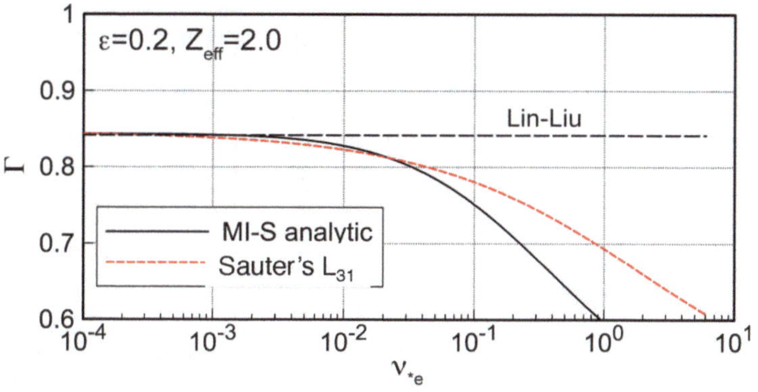

Fig. 4.10 Collisionality dependence of shielding factor by Eq. (5.39) and Sauter's fitted L_{31}. Reproduced with permission from Honda [336]. Copyright IAEA Vienna

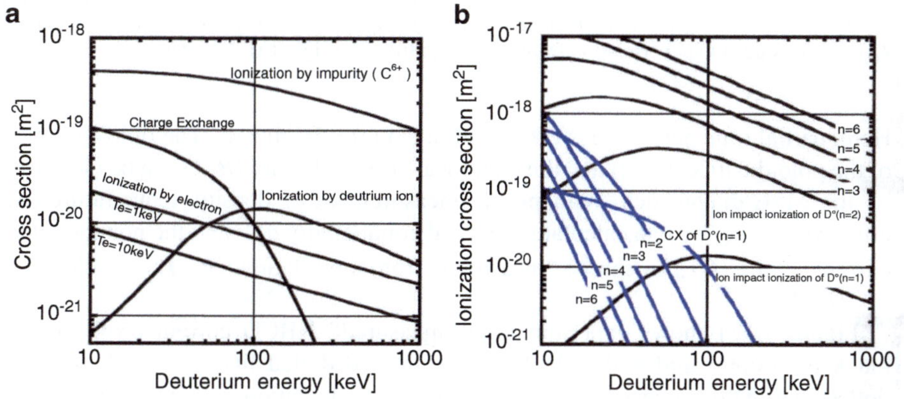

Fig. 4.11 (**a**) Ionization cross sections from ground states. (**b**) Ionization cross sections from ground and excited states. Reproduced with permission from Kikuchi [445]. Copyright IAEA, Vienna

This MSI process is important for the evaluation of local driven current density since deposition profile of fast ions is changed significantly. Various numerical codes such as ACCOME [752], ASTRA [609], NFREYA [208], NUBEAM [602] are developed to calculate beam driven current.

For a quantitative comparison of driven current density with calculation, important development to "measure" non-inductive local current density was developed by Forest [207]. In generalized Ohm's law (4.189), inductive electric field is given by,

$$\langle \boldsymbol{E} \cdot \boldsymbol{B} \rangle = \frac{\langle B_\zeta^2 \rangle}{F} \frac{\partial \psi}{\partial t}|_\phi \tag{4.191}$$

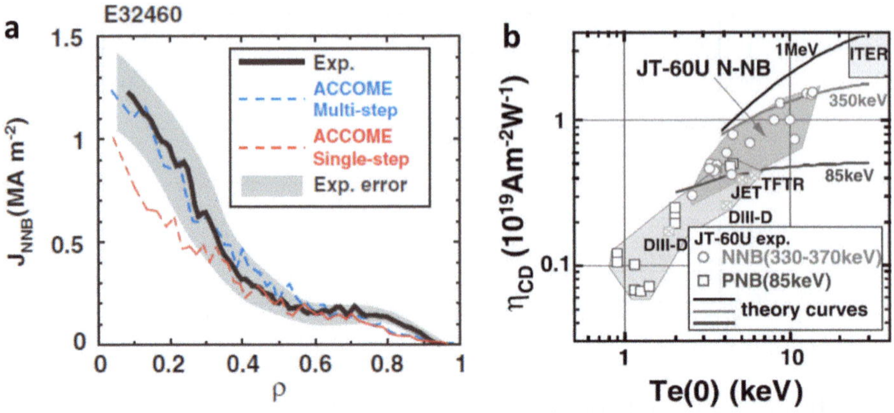

Fig. 4.12 (**a**) Comparison of experimentally measured NBCD current profile with calculations with and without multi-step ionization. After Gormezano [259] with the permission of IAEA. (**b**) Neutral beam current drive efficiency $\eta_{CD} = \bar{n}_e R I_p^{CD} / P_{CD}$ on central electron temperature $T_e(0)$. Reproduced with permission from Oikawa [586]. Copyright IAEA Vienna

Here, partial time derivative is taken with fixed toroidal flux, ϕ. The time evolution of the poloidal flux $\psi(\rho, t)$ and the total local current density $\langle \boldsymbol{B} \cdot \boldsymbol{J} \rangle$ can be measured by the MSE diagnostics. With profile measurement of density, temperature and effective charge, we can calculate electrical conductivity $\sigma_{\parallel}^{NC}$ and the non inductive current profile $\langle \boldsymbol{B} \cdot \boldsymbol{J} \rangle_{ni}$ can be "measured" from the relation, $\langle \boldsymbol{B} \cdot \boldsymbol{J} \rangle_{ni} = \langle \boldsymbol{B} \cdot \boldsymbol{J} \rangle - \sigma_{\parallel}^{NC} \langle \boldsymbol{E} \cdot \boldsymbol{B} \rangle$.

Figure 4.12a shows experimentally "measured" NBCD current profile (thick black line) compared with calculations with (red dotted line) and without (blue dotted line) MSI process [259]. It is shown that calculation with MSI agrees with the measurement.

In addition to such local measurements, systematic comparison between theory and experiments is carried out in JT-60 as shown in Fig. 4.12b showing good agreement if there is no significant MHD activities. The dependence of neutral beam current drive efficiency $\eta_{CD} = \bar{n}_e R I_p^{CD} / P_{CD}$ on the central electron temperature $T_e(0)$ is also verified showing that efficiency increases with $T_e(0)$ as shown in Fig. 4.12b [586].

4.4.5 EC Current Drive

After the comprehensive review by Fisch in 1987 [204], significant progresses are made in various RF fields. Here, recent progresses on ECCD are described due to its better accessibility to reactor plasma and its importance in stabilization of neoclassical tearing modes. A recent review by Prater [615] provided good summary of recent progress.

There are some advantages of ECRF application in reactor environment. First is the absence of accessibility problem and robust coupling irrespective of plasma conditions. Second is the high power density enabling smaller port penetration with low neutron back-streaming and tritium. Third is good controllability of local power deposition and parallel n index with steerable launcher. Fourth is availability of efficient high power Gyrotron by recent innovation [653] (see also Chap. 9).

Momentum input to electron is primary perpendicular to magnetic field by electron cyclotron damping (resonance condition given by $(\omega - l\Omega_e)/k_\parallel = v_\parallel$) of 1st or 2nd harmonic electron cyclotron waves ($l = 1, 2$). There are two types of ECRF waves for CD, O-mode and X-mode, where wave electric field is parallel and perpendicular to magnetic field $\boldsymbol{B}$, respectively [717]. The wave propagation is governed by Hamilton (or ray tracing) equation,

$$\frac{d\boldsymbol{x}}{dt} = \frac{\partial \Omega}{\partial \boldsymbol{k}}|_x, \frac{d\boldsymbol{k}}{dt} = -\frac{\partial \Omega}{\partial \boldsymbol{x}}|_k \tag{4.192}$$

Here $\omega = \Omega(\boldsymbol{k}, \boldsymbol{x}, t)$ is the local dispersion relation. We expand electron velocity distribution function as $f_e = f_{eM} + f_{e1}$ (f_{eM} is Maxwellian) and its drift kinetic equation in the toroidal geometry for a given wave amplitude is given by,

$$\frac{u_\parallel}{\gamma}\boldsymbol{b} \cdot \frac{\partial f_{e1}}{\partial \boldsymbol{x}} - C(f_{e1}) = -\frac{\partial}{\partial \boldsymbol{u}} \cdot \boldsymbol{S}_w \tag{4.193}$$

Here, $\boldsymbol{b}$ is unit vector along the magnetic field, $\boldsymbol{u} = \boldsymbol{p}/m_e = \gamma v$ is momentum per unit mass, $\gamma = (1 + u^2/c^2)^{1/2}$, $C(f_{e1})$ is the linearized electron collision term, $\boldsymbol{S}_w = \boldsymbol{D}_{QL} \cdot \partial f_{e1}/\partial \boldsymbol{u}$ is wave-induced flux in velocity space due to quasi-linear diffusion via wave-particle interaction, respectively. Wave induced flux ($\boldsymbol{S}_w$) is given by Kennel-Engelmann [426] for Landau damping ($l = 0$) and electron cyclotron damping ($l \neq 0$) as follows,

$$\boldsymbol{D}_{QL} = \sum_{n=-\infty}^{\infty} \frac{\pi}{2}\left(\frac{e}{m_e}\right)^2 \delta(\gamma\omega - k_\parallel v_\parallel - l\Omega_e)\boldsymbol{a}_n^* \boldsymbol{a}_n \tag{4.194}$$

$$\boldsymbol{a}_n = \Theta_n \left[(1 - \frac{k_\parallel u_\parallel}{\gamma\omega})\hat{\boldsymbol{u}}_\perp + \frac{k_\parallel u_\perp}{\gamma\omega}\hat{\boldsymbol{u}}_\parallel \right], \tag{4.195}$$

$$\Theta_n = \frac{E_{w+}J_{n-1} + E_{w-}J_{n+1}}{\sqrt{2}} + \frac{u_\parallel}{u_\perp} J_n E_{w\parallel} \tag{4.196}$$

Here, $\Omega_e = -eB/m_e$, $*$ means complex conjugate, $\hat{\boldsymbol{u}}_\parallel$ and $\hat{\boldsymbol{u}}_\perp$ are unit vectors in perpendicular and parallel directions, J_n is the nth order Bessel function with its argument $k_\perp v_\perp/\Omega_e$. E_{w+} and E_{w-} are left and right handed components of E_w, respectively and $E_{w\parallel}$ is parallel component of E_w. Instead of solving equation (4.193), Antonsen-Chu [23] and Taguchi [732] realized that RF induced current can be obtained by using a solution of the following adjoint equation originally used by Spitzer-Härm to obtain electrical conductivity,

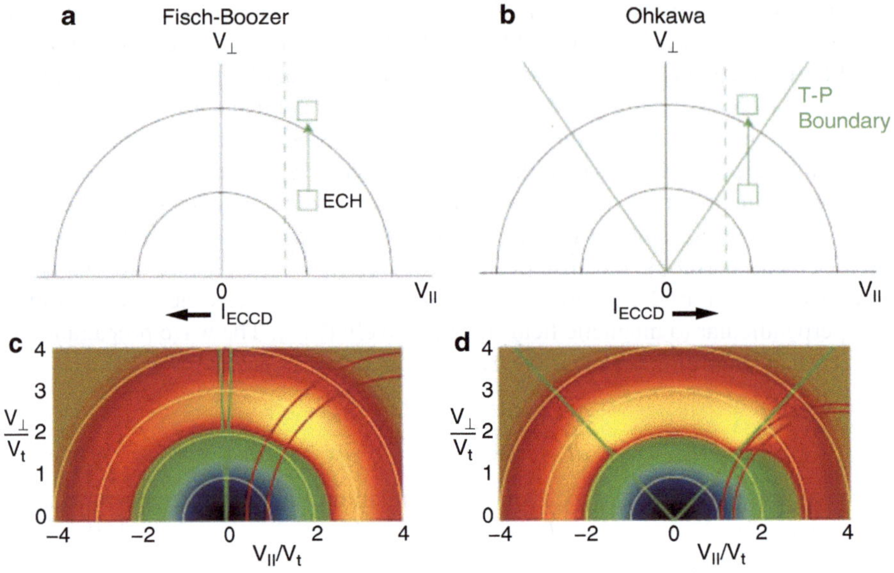

Fig. 4.13 Schematics of (**a**) Fisch-Boozer mechanism and (**b**) Ohkawa mechanism. (**c**) and (**d**) are numerical simulations of (**a**) and (**b**), respectively. Reproduced with permission from Prater [615]. Copyright AIP Publishing LLC

$$\frac{u_\parallel}{\gamma}\boldsymbol{b}\cdot\frac{\partial\chi}{\partial\boldsymbol{x}} + \frac{1}{f_{eM}}C(\chi f_{eM}) = -\frac{eu_\parallel B}{\gamma\langle B^2\rangle} \tag{4.197}$$

It is easy to show following formula for the generated current and the dissipated power by using Eqs. (4.193) and (4.197), and partial integration:

$$\langle\boldsymbol{J}\cdot\boldsymbol{B}\rangle = \langle B^2\rangle\langle\int d^3u\boldsymbol{S}_w\cdot\frac{\partial\chi}{\partial\boldsymbol{u}}\rangle \tag{4.198}$$

$$P_d = \langle\int d^3u\boldsymbol{S}_w\cdot\frac{\partial\epsilon}{\partial\boldsymbol{u}}\rangle \tag{4.199}$$

Here $\epsilon = \gamma m_e c^2$ is the electron energy, and $\partial\epsilon/\partial\boldsymbol{u} = m_e\boldsymbol{u}/\gamma$. The solution of adjoint equation (4.197) is $\chi = 0$ for $\lambda > \lambda_c$ and that for $\lambda < \lambda_c$ is given as [733]:

$$\chi(u,\lambda) = \frac{\langle B^2\rangle}{2f_c}\sigma K(u)\int_\lambda^{\lambda_c}\frac{d\lambda}{\langle\sqrt{1-\lambda B}\rangle} \tag{4.200}$$

Here, $\lambda = (u_\perp/u_\parallel)^2/B, \lambda_c = 1/B_{max}, \sigma = u_\parallel/|u_\parallel|$ and K is a solution of 1 dimensional integro-differential equation, which could be obtained using Sonnine expansion [733] or numerically using exact relativistic collision term [284].

There are two mechanisms responsible for EC current drive. One is Fisch-Boozer mechanism [202] to create asymmetry in parallel resistivity via wave-induced

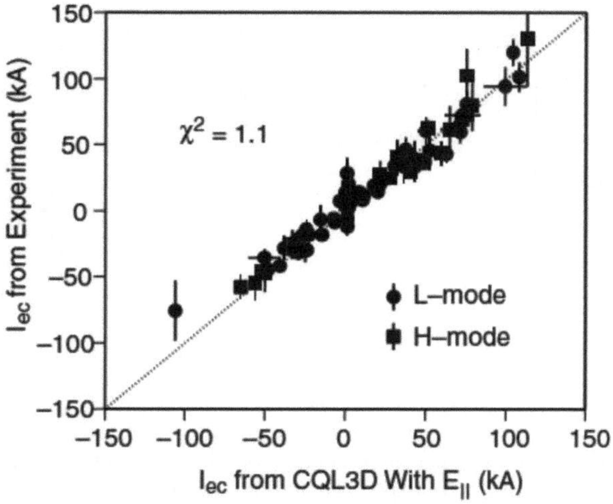

Fig. 4.14 Comparison of measured and calculated EC driven current. Reproduced with permission from Petty [610]. Copyright IAEA Vienna

velocity space diffusion. Other is Ohkawa mechanism [584] to create untrapped-trapped transition since trapped electron can not contribute to the toroidal current as shown in Fig. 4.13.

Comparison of experimental and theoretical expectation of the ECCD has been done in DIII-D [610] as shown in Fig. 4.14. The ECCD of 0.74MA is demonstrated in high electron temperature ($T_e(0) = 23$ keV) plasma in JT-60U by taking $E_{\|}$ effect [170] into account [730]. Existence of trapped electron reduces net driven current and its effect is more significant for outboard EC injection [134] and confirmed experimentally by Petty in DIII-D [611] and by Suzuki in JT-60U [729]. Figure 4.15 shows calculation and experimental results. Here $\zeta = e^3\eta_{CD}/\epsilon_0^2 kT_e$ is normalized EC current drive efficiency [515]. Experiments on TCV ECCD demonstrated that full CD plasma can be obtained [664].

4.5 Plasma Rotation in Tokamak

Toroidal rotation physics is now very important for the role of rotation shear on confinement improvement, rotational stabilization of MHD modes, and also observation of intrinsic toroidal rotation. Here we introduce neoclassical effect on rotation.

4.5.1 Neoclassical Rotations

Equation (4.34) govern the parallel flows in tokamak. Kim derived an analytic expression for the poloidal and toroidal rotations assuming impurity is in Pfirsch-

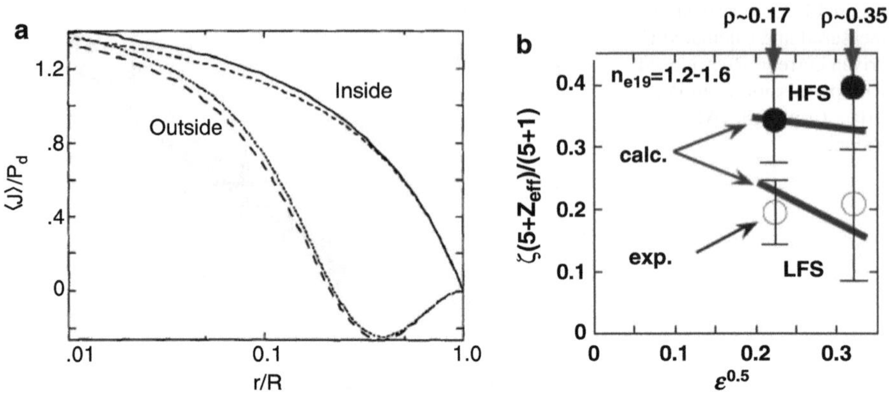

Fig. 4.15 (**a**) Calculation of effect of inboard and outboard injection on ECCD efficiency. After Cohen [134]. (**b**) Experimental study of effect of inboard and outboard injection on ECCD efficiency. Reproduced with permission from Suzuki [729]. Copyright IAEA Vienna

Schlüter regime and large mass ratio ($m_i/m_I \ll 1$) [446], while Kikuchi-Azumi solved moment equation to analyze toroidal and poloidal rotations in JT-60U [434]. The general flow balance equation (4.33) allows us to derive an expression for the toroidal flows of electrons, ions, and impurities. Here, we concentrate on effect of thermodynamic force on toroidal rotation assuming $E_\parallel = \hat{S}_\parallel = 0$ and neglect fast ion component, for simplicity. From Eq. (4.29), the local toroidal flow for species a is the summation of flux surface averaged parallel flow and Pfirsch-Schlüter flow as follows,

$$u_{a\zeta}^{(1)} = \frac{B_\zeta \langle B u_{\parallel a} \rangle}{\langle B^2 \rangle} - \left[1 - \frac{B_\zeta^2}{\langle B^2 \rangle} \right] R \left(\Phi' + \frac{P_a'}{e_a n_a} \right) \tag{4.201}$$

Here, $'$ denotes derivative with respect to ψ. From Eq. (4.34), we obtain following expression of flux surface averaged parallel flows,

$$\langle B u_{\parallel a} \rangle = -F \sum_{b=1}^{3} \left[\alpha_{ab} \left(\Phi' + \frac{P_b'}{e_b n_b} \right) + \frac{\alpha_{a,b+3}}{e_b} T_b' \right] \tag{4.202}$$

The impurity (for example, Carbon) toroidal rotation can be measured by the charge exchange recombination spectroscopy (CXRS), which can be used to determine radial electric field using (4.201) if density and temperature profiles are known as follows,

$$\frac{d\Phi}{d\psi} = -\frac{1}{\sum_b \alpha_{Ib}^*} \left[\frac{\langle B^2 \rangle u_{\zeta I}}{R B_\zeta^2} + \sum_{b=1}^{3} \left(\frac{\alpha_{Ib}^* P_b'}{e_b n_b} + \frac{\alpha_{I,b+3} T_b'}{e_b} \right) \right] \tag{4.203}$$

Here, $\alpha^*_{Ib} = \alpha_{Ib} + (\langle B^2\rangle/B_\zeta^2 - 1)\delta_{ab}$. This radial electric field profile can be used to discuss effect of radial electric field shear on transport reduction discussed in Sect. 7.4. Once radial electric field is known, we can calculate parallel flows of electron, ion, impurity from Eq. (4.201). If we calculate these flows for typical high temperature plasma, we find significant difference between ion and impurity toroidal rotations of several 10 km/s in high β_p plasma as shown by Kikuchi [434].

Toroidal and poloidal rotations of ion and impurity are analyzed analytically by Kim [446]. Since electron inertia is small, electron contribution to ion and impurity momentum balance can be neglected and flow relations of ion and impurity are given as follows.

$$\hat{\boldsymbol{\mu}}^I \cdot \boldsymbol{u}_\theta^I = \hat{\boldsymbol{L}}_{II} \cdot \boldsymbol{u}_\parallel^I + \hat{\boldsymbol{L}}_{Ii} \cdot \boldsymbol{u}_\parallel^i \tag{4.204}$$

$$\hat{\boldsymbol{\mu}}^i \cdot \boldsymbol{u}_\theta^i = \hat{\boldsymbol{L}}_{iI} \cdot \boldsymbol{u}_\parallel^I + \hat{\boldsymbol{L}}_{ii} \cdot \boldsymbol{u}_\parallel^i \tag{4.205}$$

$$\hat{\boldsymbol{u}}_\theta^a = \boldsymbol{u}_\parallel^a - \boldsymbol{V}^a, \tag{4.206}$$

where last equation for $\hat{\boldsymbol{u}}_\theta^a$ is obtained from Eq. (4.27) and $\hat{\boldsymbol{u}}_\theta^a$, $\boldsymbol{u}_\parallel^a$, $\boldsymbol{V}^a$, $\hat{\boldsymbol{\mu}}^a$, $\hat{\boldsymbol{L}}_{ab}$ are defined as follows,

$$\hat{\boldsymbol{u}}_\theta^a = \begin{bmatrix} \langle B^2\rangle \hat{u}_{a\theta} \\ \frac{2\langle B^2\rangle \hat{q}_{a\theta}}{5P_a} \end{bmatrix}, \boldsymbol{u}_\parallel^a = \begin{bmatrix} \langle Bu_{\parallel a}\rangle \\ \frac{2\langle Bq_{\parallel a}\rangle}{5P_a} \end{bmatrix}, \boldsymbol{V}^a = \begin{bmatrix} BV_{1a} \\ BV_{2a} \end{bmatrix} \tag{4.207}$$

$$\hat{\boldsymbol{\mu}}^a = \begin{bmatrix} \hat{\mu}_{a1} & \hat{\mu}_{a2} \\ \hat{\mu}_{a2} & \hat{\mu}_{a3} \end{bmatrix}, \hat{\boldsymbol{L}}_{ab} = \begin{bmatrix} \hat{l}_{11}^{ab} & -\hat{l}_{12}^{ab} \\ -\hat{l}_{21}^{ab} & \hat{l}_{22}^{ab} \end{bmatrix} \tag{4.208}$$

Since impurity collisionality is given as $\nu_I^* = (n_I Z_I^4/n_i Z_i^4)\nu_i^*$ considering fast equipartition between ion and impurity ($T_i \approx T_I$), impurity may be in the Pfirsch-Schlüter regime (negligible impurity viscous force: $\hat{\boldsymbol{\mu}}^I \cdot \boldsymbol{u}_\theta^I \approx 0$) while bulk ion is deeply collisionless regime, $\nu_I^* \gg \nu_i^*$. Therefore, impurity parallel flow can be given as $\boldsymbol{u}_\parallel^I = -\hat{\boldsymbol{L}}_{II}^{-1} \hat{\boldsymbol{L}}_{Ii} \cdot \boldsymbol{u}_\parallel^i$. Using large impurity mass approximation $m_I \gg m_i$, substitution into (4.205) gives following ion momentum balance equation.

$$\hat{\boldsymbol{\mu}}^i \cdot \hat{\boldsymbol{u}}_\theta^i = -\begin{bmatrix} 0 & 0 \\ 0 & \gamma \end{bmatrix} \cdot \boldsymbol{u}_\parallel^i \tag{4.209}$$

Here, $\gamma = \sqrt{2} + \alpha$, $\alpha = n_I Z_I^2/n_i Z_i^2$ (Here $\beta = O((m_i/m_I)^2)$ term of [446] is neglected). Using the relation $\boldsymbol{u}_\parallel^i = \hat{\boldsymbol{u}}_\theta^i + \boldsymbol{V}^i$, we obtain following expression.

$$\begin{bmatrix} \hat{\mu}_{i1} & \hat{\mu}_{i2} \\ \hat{\mu}_{i2} & \hat{\mu}_{i3} + \gamma \end{bmatrix} \hat{\boldsymbol{u}}_\theta^i = -\begin{bmatrix} 0 & 0 \\ 0 & \gamma \end{bmatrix} \boldsymbol{V}^i \tag{4.210}$$

This equation indicates that $\hat{u}_\theta$-driven ion viscous force is balanced against $\hat{q}_\theta$-driven ion viscous force so that total parallel ion viscous force becomes zero. But, the ion poloidal heat flow is driven by the thermal force (dT_i/dr) through the parallel heat friction. Equation (4.210) leads to following analytic expressions for $\hat{\boldsymbol{u}}^i_\theta, \boldsymbol{u}^i_\|, \boldsymbol{u}^I_\|, \hat{\boldsymbol{u}}^I_\theta$.

$$\hat{\boldsymbol{u}}^i_\theta = \begin{bmatrix} 0 & K_1 \\ 0 & -\frac{\hat{\mu}_{i1}}{\hat{\mu}_{i2}} K_1 \end{bmatrix} \boldsymbol{V}^i, \boldsymbol{u}^i_\| = \begin{bmatrix} 1 & K_1 \\ 0 & K_2 \end{bmatrix} \boldsymbol{V}^i \tag{4.211}$$

$$\boldsymbol{u}^I_\| = \begin{bmatrix} 1 & K_1 + 1.5K_2 \\ 0 & 0 \end{bmatrix} \boldsymbol{V}^i, \hat{\boldsymbol{u}}^I_\theta = \boldsymbol{u}^I_\| - \boldsymbol{V}^I \tag{4.212}$$

$$K_1 = \frac{\gamma \hat{\mu}_{i2}}{D}, K_2 = \frac{\hat{\mu}_{i1}\hat{\mu}_{i3} - \hat{\mu}_{i2}^2}{D}, D = \hat{\mu}_{i1}(\hat{\mu}_{i3} + \gamma) - \hat{\mu}_{i2}^2 \tag{4.213}$$

If we use tokamak ordering $B \approx B_\zeta$ and $\hat{u}_{a\theta} \approx u_\theta / B_\theta$ by neglecting poloidal variation, we obtain following form of $u_{\|i}, u_{\|I}, u_{\theta i}, u_{\theta I}$.

$$u_{\|i} \approx \frac{-1}{B_\theta}\left[\frac{d\Phi}{dr} + \frac{1}{eZ_i n_i}\frac{dP_i}{dr} + \frac{K_1}{eZ_i}\frac{dT_i}{dr}\right] \tag{4.214}$$

$$u_{\|I} \approx \frac{-1}{B_\theta}\left[\frac{d\Phi}{dr} + \frac{1}{eZ_i n_i}\frac{dP_i}{dr} + \frac{K_1 + 1.5K_2}{eZ_i}\frac{dT_i}{dr}\right] \tag{4.215}$$

$$u_{\theta i} \approx -\frac{K_1}{eZ_i B_\zeta}\frac{dT_i}{dr} \tag{4.216}$$

$$u_{\theta I} \approx -\frac{K_1 + 1.5K_2}{eZ_i B_\zeta}\frac{dT_i}{dr} - \frac{1}{eZ_I n_I B_\zeta}\frac{dP_I}{dr} + \frac{1}{eZ_i n_i B_\zeta}\frac{dP_i}{dr} \tag{4.217}$$

All terms proportional to dT_i/dr in above 4 equations originates from heat flow balance equation.

If we neglect these terms,we see $u_{\theta i} \approx 0$ due to strong viscous damping of poloidal rotation. The $u_{\|i}$ is determined to satisfy $u_{\theta i} \approx 0$ (poloidal component of parallel flow must compensate poloidal component of $E_r \times B$ and ion diamagnetic flows). Also, $u_{\|i} \approx u_{\|I}$ due to negligible impurity parallel viscous force. Then impurity parallel flow cancels poloidal component of $E_r \times B$ and ion (not impurity) diamagnetic flows. Then, impurity poloidal flow is difference of impurity and ion diamagnetic flows ($u_{\theta I} \approx [(dP_i/dr)/(Z_i n_i) - (dP_I/dr)/(Z_I n_I)]/(eB_\zeta)$).

Retaining dT_i/dr terms, we see ion and impurity toroidal rotation proportional to dT_i/dr and $1/B_\theta$ (large at high β_p).

$$\Delta u_\| = u_{\|I} - u_{\|i} = -\frac{1.5K_2}{eZ_i B_\theta}\frac{dT_i}{dr} \tag{4.218}$$

If the total toroidal momentum is zero ($n_i m_i u_{\|i} + n_I m_I u_{\|I} = 0$), this equation together with (4.218) implies positive toroidal rotation for ion and negative toroidal rotation for impurity as follows,

$$u_{\|I} = -\frac{n_i m_i}{n_I m_I} u_{\|i} = \frac{-n_i m_i}{n_i m_i + n_I m_I}\frac{1.5K_2}{eZ_i B_\theta}\frac{dT_i}{dr} \tag{4.219}$$

Since $n_i m_i / n_I m_I \gg 1$ holds in most case, impurity rotates strongly in counter direction and bulk ion rotates weakly in co direction ($|u_{\parallel I}| \gg |u_{\parallel i}|$) if total toroidal momentum is zero. Since there is very little expected momentum source, Ohmically heated plasma is ideal to test strong counter impurity rotation. In 1991 Sherwood conference, Kim discussed the impurity toroidal rotation (nearly parallel flow) in Ohmically heated plasma in JFT-2M implying that its value of -12.5 km/s is consistent with this temperature gradient-driven counter impurity toroidal rotation from Eq. (4.227). This suggests that there is no significant hidden torque drive in Ohmically heated plasma. But later many experiments showed strong intrinsic rotation driven by the hidden torque as will be discussed in Sect. 7.6.

4.5.2 Neoclassical Toroidal Viscosity

One important property of axisymmetric system is the conservation of total toroidal momentum. Total toroidal angular momentum balance equation is given as follows,

$$\sum_a m_a \langle n_a R \frac{du_{a\zeta}}{dt} \rangle = \sum_a \langle R^2 \nabla\zeta \cdot (\nabla \cdot \boldsymbol{\Pi}_a + \boldsymbol{M}_a) \rangle \tag{4.220}$$

Noting viscous tensor $\boldsymbol{\Pi}_a$ is a symmetric tensor for axisymmetric plasma and $\nabla(R^2\nabla\zeta)$ is antisymmetric tensor [328], flux surface average of toroidal viscous force is shown to be zero, namely $\langle R^2\nabla\zeta \cdot \nabla \cdot \boldsymbol{\Pi}_a \rangle = 0$. This means that toroidal drag force by magnetic field variation is zero for axisymmetric system. When symmetry is broken by the application of non-axisymmetric field, $\langle R^2\nabla\zeta \cdot \nabla \cdot \boldsymbol{\Pi}_a \rangle$ becomes non zero. This drag force is called Neoclassical Toroidal Viscosity (NTV). The symmetry breaking produces offset toroidal rotation and the force.

We first describe the offset toroidal rotation. The 0th order ion radial force balance is:

$$0 = eZ_i n_i (\boldsymbol{E} + \boldsymbol{u}_i \times \boldsymbol{B}) - \nabla P_i \tag{4.221}$$

In the flux coordinates (ψ, θ, ζ), the magnetic field is expressed as $\boldsymbol{B} = \nabla\psi \times \nabla(q\theta - \zeta)$. The radial component of above equation can be obtained by taking inner product with tangent vector $\partial \boldsymbol{x}/\partial\psi$ and using the identity $\partial \boldsymbol{x}/\partial\psi \cdot \nabla\psi = 1$, where $\boldsymbol{x}$ is position vector.

$$\boldsymbol{u}_i \cdot \nabla\zeta = -\left[\frac{d\Phi}{d\psi} + \frac{1}{eZ_i n_i}\frac{dP_i}{d\psi}\right] + q\boldsymbol{u}_i \cdot \nabla\theta \tag{4.222}$$

In the tokamak plasma with symmetry breaking, the electrostatic potential Φ is determined so that non-ambipolar flux $\langle \boldsymbol{\Gamma}_{na} \cdot \nabla V \rangle$ becomes zero. In Hamada coordinates, the non-ambipolar flux $\langle \boldsymbol{\Gamma}_{na} \cdot \nabla V \rangle$ is related to toroidal viscous force as $\langle \boldsymbol{\Gamma}_{na} \cdot \nabla V \rangle = (V'(\psi)^2 / e_a q) \langle \boldsymbol{B}_t \cdot \nabla \cdot \boldsymbol{\Pi}_a \rangle$ [678]. In the collisionless regime, ion viscous force is larger than that for electron by a factor of $(m_i/m_e)^{1/2}$. The zero non-ambipolar flux condition is then given by $\langle \boldsymbol{B}_t \cdot \nabla \cdot \boldsymbol{\Pi}_i \rangle = 0$ for electron ion plasma. Shaing [681] derived following relation for the ion in collisionless regime,

$$\frac{d\Phi}{d\psi} + \frac{1}{eZ_i n_i}\frac{dP_i}{d\psi} = -\frac{\lambda_2}{eZ_i\lambda_1}\frac{dT_i}{d\psi} \tag{4.223}$$

Here λ_1 and λ_2 are numerical constants in [681]. Even if we include impurity, we may be able to assume impurity is in a Pfirsch-Schlüter regime (namely impurity parallel viscosity is small) while ion is deeply in collisionless regime as in previous section. Then (4.223) is still zero non-ambipolar flux condition. Substituting (4.223) into (4.222), we obtain following expression of offset toroidal rotation,

$$\boldsymbol{u}_{i0} \cdot \nabla\zeta = \frac{\lambda_2}{eZ_i\lambda_1}\frac{dT_i}{d\psi} + q\boldsymbol{u}_{i0} \cdot \nabla\theta, \tag{4.224}$$

where offset poloidal rotation $\boldsymbol{u}_{i0} \cdot \nabla\theta$ is given by,

$$\boldsymbol{u}_{i0} \cdot \nabla\theta = -\frac{K_1 F(\boldsymbol{B} \cdot \nabla\theta)}{eZ_i\langle B^2\rangle}\frac{dT_i}{d\psi} \tag{4.225}$$

Since measurement of toroidal rotation is made using impurity toroidal rotation, actual measurement of offset toroidal rotation of the impurity is given by Kikuchi [444]. In the cylindrical coordinates, offset toroidal rotations of ion and impurity are:

$$u_{i\zeta 0} = \frac{3.54 - K_1}{eZ_i B_\theta}\frac{dT_i}{dr} \tag{4.226}$$

$$u_{I\zeta 0} = \frac{3.54 - 1.5K_2 - K_1}{eZ_i B_\theta}\frac{dT_i}{dr} \tag{4.227}$$

The offset rotation is confirmed experimentally in DIII-D [246] as shown in Fig. 4.16b. Since K_1 and K_2 are order of 0.3–1.2 in collisionless tokamak plasma as given by Kim [446], agreement with collisionless ion $(1/\nu)$ regime becomes better when we include finite poloidal flow correction and impurity effect.

Now we discuss drag force due to neoclassical toroidal viscosity. The 3D non-resonant field can cause increased radial excursion of ions, which produces NTV torque. This can be enhanced when the toroidal rotation frequency resonates with natural frequency of the ions. If the rotation frequency is much lower than the ion bounce frequency, the E_r/B_p toroidal precession drift (PD) play the key role, which is relatively small, for example, less than several kHz. At higher toroidal rotation frequency regime, the toroidal rotation frequency can resonate with ion bounce frequency. The NTV resonate with trapped ion bounce motion may be called BH-NTV. The resonance condition is given by $\ell\omega_b \sim n\omega_E$, where ω_b is bounce frequency, $\omega_E \equiv E_r/B_p/R$ is the electric field-induced precession drift frequency, and ℓ, n are integers.

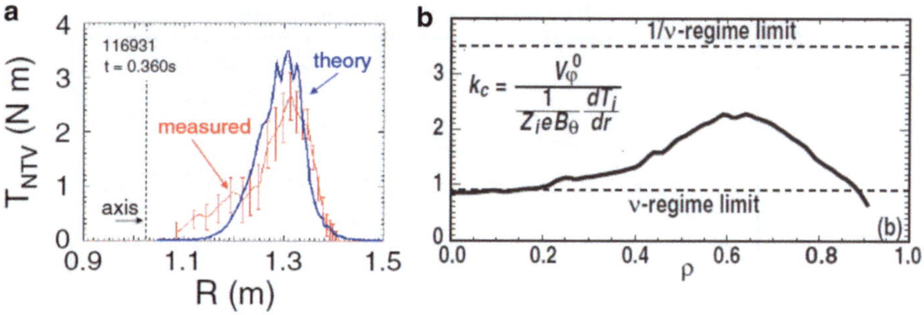

Fig. 4.16 (**a**) Observation of neoclassical toroidal viscosity (NTV) in NSTX [860]. (**b**) Observation of offset toroidal rotation due to NTV in DIII-D [246]. Reproduced with permission from Zhu [860] and Garofalo [246]. Copyright American Physical Society

The NTV has parametric dependences on the collisionality ν, and the toroidal precession drift $E_r/B_p \equiv R\omega_E$. Collisionality dependence includes well-known $1/\nu$ regime as clarified by Shaing [681] in relatively high collisionality regime and $\sqrt{\nu}$ regime in very low collisionality regime also by Shaing [682] while it can have ν dependence if $\omega_E \sim 0$.

The NTV drag in $1/\nu$ regime is first observed by W. Zhu in NSTX [860] as shown in Fig. 4.16a. The observed torque is consistent with the calculation including the resonant field amplification (RFA) effect identified by Boozer [72]. Sun [724] gives general solution of NTV torque by non-axisymmetric perturbation in the collision less regime and showed peaking against the toroidal rotation frequency whose result can be given as shown in Fig. 4.17. Since the NTV in low toroidal rotation resonates with ion precession drift, it should have some peaking against the rotation frequency. Cole [136] showed that the observed NTV at low toroidal rotation frequency in DIII-D has a peak in toroidal rotation consistent with Sun's result. Satake [659] developed a δf Monte Carlo simulation code to check $1/\nu$ dependence. Sun [725] showed the $1/\nu$ dependence in the JET experiment.

Kim [451] shows that the toroidal rotation frequency (ω_p) can have a bounce harmonic resonance ($l\omega_b = n\omega_p$), where strong breaking occurs if the plasma rotation resonates $\omega_p \sim l\omega_b/n$. J.K. Park [606] gives an satisfactory agreement between numerical computations and experiments in KSTAR.

Shaing [683] gives an analytical connection formula of NTV torque with appropriate boundary condition in the super-banana plateau regime valid for all collisionality regime, which has been refined by Sun [726].

The NTV can be induced also by exciting the internal kink mode. Seol [676] reported that central ECH can destabilize the internal kink mode and can break the toroidal rotation in KSTAR.

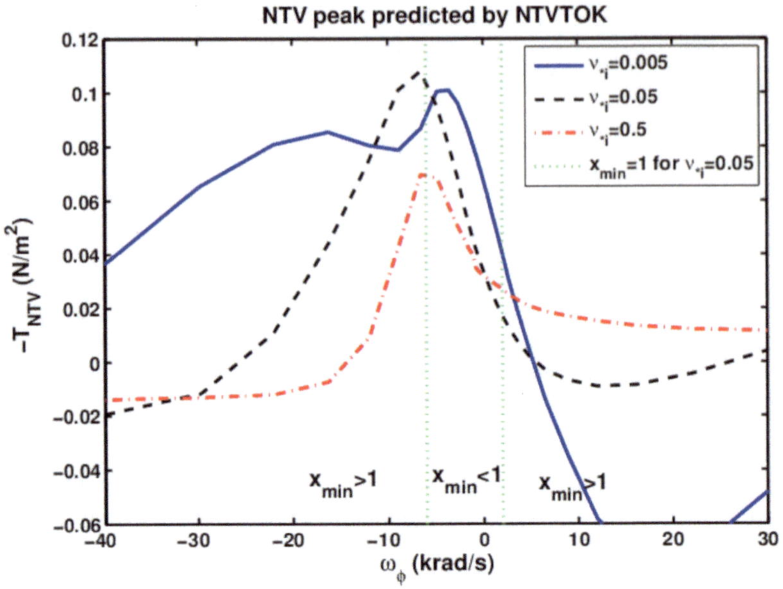

Fig. 4.17 Dependence of NTV torque density $-T_{NTV}$ on the toroidal rotation frequency ω_ϕ. Plotted from Sun [724] (courtesy of Sun)

Column 4-5: Some Basic Quantities

Frequencies: Basic frequencies in plasma physics are given as follows, while n_{e20} is electron density in $10^{20}m^{-3}$ and n_{i20} is ion density in $10^{20}m^{-3}$.
Electron cyclotron freq.: $\Omega_e(s^{-1}) = \frac{eB}{m_e} = 1.76 \times 10^{11} B(T)$
Ion cyclotron freq.: $\Omega_i(s^{-1}) = \frac{eZ_iB}{m_i} = 9.55 \times 10^7 \frac{Z_iB(T)}{A_i}$
Electron plasma freq.: $\omega_{pe}(s^{-1}) = \sqrt{\frac{n_e e^2}{m_e \varepsilon_0}} = 5.64 \times 10^{11} \sqrt{n_{e20}}$
Ion plasma freq.: $\omega_{pi}(s^{-1}) = \sqrt{\frac{n_i Z_i^2 e^2}{m_i \varepsilon_0}} = 1.32 \times 10^{10} Z_i \sqrt{\frac{n_{i20}}{A_i}}$
Drift frequency: $\omega_{*a}(s^{-1}) = \frac{k_\perp T_a}{e_a B n_a} \frac{dn_a}{dr}$
Thermal drift frequency: $\omega_{*Ta}(s^{-1}) = \frac{k_\perp}{e_a B} \frac{dT_a}{dr}$
Diamagnetic frequency: $\omega_{*pa}(s^{-1}) = \frac{k_\perp}{e_a B n_a} \frac{dp_a}{dr}$
Toroidal drift frequency: $\omega_{Da}(s^{-1}) = \boldsymbol{k} \cdot \boldsymbol{v}_{Da} = \frac{-2k_\perp T_a}{e_a BR}$

Speed: Basic speeds in plasma physics are given as follows,
Electron thermal speed: $v_{Te}(m/s) = \sqrt{\frac{2T_e}{m_e}} = 5.9 \times 10^5 \sqrt{T_e(eV)}$

(continued)

Ion thermal speed: $v_{Ti}(m/s) = \sqrt{\frac{2T_i}{m_i}} = \times 10^4\sqrt{T_i(eV)}$
Alfven speed: $v_A(m/s) = \frac{B}{\sqrt{\mu_0\rho}} = 2.18 \times 10^6 \frac{B(T)}{\sqrt{A_i n_{i20}}}$
$E \times B$ drift speed: $v_E(m/s) = \frac{E}{B} = \frac{E(V/m)}{B(T)}$
Polarization drift speed: $v_p(m/s) = \frac{m_a}{eZ_aB^2}\frac{dE}{dt}$

Lengths: Basic lengths of plasma physics are given as follows.
Debye length: $\lambda_D(m) = \sqrt{\varepsilon_0 T_e/n_e e^2} = 2.35 \times 10^5\sqrt{T_e(keV)/n_e(m^{-3})}$
Larmor radius (e, th.): $\rho_{Te}(m) = \frac{\sqrt{2m_eT_e}}{eB} = 1.07 \times 10^{-4}\frac{\sqrt{T_e(keV)}}{B(T)}$
Larmor radius (ion, th.): $\rho_{Ti}(m) = \frac{\sqrt{2m_iT_i}}{eZ_iB} = 4.57 \times 10^{-3}\frac{\sqrt{A_iT_i(keV)}}{Z_iB(T)}$
Electron mean free path: $\lambda_e = v_{Te}\tau_e = 1.44 \times 19^{23}\frac{T_e(keV)^2}{n_e(m^{-3})ln\Lambda}$

Dimensionless parameters

Collisionality: $\nu_i^* = \frac{Rq}{\varepsilon^{1.5}v_{Ti}\tau_{ii}} = 4.9 \times 10^{-12}\frac{Rqn_i(m^{-3})Z_i^4 ln\Lambda}{\varepsilon^{1.5}T_i(keV)^2}$
Normalized gyro: $\rho_i^* = \frac{\rho_i}{a} = \frac{\sqrt{2m_iT_i}}{eZ_iB} = 4.57 \times 10^{-3}\frac{\sqrt{A_iT_i(keV)}}{Z_iB(T)}$
Toroidal beta: $\beta = \frac{nT}{B^2/2\mu_0} = 4.03 \times 10^{-2}\frac{T_i(keV)n(10^{20}m^{-3})}{B(T)^2}$

Exercise 4.1. Show that the cross field particle and heat fluxes can be decomposed into classical and neoclassical fluxes as:

$$\begin{bmatrix}\langle n_a \boldsymbol{u}_{a\perp}\cdot\nabla\psi\rangle \\ \langle \boldsymbol{q}_{a\perp}\cdot\nabla\psi/T_a\rangle\end{bmatrix} = \begin{bmatrix}\Gamma_a^{cl} + \Gamma_a^{NC} \\ (q_a^{cl} + q_a^{NC})/T_a\end{bmatrix} \tag{4.228}$$

$$\text{where, } \begin{bmatrix}\Gamma_a^{cl} \\ q_a^{cl}/T_a\end{bmatrix} = \begin{bmatrix}\langle(\boldsymbol{B}\times\nabla\psi)\cdot\boldsymbol{F}_{a1}/e_aB^2\rangle \\ \langle(\boldsymbol{B}\times\nabla\psi)\cdot\boldsymbol{F}_{a2}/e_aB^2\rangle\end{bmatrix}$$

$$\begin{bmatrix}\Gamma_a^{NC} \\ q_a^{NC}/T_a\end{bmatrix} = \begin{bmatrix}\langle(\boldsymbol{B}\times\nabla\psi)\cdot(\nabla P_a + \nabla\cdot\boldsymbol{\Pi}_a - e_an_a\boldsymbol{E})/e_aB^2\rangle \\ \langle(\boldsymbol{B}\times\nabla\psi)\cdot((5/2)n_a\nabla T_a + \nabla\cdot\boldsymbol{\Theta}_a)/e_aB^2\rangle\end{bmatrix}$$

Answer. See Kikuchi [443].

Exercise 4.2. Show that the neoclassical particle flux is decomposed into Pfirsch-Schlüter, banana-plateau, and electric fluxes as:

$$\begin{bmatrix}\Gamma_a^{NC} \\ q_a^{NC}/T_a\end{bmatrix} = \begin{bmatrix}\Gamma_a^{ps} + \Gamma_a^{bp} + \Gamma_a^{E} \\ (q_a^{ps} + q_a^{bp})/T_a\end{bmatrix} \tag{4.229}$$

$$\text{where, } \begin{bmatrix}\Gamma_a^{ps} \\ q_a^{ps}/T_a\end{bmatrix} = -\frac{F(\psi)}{e_a}\left\langle\left(\frac{1}{B^2} - \frac{1}{\langle B^2\rangle}\right)\begin{bmatrix}\boldsymbol{B}\cdot\boldsymbol{F}_{1a} \\ \boldsymbol{B}\cdot\boldsymbol{F}_{1a}\end{bmatrix}\right\rangle$$

$$\begin{bmatrix} \Gamma_a^{bp} \\ q_a^{bp}/T_a \end{bmatrix} = -\frac{F(\psi)}{e_a} \begin{bmatrix} \langle \boldsymbol{B} \cdot \nabla \cdot \boldsymbol{\Pi}_a \rangle \\ \langle \boldsymbol{B} \cdot \nabla \cdot \boldsymbol{\Theta}_a \rangle \end{bmatrix}$$

$$\Gamma_a^E = \langle n_a \boldsymbol{B} \cdot \boldsymbol{E} \rangle F(\psi)/\langle B^2 \rangle + \langle R^2 \nabla \zeta \cdot n_a \partial \boldsymbol{A}/\partial t \rangle$$

Answer. See Kikuchi [443].

Exercise 4.3. Let define radial coordinate $\rho = (\phi/\phi_a)^{1/2} a$. Show that the summation of classical and Pfirsch-Schlüter fluxes are given by,

$$\begin{bmatrix} \Gamma_a^{cl} + \Gamma_a^{ps} \\ (q_a^{cl} + q_a^{ps})/T_a \end{bmatrix} = \frac{\langle R^2 \rangle - F^2/\langle B^2 \rangle}{\psi'(\rho)\tau_{aa}} \sum_b \frac{m_a n_a}{e_a e_b} \begin{bmatrix} \hat{l}_{11}^{ab} & -\hat{l}_{12}^{ab} \\ -\hat{l}_{21}^{ab} & \hat{l}_{22}^{ab} \end{bmatrix} \begin{bmatrix} P_b'(\rho)/n_b \\ T_b'(\rho) \end{bmatrix} \tag{4.230}$$

Answer. See Kikuchi [443].

Exercise 4.4. Show that the banana-plateau fluxes are given by,

$$\begin{bmatrix} \Gamma_a^{bp} \\ q_a^{bp}/T_a \end{bmatrix} = -\sum_b \begin{bmatrix} K_{11}^{ab} & K_{12}^{ab} \\ K_{21}^{ab} & K_{22}^{ab} \end{bmatrix} \begin{bmatrix} P_b'(\rho)/n_b \\ T_b'(\rho) \end{bmatrix} + \begin{bmatrix} g_{1a} \\ g_{2a} \end{bmatrix} \langle B E_{\parallel} \rangle \tag{4.231}$$

Answer. See Kikuchi [443].

Exercise 4.5. Derive (4.183) from the following θ dependent FP equation.
$\frac{\partial f}{\partial t} - \frac{B_\theta v_\parallel}{r B_\zeta} \frac{\partial f}{\partial \theta} = \frac{1}{\tau_{se}} [\frac{1}{v^2} \frac{\partial}{\partial v} \{(v_c^3 + v^3) f\} + \frac{\beta B_{min}}{B} \frac{v_c^3}{v^3} \frac{|v_\parallel|}{\eta v} \frac{\partial}{\partial \eta} \{(1 - \eta^2) \frac{|v_\parallel|}{\eta v} \frac{\partial f}{\partial \eta}\}] + S.$

Hint: Use expansion $f = f_0 + f_1$ in power of τ_B/τ_{se} and apply solvability condition, where τ_B is the bounce period of fast ion. For detail, see Cordey [148].

Exercise 4.6. Derive the relation (4.158).

Answer. Read Helander-Sigmar Sect. 11.2 to reach [317](11.23).

Chapter 5
Low Frequency Collective Motions in Tokamak

Abstract Low frequency collective motions play major roles in turbulent transport and also MHD instabilities in tokamaks. In this chapter, we introduce drift waves, shear Alfven waves, wave-particle interactions using fluid and gyro kinetic equations as important collective motions in tokamak. Electrostatic drift waves such as slab drift wave including ion temperature gradient (ITG) drift instability, and toroidal drift waves including ITG and trapped electron mode (TEM) instabilities are introduced in Sect. 5.1. The shear Alfven wave, kinetic Alfven wave and drift Alfven wave are introduced in Sect. 5.2. Classical and modern gyrokinetic theories are introduced in Sect. 5.3. Numerical calculations of toroidal ITG mode, TEM and the electron temperature gradient (ETG) mode based on gyrokinetic theories are described in Sect. 5.4. Columns includes derivation of guiding center Lagrangian, Non-canonical mechanics, Lagrange tensor, and Lie transformation.

Further Readings:

Book: Kadomtsev (1976) [409] is nice book on collective phenomena in plasmas. Kadomtsev (1965) includes some of important analysis on drift Alfven waves. Chen (1987) [110] is well written to understand classical gyro kinetic equation. Hazeltine-Waelbloeck (2004) [309] is useful for basic drift wave. Weiland (2000) [815] is useful to understand fluid views of drift waves. You may read Flanders (1989) [206], Arnolds (1978) [26], Frankel (2004) [209] for mathematical bases.

Review: Kadomtsev-Pogutse [408] is useful as introduction to trapped particle dynamics. Horton (1999) [341] is good survey of fundamentals of drift waves. Cary-Littlejohn [99] is nice paper on introduction of Lie transform. Brizard-Hahm (2007) [82] is comprehensive report on modern gyrokinetic theory. Cary-Brizard (2009) [100] is useful to understand Lie transform.

M. Kikuchi, M. Azumi, *Frontiers in Fusion Research II*,
DOI 10.1007/978-3-319-18905-5_5

5.1 Electrostatic Drift Waves

5.1.1 Density Gradient Drift Waves

We first discuss a drift wave in a simple geometry shown in Fig. 5.1a. The density gradient is in the negative x direction ($dn/dx < 0$) and the magnetic field is in the z direction. We assume uniform temperature, i.e. $dT/dx = 0$. We consider the drift wave driven by the density gradient dn/dx. The $\partial\tilde{\Phi}/\partial y$ produces $E \times B$ drift motion in the x direction ($\tilde{u}_{Ex} = \tilde{E}_y/B = -(\partial\tilde{\Phi}/\partial y)/B$). This $\tilde{u}_{Ex}$ produces the ion density fluctuation $\tilde{n}_i$. Taking finite $k_\parallel$ ($\partial/\partial z \neq 0$) in the parallel ion motion to include sound wave dynamics, the ion continuity equation ($\partial n_i/\partial t + \nabla(n_i \boldsymbol{u}_i) = 0$) and the parallel ion momentum balance equation is given as follows,

$$\frac{\partial \tilde{n}_i}{\partial t} + n_i \frac{\partial \tilde{u}_{\parallel i}}{\partial z} + \tilde{u}_{Ex} \frac{dn_i}{dx} = 0 \tag{5.1}$$

$$m_i n_i \frac{\partial \tilde{u}_{\parallel i}}{\partial t} = -\frac{\partial \tilde{p}_i}{\partial z} - e_i n_i \frac{\partial \tilde{\Phi}}{\partial z} \tag{5.2}$$

We note $\nabla \cdot \tilde{\boldsymbol{u}}_E = -\nabla \cdot \left(\nabla\tilde{\Phi} \times \boldsymbol{B}/B^2\right) = 0$ in uniform B. For $\sim e^{-i\omega t + ik_y y + ik_\parallel z}$, $\tilde{u}_{Ex}(dn_i/dx) = in_i\omega_{*e}(e\tilde{\Phi}/T_e)$, where $e_i = eZ_i$, $\omega_{*e} = k_y T_e/eL_n B$, $L_n = -n_e/(dn_e/dx)$ and Eqs. (5.2) and (5.3) becomes:

$$\frac{\tilde{n}_i}{n_i} = \frac{\omega_{*e}}{\omega}\frac{e\tilde{\Phi}}{T_e} + \frac{k_\parallel \tilde{u}_{\parallel i}}{\omega} \tag{5.3}$$

$$\tilde{u}_{\parallel i} = \frac{k_\parallel}{\omega}\frac{\tilde{p}_i + e_i n_i \tilde{\Phi}}{m_i n_i} \tag{5.4}$$

The adiabatic ion response is given by $d(p_i n_i^{-\gamma_i})/dt = 0$ ($\gamma_i = 5/3$), which gives $\tilde{p}_i = \gamma_i T_i \tilde{n}_i$. Eliminating $\tilde{u}_{iz}$ and assuming ion adiabatic response, we obtain:

$$\frac{\tilde{n}_i}{n_i} = \frac{\omega\omega_{*e} + k_\parallel^2 c_{se}^2}{\omega^2 - k_\parallel^2 c_{si}^2}\frac{e\tilde{\Phi}}{T_e} \tag{5.5}$$

Here, $c_{se}^2 = Z_i T_e/m_i$, $c_{si}^2 = \gamma_i T_i/m_i$ and ω_{*e} is the electron diamagnetic frequency.

The electron dynamics is dominated by the parallel momentum balance as $m_e du_{\parallel e}/dt = e\partial\Phi/\partial z - \partial p_e/n_e\partial z$. For low frequency phenomena ($\omega \ll \Omega_i$), we can neglect small electron mass. Since parallel electron heat conduction is sufficiently high, we can also assume isothermal electron ($\partial T_e/\partial z = 0$) so that $e\partial\Phi/\partial z - (T_e/n_e)(\partial n_e/\partial z) = 0$. This gives Boltzmann relation, $n_e/n_{e0} = exp[e\Phi/T_e]$. Expanding $n_e = n_{e0} + \tilde{n}_e$ and $\Phi = \Phi_0 + \tilde{\Phi}$, we have electron Boltzmann response:

$$\tilde{n}_e/n_{e0} = e\tilde{\Phi}/T_e \tag{5.6}$$

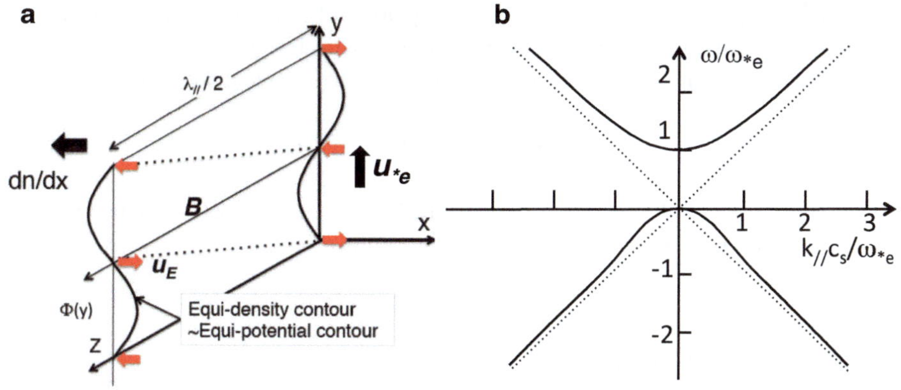

Fig. 5.1 (**a**) Geometry of the drift wave. Modified from F.F. Chen [108]. (**b**) Dispersion relation of the electrostatic drift wave

Substitution of above equation and Boltzmann relation ($\tilde{n}_e/n_e = e\tilde{\Phi}/T_e$) to charge neutrality condition ($\tilde{n}_e/n_e = \tilde{n}_i/n_i$) gives following dispersion relation.

$$\omega^2 = \omega\omega_{*e} + k_\parallel^2 c_s^2 \tag{5.7}$$

Here, $c_s^2 = (\gamma_i T_i + Z_i T_e)/m_i$ and c_s is a sound speed. The dispersion relation $\omega = \omega(k_\parallel)$ is shown in Fig. 5.1b. Upper branch is a drift wave propagating in the electron diamagnetic drift direction. Even at $k_\parallel = 0$, the dispersion relation $\omega = \omega_{*e}$ implies that the phase velocity $u_{ph} = \omega/k_\perp = u_{*e}$. At finite $k_\parallel$, it couples to sound wave. Lower branch is a drift wave propagating in the ion diamagnetic drift direction. The dispersion relation at low $k_\parallel$ is approximately given by $\omega = -k_\parallel^2 c_s^2/\omega_{*e}$. This mode does not have real frequency at $k_\parallel = 0$ ($\omega(0) = 0$), i.e. it is the convective cell, which can be destabilized by the free energy from the ion temperature gradient.

Due to large ion inertia, the polarization drift slightly modifies the dispersion relation (5.7). The ion polarization drift $\boldsymbol{u}_{pi} = (m_i/e_i B^2)(d\boldsymbol{E}/dt)$ is given by,

$$\tilde{u}_{piy} = -\frac{m_i \omega k_\perp}{e_i B^2}\tilde{\Phi} \tag{5.8}$$

Since $\nabla \cdot \boldsymbol{u}_{pi} \sim \frac{\partial u_{piy}}{\partial y} \neq 0$, ion continuity equation may be modified as,

$$\frac{\partial \tilde{n}_i}{\partial t} + n_i \left(\frac{\partial \tilde{u}_{iz}}{\partial z} + \frac{\partial \tilde{u}_{piy}}{\partial y} \right) + \tilde{v}_{Ex}\frac{dn_i}{dx} = 0 \tag{5.9}$$

$$\text{or,}\ \frac{\tilde{n}_i}{n_i} = \left[\frac{\omega_{*e}}{\omega} - k_\perp^2 \rho_s^2 \right] \frac{e\tilde{\Phi}}{T_e} + \frac{k_\parallel \tilde{u}_{\parallel i}}{\omega} \tag{5.10}$$

Here, $\rho_s^2 = c_{se}^2/\Omega_i^2$. Together with the ion equation of motion (5.4) and the electron response is Boltzmann (5.6) and using ion adiabatic law $\tilde{p}_i = \gamma_i T_i \tilde{n}_i$, the charge neutrality condition $\tilde{n}_e/n_e = \tilde{n}_i/n_i$ gives modified drift wave dispersion relation.

$$\omega^2(1 + k_\perp^2 \rho_s^2) = \omega\omega_{*e} + k_\parallel^2 c_s^2 \tag{5.11}$$

Here $c_s^2 = (\gamma_i T_i + Z_i T_e)/m_i$ is a sound speed. The effect of polarization drift appears as $k_y^2 \rho_s^2$, which is related to the finite Larmor radius effect.

5.1.2 Slab ITG Mode

The drift wave driven by dn/dx is stable but can be destabilized by the dT_i/dx called the Ion Temperature Gradient (ITG) mode even in a uniform magnetic field. We concentrate on ion response since electron response is Boltzmann relation, $\tilde{n}_e/n_e = e\tilde{\Phi}/T_e$. Eliminating $\tilde{u}_{\parallel i}$ from (5.3) and (5.4) with $\sim e^{-i\omega t + ik_y y + ik_\parallel z}$, $\tilde{u}_{Ex} = -ik_y\tilde{\Phi}/B$:

$$\frac{\tilde{n}_i}{n_i} - \frac{\omega_{*e}}{\omega}\frac{e\tilde{\Phi}}{T_e} = \frac{T_i k_\parallel^2}{m_i \omega^2}\left(\frac{\tilde{p}_i}{p_i} + \frac{e_i\tilde{\Phi}}{T_i}\right) \tag{5.12}$$

Instead of the ion adiabatic law $\tilde{p}_i = \gamma_i T_i \tilde{n}_i$, we use ion energy balance equation, $\frac{3}{2}\left(\frac{\partial}{\partial t} + \boldsymbol{u}_i \cdot \nabla\right) p_i + \frac{5}{2} p_i \nabla \cdot \boldsymbol{u}_i = -\nabla \cdot \boldsymbol{q}_i$ where $\boldsymbol{u}_i = \boldsymbol{u}_{\parallel i} + \boldsymbol{u}_{*i} + \boldsymbol{u}_E$ and $\boldsymbol{q}_i = \boldsymbol{q}_{*i} + \boldsymbol{q}_{\parallel i}$. In an uniform B plasma, we have $\nabla \cdot \boldsymbol{u}_E = 0$, $\nabla \cdot \boldsymbol{q}_{*i} + \frac{5}{2} p_i \nabla \cdot \boldsymbol{u}_{*i} = 0$, and $\boldsymbol{u}_{*i} \cdot \nabla p_i = 0$ from (5.37). Neglecting $q_{\parallel i}$, we have $\frac{3}{2}(\frac{\partial p_i}{\partial t} + \boldsymbol{u}_E \cdot \nabla p_i) + \frac{5}{2} p_i \frac{\partial u_{\parallel i}}{\partial z} = 0$. Using (5.3),

$$\frac{\tilde{p}_i}{p_i} + \frac{\omega_{*pi}}{\omega}\frac{e_i\tilde{\Phi}}{T_i} = \frac{5}{3}\left(\frac{\tilde{n}_i}{n_i} - \frac{\omega_{*e}}{\omega}\frac{e\tilde{\Phi}}{T_e}\right) \tag{5.13}$$

where $\omega_{*pi} = k_y (dP_i/dx)/e_i n_i B$. Eliminating $\tilde{p}_i/p_i$ from (5.12) and (5.13), we have:

$$\frac{\tilde{n}_i}{n_i} = \left[\frac{\omega_{*e}}{\omega} + \frac{(1 - \frac{\omega_{*pi}}{\omega}) k_\parallel^2 c_{se}^2}{\omega^2 - k_\parallel^2 c_{si}^2}\right]\frac{e\tilde{\Phi}}{T_e} = \frac{\omega\omega_{*e} + k_\parallel^2 c_{se}^2 \left[1 - \frac{\omega_{*i}}{\omega}(\eta_i - \frac{2}{3})\right]}{\omega^2 - k_\parallel^2 c_{si}^2}\frac{e\tilde{\Phi}}{T_e} \tag{5.14}$$

Here, we used $\omega_{*pi} = \omega_{*i}(1 + \eta_i)$, $\omega_{*i} = -k_y T_i/e_i L_n B = -\omega_{*e}/\tau_e Z_i$, $\eta_i = L_n/L_{Ti}$, $L_{Ti} = -T_i/(dT_i/dx)$, and $\tau_e = T_e/T_i = 1/\tau_i$. Using the Boltzmann relation for the electron $\tilde{n}_e/n_e = e\tilde{\Phi}/T_e$, we have following ion temperature gradient (ITG) mode dispersion relation from the charge neutrality condition $\tilde{n}_i/n_i = \tilde{n}_e/n_e$.

$$\omega(\omega - \omega_{*e}) = \left[1 + (5/3Z_i\tau_e) - (\omega_{*i}/\omega)(\eta_i - 2/3)\right] k_\parallel^2 c_{se}^2 \tag{5.15}$$

This cubic equation is obtained by Horton in 1972 [340] and has one real solution $\omega(k_\parallel)$ corresponds to drift wave and two complex conjugate convective cell/sound wave solution. The latter solution becomes unstable when $|dT_i/dr|$ is sufficiently large. The instability appears at $\omega \ll |\omega_{*e}|, |\omega_{*i}|$. The dispersion relation becomes,

$$\omega^2 = -(k_\parallel^2 c_{se}^2/\tau_e Z_i)\,(\eta_i - 2/3) \tag{5.16}$$

Therefore, electrostatic ITG mode in uniform magnetic field becomes unstable when $\eta_i > \frac{2}{3}$. In the limit of flat density $dn/dx = 0$, Eq. (5.15) leads to,

$$\omega^2 = k_\parallel^2 c_{se}^2\,[1 + 5/3\tau_e Z_i - \omega_{*Ti}/\omega] \tag{5.17}$$

Here, $\omega_{*Ti} = k_y(dT_i/dx)/e_iB$. This cubic equation in ω gives instability (A.108):

$$|\omega_{*Ti}| \geq 2[(1 + 5/3\tau_e Z_i)/3]^{3/2} k_\parallel c_{se} \tag{5.18}$$

5.1.3 Toroidal ITG and ETG Modes

Fluid equation of motion for species a is given by,

$$m_a n_a \frac{d\boldsymbol{u}_a}{dt} = e_a n_a\,(\boldsymbol{E} + \boldsymbol{u}_a \times \boldsymbol{B}) - \nabla p_a + \nabla \cdot \boldsymbol{\pi}_a + \boldsymbol{F}_a, \tag{5.19}$$

where $d/dt = \partial/\partial t + \boldsymbol{u}_a \cdot \nabla$ is the total derivative, $\boldsymbol{\pi}_a$ is stress tensor, $\boldsymbol{F}_a$ is force acting on fluid a including the friction force. The stress tensor $\boldsymbol{\pi}_a$ has parallel and perpendicular components. We use the electrostatic approximation in this section.

Taking the cross product $\boldsymbol{b}\times$ (5.19) with $\boldsymbol{b} = \boldsymbol{B}/B$, we obtain following equation for perpendicular drifts.

$$\boldsymbol{u}_{\perp a} = \boldsymbol{u}_E + \boldsymbol{u}_{*a} + \boldsymbol{u}_{\pi a} + \boldsymbol{u}_{Fa} + \boldsymbol{u}_{pa} \tag{5.20}$$

where, $\boldsymbol{u}_E$, $\boldsymbol{u}_{*a}$, $\boldsymbol{u}_{\pi a}$, $\boldsymbol{u}_{Fa}$, $\boldsymbol{u}_{pa}$ are called $E \times B$ drift, diamagnetic drift, stress tensor drift, force drift, and polarization drift respectively, and are given as follows,

$$\boldsymbol{u}_E = (\boldsymbol{E} \times \boldsymbol{b})/B = (\boldsymbol{b} \times \nabla \Phi)/B \tag{5.21}$$

$$\boldsymbol{u}_{*a} = (\boldsymbol{b} \times \nabla p_a)/e_a n_a B \tag{5.22}$$

$$\boldsymbol{u}_{\pi a} = (\boldsymbol{b} \times \nabla \cdot \boldsymbol{\pi}_a)/e_a n_a B \tag{5.23}$$

$$\boldsymbol{u}_{Fa} = -(\boldsymbol{b} \times \boldsymbol{F}_a)/e_a n_a B \tag{5.24}$$

$$\boldsymbol{u}_{pa} = m_a \boldsymbol{b} \times (d\boldsymbol{u}_a/dt)/e_a B \tag{5.25}$$

Since polarization drift is proportional to mass, electron polarization is negligible ($u_{pe} \sim 0$) and the ion polarization drift (u_{pi}) is finite. But, the ion polarization drift $\boldsymbol{u}_{pi}$ is order down in ω/Ω_i compared with $E \times B$ and diamagnetic drifts noting $\omega/\Omega_i \ll 1$ as seen from Eq. (5.25). Force drift is neglected below for simplicity unless otherwise specified. Continuity equation for species a is given by,

$$\frac{\partial n_a}{\partial t} + \nabla \cdot (n_a \boldsymbol{u}_a) = 0 \tag{5.26}$$

Using the expression for perpendicular flow (5.20) and including parallel flow term, continuity equation for species a is given as follows,

$$\frac{\partial n_a}{\partial t} + \nabla \cdot [n_a(\boldsymbol{u}_E + \boldsymbol{u}_{*a} + \boldsymbol{u}_{\pi a} + \boldsymbol{u}_{pa} + \boldsymbol{u}_{\|a})] = 0 \tag{5.27}$$

where, $\boldsymbol{u}_E$, $\boldsymbol{u}_{*a}$, $\boldsymbol{u}_{\pi a}$, $\boldsymbol{u}_{pa}$, $\boldsymbol{u}_{\|a}$ are $E \times B$ drift, diamagnetic drift, stress tensor drift, polarization drift, and parallel flow, respectively. For electron continuity equation, it is noted that electron polarization drift and stress tensor drift are negligible.

Neglecting stress tensor and heat source terms, the energy balance equation is:

$$\frac{3}{2}\left(\frac{\partial}{\partial t} + \boldsymbol{u}_a \cdot \nabla\right) p_a + \frac{5}{2} p_a \nabla \cdot \boldsymbol{u}_a = -\nabla \cdot \boldsymbol{q}_a \tag{5.28}$$

where $\boldsymbol{q}_a$ is random heat flux.

The divergence of drift appears in many cases, such as above continuity equations. We have following relations, called the curvature relations [815].

$$\nabla \cdot \boldsymbol{u}_E = \frac{e_a}{T_a} \boldsymbol{u}_{da} \cdot \nabla \Phi \tag{5.29}$$

$$\nabla \cdot (n_a \boldsymbol{u}_{*a}) = \frac{1}{T_a} \boldsymbol{u}_{da} \cdot \nabla p_a \tag{5.30}$$

$$\nabla \cdot \boldsymbol{q}_{*a} = \frac{5}{2} n_a (\boldsymbol{u}_{da} - \boldsymbol{u}_{*a}) \cdot \nabla T_a \tag{5.31}$$

$$\text{where, } \boldsymbol{u}_{da} = (T_a/e_a B) \boldsymbol{b} \times \boldsymbol{\kappa}_2 \, , \, \boldsymbol{\kappa}_2 \equiv \boldsymbol{\kappa} + \nabla ln B \tag{5.32}$$

Here, $\boldsymbol{\kappa} \equiv (\boldsymbol{b} \cdot \nabla)\boldsymbol{b}$ is the curvature of the magnetic field defined in (2.6).

Considering $\nabla \cdot (\boldsymbol{b} \times \nabla \Phi) = \nabla \Phi \cdot \nabla \times \boldsymbol{b}$ and $(\boldsymbol{b} \times \nabla \Phi) \cdot \nabla(1/B) = (1/B) \nabla \Phi \cdot (\boldsymbol{b} \times \nabla ln B)$, the divergence of $\boldsymbol{u}_E = -\nabla \Phi \times \boldsymbol{b}/B$ drift is given by,

$$\nabla \cdot \boldsymbol{u}_E = \frac{1}{B} \nabla \cdot (\boldsymbol{b} \times \nabla \Phi) + (\boldsymbol{b} \times \nabla \Phi) \cdot \nabla \frac{1}{B} = \frac{1}{B} \nabla \Phi \cdot (\nabla \times \boldsymbol{b} + \boldsymbol{b} \times \nabla ln B) \tag{5.33}$$

Using Eq. (2.7), we have,

$$\nabla \cdot \boldsymbol{u}_E = -\boldsymbol{\kappa}_2 \cdot \boldsymbol{u}_E + \frac{(\boldsymbol{b} \cdot \nabla \Phi)(\boldsymbol{b} \cdot \nabla \times \boldsymbol{b})}{B} = -\boldsymbol{\kappa}_2 \cdot \boldsymbol{u}_E + \frac{\mu_0 J_\| \boldsymbol{b} \cdot \nabla \Phi}{B^2}, \tag{5.34}$$

where we use $\boldsymbol{b} \cdot \nabla \times \boldsymbol{b} = \boldsymbol{b} \cdot \nabla \times (\boldsymbol{B}/B) = \mu_0 J_\parallel / B$. The last term of RHS is negligible since it is $O(k_\parallel / k_\perp qA)$, where q and $A = R/a$ are safety factor and aspect ratio, respectively. Substituting $\boldsymbol{u}_E = -\nabla\Phi \times \boldsymbol{b}/B$, we arrive at (5.29).

For diamagnetic drift, we have similar procedures to show following formula.

$$\nabla \cdot (n_a \boldsymbol{u}_{*a}) = -\boldsymbol{\kappa}_2 \cdot (n_a \boldsymbol{u}_{*a}) + \frac{\mu_0 J_\parallel}{e_a B^2} \boldsymbol{b} \cdot \nabla p_a \tag{5.35}$$

If we neglect second term of RHS, we arrive at (5.30). The drift $u_{da} \sim -2T_a/e_a BR$ is a toroidal drift velocity of component a. The perpendicular heat flow $\boldsymbol{q}_{\perp a}$ is a diamagnetic thermal flow given by,

$$\boldsymbol{q}_{\perp a} = \frac{5}{2} \frac{p_a}{e_a B} \boldsymbol{b} \times \nabla T_a \equiv \boldsymbol{q}_{*a} \tag{5.36}$$

Similarly, we arrive at (5.31) for the divergence of $\boldsymbol{q}_{*a}$.

In the uniform magnetic field, the toroidal drift disappears and we have:

$$\nabla \cdot \boldsymbol{u}_E = 0\,, \; \nabla \cdot (n_a \boldsymbol{u}_{*a}) = 0\,, \; \nabla \cdot \boldsymbol{q}_{*a} = -\frac{5}{2} n_a \boldsymbol{u}_{*a} \cdot \nabla T_a \tag{5.37}$$

To understand the toroidal ITG dynamics, we start from the perturbed ion continuity equation is:

$$\partial \tilde{n}_i / \partial t + n_i \nabla \cdot \tilde{\boldsymbol{u}}_E + \tilde{\boldsymbol{u}}_E \cdot \nabla n_i + \nabla \cdot \delta (n_i \boldsymbol{u}_{*i}) + n_i \nabla \cdot \tilde{\boldsymbol{u}}_{\parallel i} = 0 \tag{5.38}$$

Here, we neglected stress tensor drift and polarization drift and using the curvature relations (5.29) and (5.30), we have $\nabla \cdot \tilde{\boldsymbol{u}}_E = -ie\omega_{de} \tilde{\Phi}/T_e$, $\nabla \cdot \delta(n_i \boldsymbol{u}_{*i}) = i\omega_{di} \tilde{p}_i / T_i$, where $\omega_{di} = \boldsymbol{u}_{di} \cdot \boldsymbol{k}$. In the strong ballooning limit where the mode is localized in the bad curvature region, we may approximate $\omega_{di} = u_{di} k_\perp$, while $\boldsymbol{u}_{di}$ is in the vertical direction. This approximation is called the strong ballooning limit. Then, we have $\omega_{di} = -2k_\perp T_i / e_i BR$ and $\omega_{de} = 2k_\perp T_e / eBR = -\tau_e Z_i \omega_{di}$. The perturbed ion continuity equation is:

$$\omega \frac{\tilde{n}_i}{n_i} = -\omega_{de} \frac{e\tilde{\Phi}}{T_e} + \omega_{*e} \frac{e\tilde{\Phi}}{T_e} + \omega_{di} \frac{\tilde{p}_i}{p_i} + k_\parallel \tilde{u}_{\parallel i} \tag{5.39}$$

Taking $\boldsymbol{b}\cdot$ (5.19) gives (5.4). Eliminating $\tilde{u}_{\parallel i}$ from (5.4) and (5.39), we have:

$$\frac{\tilde{n}_i}{n_i} - \frac{\omega_{*e} - \omega_{de}}{\omega} \frac{e\tilde{\Phi}}{T_e} = \frac{T_i k_\parallel^2}{m_i \omega^2} \left(\frac{\tilde{p}_i}{p_i} + \frac{e_i \tilde{\Phi}}{T_i} \right) + \frac{\omega_{di}}{\omega} \frac{\tilde{p}_i}{p_i}, \tag{5.40}$$

which agrees with (5.12) in $\omega_{di}, \omega_{de} \to 0$ limit.

Before discussion of the toroidal ITG, we first consider the stable drift wave with uniform temperature plasma and assume the ion adiabatic response, $\tilde{p}_i = \gamma_i T_i \tilde{n}_i$. In this case, we have:

$$\frac{\tilde{n}_i}{n_i} = \frac{\omega(\omega_{*e} - \omega_{de}) + k_\parallel^2 c_{se}^2}{\omega(\omega - \gamma_i \omega_{di}) - k_\parallel^2 c_{si}^2} \frac{e\tilde{\Phi}}{T_e} \tag{5.41}$$

This ion response has a correct asymptotic form (5.5) with $\omega_{di}, \omega_{de} \rightarrow 0$. If we assume Boltzmann relation for electron, $\tilde{n}_e/n_e = e\tilde{\Phi}/T_e$, we have following dispersion relation.

$$\omega[\omega - (\omega_{*e} - \omega_{de} + \gamma_i \omega_{di})] = k_\parallel^2 c_s^2 \tag{5.42}$$

Thus, the toroidal effect simply shifts the electron diamagnetic frequency and the drift wave and convective cell is stable with $dT_i/dr = 0$.

Now, we consider $dT_i/dr \neq 0$ ITG drift wave and we use ion pressure balance equation (5.28) instead of the ion adiabatic law. Substituting the curvature relation (5.31) into (5.28) and retaining only $\boldsymbol{u}_E$, $\boldsymbol{u}_{*i}$ and $\boldsymbol{u}_{\parallel i}$ as flow components and neglecting $\nabla \cdot \boldsymbol{q}_{\parallel i}$, we have following leading order equation (we neglect $5/3 p_i \nabla \cdot (\boldsymbol{u}_{\pi i} + \boldsymbol{u}_{pi})$).

$$\frac{\partial p_i}{\partial t} + \boldsymbol{u}_E \cdot \nabla p_i + \frac{5}{3} p_i \nabla \cdot (\boldsymbol{u}_E + \boldsymbol{u}_{*i} + \boldsymbol{u}_{\parallel i}) + \frac{5}{3} n_i (\boldsymbol{u}_{di} - \boldsymbol{u}_{*i}) \cdot \nabla T_i = 0 \tag{5.43}$$

Here, we used $\boldsymbol{u}_{*i} \cdot \nabla p_i = 0$. Using $\boldsymbol{u}_{*i} \cdot \nabla p_i = 0$, we also have

$$p_i \nabla \cdot \boldsymbol{u}_{*i} - n_i \boldsymbol{u}_{*i} \cdot \nabla T_i = T_i \nabla \cdot (n_i \boldsymbol{u}_{*i}) = \boldsymbol{u}_{di} \cdot \nabla p_i \tag{5.44}$$

since $n_i \boldsymbol{u}_{di} \cdot \nabla T_i = \boldsymbol{u}_{di} \cdot \nabla p_i - T_i \boldsymbol{u}_{di} \cdot \nabla n_i$, we have

$$\frac{\partial p_i}{\partial t} + \boldsymbol{u}_E \cdot \nabla p_i + \frac{5}{3} p_i \nabla \cdot \boldsymbol{u}_{\parallel i} + \frac{5}{3} \boldsymbol{u}_{di} \cdot [e_i n_i \nabla \Phi + 2 \nabla p_i - T_i \nabla n_i] = 0 \tag{5.45}$$

Equation (5.4) is still valid, and the linearized equation for $\tilde{p}_i$ is obtained as,

$$\frac{\tilde{p}_i}{p_i} + \frac{\omega_{di}}{\omega} \frac{e_i \tilde{\Phi}}{T_i} = \frac{k_\parallel^2 c_{si}^2}{\omega^2} \left(\frac{\tilde{p}_i}{p_i} + \frac{e_i \tilde{\Phi}}{T_i} \right) + \frac{5\omega_{di}}{3\omega} \left(\frac{e_i \tilde{\Phi}}{T_i} + 2\frac{\tilde{p}_i}{p_i} - \frac{\tilde{n}_i}{n_i} \right) \tag{5.46}$$

Eliminating $\tilde{p}_i/p_i$ from (5.40) and (5.46), we obtain:

$$\frac{\tilde{n}_i}{n_i} = \frac{\omega(\omega_{*e} - \omega_{de}) + (\eta_i - \frac{7}{3} + \frac{10}{3}\epsilon_n)\omega_{di}\omega_{*e} + k_\parallel^2 c_{se}^2 [1 - \frac{\omega_{*i}}{\omega}(\eta_i + \frac{10}{3}\epsilon_n - \frac{2}{3})]}{\omega^2 - \frac{10}{3}\omega\omega_{di} + \frac{5}{3}\omega_{di}^2 - k_\parallel^2 c_{si}^2 (1 - \frac{\omega_{di}}{\omega})} \frac{e\tilde{\Phi}}{T_e} \tag{5.47}$$

Here, we used $\epsilon_n \equiv \omega_{di}/2\omega_{*i} = L_n/R$ and $\omega_{*e} = -\tau_e Z_i \omega_{*i}$. This ion density response has following correct asymptotic forms.

$$\frac{\tilde{n}_i}{n_i} = \begin{cases} -\frac{e_i\tilde{\Phi}}{T_i} \ (\omega_{di}, \omega_{de} \to \infty) \\ \frac{\omega\omega_{*e} + k_\parallel^2 c_{se}^2 [1 - \frac{\omega_{*i}}{\omega}(\eta_i - \frac{2}{3})]}{\omega^2 - k_\parallel^2 c_{si}^2} \frac{e\tilde{\Phi}}{T_e} \ (\omega_{di}, \omega_{de} \to 0) \end{cases} \tag{5.48}$$

Assuming $\tilde{n}_e/n_e = e\tilde{\Phi}/T_e$, we obtain following toroidal ITG dispersion relation.

$$\omega^2 - c_1\omega_{*e}\omega + c_2\omega_{*e}^2 = \left(c_3 - \frac{\omega_{*i}}{\omega}c_4\right)k_\parallel^2 c_{se}^2 \tag{5.49}$$

$$c_1 = 1 - 2\epsilon_n\left(1 + \frac{10}{3\tau_e Z_i}\right), c_2 = \frac{2\epsilon_n}{\tau_e Z_i}\left(\eta_i - \frac{7}{3} + \frac{10}{3}\epsilon_n\left(1 + \frac{1}{\tau_e Z_i}\right)\right) \tag{5.50}$$

$$c_3 = 1 + \frac{5}{3\tau_e Z_i}, c_4 = \eta_i - \frac{2}{3} + \frac{10}{3}\epsilon_n + \frac{20}{3\tau_e^2 Z_i^2}\epsilon_n^2 \tag{5.51}$$

This dispersion relation agrees with (5.15) asymptotically $\epsilon_n \to 0$. At finite ϵ_n, the toroidal ITG mode can be unstable even at $k_\parallel = 0$, while the ITG mode in the uniform magnetic field can be unstable only at finite $k_\parallel$. The instability threshold condition for the toroidal ITG is given using (A.108) at $k_\parallel = 0$ as:

$$\eta_{ic} = \frac{3}{2} - \frac{\tau_e Z_i}{2} + \frac{\tau_e Z_i}{8\epsilon_n} + \epsilon_n\left(\frac{\tau_e Z_i}{2} + \frac{20}{9\tau_e Z_i}\right) \tag{5.52}$$

The electron temperature gradient can also drive an instability called the electron temperature gradient (ETG) mode similar to ITG mode. The mode is characterized by its short wavelength $\rho_e \ll \lambda \ll \rho_i$. The mode frequency ω satisfies $\Omega_i < \omega \ll \Omega_e$ and ions are unmagnetized. The ion may reach thermal equilibrium by the perpendicular motion to satisfy Boltzmann relation, $\tilde{n}_i/n_i = -e_i\tilde{\Phi}/T_i$.

5.1.4 Trapped Electron Mode/ITG

Until now, we consider the simple Boltzmann response for electrons, $\tilde{n}_e/n_e = e\tilde{\Phi}/T_e$. In tokamak, we have circulating and trapped electrons, which behave differently to the electrostatic potential perturbation. The electron density perturbation may be expressed as a summation of trapped electron response ($\tilde{n}_{et}$) and circulating electron response ($\tilde{n}_{ec}$).

$$\frac{\tilde{n}_e}{n_e} = f_t\frac{\tilde{n}_{et}}{n_{et}} + f_c\frac{\tilde{n}_{ec}}{n_{ec}} \tag{5.53}$$

Here, f_t is trapped particle fraction of electron and $f_c = 1 - f_t$ is circulating fraction of electron. Since trapped electron can not move parallel to magnetic field, we can neglect parallel motion ($k_\parallel$ terms). Then, trapped electron response $\tilde{n}_{et}$ may be given by (5.47) by exchanging ion to electron.

$$\frac{\tilde{n}_{et}}{n_{et}} = \frac{\omega(\omega_{*e} - \omega_{de}) + (\eta_e - \frac{7}{3} - \frac{10}{3}\epsilon_n)\omega_{de}\omega_{*e}}{\omega^2 - \frac{10}{3}\omega\omega_{de} + \frac{5}{3}\omega_{de}^2} \frac{e\tilde{\Phi}}{T_e} \tag{5.54}$$

Considering the circulating electron may follow Boltzmann relation, we have,

$$\frac{\tilde{n}_e}{n_e} = \left[\frac{\omega(\omega_{*e} - \omega_{de}) + (\eta_e - \frac{7}{3} - \frac{10}{3}\epsilon_n)\omega_{de}\omega_{*e}}{\omega^2 - \frac{10}{3}\omega\omega_{de} + \frac{5}{3}\omega_{de}^2} f_t + f_c \right] \frac{e\tilde{\Phi}}{T_e} \tag{5.55}$$

Using Eq. (5.47), the normalized ion density perturbation neglecting the parallel sound wave contribution is given by,

$$\frac{\tilde{n}_i}{n_i} = \frac{\omega(\omega_{*e} - \omega_{de}) + (\eta_i - \frac{7}{3} + \frac{10}{3}\epsilon_n)\omega_{di}\omega_{*e}}{\omega^2 - \frac{10}{3}\omega\omega_{di} + \frac{5}{3}\omega_{di}^2} \frac{e\tilde{\Phi}}{T_e} \tag{5.56}$$

The charge neutrality condition ($\tilde{n}_e/n_e = \tilde{n}_i/n_i$) gives following dispersion relation for ITG and trapped electron mode (TEM).

$$\frac{\omega(\omega_{*e} - \omega_{de}) + (\eta_i - \frac{7}{3} + \frac{10}{3}\epsilon_n)\omega_{di}\omega_{*e}}{\omega^2 - \frac{10}{3}\omega\omega_{di} + \frac{5}{3}\omega_{di}^2} = \frac{\omega(\omega_{*e} - \omega_{de}) + (\eta_e - \frac{7}{3} - \frac{10}{3}\epsilon_n)\omega_{de}\omega_{*e}}{\omega^2 - \frac{10}{3}\omega\omega_{de} + \frac{5}{3}\omega_{de}^2} f_t + f_c \tag{5.57}$$

Weiland [814, 815] derived the dispersion relation including the finite Larmor radius correction as follows,

$$\frac{\omega(\omega_{*e} - \omega_{de}) + (\eta_i - \frac{7}{3} + \frac{10}{3}\epsilon_n)\omega_{di}\omega_{*e} + FL}{\omega^2 - \frac{10}{3}\omega\omega_{di} + \frac{5}{3}\omega_{di}^2} = \frac{\omega(\omega_{*e} - \omega_{de}) + (\eta_e - \frac{7}{3} - \frac{10}{3}\epsilon_n)\omega_{de}\omega_{*e}}{\omega^2 - \frac{10}{3}\omega\omega_{de} + \frac{5}{3}\omega_{de}^2} f_t + f_c$$

$$FL = k_\perp^2 \rho_i^2 [\omega - \omega_{*i}(1 + \eta_i)] \left(\frac{\omega}{\omega_{*e}} + \frac{10\epsilon_n}{3\tau_e Z_i} \right) \tag{5.58}$$

This dispersion relation is useful to see qualitative picture of ITG/TEM mode diagram. Figure 5.2 shows typical result given by Garbet [243].

In the peaked density regime, the Trapped Electron Mode (TEM) can be unstable. This TEM is first predicted by Kadomtsev as candidate instability to explain anomalous electron transport [408]. And TEM can give a scaling of energy

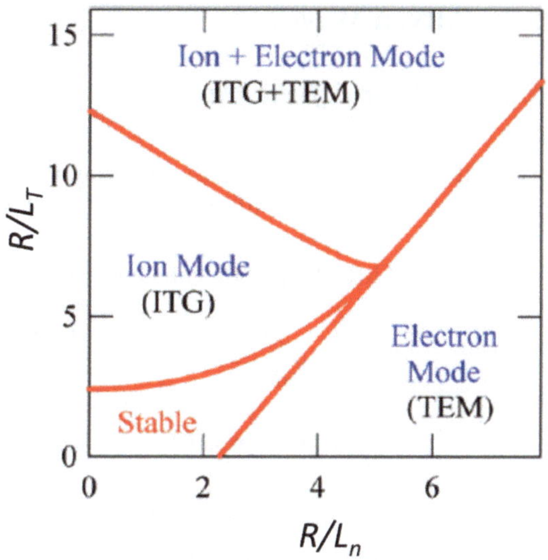

Fig. 5.2 Stability diagram in $(R/L_n, R/L_T)$ for the ITG/TEM modes under $T_e(r) = T_i(r)$, trapped electron fraction of 0.3, collisionless limit ($\nu_* = 0$) calculated using Weiland fluid model. Reproduced with permission from Garbet [243]. Copyright IOP Publishing

confinement similar to Neo-Alcator scaling ($\tau_E \sim R^2 a$). The collisionless TEM has wave length closer to ITG mode $k_\perp \rho_i \sim 0.3$.

In the flat density regime ($R/L_n < 2$), ITG mode can be unstable if there is sufficient free energy from ion temperature gradient dT_i/dr, where inward particle flux is expected. For sufficiently high R/L_T, ITG and TEM can be unstable where outward particle flux is expected. We summarize drift wave fluid quantities as follows,

$$\omega_{*a} = \frac{k_\perp T_a}{e_a B n_a}\frac{dn_a}{dr}, \quad \omega_{*Ta} = \frac{k_\perp}{e_a B}\frac{dT_a}{dr}, \quad \omega_{*pa} = \frac{k_\perp}{e_a B n_a}\frac{dp_a}{dr}, \quad \omega_{da} = -\frac{2k_\perp T_a}{e_a BR} \tag{5.59}$$

$$c_{se}^2 = \frac{Z_i T_e}{m_i}, \quad c_{si}^2 = \frac{\gamma_i T_i}{m_i}, \quad u_{*a} = \frac{T_a}{e_a B n_a}\frac{dn_a}{dr}, \quad u_{Da} = -\frac{2T_a}{e_a BR} \tag{5.60}$$

$$\gamma_i = \frac{5}{3}, \quad c_s^2 = \frac{\gamma_i T_i + Z_i T_i}{m_i}, \quad \tau_e = \frac{T_e}{T_i}, \quad \tau_i = \frac{T_i}{T_e} \tag{5.61}$$

$$L_{na} = -\frac{r}{n_a}\frac{dn_a}{dr}, \quad L_{Ta} = -\frac{r}{T_a}\frac{dT_a}{dr}, \quad \epsilon_n = \frac{L_n}{R}, \quad \eta_a = \frac{L_{na}}{L_{Ta}} \tag{5.62}$$

5.2 Alfven Waves

5.2.1 *Shear Alfven Wave*

The Alfven wave is one of most important waves in plasma physics originated from Nobel laureate Hannes Alfven (1908–1995) [10]. We show the dispersion relation of the shear Alfven wave.

Linearized MHD equation can be given as,

$$\rho\frac{\partial \boldsymbol{v}}{\partial t} = -\nabla\left(p_1 + \frac{\boldsymbol{B}\cdot\boldsymbol{B}_1}{\mu_0}\right) + \frac{1}{\mu_0}\boldsymbol{B}\cdot\nabla\boldsymbol{B}_1 \tag{5.63}$$

$$\frac{\partial \boldsymbol{B}_1}{\partial t} = \nabla\times(\boldsymbol{v}\times\boldsymbol{B}) = \boldsymbol{B}\cdot\nabla\boldsymbol{v} - \boldsymbol{v}\cdot\nabla\boldsymbol{B} - \boldsymbol{B}\nabla\cdot\boldsymbol{v} \tag{5.64}$$

If we consider the divergence-free ($\nabla\cdot\boldsymbol{v} = 0$) shear flow ($\boldsymbol{v}\perp\boldsymbol{B}$) propagating parallel to the magnetic field ($\nabla \parallel \boldsymbol{B}$), we have

$$\rho\frac{\partial \boldsymbol{v}}{\partial t} = \frac{1}{\mu_0}\boldsymbol{B}\cdot\nabla\boldsymbol{B}_1\,,\ \frac{\partial \boldsymbol{B}_1}{\partial t} = \boldsymbol{B}\cdot\nabla\boldsymbol{v} \tag{5.65}$$

If we let z a coordinate along $\boldsymbol{B}$, we have

$$\frac{\partial^2 \boldsymbol{v}}{\partial t^2} = \frac{B^2}{\mu_0\rho}\frac{\partial^2 \boldsymbol{v}}{\partial z^2} \tag{5.66}$$

This is a wave equation with a phase velocity $v_A = B/\sqrt{\mu_0\rho}$, which is called Alfven wave. This wave is called the shear Alfven wave. If we assume $e^{-i\omega t+ik_\parallel z}$, we have following dispersion relation of the shear Alfven wave.

$$\omega^2 = k_\parallel^2 v_A^2 \tag{5.67}$$

Another way to derive shear Alfven wave dispersion relation uses

$$\nabla\cdot\boldsymbol{j} = ik_\parallel j_\parallel + ik_\perp j_\perp = 0 \tag{5.68}$$

Since $E_\parallel = 0$ for ideal plasma, we have $A_\parallel = (k_\parallel/\omega)\Phi$. Using the Ampere's law $k_\perp^2 A_\parallel = \mu_0 j_\parallel$, we have

$$\mu_0 j_\parallel = (k_\perp^2 k_\parallel/\omega)\Phi \tag{5.69}$$

Using the ion polarization current ($j_p = e_i n_i(\partial\boldsymbol{E}/\partial t)m_i/e_i B^2$) for the perpendicular current, we have

$$j_\perp = j_p = -m_i n_i \omega k_\perp \Phi/B^2 \tag{5.70}$$

Fig. 5.3 From left, H. Alfven, A. Hasegawa, Liu Chen. AH and LC are recipients of APS J.C. Maxwell Prize and EPS H. Alfven Prize due to their outstanding contribution to the plasma physics

The $\nabla \cdot \boldsymbol{j} = 0$ gives the shear Alfven wave dispersion relation $\omega^2 = k_\parallel^2 v_A^2$.

It is important to note that the ion polarization drift plays an essential role in the dynamics and the dispersion relation of the shear Alfven wave.

5.2.2 *Kinetic Alfven Wave*

We now introduce an important kinetic MHD wave called kinetic Alfven wave originated from A. Hasegawa and L. Chen [293, 294] (Fig. 5.3).

In the low beta plasma such as tokamak plasma, we consider the bending of the magnetic field as the electromagnetic effect and can neglect the compressional Alfven wave (magneto sonic wave), $\tilde{B}_\parallel = (\nabla \times \tilde{\boldsymbol{A}})_\parallel = \partial \tilde{A}_y/\partial x - \partial \tilde{A}_x/\partial y = 0$. This means $\tilde{\boldsymbol{A}}_\perp$ has a scalar potential, $\tilde{\boldsymbol{A}}_\perp = \nabla_\perp \tilde{G}$. Then, the electric field $\tilde{\boldsymbol{E}}$ is given as,

$$\tilde{\boldsymbol{E}} = -\nabla \tilde{\Phi} - \frac{\partial \tilde{\boldsymbol{A}}}{\partial t} = -\nabla_\perp \tilde{\phi} - \frac{\partial \tilde{\psi}}{\partial z}\boldsymbol{e}_z = -\nabla \tilde{\phi} + \frac{\partial(\tilde{\phi} - \tilde{\psi})}{\partial z}\boldsymbol{e}_z \tag{5.71}$$

$$\tilde{\phi} = \tilde{\Phi} + \partial \tilde{G}/\partial t\,,\ \tilde{\psi} = \tilde{\Phi} + \int (\partial \tilde{A}_\parallel/\partial t) dz \tag{5.72}$$

According to Kadomtsev [407, p. 82], $\tilde{\phi}$ and $\tilde{\psi}$ are called perpendicular and parallel potentials, respectively. The Maxwell's induction law $\partial \boldsymbol{B}/\partial t = -\nabla \times \boldsymbol{E}$ can be expressed in terms of these scalar potentials as,

$$\frac{\partial \tilde{\boldsymbol{B}}}{\partial t} = (\boldsymbol{e}_z \times \nabla_\perp)\frac{\partial}{\partial z}(\tilde{\phi} - \tilde{\psi}) \tag{5.73}$$

Assuming $e^{-i\omega t + ik_\parallel z + ik_\perp y}$, $k_\perp = k_y$ and $k_x = 0$, we have $\omega \tilde{B}_x = -ik_y k_\parallel (\tilde{\phi} - \tilde{\psi})$.

The relation between $\tilde{\phi}$ and $\tilde{\psi}$ can be obtained by $\nabla \cdot \tilde{\boldsymbol{j}} = 0$. We can obtain expression for $\tilde{j}_\parallel$ by using $\omega \tilde{B}_x = -ik_y k_\parallel (\tilde{\phi} - \tilde{\psi})$ and $\mu_0 \tilde{j}_\parallel = -ik_y \tilde{B}_x$ ($\mu_0 \boldsymbol{j} = \nabla \times \boldsymbol{B}$)

and that for the perpendicular current $\tilde{j}_\perp$ by the polarization current (5.70) with $\Phi \rightarrow \tilde{\phi}$.

$$\tilde{j}_\parallel = -\frac{k_\parallel k_\perp^2}{\mu_0 \omega}(\tilde{\phi} - \tilde{\psi}) \, , \tilde{j}_\perp = \tilde{j}_p = \frac{m_i n_i \omega k_\perp}{B^2} \tilde{\phi} \tag{5.74}$$

Substituting these expressions into $k_\parallel \tilde{j}_\parallel + k_\perp \tilde{j}_\perp = 0$, we obtain:

$$(k_\parallel^2 v_A^2 - \omega^2)\tilde{\phi} = k_\parallel^2 v_A^2 \tilde{\psi} \tag{5.75}$$

We consider the electron parallel response in an uniform plasma. Obviously, the electron should follow the Boltzmann response against parallel potential $\tilde{\psi}$.

$$\frac{\tilde{n}_e}{n_e} = \frac{e\tilde{\psi}}{T_e} \tag{5.76}$$

Similar to (5.10), the ion continuity equation including polarization drift is expressed in a uniform plasma using the perpendicular potential $\tilde{\phi}$ as,

$$\frac{\tilde{n}_i}{n_i} = -k_\perp^2 \rho_s^2 \frac{e\tilde{\phi}}{T_e} + \frac{k_\parallel \tilde{u}_{\parallel i}}{\omega}, \tag{5.77}$$

where $\rho_s = c_{se}/\Omega_i$.

Similar to (5.4), the parallel ion flow is expressed using parallel potential $\tilde{\psi}$ as,

$$\tilde{u}_{\parallel i} = \frac{k_\parallel}{\omega} \frac{\tilde{p}_i + e_i n_i \tilde{\psi}}{m_i n_i} \tag{5.78}$$

Using the adiabatic response $\tilde{p}_i = \gamma_i T_i \tilde{n}_i$, we have

$$\frac{\tilde{n}_i}{n_i} = \frac{-k_\perp^2 \rho_s^2 \omega^2}{\omega^2 - k_\parallel^2 c_{si}^2} \frac{e\tilde{\phi}}{T_e} + \frac{k_\parallel^2 c_{se}^2}{\omega^2 - k_\parallel^2 c_{si}^2} \frac{e\tilde{\psi}}{T_e} \tag{5.79}$$

Using (5.75), we have:

$$\frac{\tilde{n}_i}{n_i} = \frac{-k_\parallel^2 v_A^2 k_\perp^2 \rho_s^2 \omega^2 + k_\parallel^2 c_{se}^2 (k_\parallel^2 v_A^2 - \omega^2)}{(\omega^2 - k_\parallel^2 c_{si}^2)(k_\parallel^2 v_A^2 - \omega^2)} \frac{e\tilde{\psi}}{T_e} \tag{5.80}$$

The charge neutrality condition $\tilde{n}_e/n_e = \tilde{n}_i/n_i$ gives following dispersion relation.

$$(\omega^2 - k_\parallel^2 c_s^2)(k_\parallel^2 v_A^2 - \omega^2) = -k_\parallel^2 v_A^2 k_\perp^2 \rho_s^2 \omega^2, \tag{5.81}$$

where $c_s^2 = c_{se}^2 + c_{si}^2$ is the sound speed.

This dispersion relation shows that the shear Alfven wave can couple to the slow ion sound wave due to finite Larmor radius effect. For $\beta = c_s^2/v_A^2 \ll 1$, we find the slow ion sound wave dispersion relation as $\omega^2 = k_\parallel^2 c_s^2/(1 + k_\perp^2 \rho_s^2)$. On the other hand, for the Shear Alfven wave, we find

$$\omega^2 = k_\parallel^2 v_A^2 (1 + k_\perp^2 \rho_s^2) \tag{5.82}$$

This finite Larmor radius correction to the shear Alfven wave plays an essential role near the shear Alfven resonance and is called the Kinetic Alfven Wave (KAW).

A. Hasegawa and L. Chen [294] derived the dispersion relation of the kinetic Alfven wave from kinetic equation as $\omega^2 = k_\parallel^2 v_A^2 [k_\perp^2 \rho_i^2/(1 - I_0(k_\perp^2 \rho_i^2) e^{-k_\perp^2 \rho_i^2}) + k_\perp^2 \rho_i^2 \tau_e]$. This dispersion relation is reduced to $\omega^2 = k_\parallel^2 v_A^2 [1 + k_\perp^2 \rho_i^2 (T_e/T_i + 3/4)]$ when $k_\perp^2 \rho_i^2 \ll 1$ and to $\omega^2 = k_\parallel^2 v_A^2 k_\perp^2 \rho_i^2 (1 + \tau_e)$ when $k_\perp^2 \rho_i^2 \gg 1$.

5.2.3 *Drift Alfven Wave*

The drift wave introduced in Sect. 5.1.1 can be coupled to the shear Alfven wave in medium beta ($m_e/m_i \ll \beta \ll 1$) and to the compressional Alfven wave at high beta ($\beta \sim 1$). Since the parallel phase velocity of the shear Alfven wave $\omega/k_\parallel = v_A = B/\sqrt{\mu_0 m_i n_i} = v_{Te}(m_e/m_i\beta_e)^{1/2}$ in hydrogen plasmas, parallel phase velocity becomes comparable to the electron thermal speed at $\beta_e \sim m_e/m_i$. When $\beta_e \sim (m_e/m_i)^{1/2}$, the phase velocity of the shear Alfven wave satisfies $v_{Ti} \ll \omega/k_\parallel \ll v_{Te}$, where the drift wave becomes important. Tokamak operates at medium beta and the coupling to the shear Alfven wave is important.

We consider the same configuration in Fig. 5.1a to consider the coupling of the density gradient drift wave to the shear Alfven wave (i.e. $dT_i/dr = 0$).

At first, we consider the relation between $\tilde{\psi}$ and $\tilde{\phi}$ to include diamagnetic current. The fluid equation of motion is given by,

$$m_i n_i \frac{d\boldsymbol{u}}{dt} = -\nabla p + \boldsymbol{j} \times \boldsymbol{B} \tag{5.83}$$

This gives the perpendicular current as a combination of the diamagnetic and polarization current for uniform magnetic field ($\boldsymbol{b} = \boldsymbol{e}_z$).

$$\boldsymbol{j}_\perp = \boldsymbol{j}_* + \boldsymbol{j}_p, \text{ where } \boldsymbol{j}_* = \frac{\boldsymbol{e}_z}{B} \times \nabla p \,, \boldsymbol{j}_p = \frac{m_i n_i}{B} \boldsymbol{e}_z \times \frac{d\boldsymbol{u}_\perp}{dt} \tag{5.84}$$

The perturbed parallel current $\tilde{j}_\parallel$ is given by (5.74). Using the vector formula (A.11), we find $\nabla \cdot \boldsymbol{j}_* = 0$. Using $d\boldsymbol{u}_\perp/dt \sim \partial \boldsymbol{u}_\perp/\partial t + (\boldsymbol{u}_{*i} \cdot \nabla)\boldsymbol{u}_\perp$ and $\omega_{*i} = k_\perp u_{*i}$, the polarization current is given by,

$$\boldsymbol{j}_p = -\frac{i(\omega - \omega_{*i})}{\mu_0 v_A^2} B \boldsymbol{e}_z \times \boldsymbol{u}_\perp = \frac{(\omega - \omega_{*i}) k_\perp}{\mu_0 v_A^2} \tilde{\phi} \boldsymbol{e}_x \tag{5.85}$$

Substituting these expressions into $\nabla \cdot \boldsymbol{j} = 0$ $(k_\parallel \tilde{j}_\parallel + k_\perp \tilde{j}_\perp = 0)$, we obtain:

$$(k_\parallel^2 v_A^2 - \omega(\omega - \omega_{*i}))\tilde{\phi} = k_\parallel^2 v_A^2 \tilde{\psi} \tag{5.86}$$

The isothermal electron parallel equation of motion at low frequency $\omega \ll \Omega_i$ is:

$$0 = -en_e \tilde{E}_\parallel - ik_\parallel T_e \tilde{n}_e + en_e u_{*e} \tilde{B}_x \tag{5.87}$$

Substituting $E_\parallel = -ik_\parallel \tilde{\psi}$ and $\tilde{B}_x = (ik_\parallel k_\perp/\omega)(\tilde{\phi} - \tilde{\psi})$, we have

$$\frac{\tilde{n}_e}{n_e} = \frac{e}{T_e}\left[\tilde{\psi} + \frac{\omega_{*e}}{\omega}(\tilde{\phi} - \tilde{\psi})\right] \tag{5.88}$$

Similar to (5.10), the ion continuity equation including polarization drift is expressed using the perpendicular potential $\tilde{\phi}$.

$$\frac{\tilde{n}_i}{n_i} = \left[\frac{\omega_{*e}}{\omega} - k_\perp^2 \rho_s^2\right]\frac{e\tilde{\phi}}{T_e} + \frac{k_\parallel \tilde{u}_{\parallel i}}{\omega} \tag{5.89}$$

From the ion parallel equation of motion, $-i\omega m_i n_i \tilde{u}_i = e_i n_i E_\parallel - ik_\parallel \tilde{p}_i - e_i n_i u_{*i} \tilde{B}_x$:

$$\tilde{u}_{\parallel i} = \frac{k_\parallel}{\omega}\frac{\tilde{p}_i + e_i n_i \tilde{\psi} + e_i n_i \frac{\omega_{*i}}{\omega}(\tilde{\phi} - \tilde{\psi})}{m_i n_i} \tag{5.90}$$

Substituting (5.90) into (5.89) and assuming adiabatic response $\tilde{p}_i = \gamma_i T_i \tilde{n}_i$:

$$(\omega^2 - k_\parallel^2 c_{si}^2)\frac{\tilde{n}_i}{n_i} = \left(\omega\omega_{*e} - k_\perp^2 \rho_s^2 \omega^2 + k_\parallel^2 c_{se}^2 \frac{\omega_{*i}}{\omega}\right)\frac{e\tilde{\phi}}{T_e} + k_\parallel^2 c_{se}^2 (1 - \frac{\omega_{*i}}{\omega})\frac{e\tilde{\psi}}{T_e} \tag{5.91}$$

Using the charge neutrality condition $\tilde{n}_e/n_e = \tilde{n}_i/n_i$ and (5.88), we have

$$(\omega^2 - k_\parallel^2 c_{si}^2)\left[\tilde{\psi} + \frac{\omega_{*e}}{\omega}(\tilde{\phi} - \tilde{\psi})\right] = \left(\omega\omega_{*e} - k_\perp^2 \rho_s^2 \omega^2 + k_\parallel^2 c_{se}^2 \frac{\omega_{*i}}{\omega}\right)\tilde{\phi} + k_\parallel^2 c_{se}^2 (1 - \frac{\omega_{*i}}{\omega})\tilde{\psi}$$

Using $c_{si}^2 \omega_{*e} = -c_{se}^2 \omega_{*i}$, we have

$$[\omega(\omega - \omega_{*e}) - k_\parallel^2 c_s^2]\tilde{\psi} = -k_\perp^2 \rho_s^2 \omega^2 \tilde{\phi} \tag{5.92}$$

$$(k_\parallel^2 v_A^2 - \omega(\omega - \omega_{*i}))\tilde{\phi} = k_\parallel^2 v_A^2 \tilde{\psi}, \text{: (5.86)}$$

where $c_s^2 = c_{se}^2 + c_{si}^2$. Combining Eqs. (5.86) and (5.92), we have following dispersion relation of the drift Alfven wave.

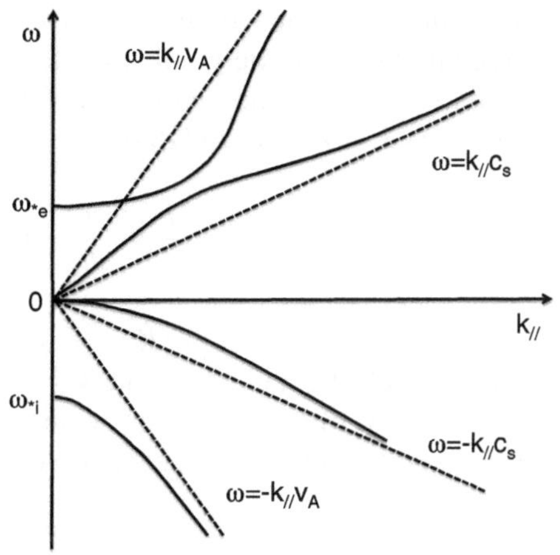

Fig. 5.4 The dispersion relation of drift Alfven wave

$$\left[\omega(\omega-\omega_{*e})-k_\parallel^2c_s^2\right]\left[\omega(\omega-\omega_{*i})-k_\parallel^2v_A^2\right]=k_\perp^2\rho_s^2k_\parallel^2v_A^2\omega^2 \tag{5.93}$$

More detailed kinetic calculation of the drift Alfven wave gives $\omega^2 \to \omega(\omega-\omega_{*i})$ in the right hand side. This electromagnetic drift wave has four branches. Shape of this dispersion relation is given in Fig. 5.4 for a case of $v_A > c_s$. The parallel phase speed $v_{\parallel ph} = \omega/k_\parallel$ may reach Alfven speed v_A and the condition $v_{\parallel ph} < v_{the}$ implies $v_A < v_{the}$. The condition $v_A < v_{the}$ is equivalent to $\beta > m_e/m_i$.

5.3 Gyro Kinetic Theory of Drift Waves

5.3.1 Classical Gyrokinetic Theory

Classical gyrokinetic theory by the direct averaging of Vlasov-Maxwell equation over the gyro motion is developed by Rutherford-Frieman [642], Taylor-Hastie [755], Catto [101], Antonsen-Lane [22], and by Frieman-Chen [213].

5.3.1.1 Linear Gyrokinetic Theory

The basic equation is Vlasov-FP equation (4.1). Using smallness parameter $\epsilon = \rho_i/a \ll 1$, we use the gyro kinetic ordering: $\omega/\Omega_i \sim \epsilon$, $k_\parallel\rho_i \sim \epsilon$, $k_\perp\rho_i \sim \epsilon^0$. We derive electrostatic linearized Gyrokinetic equation neglecting the collision term. Expanding $f_a = f_{aM} + f_{a1}$, we have:

$$\frac{df_{a1}}{dt} \equiv \left[\frac{\partial}{\partial t} + \boldsymbol{v}\cdot\frac{\partial}{\partial \boldsymbol{x}} + \frac{e_a}{m_a}\boldsymbol{v}\times\boldsymbol{B}\cdot\frac{\partial}{\partial \boldsymbol{v}}\right]f_{a1} = \frac{e_a}{m_a}\nabla\tilde{\Phi}\cdot\frac{\partial f_{aM}}{\partial \boldsymbol{v}} \tag{5.94}$$

Here d/dt is the total time derivative along the unperturbed trajectories of the charged particles. In the guiding center coordinates $(\boldsymbol{r}, K, \mu, \varphi)$ where the particle position $\boldsymbol{x}$ is given by the summation of the guiding center position $\boldsymbol{r}$ and the gyration vector $\boldsymbol{\rho}_a$ as $\boldsymbol{x} = \boldsymbol{r} + \boldsymbol{\rho}_a$ and d/dt is given by:

$$\frac{d}{dt} = \frac{\partial}{\partial t} + \boldsymbol{v}_g \cdot \frac{\partial}{\partial \boldsymbol{r}} - \Omega_a \frac{\partial}{\partial \varphi} \tag{5.95}$$

Here, we used $\dot{\mu} = 0, \dot{K} = 0$ for the unperturbed orbit assuming zero equilibrium electrostatic potential $\Phi_0 = 0$. Transformation formula from particle phase space $(\boldsymbol{x}, \boldsymbol{v})$ to the guiding center coordinates $(\boldsymbol{r}, K, \mu, \varphi)$ is given in Column 5-1. Substituting (5.111)–(5.113) into (5.94), we have

$$\frac{df_{a1}}{dt} = e_a \frac{\partial \tilde{\Phi}}{\partial \boldsymbol{r}} \cdot \left[\boldsymbol{v} \frac{\partial}{\partial K} + \frac{\boldsymbol{b}}{\Omega_a} \times \frac{\partial}{\partial \boldsymbol{r}} \right] f_{aM}, \tag{5.96}$$

where the zero order distribution function f_{aM} is a constant of motion and is given by the canonical Maxwellian satisfying $\partial f_{aM}/\partial \varphi = \partial f_{aM}/\partial \mu = 0$.

$$f_{aM} = \frac{n_a(\psi_c)}{(2\pi T_a(\psi_c)/m_a)^{3/2}} e^{-K/T_a(\psi_c)}, \tag{5.97}$$

where $\psi_c = \psi + m_a R v_\zeta / e_a$ is the normalized canonical angular momentum.

Since $\tilde{\Phi}$ is independent of $\boldsymbol{v}$, we have $\partial \tilde{\Phi}/\partial \boldsymbol{v} = 0$. Using (5.113), we have

$$\left[\boldsymbol{v} \frac{\partial}{\partial K} + \boldsymbol{v}_\perp \frac{\partial}{B \partial \mu} + \frac{\boldsymbol{b} \times \boldsymbol{v}_\perp}{v_\perp^2} \frac{\partial}{\partial \varphi} + \frac{\boldsymbol{b}}{\Omega_a} \times \frac{\partial}{\partial \boldsymbol{r}} \right] \tilde{\Phi} = 0 \tag{5.98}$$

Taking $\boldsymbol{b} \times \boldsymbol{v}_\perp / v_\perp$ component of Eq. (5.98) and using the vector formula $(\boldsymbol{a} \times \boldsymbol{b}) \cdot (\boldsymbol{c} \times \boldsymbol{d}) = (\boldsymbol{a} \cdot \boldsymbol{c})(\boldsymbol{b} \cdot \boldsymbol{d}) - (\boldsymbol{a} \cdot \boldsymbol{d})(\boldsymbol{b} \cdot \boldsymbol{c})$ (A.3), we have

$$\boldsymbol{v}_\perp \cdot \frac{\partial \tilde{\Phi}}{\partial \boldsymbol{r}} = -\Omega_a \frac{\partial \tilde{\Phi}}{\partial \varphi}$$

Substituting this expression into (5.96), we have:

$$\frac{df_{a1}}{dt} = e_a \left[v_\parallel \boldsymbol{b} \cdot \frac{\partial \tilde{\Phi}}{\partial \boldsymbol{r}} \frac{\partial}{\partial K} - \Omega_a \frac{\partial \tilde{\Phi}}{\partial \varphi} \frac{\partial}{\partial K} + \nabla \tilde{\Phi} \cdot \frac{\boldsymbol{b}}{\Omega_a} \times \frac{\partial}{\partial \boldsymbol{r}} \right] f_{aM} \tag{5.99}$$

Here, we used $\partial f_{aM}/\partial \mu = 0$ for the isotropic distribution function (5.97). We divide f_{a1} into adiabatic and non-adiabatic parts as follows,

$$f_{a1} = e_a \tilde{\Phi} \frac{\partial f_{aM}}{\partial K} + g_a \tag{5.100}$$

Total time derivative of the adiabatic part $e_a \tilde{\Phi} \partial f_{aM}/\partial K$ is calculated as,

$$e_a \left[\frac{\partial}{\partial t} + \boldsymbol{v}_g \cdot \frac{\partial}{\partial \boldsymbol{r}} - \Omega_a \frac{\partial}{\partial \varphi} \right] \tilde{\Phi} \frac{\partial f_{aM}}{\partial K} = e_a \left[\frac{\partial \tilde{\Phi}}{\partial t} + v_\parallel \boldsymbol{b} \cdot \frac{\partial \tilde{\Phi}}{\partial \boldsymbol{r}} - \Omega_a \frac{\partial \tilde{\Phi}}{\partial \varphi} \right] \frac{\partial f_{aM}}{\partial K},$$

where we neglected higher order term $\boldsymbol{v}_d \cdot \partial \tilde{\Phi}/\partial \boldsymbol{r}$ as compared with $\partial \tilde{\Phi}/\partial t$.

The linearized Vlasov equation for non-adiabatic part g_a is given as follows,

$$\left[\frac{\partial}{\partial t} + \boldsymbol{v}_g \cdot \frac{\partial}{\partial \boldsymbol{r}} - \Omega_a \frac{\partial}{\partial \varphi} \right] g_a = e_a \left[-\frac{\partial f_{aM}}{\partial K} \frac{\partial \tilde{\Phi}}{\partial t} + \frac{\partial \tilde{\Phi}}{\partial \boldsymbol{r}} \cdot \frac{\boldsymbol{b}}{\Omega_a} \times \frac{\partial f_{aM}}{\partial \boldsymbol{r}} \right] \tag{5.101}$$

We define the gyro phase average of $A(\boldsymbol{x}, E, \mu, \varphi, t)$:

$$< A >= \oint \frac{d\varphi}{2\pi} A(\boldsymbol{x}, E, \mu, \varphi, t), \tag{5.102}$$

where $E = K + e_a \Phi$. Taking the gyro phase average of (5.101), we obtain:

$$\left[\frac{\partial}{\partial t} + \boldsymbol{v}_g \cdot \frac{\partial}{\partial \boldsymbol{r}} \right] \langle g_a \rangle = e_a \left[-\frac{\partial f_{aM}}{\partial K} \frac{\partial \langle \tilde{\Phi} \rangle}{\partial t} + \frac{\partial \langle \tilde{\Phi} \rangle}{\partial \boldsymbol{r}} \cdot \frac{\boldsymbol{b}}{\Omega_a} \times \frac{\partial f_{aM}}{\partial \boldsymbol{r}} \right] \tag{5.103}$$

We use following eikonal approximation perpendicular to the magnetic field.

$$\langle g_a \rangle = \hat{g}_a exp \left[i \int^{\boldsymbol{r}_\perp} \boldsymbol{k}_\perp \cdot d\boldsymbol{r}_\perp - i\omega t \right]$$

We define the normal vector as $\boldsymbol{n} = \nabla\psi/|\nabla\psi|$ and the bi-normal vector by $\boldsymbol{s} = \boldsymbol{b} \times \boldsymbol{n}$ so that $(\boldsymbol{b}, \boldsymbol{n}, \boldsymbol{s})$ forms the right-handed orthogonal set. The bi-normal vector $\boldsymbol{s} = \boldsymbol{b} \times \boldsymbol{n}$ is a unit vector on the flux surface perpendicular to the magnetic field.

Since the driving force $\partial f_{aM}/\partial \boldsymbol{r}$ is mainly directed to $\boldsymbol{n} = \nabla\psi/|\nabla\psi|$, we may approximate $\boldsymbol{b} \times \partial f_{aM}/\partial \boldsymbol{r} \sim (\boldsymbol{b} \times \boldsymbol{n})(\boldsymbol{n} \cdot \partial f_{aM}/\partial \boldsymbol{r}) = \boldsymbol{s}(\boldsymbol{n} \cdot \partial f_{aM}/\partial \boldsymbol{r})$. Using the eikonal expression, we define wave number in the bi-normal direction k_s by $ik_s = \boldsymbol{s} \cdot \partial/\partial \boldsymbol{r}$. Using these definition, we define ω_{*a} as follows,

$$\omega_{*a} = \frac{m_a T_a k_s}{e_a B} \left[\boldsymbol{n} \cdot \frac{\partial ln f_{aM}}{\partial \boldsymbol{r}} \right] = \omega_n \left[1 + \eta_a \left(\frac{K}{T_a} - \frac{3}{2} \right) \right] \tag{5.104}$$

$$\omega_n = (T_a k_s / e_a B) \boldsymbol{n} \cdot \nabla ln n_a \,, \, \eta_a = \frac{dln n_a}{dln T_a}$$

Using these quantities, the gyro-phase averaged linearized kinetic equation (5.103) in the eikonal approximation is given as follows.

$$\left[-i\omega + v_{\|}\frac{\partial}{\partial r_{\|}} + i\boldsymbol{k}_{\perp}\cdot\boldsymbol{v}_d\right]\hat{g}_a = -i\frac{e_a}{T_a}f_{aM}(\omega-\omega_*)\hat{\Phi} \tag{5.105}$$

Now we calculate gyro phase averaged electrostatic potential in the eikonal form, $\hat{\Phi}$. Consider the Fourier representation of $\tilde{\Phi}$ in space perpendicular to the magnetic field.

$$\tilde{\Phi}(\boldsymbol{x}) = \int d\boldsymbol{k}_{\perp}e^{i\boldsymbol{k}_{\perp}\cdot\boldsymbol{x}}\Phi(\boldsymbol{k}_{\perp}) = \int d\boldsymbol{k}_{\perp}e^{i\boldsymbol{k}_{\perp}\cdot(\boldsymbol{r}+\boldsymbol{\rho}_a)}\Phi(\boldsymbol{k}_{\perp}) \tag{5.106}$$

The gyro phase average is given by,

$$\langle\tilde{\Phi}(\boldsymbol{x})\rangle = \int d\boldsymbol{k}_{\perp}e^{i\boldsymbol{k}_{\perp}\cdot\boldsymbol{r}}\Phi(\boldsymbol{k}_{\perp})\langle e^{i\boldsymbol{k}_{\perp}\cdot\boldsymbol{\rho}_a}\rangle = \int d\boldsymbol{k}_{\perp}e^{i\boldsymbol{k}_{\perp}\cdot\boldsymbol{r}}\Phi(\boldsymbol{k}_{\perp})J_0(k_{\perp}\rho_a)$$

$$\text{where we use: } \langle e^{i\boldsymbol{k}\cdot\boldsymbol{\rho}_a}\rangle = \int_0^{2\pi}\frac{d\varphi}{2\pi}e^{ik_{\perp}\rho_a sin\varphi} = J_0(k_{\perp}\rho_a) \tag{5.107}$$

Since we are using eikonal form $\langle\tilde{\Phi}\rangle = \hat{\Phi}exp\left[i\int^{r_\perp}\boldsymbol{k}_{\perp}\cdot d\boldsymbol{r}_{\perp} - i\omega t\right]$, we have

$$\hat{\Phi} = \Phi(\boldsymbol{k}_{\perp})J_0(k_{\perp}\rho_a)$$

Thus, we obtain following classical linear gyrokinetic equation.

$$\left[-i\omega + v_{\|}\frac{\partial}{\partial r_{\|}} + i\boldsymbol{k}_{\perp}\cdot\boldsymbol{v}_d\right]\hat{g}_a = -i\frac{e_a}{T_a}f_{aM}(\omega-\omega_*)\Phi(\boldsymbol{k}_{\perp})J_0(k_{\perp}\rho_a) \tag{5.108}$$

Column 5-1: Guiding Center Coordinate Transformation

In both drift and gyro kinetic theories, coordinate transformation from the particle coordinates $(\boldsymbol{x}, \boldsymbol{v})$ to the guiding center coordinates $(\boldsymbol{r}, K, \mu, \varphi)$ is necessary. Here we give a formula given by Catto [101]. Here $K = m_a v^2/2$ is the kinetic energy, $\mu = m_a v_{\perp}^2/2B$ is the magnetic moment, and φ is the gyro phase. Some papers use $E = K + e_a\Phi$ instead of K.

$$\boldsymbol{r} = \boldsymbol{x} + \Omega_a^{-1}\boldsymbol{v}\times\boldsymbol{b} \tag{5.109}$$

$$\boldsymbol{v} = v_{\perp}(\boldsymbol{n}cos\varphi + \boldsymbol{s}sin\varphi) + v_{\|}\boldsymbol{b} \tag{5.110}$$

Here, $\boldsymbol{b} = \boldsymbol{B}/B, \boldsymbol{n} = \nabla\psi/|\nabla\psi|$, and $\boldsymbol{s} = \boldsymbol{b}\times\boldsymbol{n}$. The coordinate transformation is given by,

(continued)

$$\frac{\partial}{\partial t} \to \frac{\partial}{\partial t} \tag{5.111}$$

$$\frac{\partial}{\partial \boldsymbol{x}} \to \frac{\partial}{\partial \boldsymbol{r}} - \left[\frac{\partial(\boldsymbol{b}/\Omega_a)}{\partial \boldsymbol{x}} \times \boldsymbol{v}\right] \cdot \frac{\partial}{\partial \boldsymbol{r}} + \frac{\partial \mu}{\partial \boldsymbol{x}}\frac{\partial}{\partial \mu} + \frac{\partial \varphi}{\partial \boldsymbol{x}}\frac{\partial}{\partial \varphi} \tag{5.112}$$

$$\frac{\partial}{\partial \boldsymbol{v}} \to m_a \boldsymbol{v}\frac{\partial}{\partial K} + \frac{m_a \boldsymbol{v}_\perp}{B}\frac{\partial}{\partial \mu} + \frac{\boldsymbol{b} \times \boldsymbol{v}_\perp}{v_\perp^2}\frac{\partial}{\partial \varphi} + \frac{\boldsymbol{b}}{\Omega_a} \times \frac{\partial}{\partial \boldsymbol{r}} \tag{5.113}$$

We define the phase angle φ in the $(\boldsymbol{n}, \boldsymbol{s})$ plane perpendicular to the magnetic field by $\boldsymbol{v}_\perp = v_\perp[\boldsymbol{n} cos\varphi + \boldsymbol{s} sin\varphi]$, which leads to $\boldsymbol{n} \cdot \boldsymbol{v} = v_\perp cos\varphi$ and $\boldsymbol{s} \cdot \boldsymbol{v} = v_\perp sin\varphi$.

$$\frac{\partial \varphi}{\partial \boldsymbol{x}} = \frac{v_\parallel}{v_\perp}\frac{\partial \boldsymbol{b}}{\partial \boldsymbol{x}} \cdot (-\boldsymbol{s} cos\varphi + \boldsymbol{n} sin\varphi) + \left(\frac{\partial}{\partial \boldsymbol{x}}\boldsymbol{s}\right) \cdot \boldsymbol{n} \tag{5.114}$$

$$\frac{\partial \mu}{\partial \boldsymbol{x}} = -\frac{\mu}{B}\frac{\partial B}{\partial \boldsymbol{x}} - \frac{v_\parallel}{B}\frac{\partial \boldsymbol{b}}{\partial \boldsymbol{x}} \cdot \boldsymbol{v}_\perp \tag{5.115}$$

We also note

$$\frac{e_a \boldsymbol{v} \times \boldsymbol{B}}{m_a} \cdot \frac{\partial}{\partial \boldsymbol{v}} = -\Omega_a \frac{\partial}{\partial \varphi} \tag{5.116}$$

$$(\Omega_a \boldsymbol{v} \times \boldsymbol{b}) \cdot \left(\Omega_a^{-1} \boldsymbol{b} \times \frac{\partial}{\partial \boldsymbol{r}}\right) = -\boldsymbol{v}_\perp \cdot \frac{\partial}{\partial \boldsymbol{r}} \tag{5.117}$$

Then, we have following transformation for the Vlasov propagator.

$$\frac{\partial}{\partial t} + \boldsymbol{v} \cdot \frac{\partial}{\partial \boldsymbol{x}} + \frac{e_a \boldsymbol{v} \times \boldsymbol{B}}{m_a} \cdot \frac{\partial}{\partial \boldsymbol{v}} = \frac{\partial}{\partial t} - \Omega_a \frac{\partial}{\partial \varphi} + v_\parallel \boldsymbol{b} \cdot \frac{\partial}{\partial \boldsymbol{r}}$$
$$+\boldsymbol{v} \cdot \left[\frac{\partial \varphi}{\partial \boldsymbol{x}}\frac{\partial}{\partial \varphi} + \frac{\partial \mu}{\partial \boldsymbol{x}}\frac{\partial}{\partial \mu} - \left(\frac{\partial}{\partial \boldsymbol{x}}(\Omega_a^{-1}\boldsymbol{b})\right) \cdot \left(\boldsymbol{v} \times \frac{\partial}{\partial \boldsymbol{r}}\right)\right] \tag{5.118}$$

Exercise 5.1. Derive relations $\partial \boldsymbol{r}/\partial \boldsymbol{x}|_{\boldsymbol{v}} = 1 - \partial(\boldsymbol{b}/\Omega_a)/\partial \boldsymbol{x} \times \boldsymbol{v}$, $\partial K/\partial \boldsymbol{v}|_x = m_a \boldsymbol{v}$, $\partial \mu/\partial m_a \boldsymbol{v}|_x = \boldsymbol{v}_\perp/B$, $(\partial \boldsymbol{r}/\partial \boldsymbol{v})\partial/\partial \boldsymbol{r} = (\boldsymbol{b}/\Omega_a) \times \partial/\partial \boldsymbol{r}$.

Answer. Use (5.109), $K = m_a v^2/2$, $\mu = m_a v_\perp^2/2B$.

5.3.1.2 Bounce-Averaged Drift Kinetic Equation

While we have to solve the gyro kinetic equation (5.108) to include finite Larmor radius (FLR) effect for the ion dynamics, we can simplify the gyro kinetic equation (5.108) for the electron dynamics by neglecting FLR effect.

If the drift wave frequency ω is much smaller than the bounce frequency of the trapped electron ω_{be}: ($\omega \leq \omega_{*e} \sim O((m_e/m_i)^{1/2}\omega_{be})$, we can use an averaged drift kinetic equation over the bounce motion, which is called the bounce-averaged drift kinetic equation. We show its derivation in this section.

Our starting point is a linearized gyro kinetic equation in the $k_\perp \rho_e \to 0$ for a perturbations of the form $e^{-i\omega t + in\zeta}$ as follows,

$$\left[\frac{\partial}{\partial t} + \boldsymbol{v}_g \cdot \frac{\partial}{\partial \boldsymbol{r}}\right] g_e = i\frac{e}{T_e} f_{eM}(\omega - \omega_*)\tilde{\Phi} \tag{5.119}$$

$$\omega_* = \omega_n \left[1 + \eta\left(\frac{K}{T_e} - \frac{3}{2}\right)\right], \omega_n = \frac{T_e k_\zeta}{eB_p} \boldsymbol{n} \cdot \nabla ln n_e \tag{5.120}$$

Here ζ is a toroidal angle and $\langle\rangle$ for gyro-phase average is suppressed and the toroidal wave number $k_\zeta = n/R$ is approximated by the toroidal projection of the bi-normal wave number k_s as $k_\zeta = -k_s B_p/B$. Averaging over the trapped electron bounce motion, remaining motion is a slow toroidal drift, called as the toroidal precession drift, whose toroidal precession frequency $\langle\dot{\zeta}\rangle_b$ is give in [443],

$$\langle\dot{\zeta}\rangle_b = \frac{1}{e_a} \frac{\partial J/\partial\psi}{\partial J/\partial K}, \tag{5.121}$$

where $J = \int m_a v_\parallel dl_\parallel$, ψ and K are longitudinal adiabatic invariant, poloidal flux, and particle kinetic energy, respectively. This means $\langle \boldsymbol{v}_g \cdot \partial/\partial\boldsymbol{r}\rangle_b = \langle\dot{\zeta}\rangle_b \partial/\partial\zeta$, where $\langle A\rangle_b$ is the bounce average of A.

The bounce-average of the drift kinetic equation (5.119) is given as,

$$\left[\frac{\partial}{\partial t} + \langle\dot{\zeta}\rangle_b \frac{\partial}{\partial\zeta}\right] \langle g_e\rangle_b = i\frac{e}{T_e} f_{eM}(\omega - \omega_*)\langle\tilde{\Phi}\rangle_b \tag{5.122}$$

If we take the difference (5.119)-(5.122), the 0-th order drift kinetic equation is given by $\left[\boldsymbol{v}_g \cdot (\partial/\partial\boldsymbol{r}) - \langle\dot{\zeta}\rangle_b(\partial/\partial\zeta)\right] g_{e0} = 0$.

This equation simply means that g_{e0} is constant along the bounce banana orbit subtracting slow precession drift. The first order equation is given by,

$$\left[\boldsymbol{v}_g \cdot \frac{\partial}{\partial\boldsymbol{r}} - \langle\dot{\zeta}\rangle_b \frac{\partial}{\partial\zeta}\right] g_{e1} - i(\omega - n\langle\dot{\zeta}\rangle_b) g_{e0} = i\frac{e}{T_e} f_{eM}(\omega - \omega_*)\tilde{\Phi} \tag{5.123}$$

Taking the bounce average of above equation, we obtain the solution of g_{e0}.

$$g_{e0} = -\frac{e\langle\tilde{\Phi}\rangle_b}{T_e} \frac{\omega - \omega_*}{\omega - \omega_{De}} f_{eM} \tag{5.124}$$

Here, $\langle\tilde{\Phi}\rangle_b$ is the bounce average of the electrostatic potential, and $\omega_{De} = n\langle\dot{\zeta}\rangle_b$ is a toroidal precession drift frequency, n is the toroidal mode number.

5.3.2 Modern Gyro Kinetic Theory

Classical Gyrokinetic theory using Bessel function is relatively easy to understand but do not have exact conservation properties required from Liouville theorem. Modern gyrokinetic theory [82] provides such a framework by using Lagrangian approach. Use of Lagrangian guarantees the Liouville theorem in the gyro center phase space. Cary-Littlejohn [99] provides a good guidance to the Lie transformation. Here, we note that gyro-phase averaged Lagrangian can be obtained by the direct averaging of the charged particle Lagrangian (see Column 5-2).

Column 5-2: Guiding Center Lagrangian; Kikuchi [443]
The single particle Lagrangian in non-canonical coordinates $z = (\boldsymbol{x}, \boldsymbol{v})$ is:

$$L(\boldsymbol{x}, \boldsymbol{v}, t) = (e_a \boldsymbol{A} + m_a \boldsymbol{v}) \cdot \dot{\boldsymbol{x}} - h(\boldsymbol{x}, \boldsymbol{v}, t) \tag{5.125}$$

$$h(\boldsymbol{x}, \boldsymbol{v}, t) = \frac{1}{2} m_a v^2 + e_a \Phi(\boldsymbol{x}, t) \tag{5.126}$$

Let $\boldsymbol{x}$, $\boldsymbol{r}$ and $\boldsymbol{\rho}_a = \rho_a(\boldsymbol{e}_x cos\varphi + \boldsymbol{e}_y sin\varphi)$ are the charged particle position, the guiding center positions and the gyration vector, respectively. Here, $\rho_a = v_\perp/\Omega_a$ is the Larmor radius, φ is a gyro phase, and vectors $\boldsymbol{e}_x, \boldsymbol{e}_y$ are the orthogonal unit vectors perpendicular to the magnetic field, which comprises right handed coordinate system by $(\boldsymbol{e}_x, \boldsymbol{e}_y, \boldsymbol{b})$. We have $\boldsymbol{x} = \boldsymbol{r} + \boldsymbol{\rho}_a$ and $\boldsymbol{v} = v_\parallel \boldsymbol{b} + v_\perp(\boldsymbol{e}_x sin\varphi - \boldsymbol{e}_y cos\varphi)$. We expand $\boldsymbol{A}(\boldsymbol{x}) \cdot \dot{\boldsymbol{x}}$ as:

$$\boldsymbol{A}(\boldsymbol{x}) \cdot \dot{\boldsymbol{x}} \sim \left[\boldsymbol{A}(\boldsymbol{r}) + \frac{\partial \boldsymbol{A}}{\partial x} \rho_a cos\varphi + \frac{\partial \boldsymbol{A}}{\partial y} \rho_a sin\varphi\right] \cdot [\dot{\boldsymbol{r}} + \rho_a \dot{\varphi}(-\boldsymbol{e}_x sin\varphi + \boldsymbol{e}_y cos\varphi)] \tag{5.127}$$

Taking the gyro-phase average $\langle\rangle$ and using $\frac{\partial A_y}{\partial x} - \frac{\partial A_x}{\partial y} = B$, we have

$$\langle \boldsymbol{A} \cdot \dot{\boldsymbol{x}} \rangle \sim \boldsymbol{A}(\boldsymbol{r}) \cdot \dot{\boldsymbol{r}} + \frac{1}{2} B \rho_a^2 \dot{\varphi} \tag{5.128}$$

Since $\boldsymbol{v} \cdot \dot{\boldsymbol{x}} = (v_\parallel \boldsymbol{b} + v_\perp(\boldsymbol{e}_x sin\varphi - \boldsymbol{e}_y cos\varphi))(\dot{\boldsymbol{r}} + \rho_a \dot{\varphi}(-\boldsymbol{e}_x sin\varphi + \boldsymbol{e}_y cos\varphi))$,

$$\langle \boldsymbol{v} \cdot \dot{\boldsymbol{x}} \rangle = v_\parallel \boldsymbol{b} \cdot \dot{\boldsymbol{r}} - v_\perp \rho_a \dot{\varphi} \tag{5.129}$$

The gyro-phase average of the Hamiltonian h is simply given by,

$$h(\boldsymbol{r}, v_\parallel, \mu, t) = \frac{1}{2} m_a v_\parallel^2 + \mu B(\boldsymbol{r}) + e_a \Phi(\boldsymbol{r}) \ , \ \mu = \frac{m_a v_\perp^2}{2B} \tag{5.130}$$

(continued)

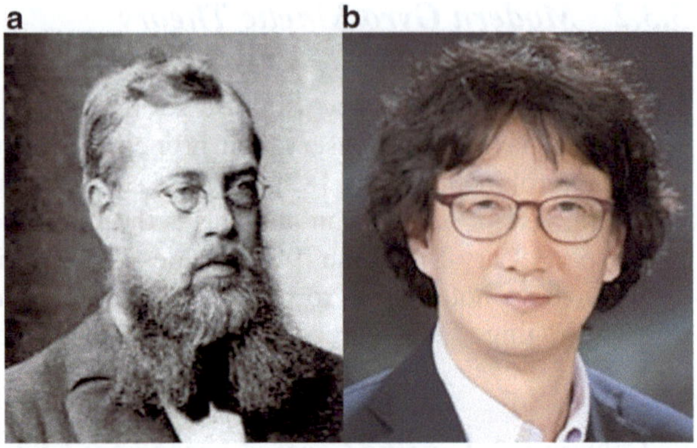

Fig. 5.5 (**a**) Sophus Lie (1842–1899), (**b**) T.S. Hahm. TSH is a professor at Seoul National University and formerly worked at Princeton Plasma Physics Laboratory, who made number of significant theoretical works including gyro kinetic theory

Therefore, the guiding center Lagrangian is given by,

$$L = (e_a \boldsymbol{A}(\boldsymbol{r}, t) + m_a v_\parallel \boldsymbol{b}) \cdot \dot{\boldsymbol{r}} - (e_a \mu / m_a)\dot{\varphi} - h(\boldsymbol{r}, v_\parallel, \mu, t) \quad (5.131)$$

This form of the gyro-phase averaged Lagrangian is derived by R. Littlejohn using Lie transformation in a series of paper by R. Littlejohn [518–522].

Modern gyrokinetic theory in the turbulent toroidal plasma is developed by T.S. Hahm [275, 276] (Fig. 5.5) using the non-canonical mechanics and Lie transformation. Its application to nonlinear ITG simulation is given by Parker [607]. In the turbulent plasma, the drift wave fluctuation with $k_\perp \rho_i \sim 1$ breaks the conservation of the magnetic moment and the Lie transformation is effective to recover its conservation property as shown in the earlier work in slab geometry [178]. Fundamentals of the modern gyrokinetic theory can be seen in Brizard-Hahm [82] and Cary-Brizard [100]. Column 5-3 gives short summary of the non-canonical mechanics.

Column 5-3: Non-canonical Mechanics; Kikuchi [443]

While Hamilton mechanics is powerful, the set of transformation is restricted to canonical transformation. To analyze gyro motion, it is useful to use non canonical coordinates. The guiding center coordinates $(\boldsymbol{R}, E, \mu, \varphi)$ discussed in Sect. 4.2 is such an example of useful non-canonical coordinates. Consider 6 dimensional non-canonical variables z is given as $z = z(\boldsymbol{q}, \boldsymbol{p})$. When the Lagrangian L is given as $L = \boldsymbol{p} \cdot \dot{\boldsymbol{q}} - H(\boldsymbol{q}, \boldsymbol{p})$ in canonical form, the Lagrangian in non-canonical form is given as,

(continued)

$$L(z, \dot{z}, t) = \sum_{i=1}^{6} \gamma_i \dot{z}^i - h \tag{5.132}$$

$$\gamma_i(z,t) = \boldsymbol{p} \cdot \frac{\partial \boldsymbol{q}}{\partial z^i}\,,\ h(z,t) = H(\boldsymbol{q}(z,t), \boldsymbol{p}(z,t), t) - \boldsymbol{p} \cdot \frac{\partial \boldsymbol{q}}{\partial t} \tag{5.133}$$

Due to the coordinate invariance of the variational principle, the variational principle to give the equation of motion is given by,

$$\delta S = \delta \int L(z, \dot{z}, t) dt = 0 \tag{5.134}$$

Addition of gauge term dG/dt to Lagrangian L does not change the equation of motion, but transform $\gamma_i \to \gamma_i + \partial G/\partial z^i$ and $h \to h - \partial G/\partial t$.

The Euler-Lagrange equation to extremize the action integral S is given by,

$$\frac{d}{dt}\left(\frac{\partial L}{\partial \dot{z}^i}\right) - \frac{\partial L}{\partial z^i} = 0 \tag{5.135}$$

This leads to following equation of motion.

$$\omega_{ij} \frac{dz^j}{dt} = \frac{\partial h}{\partial z^i} + \frac{\partial \gamma_i}{\partial t} \tag{5.136}$$

where ω_{ij} is called the Lagrange tensor defined as:

$$\omega_{ij} = \frac{\partial \gamma_j}{\partial z^i} - \frac{\partial \gamma_i}{\partial z^j} = \frac{\partial (\boldsymbol{p} \cdot \frac{\partial \boldsymbol{q}}{\partial z^j})}{\partial z^i} - \frac{\partial (\boldsymbol{p} \cdot \frac{\partial \boldsymbol{q}}{\partial z^i})}{\partial z^j} = \frac{\partial \boldsymbol{p}}{\partial z^i} \cdot \frac{\partial \boldsymbol{q}}{\partial z^j} - \frac{\partial \boldsymbol{p}}{\partial z^j} \cdot \frac{\partial \boldsymbol{q}}{\partial z^i} \tag{5.137}$$

The non-canonical mechanics in Column 5-3 can be described by using the language of differential form given in Appendix A.4. The one form $\gamma = Ldt = \boldsymbol{p} \cdot d\boldsymbol{q} - hdt$ is called the fundamental one-form of Poincare-Cartan. We use the terminology of "Lagrangian one form" or "one form" in short.

Single particle Lagrangian is given by $L = \boldsymbol{p} \cdot \dot{\boldsymbol{q}} - h(\boldsymbol{q}, \boldsymbol{p})$, where h is the Hamiltonian. Let $z = z(\boldsymbol{q}, \boldsymbol{p}, t)$ is a non-canonical coordinate system. The Lagrangian one form $\gamma = Ldt$ in the non-canonical coordinates z is given by,

$$\gamma = \gamma_\mu dz^\mu = \gamma_i dz^i - hdt\,, (\mu = 0, 6, i = 1, 6), \tag{5.138}$$

where $\gamma_0 = -h$ and $dz^0 = dt$ and γ_i and h are given by (5.133).

The Euler-Lagrange equation for the non-canonical coordinate z^j is given by $d\gamma = 0$ (equivalent to the variational principle $\delta \int L dt = 0$). Using the algebra of the differential form, we obtain:

$$0 = d\gamma = d(\gamma_i dz^i - hdt) = \left(\frac{\partial \gamma_j}{\partial z^i} - \frac{\partial \gamma_i}{\partial z^j}\right) dz^i dz^j - \left(\frac{\partial h}{\partial z^i} + \frac{\partial \gamma_i}{\partial t}\right) dz^i dt \quad (5.139)$$

Using the Lagrange tensor $\omega_{ij} = \partial\gamma_j/\partial z^i - \partial\gamma_i/\partial z^j$, we have:

$$\omega_{ij}\frac{dz^j}{dt} = \frac{\partial h}{\partial z^i} + \frac{\partial \gamma_i}{\partial t} \quad (5.140)$$

In the standard gyrokinetic ordering, we use the ordering: $\omega/\Omega_i \sim k_\parallel/k_\perp \sim e\Phi/T_e \sim \rho_i/L_n = O(\epsilon)$. Under this ordering, we can use gyro-phase averaged Lagrangian (5.131) as a 0-th order Lagrangian. Then the Lagrangian one-form in the guiding center coordinates $(\boldsymbol{r}, v_\parallel, M_0, \varphi)$ is given by

$$\gamma_0 = (e_a \boldsymbol{A}(\boldsymbol{r}) + m_a v_\parallel \boldsymbol{b}) \cdot d\boldsymbol{r} - M_0 d\varphi - h_0(\boldsymbol{r}, v_\parallel, M_0)dt \quad (5.141)$$

$$h_0 = \frac{1}{2} m_a v_\parallel^2 + \mu B \quad (5.142)$$

Here, $M_0 = e_a\mu/m_a$ is used instead of μ, $\boldsymbol{b} = \boldsymbol{B}/B$ and we assume no 0-th order electrostatic potential and the equilibrium field has no explicit time dependence.

We define the non-canonical coordinates $z = (t, r_1, r_2, r_3, v_\parallel, M_0, \varphi)$. While γ_0 is independent of gyro phase φ, the electrostatic perturbation Φ due to the drift wave adds 1st order perturbation to the one-form, by which γ depends on φ.

The perturbed one-form γ_1 is given by,

$$\gamma_1 = -e_a \Phi(\boldsymbol{r} + \boldsymbol{\rho})dt, \quad (5.143)$$

which means $\gamma_{1t} = -e_a\Phi$, $\gamma_{1i} = 0$ $(i = 1, 6)$ and we can assume $\gamma_2 = 0$. This results in the breakdown of conservation of the magnetic moment μ (and hence M_0).

Modern gyrokinetic theory tries to recover the conservation of the magnetic moment through coordinate transformation, $z \to \boldsymbol{Z}$. In our case,

$$z = (t, \boldsymbol{r}, v_\parallel, M_0, \varphi) \to \boldsymbol{Z} = (t, \boldsymbol{R}, U, M, \bar{\varphi}) \quad (5.144)$$

We try to find the Lagrangian one form in the new coordinates to have following gyro-phase independent form.

$$\Gamma = \Gamma_i(\boldsymbol{Z})dZ^i - H(\boldsymbol{Z})dt + dS(\boldsymbol{Z}) \text{ , in } \boldsymbol{Z} = (t, Z^i)$$

We consider the new coordinate system $\boldsymbol{Z} = \boldsymbol{Z}(z)$ (we do not transform 't'). Under this coordinate transform, we have $\gamma = \gamma_\nu dz^\nu = \Gamma_\mu dZ^\mu$. This means,

$$\Gamma_\mu = \gamma_\nu \frac{\partial z^\nu}{\partial Z^\mu} \tag{5.145}$$

Using the Lie transform formula shown in Column 5-5, the 1st order one-form Γ_1 is related as (5.206).

$$\Gamma_1 = dS_1 - L_1\gamma_0 + \gamma_1 \tag{5.146}$$

To remove gyro phase dependence in new coordinates $\boldsymbol{Z}$, we consider simple form $\Gamma_1 = \Gamma_{1t}dt$, i.e. $\Gamma_{1i} = 0$ $(i = 1, 6)$. Taking the i-th component of (5.146):

$$0 = (dS_1)_i - (L_1\gamma_0)_i + 0 \tag{5.147}$$

Using (5.204) and the definition of the Lagrange bracket (5.137), we have

$$(L_1\gamma_0)_i = g_1^j \left[\frac{\partial \gamma_{0i}}{\partial z^j} - \frac{\partial \gamma_{0j}}{\partial z^i} \right] = g_1^j \omega_{ji} = -\omega_{ij} g_1^j \tag{5.148}$$

$$\omega_{ij} = [z^i, z^j] \equiv \frac{\partial \boldsymbol{p}}{\partial z^i} \cdot \frac{\partial \boldsymbol{q}}{\partial z^j} - \frac{\partial \boldsymbol{p}}{\partial z^j} \cdot \frac{\partial \boldsymbol{q}}{\partial z^i} \tag{5.149}$$

Since $(dS_1)_i = \partial S_1 / \partial z^i$, (5.147) gives

$$\omega_{ij} g_1^j = -\frac{\partial S_1}{\partial z^i} \tag{5.150}$$

The inverse matrix of the Lagrange tensor $\omega_{ij}^{-1} \equiv J_{ij}$ is called the Poisson tensor.

$$J_{ij} = \{z^i, z^j\} \equiv \frac{\partial z^i}{\partial \boldsymbol{q}} \cdot \frac{\partial z^j}{\partial \boldsymbol{p}} - \frac{\partial z^j}{\partial \boldsymbol{q}} \cdot \frac{\partial z^i}{\partial \boldsymbol{p}} \tag{5.151}$$

Therefore, we obtain following formula for the generating function.

$$g_1^i = -J_{ij} \frac{\partial S_1}{\partial z^j} = -\{z^i, z^j\} \frac{\partial S_1}{\partial z^j} = \{S_1, z^i\} \tag{5.152}$$

Taking the t component of (5.146), we have

$$\Gamma_{1t} = \frac{\partial S_1}{\partial t} - (L_1\gamma_0)_t + \gamma_{1t} \tag{5.153}$$

Noting that t is not transformed $(g_1^0 = 0)$ and the equilibrium field is stationary $(\partial \gamma_{0j} / \partial t = 0)$, we have

$$(L_1\gamma_0)_t = g_1^j \left[\frac{\partial \gamma_{0t}}{\partial z^j} - \frac{\partial \gamma_{0j}}{\partial t} \right] = -\{S_1, z^j\} \frac{\partial h_0}{\partial z^j} = -\{S_1, h_0\} \tag{5.154}$$

Here we use (5.152). Substituting this equation and using $\Gamma_{1t} = -H_1$, $\gamma_{1t} = -h_1 = -e_a\Phi$, Eq. (5.153) reads

$$-H_1 = \frac{\partial S_1}{\partial t} + \{S_1, h_0\} - e_a\Phi = \frac{dS_1}{dt} - e_a\Phi \tag{5.155}$$

Since we are looking for a solution where $\Gamma_{1t} = -H_1$ do not have any gyro-phase dependence, we take a gyro-phase average $\langle(5.155)\rangle$ to obtain:

$$H_1 = \langle e_a\Phi\rangle \tag{5.156}$$

Here, we intend to absorb gyro-phase dependent part by S_1. Defining $\tilde{\Phi} = \Phi - \langle\Phi\rangle$, we have following equation to determine S_1.

$$\frac{dS_1}{dt} = \frac{\partial S_1}{\partial t} + \{S_1, h_0\} = e_a\tilde{\Phi} \tag{5.157}$$

Using the ordering $\{S_1, h_0\} \sim \Omega_a\partial S_1/\partial\varphi$ and $\partial S_1/\partial t \sim O(\omega/\Omega_a)$, we have

$$S_1 = -e_a\int\tilde{\Phi}dt \sim -\frac{e_a}{\Omega_a}\int\tilde{\Phi}d\varphi \tag{5.158}$$

The second order differential one-form Γ_2 is related as (5.207),

$$\Gamma_2 = dS_2 - L_2\gamma_0 + \gamma_2 - L_1\gamma_1 + \frac{1}{2}L_1^2\gamma_0 \tag{5.159}$$

The i-th component of Eq. (5.159) is given as,

$$\Gamma_{2i} = (dS_2)_i - (L_2\gamma_0)_i + \gamma_{2i} - (L_1\gamma_1)_i + \frac{1}{2}(L_1[dS_1 + \gamma_1 - \Gamma_1])_i \tag{5.160}$$

We can make procedures similar to g_1^i noting $L_1dS_1 \equiv 0$, $\Gamma_{2i} = \Gamma_{1i} = \gamma_{2i} = \gamma_{1i} = 0$ $(i = 1, 6)$ to obtain following expression for g_2^i:

$$g_2^i = \{S_2, z^i\} \tag{5.161}$$

The t-th component of Eq. (5.159) is given as,

$$\Gamma_{2t} = (dS_2)_t - (L_2\gamma_0)_t + \gamma_{2t} - (L_1\gamma_1)_t + \frac{1}{2}(L_1[dS_1 + \gamma_1 - \Gamma_1])_t \tag{5.162}$$

Noting $\gamma_{1t} = -e_a\Phi$, $\gamma_{2t} = -h_2 = 0$, we have

$$-H_2 = \frac{\partial S_2}{\partial t} + \{S_2, h_0\} + \frac{1}{2}\{S_1, h_1\} + \frac{1}{2}\{S_1, H_1\} \tag{5.163}$$

We can make H_2 to be gyro-phase independent by the appropriate choice of S_2 similar to S_1 in the first order relation and H_2 can be given as follows by noting that H_1 has no gyro-phase dependence.

$$H_2 = -\frac{1}{2}\langle\{S_1, h_1\}\rangle \tag{5.164}$$

The actual form of H_2 in terms of Φ is given by Dubin [178] as follows.

$$H_2 = -\frac{e_a^2}{2\Omega_a}\left[\frac{\partial\langle\tilde{\Phi}^2\rangle}{\partial M_0} + \frac{1}{\Omega_a}\langle\frac{\partial\psi_1}{\partial \boldsymbol{r}}\cdot \boldsymbol{b}\times\frac{\partial\tilde{\Phi}}{\partial \boldsymbol{r}}\rangle\right] \tag{5.165}$$

$$\text{where, } \psi_1 = \int^{\varphi}\tilde{\Phi}d\varphi$$

Adding $\Gamma_1 = -H_1$ and $\Gamma_2 = -H_2$ to the Lagrangian differential one-form, we have

$$\Gamma = [e_a\boldsymbol{A}(\boldsymbol{R}) + m_aU\boldsymbol{b}(\boldsymbol{R})]\cdot d\boldsymbol{R} + Md\bar{\varphi} - Hdt \tag{5.166}$$

$$H = \frac{1}{2}m_aU^2 + M\Omega_a(\boldsymbol{R}) + H_* \tag{5.167}$$

Here $H_* = H_1 + H_2$.

Then the Euler-Lagrange equation in new coordinates $\boldsymbol{Z}$ (to be obtained from $d\Gamma = 0$) can be obtained using the Lagrange tensor form in Column 5-4:

$$\frac{dM}{dt} = 0 \tag{5.168}$$

$$\frac{d\bar{\varphi}}{dt} = \Omega_a + \frac{\partial H_*}{\partial M} \tag{5.169}$$

$$\boldsymbol{b}\cdot\frac{d\boldsymbol{R}}{dt} = U \tag{5.170}$$

$$\frac{d\boldsymbol{R}}{dt} = U\boldsymbol{b} + \frac{\boldsymbol{b}\times}{e_aB^*}\left[M\frac{\partial\Omega_a}{\partial\boldsymbol{R}} + m_aU^2\boldsymbol{b}\cdot\frac{\partial}{\partial\boldsymbol{R}}\boldsymbol{b} + \frac{\partial H_*}{\partial\boldsymbol{R}}\right] \tag{5.171}$$

$$m_a\frac{dU}{dt} = -\frac{1}{B^*}\boldsymbol{B}^*\cdot\left(M\frac{\partial\Omega_a}{\partial\boldsymbol{R}} + \frac{\partial H_*}{\partial\boldsymbol{R}}\right), \tag{5.172}$$

where $e_a\boldsymbol{B}^* = e_a\boldsymbol{B} + m_aU\partial/\partial\boldsymbol{R}\times\boldsymbol{b}$ and $B^* = \boldsymbol{b}\cdot\boldsymbol{B}^*$.

These equations are almost same with the guiding center equation of motion using the Littlejohn's variational principle [443] in new coordinates $\boldsymbol{Z}$ except some corrections to the Hamiltonian.

Then, the Gyrokinetic Vlasov-Poisson equations for the velocity distribution function in the gyro center coordinates $F(\boldsymbol{Z})$ can be written as [275]:

$$\frac{dF}{dt} + \frac{dU}{dt}\frac{\partial F}{\partial U} + \frac{d\boldsymbol{R}}{dt} \cdot \frac{\partial F}{\partial \boldsymbol{R}} = 0 \tag{5.173}$$

$$\nabla^2 \Phi(\boldsymbol{x}, t) = -\frac{e_i}{\epsilon_0} \int \left[F(\boldsymbol{Z}) + \frac{e_i}{\Omega_i} \left(\tilde{\Phi} \frac{\partial F}{\partial M} + \frac{1}{\Omega_i} \frac{\partial \psi_1}{\partial \boldsymbol{R}} \cdot \boldsymbol{b} \times \frac{\partial F}{\partial \boldsymbol{R}} \right) \right]$$
$$\times \delta(\boldsymbol{R} + \rho_{\boldsymbol{R}} - \boldsymbol{x}) d^6 \boldsymbol{Z} - e n_e, \tag{5.174}$$

where $d^6\boldsymbol{Z} = B^* d^3\boldsymbol{R} dU dM d\bar{\varphi}$ and B^* is the Jacobian. The terms in () is called the polarization density and n_e may be calculated by the drift kinetic equation since finite Larmor radius effect is very small for the electron.

Since Jacobian B^* gives the density of the phase space volume element, Liouville theorem in phase space may be expressed as:

$$\frac{\partial B^*}{\partial t} + \frac{\partial}{\partial Z^i}(B^* \dot{Z}^i) = 0 \tag{5.175}$$

Using this equation, the Vlasov equation can be written in the following conservation form [275].

$$\frac{\partial F B^*}{\partial t} + \frac{\partial}{\partial \boldsymbol{R}} \cdot \left[F \left(U \boldsymbol{B}^* + \frac{M \boldsymbol{b}}{e_i} \times \frac{\partial \Omega_i}{\partial \boldsymbol{R}} + \frac{\boldsymbol{b}}{e_i} \times \frac{\partial H_*}{\partial \boldsymbol{R}} \right) \right]$$
$$- \frac{1}{m_i} \frac{\partial}{\partial U} \left[F \boldsymbol{B}^* \cdot \left(M \frac{\partial \Omega_i}{\partial \boldsymbol{R}} + \frac{\partial H_*}{\partial \boldsymbol{R}} \right) \right] = 0 \tag{5.176}$$

Column 5-4: Non-vanishing Components of Lagrange Tensor; Hahm [275]
The Lagrange tensor ω_{ij} in Lie transformed coordinates $\boldsymbol{Z} = (\boldsymbol{R}, U, M, \bar{\varphi})$ is defined as:

$$\omega_{ij} = \frac{\partial \gamma_j}{\partial Z^i} - \frac{\partial \gamma_i}{\partial Z^j} \tag{5.177}$$

The non-vanishing components of the Lagrange tensor are:

$$\omega_{R_i R_j} = e_a \epsilon_{ijk} B_k^* \tag{5.178}$$

$$\omega_{\boldsymbol{R}U} = -m_a \boldsymbol{b} \tag{5.179}$$

$$\omega_{tM} = \Omega_a + \frac{\partial H_*}{\partial M} \tag{5.180}$$

$$\omega_{tU} = m_a U \tag{5.181}$$

$$\omega_{t\boldsymbol{R}} = M \frac{\partial \Omega_a}{\partial \boldsymbol{R}} + \frac{\partial H_*}{\partial \boldsymbol{R}} \tag{5.182}$$

(continued)

$$\omega_{\varphi M} = -1, \tag{5.183}$$

where

$$H_* = H_1 + H_2 = e_a\langle\Phi\rangle - \frac{e_a^2}{2\Omega_a}\left[\frac{\partial\langle\tilde{\Phi}\rangle}{\partial M} + \frac{1}{\Omega_a}\langle\frac{\partial\psi_1}{\partial \boldsymbol{r}}\cdot\boldsymbol{b}\times\frac{\partial\tilde{\Phi}}{\partial\boldsymbol{r}}\rangle\right] \tag{5.184}$$

$$\psi_1 = \int^{\varphi}\Phi d\varphi \tag{5.185}$$

Column 5-5: Lie Transformation; Cary-Littlejohn [99]
We consider a near-identity coordinate transformation from $z = \{z^\mu\}$ to $\mathbf{Z} = \{Z^\mu\}$.

$$Z^\mu = Z^\mu(z,\epsilon) = z^\mu + \epsilon Z_1^\mu + \epsilon^2 Z_2^\mu + --, \tag{5.186}$$

where ϵ is a smallness parameter and $Z^\mu(z,\epsilon=0) = z^\mu$. Lie transformation is a special coordinate transformation specified by the generating function g^μ:

$$\partial Z^\mu(z,\epsilon)/\partial\epsilon = g^\mu(\mathbf{Z}), \tag{5.187}$$

where g^μ depends on $\mathbf{Z}$ and does not explicitly depend on ϵ. Using the inverse transformation $z = z(\mathbf{Z},\epsilon)$, we have following identity transformation,

$$z^\mu(\mathbf{Z}(z,\epsilon),\epsilon) = z^\mu \tag{5.188}$$

Taking the partial derivative $\partial/\partial\epsilon$ and using (5.187), we have

$$\frac{\partial z^\mu}{\partial Z^\nu}\frac{\partial Z^\nu}{\partial\epsilon} + \frac{\partial z^\mu}{\partial\epsilon} = 0\text{ , i.e. }\frac{\partial z^\mu}{\partial\epsilon} = -g^\mu\frac{\partial z^\mu}{\partial Z^\nu} \tag{5.189}$$

If the scalar $s(z)$ at z is expressed as $S(\mathbf{Z})$ at $\mathbf{Z}$, S have epsilon dependence due to coordinate transformation as $S(\mathbf{Z},\epsilon)$. The scalar conserve its value as $s(z) = S(\mathbf{Z},\epsilon)$. Taking $\partial/\partial\epsilon$, we obtain

$$\frac{\partial S(\mathbf{Z},\epsilon)}{\partial\epsilon} = -g^\mu(\mathbf{Z})\frac{\partial S(\mathbf{Z},\epsilon)}{\partial Z^\mu} \tag{5.190}$$

(continued)

For the scalar, we define an differential operator L as,

$$L \equiv g^{\mu} \frac{\partial}{\partial y^{\mu}}, \text{ where } y^{\mu} = z^{\mu} \text{ or } Z^{\mu} \tag{5.191}$$

Equation (5.190) is expressed as

$$\frac{\partial S(\mathbf{y}, \epsilon)}{\partial \epsilon} = -LS(\mathbf{y}, \epsilon) \tag{5.192}$$

Considering the Taylor expansion of $S(\mathbf{y}, \epsilon)$, we find following relation.

$$S(\mathbf{y}, \epsilon) = S(\mathbf{y}, 0) + \sum_{n=1}^{\infty} \frac{\partial^n S(\mathbf{y}, \epsilon = 0)}{n! \partial \epsilon^n} \epsilon^n = \sum_{n=0}^{\infty} \frac{\epsilon^n (-L)^n}{n!} s(\mathbf{y}) = e^{-\epsilon L} s(\mathbf{y}) \tag{5.193}$$

Using this exponential operator $e^{-\epsilon L}$, we have following coordinate transformation.

$$z^{\mu} = e^{-\epsilon L} Z^{\mu}, \; Z^{\mu} = e^{\epsilon L} z^{\mu} \tag{5.194}$$

It should be noted that the transformation law of the scalar $S(\mathbf{y}, \epsilon) = e^{-\epsilon L} s(\mathbf{y})$ is opposite to this coordinate transformation.

We now consider the Lie transformation of the differential one form $\gamma = \gamma_{\mu} dz^{\mu}$. This differential one-form may be expressed in the new coordinates $Z^{\mu} = Z^{\mu}(z)$ as $\gamma_{\mu} dz^{\mu} = \Gamma_{\mu} dZ^{\mu}$. This gives rise to a following transformation law for the μ component of the differential one-form Γ,

$$\Gamma_{\mu}(\mathbf{Z}, \epsilon) = \frac{\partial z^{\nu}}{\partial Z^{\mu}} \gamma_{\nu}(z(\mathbf{Z}, \epsilon)) \tag{5.195}$$

Taking the partial derivative $\partial / \partial \epsilon$ of above equation and using $\partial z^{\mu} / \partial \epsilon = -g^{\mu} \partial z^{\mu} / \partial Z^{\nu}$, we have

$$\begin{aligned}
\frac{\partial \Gamma_{\mu}}{\partial \epsilon} &= \left[\frac{\partial}{\partial \epsilon} \frac{\partial z^{\nu}}{\partial Z^{\mu}} \right] \gamma_{\nu} + \frac{\partial z^{\nu}}{\partial Z^{\mu}} \frac{\partial}{\partial \epsilon} \gamma_{\nu} \\
&= -\left[\frac{\partial}{\partial Z^{\mu}} [g^{\lambda} \frac{\partial z^{\nu}}{\partial Z^{\lambda}}] \right] \gamma_{\nu} - g^{\lambda} \frac{\partial z^{\nu}}{\partial Z^{\mu}} \frac{\partial \gamma_{\nu}}{\partial Z^{\lambda}} \\
&= -\frac{\partial}{\partial Z^{\mu}} \left[g^{\lambda} \frac{\partial z^{\nu}}{\partial Z^{\lambda}} \gamma_{\nu} \right] - g^{\lambda} \left[\frac{\partial z^{\nu}}{\partial Z^{\mu}} \frac{\partial \gamma_{\nu}}{\partial Z^{\lambda}} - \frac{\partial z^{\nu}}{\partial Z^{\lambda}} \frac{\partial \gamma_{\nu}}{\partial Z^{\mu}} \right]
\end{aligned}$$

(continued)

$$= -\frac{\partial}{\partial Z^\mu}[g^\lambda \Gamma_\lambda] - g^\lambda \left[\frac{\partial}{\partial Z^\lambda}(\frac{\partial z^\nu}{\partial Z^\mu}\gamma_\nu) - \frac{\partial}{\partial Z^\nu}(\frac{\partial z^\nu}{\partial Z^\lambda}\gamma_\nu) \right]$$

$$= -\frac{\partial}{\partial Z^\mu}[g^\lambda \Gamma_\lambda] - g^\lambda \left[\frac{\partial \Gamma_\mu}{\partial Z^\lambda} - \frac{\partial \Gamma_\lambda}{\partial Z^\mu} \right] \tag{5.196}$$

We define the differential operator L and g as,

$$L\Gamma_\mu = g^\lambda \left[\frac{\partial \Gamma_\mu}{\partial y^\lambda} - \frac{\partial \Gamma_\lambda}{\partial y^\nu} \right], g = g^\mu \frac{\partial}{\partial Z^\mu} \tag{5.197}$$

Using the internal product relation (see A.4), $g \cdot \Gamma = g^\lambda \frac{\partial}{\partial Z^\lambda} \cdot \Gamma_\lambda dZ^\lambda = g^\lambda \Gamma_\lambda$, following one form relation is obtained from (5.196).

$$\frac{\partial \Gamma_\mu}{\partial \epsilon} dZ^\mu = -\frac{\partial (g \cdot \Gamma)}{\partial Z^\mu} dZ^\mu - L\Gamma_\mu dZ^\mu, \tag{5.198}$$

which can be expressed as:

$$\frac{\partial \Gamma}{\partial \epsilon} = -d(g \cdot \Gamma) - L\Gamma \tag{5.199}$$

The first term is a total derivative (or Gauge term). Noting that product of two operator is always vanish, i.e. $Ld = 0, dL = 0$, we have

$$\frac{\partial^n \Gamma}{\partial \epsilon^n} = (-dg)^n \cdot \Gamma + (-L)^n \Gamma \tag{5.200}$$

Noting that $\Gamma(\epsilon = 0) = \gamma$, we have

$$\Gamma = \sum_{n=0}^{\infty} \frac{\partial^n \Gamma}{\partial \epsilon^n}\bigg|_{\varepsilon=0} \frac{\epsilon^n}{n!} = \sum_{n=0}^{\infty} (-L)^n \frac{\epsilon^n}{n!} \gamma + dS = e^{-\epsilon L}\gamma + dS \tag{5.201}$$

If we define $T_n = e^{-\epsilon^n L_n}$ and multiple transformation as $T = --T_3 T_2 T_1$:

$$\Gamma = T\gamma + dS \tag{5.202}$$

The Lie transformation (5.188) is written as $Z^\mu = T^{-1} z^\mu$, where $T = --T_3 T_2 T_1$ is a sequence of Lie transforms. Each Lie transform T_n is given by $T_n = e^{-\epsilon^n L_n}$ and,

$$\text{For scalar, } L_n \equiv g_n^\mu \frac{\partial}{\partial z^\mu} \tag{5.203}$$

(continued)

$$\text{For one-form, } [L_n\gamma]_\mu = \sum_{\nu=0}^{6} g_n^\nu \left[\frac{\partial \gamma_\mu}{\partial z^\nu} - \frac{\partial \gamma_\nu}{\partial z^\mu}\right] \tag{5.204}$$

where g_n^μ is generation function of the Lie transform T_n and satisfy the relation $\partial Z^\mu/\partial \epsilon^n = g_n^\mu(\mathbf{Z})$. With this Lie transform, scalar s is transformed as $S = Ts$.

In case of Lie transform of differential 1 form, $\gamma = \gamma_\mu dz^\mu$ transforms to $\Gamma = \Gamma_\mu dZ^\mu$ as $\Gamma = T\gamma + dS$, where dS is the gauge term and $T = -- T_3 T_2 T_1$.

Expanding $\Gamma, \gamma, T = -e^{-\epsilon^2 L_2} e^{-\epsilon L_1}$ and dS as $\Gamma = \Gamma_0 + \epsilon\Gamma_1 + \epsilon^2\Gamma_2 + --$, $\gamma = \gamma_0 + \epsilon\gamma_1 + \epsilon^2\gamma_2 + --$, $T = 1 - \epsilon L_1 + \epsilon^2\left(\frac{1}{2}L_1^2 - L_2\right) + --$, $dS = dS_0 + \epsilon dS_1 + \epsilon^2 dS_2 + --$, we obtain following relations in each order of ϵ.

$$\Gamma_0 = dS_0 + \gamma_0 \tag{5.205}$$

$$\Gamma_1 = dS_1 - L_1\gamma_0 + \gamma_1 \tag{5.206}$$

$$\Gamma_2 = dS_2 - L_2\gamma_0 + \gamma_2 - L_1\gamma_1 + \frac{1}{2}L_1^2\gamma_0 \tag{5.207}$$

We also have approximate relation between two coordinates z and $\mathbf{Z}$ using the generating function as follows,

$$Z^i = z^i + \epsilon\frac{\partial Z^i}{\partial \epsilon}|_{\epsilon=0} + O(\epsilon^2) = z^i + \epsilon g_1^i(z^i) + O(\epsilon^2) \tag{5.208}$$

Notice: In the exponential operator in general, we observe $e^{L_1}e^{L_2} \neq e^{(L_1+L_2)}$. The Lie group and Lie algebra appears to describe the difference. For smallness parameter ϵ, exponential operator has a relation $e^{\epsilon X}e^{\epsilon Y} = e^{\epsilon(X+Y)+\frac{1}{2}\epsilon^2[X,Y]+O(\epsilon^3)}$, where $[X, Y] \equiv XY - YX$ is a commutator of operators.

5.4 Linear Gyrokinetics of Drift Waves

5.4.1 Global Structure of ITG/TEM

The structure of the toroidal drift waves such as ITG/TEM and ETG are of semi-global nature. The global linear mode structure of toroidal drift wave using linear gyro kinetic equation (5.103) has been solved by Brunner [83] and the procedures are sketched in this section. For the electrostatic drift waves, eigenmode structure can be determined using the charge neutrality condition $\sum_a e_a\tilde{n}_a = 0$, where the perturbed density $\tilde{n}_a$ can be determined from the perturbed distribution function

$f_{a1} = -e_a\tilde{\Phi} f_{aM}/T_a + g_a$. Using (5.103), we may obtain $\langle g_a\rangle$ by the integration along the unperturbed orbit:

$$\langle g_a\rangle = e_a \int_{-\infty}^{t} dt' \left[-\frac{\partial f_{aM}}{\partial K}\frac{\partial\langle\tilde{\Phi}\rangle}{\partial t} + \frac{\partial\langle\tilde{\Phi}\rangle}{\partial \boldsymbol{r}} \cdot \frac{\boldsymbol{b}}{\Omega_a} \times \frac{\partial f_{aM}}{\partial \boldsymbol{r}} \right] \tag{5.209}$$

The density perturbation $\tilde{n}_a$ is given as:

$$\tilde{n}_a = -\frac{e_a\tilde{\Phi}}{T_a} n_a + \int g_a d\boldsymbol{v}, \tag{5.210}$$

where the first term in the right hand side represents Boltzmann response, which is the reason why $-e_a\tilde{\Phi} f_{aM}/T_a$ is called 'adiabatic'.

The Fourier spectrum of the density perturbation in space having the wave number vector $\boldsymbol{k}$ may be written as:

$$\tilde{n}_a(\omega,\boldsymbol{k}) = \int d\boldsymbol{x} e^{-i\boldsymbol{k}\cdot\boldsymbol{x}} \tilde{n}_a(\omega,\boldsymbol{x}) = \int d\boldsymbol{k}' M_a(\omega,\boldsymbol{k},\boldsymbol{k}')\tilde{\Phi}(\boldsymbol{k}'), \tag{5.211}$$

where $M_a(\omega,\boldsymbol{k},\boldsymbol{k}')$ is the density response function from the electrostatic potential perturbation $\tilde{\Phi}(\boldsymbol{k}')$. In the toroidal geometry with coordinates (r,θ,ζ), we may use $\boldsymbol{k} = (2\pi k/\Delta r)\nabla r + m\nabla\theta + n\nabla\zeta$, where Δr is radial width of the mode and (k,m,n) are integers. The charge neutrality condition gives rise to the linear eigen mode equations: $\sum_{k'}\sum_a M_a(\omega,\boldsymbol{k},\boldsymbol{k}')\tilde{\Phi}(\boldsymbol{k}') = 0$. Therefore, the dispersion relation is:

$$detM(\omega,\boldsymbol{k}) = 0 \text{ , where } M_{k',m',n'} = \sum_a M_a(\omega,\boldsymbol{k},\boldsymbol{k}') \tag{5.212}$$

The density response functions for trapped and passing ions $M_i(\omega,\boldsymbol{k},\boldsymbol{k}')$ are separately calculated while that for the electron is calculated using Eq. (5.122) for non-adiabatic electron response due to trapped electron. Obviously the non-adiabatic response for the passing electron is assumed to be zero.

Based on the fact that all higher radial mode numbers are averaged out due to the FLR effect, the maximum radial mode number is given by $k_{max} = \Delta\rho/2\pi\rho_i$ so that only $|k| \le k_{max}$ are considered. For a fixed toroidal mode number n, many poloidal modes are coupled each other and their range is determined by the analytical estimate by Connor-Wilson [146].

The solution of Eq. (5.212) is searched in the complex ω plane using the Davies algorithm [152] (higher order Nyquist algorithm in which the contour integral path is chosen from the estimated position of zero's from the equi-contour plot of $|M(\omega)|$).

Eigenmode characteristics of the linear gyro kinetic drift waves are calculated and discussed in detail for a circular large tokamak (maximum $\eta_i = 4.0$ at $q =$

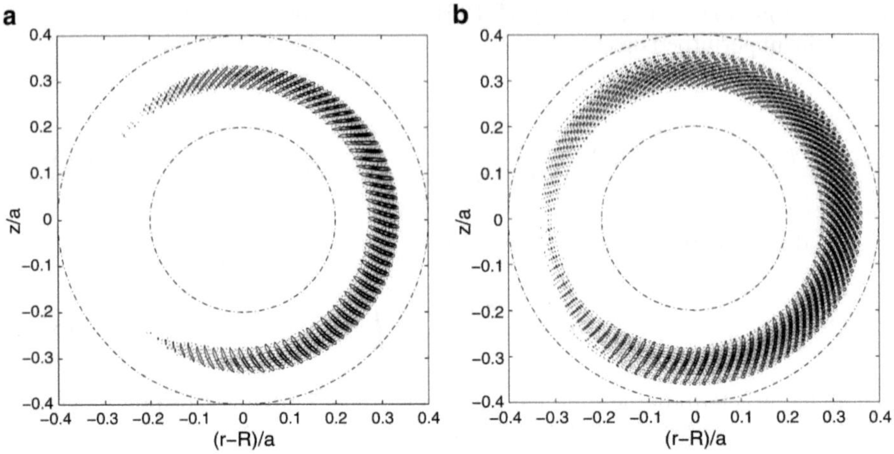

Fig. 5.6 Global linear eigenmode structures of toroidal ITG and TEM modes. Reproduced with permission from Brunner [83]. Copyright AIP Publishing LLC. (**a**) n = 64 ITG. (**b**) n = 64 TEM

$1.5, s = 0.6, T_e/T_i = 1, \epsilon_n = L_n/R = 0.29, \epsilon = 0.11$) with trapped electron dynamics [83]. A typical linear eigenmode structures of toroidal ITG and TEM are shown in Fig. 5.6. The mode structure is radially elongated and poloidally tilted due to dense overup of poloidal harmonics. This mode structure is closely related to the ideal MHD ballooning mode structure first developed by Connor-Hastie-Taylor [141]. The real frequency and the growth rate for the toroidal mode number scan are also shown in Fig. 5.7. The ITG rotates in the ion diamagnetic drift direction and the TEM in the electron diamagnetic drift direction. The ITG modes are unstable for a broad range of toroidal number $n \sim$4–100 and the growth rate is highest $n =$50–60 and higher modes have lower growth rate due to the averaging of fluctuating electrostatic potential by the finite Larmor radius effect. The ITG growth rate is strongly enhanced by including trapped electron dynamics.

Magnetic shear has an important effect on the eigenmode characteristics of the linear gyro kinetic drift waves, which is calculated for a circular large tokamak for a fixed toroidal mode number $n = 10$ which corresponds to $k_\theta \rho_i = 0.35$ at $q = 1.5$ (maximum $\eta_i = 2.5$ at $q = 1.5, T_e/T_i = 1, \epsilon_n = L_n/R = 0.25, \epsilon = r/R = 0.11$) as shown in Figs. 5.8 and 5.9 [83].

There are several modes such as three ITGs (modes $1, 2, 3$) and two TEMs (modes $4, 5$) in case of positive shear $s = 1$ (Fig. 5.8). The growth rate of most unstable ITG (mode 1) is largest at $s \sim 0.5$ and stabilized at sufficiently negative magnetic shear after transition to the TEM, while other two ITG modes remain unstable with much lower growth rates ($1/4$ of value at $s \sim 0.5$).

Two TEMs (modes $4, 5$) are also stabilized for strongly negative magnetic shear. This stabilization of TEMs is explained by the decrease of the average amplitude of the toroidal precession drift which is an instability drive of TEMs.

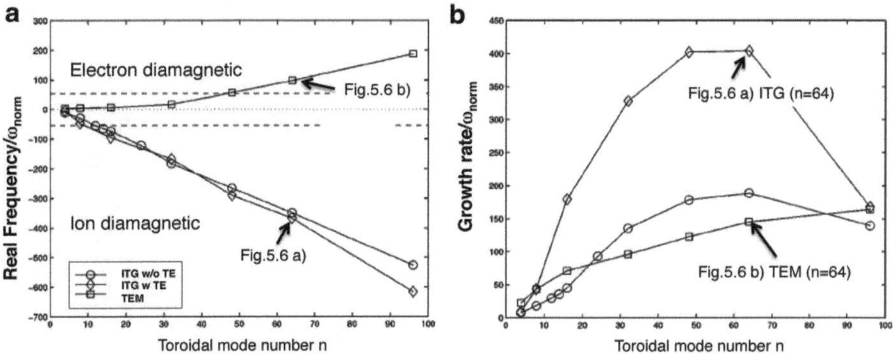

Fig. 5.7 The real frequency and the growth rate as a function of toroidal mode number for toroidal ITG (with and without trapped electron dynamics) and TEM. Reproduced with permission from Brunner [83]. Copyright AIP Publishing LLC. (**a**) Real frequency/ω_{norm}. (**b**) Growth rate/ω_{norm}

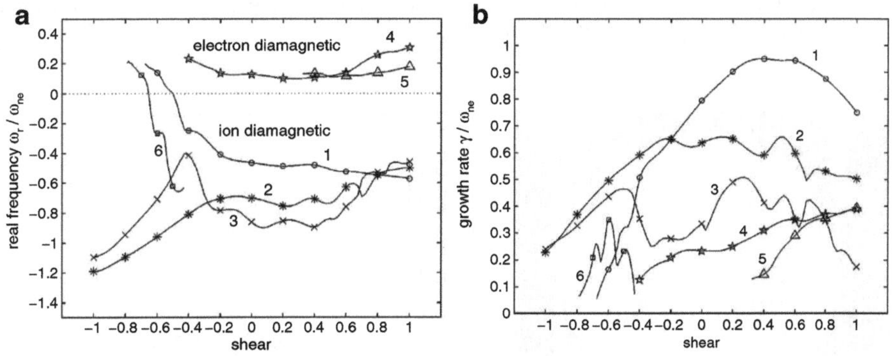

Fig. 5.8 The real frequency and the growth rates as a function of the magnetic shear. Reproduced with permission from Brunner [83]. Copyright AIP Publishing LLC

Figure 5.10 shows 2D eigenmode structure of toroidal ITG drift wave and its decomposition in poloidal harmonics from the linear gyrokinetic simulation by Idomura [366]. While mode is radially elongated, there is a poloidal angle $\theta = \theta_0$where the electrostatic potential is purely radially elongated. When the 2D mode structure is Fourier decomposed in poloidal harmonics, mode amplitudes are peaked at respective resonant surface and the each mode has relatively small radial envelope while overall envelope is much wider.

In the axisymmetric system, the electrostatic potential $\tilde{\Phi}$ can be expressed as a summation of poloidal harmonics with harmonic amplitude $\tilde{\Phi}_l(q)$ constructed by the eigenfunction for $l = 0$, $\tilde{\Phi}_0(nq - m)$ peaked at $q = m/n$ and assuming translational symmetry similar to Bloch function in crystal lattice [459, 851].

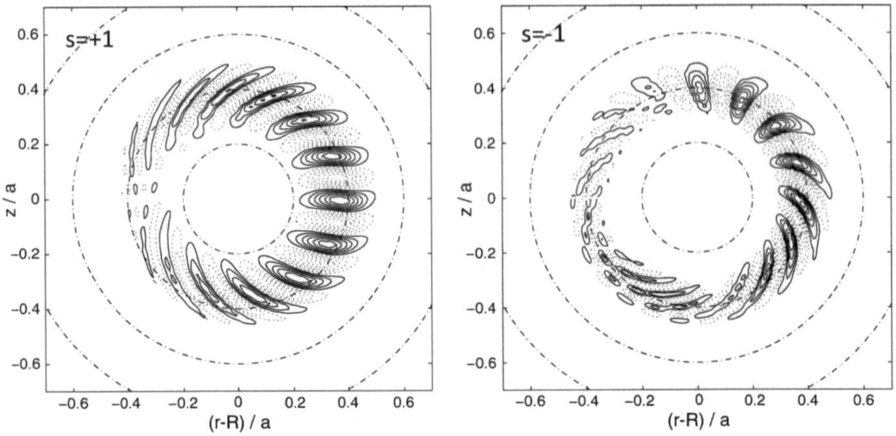

Fig. 5.9 Global linear eigenmode structure of most unstable modes for $s = 1$ and $s = -1$. Both are toroidal ITG mode rotating in the ion diamagnetic drift direction. Reproduced with permission from Brunner [83]. Copyright AIP Publishing LLC

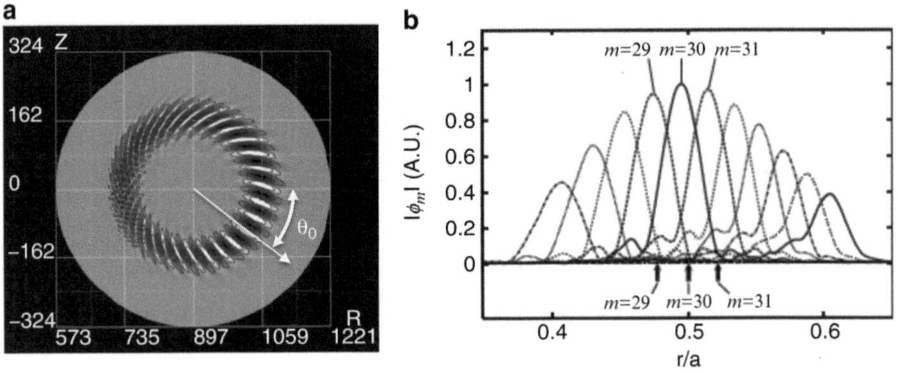

Fig. 5.10 (**a**) Eigenmode structure in poloidal cross section. (**b**) Radial harmonics of toroidal ITG with $n = 15$ and $a/\rho_i = 324$. Reproduced with permission from Idomura [366]. Copyright IAEA Vienna

$$\tilde{\Phi}(r,\theta,\zeta) = \sum_{l=-\infty}^{+\infty} \tilde{\Phi}_l(q)e^{i(m+l)\theta}e^{-in\zeta} \tag{5.213}$$

$$\tilde{\Phi}_l(q) = a(\Delta q)\tilde{\Phi}_0(nq - m - l) \tag{5.214}$$

where, each harmonics is located in the region where $q = (m+l)/n$. Here, q is safety factor used as radial coordinate, $\Delta q = l/n$, $a(\Delta q)$ is slowly varying amplitude corresponding to envelope of modes in Fig. 5.10. This is called Ballooning eigenfunction, originally developed for ideal MHD Ballooning mode. This translational symmetry holds except the mode edge. This translational symmetry comes from the dense overlap of poloidal harmonics and breaks down in case magnetic shear is very weak, similar to infernal mode discussed in Sect. 8.4.3.

5.4.2 *Electron Temperature Gradient (ETG) Mode*

In this section, we show some calculations based on the modern gyrokinetic theory.

The gyro kinetic Vlasov-Poisson system for slab ITG/ETG in the gyro-averaged coordinates $\boldsymbol{Z} = (t, \boldsymbol{R}, v_\parallel, \mu, \theta)$ is derived by treating the magnetic shear as a perturbation by Idomura [365] ($\langle\rangle$ stands for gyro-phase average):

$$\frac{\partial F_a}{\partial t} + \left[-\frac{\partial \langle \Phi \rangle}{\partial \boldsymbol{R}} \times \frac{\boldsymbol{b}}{B} + \frac{v_\parallel}{B} \langle \boldsymbol{B} \rangle \right] \cdot \frac{\partial F_a}{\partial \boldsymbol{R}} - \frac{e_a}{m_a B} \langle \boldsymbol{B} \rangle \cdot \frac{\partial \langle \Phi \rangle}{\partial \boldsymbol{R}} \frac{\partial F_a}{\partial v_\parallel} = 0 \quad (5.215)$$

The distribution function F_a and the electrostatic potential $\langle \Phi \rangle$ are Fourier expanded as $e^{-i\omega t + i\boldsymbol{k}\cdot\boldsymbol{R}}$. After velocity space integration, density perturbation is expressed in terms of the Fourier series of potential perturbation and the charge neutrality condition gives following equation.

$$\sum_{k_l} L_{k_m,k_l}(\omega) \Phi_{k_l} = 0, \quad (5.216)$$

where L_{k_m,k_l} is the element of the complex matrix $L(\omega)$ similar to M in previous section. The dispersion relation is given by $detL = 0$, which is solved using Davies method as well. Figure 5.11 shows a comparison of linear growth rates of slab ITG and ETG [365].

In the typical fusion relevant tokamak plasmas, the electron gyro radius ρ_e can be smaller than the Debye length λ_e, i.e. $\lambda_{De}^2/\rho_{te}^2(\sim B^2/n_e) \geq 1$. So the Debye shielding effect significantly modify the ETG mode structure. So the shift of unstable region in $k_y \rho_{ti}$ from ITG to ETG may not scale by ρ_{ti}/ρ_{te} but by ρ_{ti}/λ_{De} as shown for the normal shear case in Fig. 5.11, while negative shear case is more complex.

While ETG has a finer spatial scale than ITG, ETG can play a significant role in the electron heat transport due to radially elongated structure and also due to high threshold for the $E \times B$ shear stabilization.

The Gyrokinetic Vlasov-Poisson equation for toroidal ETG in the gyro-averaged coordinates, $\boldsymbol{Z} = (t; \boldsymbol{R}, v_\parallel, \mu)$ is given as [367]:

$$\frac{\partial F_a}{\partial t} + \frac{d\boldsymbol{R}}{dt} \cdot \frac{\partial F_a}{\partial \boldsymbol{R}} + \frac{dv_\parallel}{dt} \frac{\partial F_a}{\partial v_\parallel} = 0 \quad (5.217)$$

$$\frac{d\boldsymbol{R}}{dt} = v_\parallel \boldsymbol{b} + \frac{1}{e_a B^*} \boldsymbol{b} \times \left[e_a \frac{\partial \langle \Phi \rangle}{\partial \boldsymbol{R}} + m_a v_\parallel^2 \boldsymbol{b} \cdot \frac{\partial}{\partial \boldsymbol{R}} \boldsymbol{b} + \mu B \frac{\partial lnB}{\partial \boldsymbol{R}} \right] \quad (5.218)$$

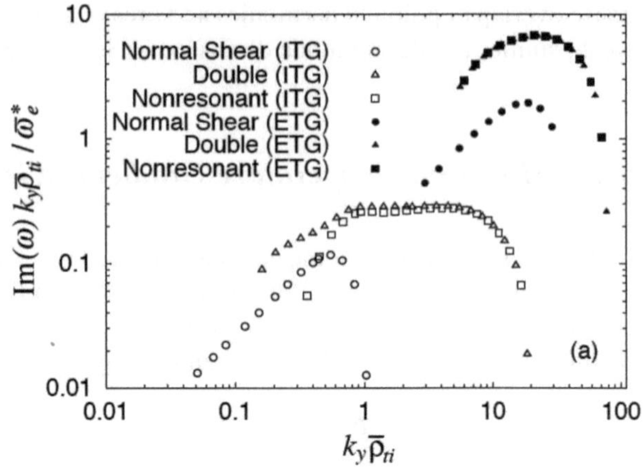

Fig. 5.11 The growth rates of slab ITG and ETG as a function of $k_\perp \rho_i$. 'Double' means there are two resonant surfaces. 'Non-resonant' means there is no resonant surface. Shift in $k_y \rho_{ti}$ from ITG to ETG unstable region is $\sim \rho_{ti}/\lambda_{De} \sim 24$ for normal shear case. Reproduced with permission from Idomura [365]. Copyright IAEA Vienna

$$\frac{dv_\parallel}{dt} = -\frac{\boldsymbol{B}^*}{m_a B^*} \cdot \left[e_a \frac{\partial \langle \Phi \rangle}{\partial \boldsymbol{R}} + m_a v_\parallel^2 \boldsymbol{b} \cdot \frac{\partial}{\partial \boldsymbol{R}} \boldsymbol{b} + \mu B \frac{\partial ln B}{\partial \boldsymbol{R}} \right] \tag{5.219}$$

$$\delta n_a = \int \delta f_a(\boldsymbol{R}, v_\parallel, \mu, t) \delta(\boldsymbol{R} + \boldsymbol{\rho}_a - \boldsymbol{x}) d^6 \boldsymbol{Z} - \frac{e_a n_a}{T_a} \sum_{\boldsymbol{k}} \left[1 - I_0(k_\perp^2 \rho_{ta}^2) e^{-k_\perp^2 \rho_{ta}^2} \right] \Phi_{\boldsymbol{k}} e^{i\boldsymbol{k}\cdot\boldsymbol{x}} \tag{5.220}$$

$$-\nabla^2 \Phi = \frac{1}{\epsilon_0} \sum_a e_a \delta n_a(\boldsymbol{x}) \tag{5.221}$$

Here, $\delta f_a = F_a - F_{a0}$ is the perturbed distribution function and F_{a0} is an equilibrium distribution function.

The ETG mode is characterized by a short wavelength, $k_\theta \rho_{te} \sim 1$ and $k_\theta \rho_{ti} \gg 1$, where k_θ is the poloidal wave number. In this mode, kinetic ion response vanishes and the ion response becomes adiabatic [365, 399].

In order to calculate accurate eigenmodes efficiently, Idomura developed quasi-ballooning mode representation and applied to both ITG [366] and ETG [367] mode analyses. This quasi-ballooning mode representation is originally developed for linear MHD solver by Grüber [269] and is applicable not only to high n modes but also medium n modes without any approximation. In the quasi-ballooning mode representation, Φ is expressed as,

$$\Phi(r, \theta, \zeta) = \sum_n \Phi_n(r, \theta) e^{-in\zeta + iS(\theta)} \tag{5.222}$$

$$S(\theta) = nq(r_s)\chi(\theta) \tag{5.223}$$

$$\chi(\theta) = \frac{1}{q(r_s)} \int_0^\theta \frac{\boldsymbol{B} \cdot \nabla \zeta}{\boldsymbol{B} \cdot \nabla \theta'} |_{r=r_s} d\theta' \tag{5.224}$$

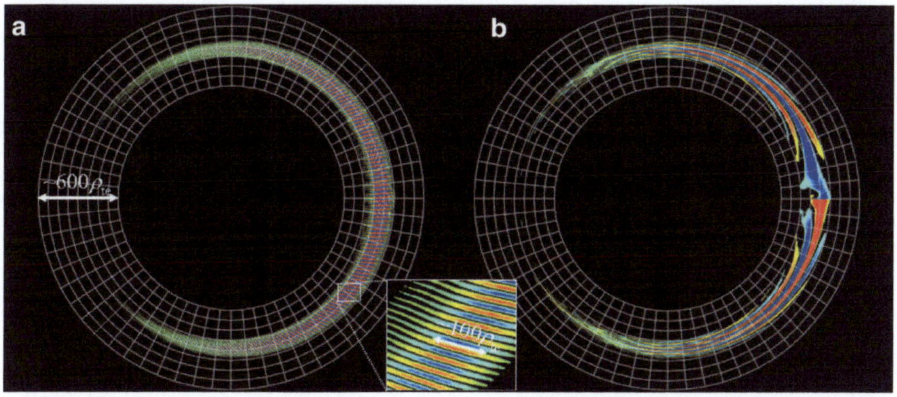

Fig. 5.12 (**a**) Linear eigenmode structure of Toroidal ETG mode showing typical ballooning mode structure. (**b**) The envelope function $\Phi_n(r,\theta)$, which has jump at $\theta = 0$ but the real eigen function do not have such gap. Reproduced with permission from Idomura [367]. Copyright IAEA Vienna

Here, $\Phi_n(r,\theta)$ becomes a slowly varying function, and q is safety factor, χ is poloidal straight field line angle at $r = r_s$ and r_s is radius of reference magnetic surface where drift waves may be most unstable (namely, radius where dT/dr is largest). In order to satisfy periodic boundary condition in the poloidal direction, $\Phi_n(r,\theta)$ is solved under the boundary condition of $\Phi_n(r,0) = \Phi_n(r,2\pi)e^{iS(2\pi)}$.

The use of quasi-ballooning mode representation for electrostatic potential Φ becomes essential for the ETG mode. Figure 5.12 shows typical linear eigenmode structure of toroidal ETG in the poloidal cross section. Mode shows ballooning character having large amplitude in the outboard and has radially elongated structure similar to eigenfunction of toroidal ITG but is much finer in poloidal direction with typical radial length $\sim 100\rho_e$.

This radially elongated mode structure is coming from strong mode over up in the radial direction as a typical characteristics of ballooning mode, which is found originally in the MHD ballooning mode theory by Connor-Hastie-Taylor [141] and Zakharov [851] (see Chap. 6).

Figure 5.13 shows safety factor q and the magnetic shear $s = rdq/dr/q$ profiles used in the Gyrokinetic calculation of ETG modes by Idomura [367] and the radial mode over up of poloidal harmonics for positive shear and negative shear cases. As we expected, poloidal harmonics densely overlap radially for the positive magnetic shear case. In the negative magnetic shear case, however, only a few poloidal harmonics are destabilized near the q_{min} surface. But non-resonant mode, which do not have resonant surface in the plasma can be excited and plays an important role as observed numerically by Candy [97]. Guttenfelder [274] showed ETG accounts for NSTX electron transport.

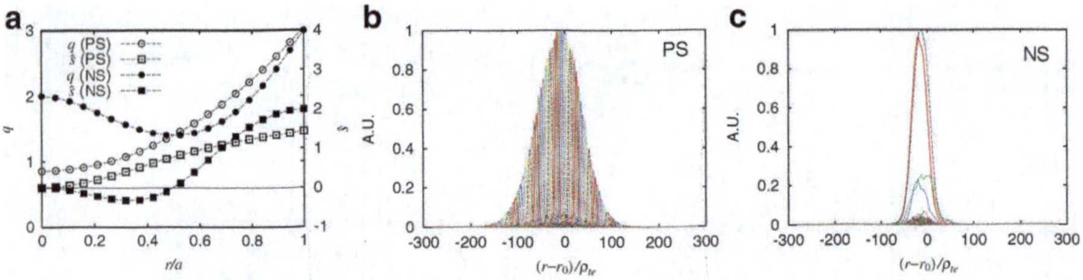

Fig. 5.13 (**a**) Safety factor q and the magnetic shear $s = rdq/dr/q$ profiles in positive shear and negative shear cases. (**b**) Amplitude profiles of various poloidal modes of toroidal ETG for positive magnetic shear case ($a = 2046\rho_{te}$). (**c**) Amplitude profiles of various poloidal modes of toroidal ETG for negative magnetic shear case ($a = 2046\rho_{te}$). Reproduced with permission from Idomura [367]. Copyright IAEA Vienna

Chapter 6
Fundamentals of Ballooning Modes in Tokamak

Abstract Tokamak flux surfaces are dominated by irrational surfaces, which brings the mathematical problem how to satisfy double periodicity in toroidal and poloidal angles for flute like perturbation expressed by the eikonal form. Ballooning transform is essential method to analyze 2D mode structure in toroidal geometry not only to ideal Ballooning mode, but also for high n toroidal drift waves and Alfven eigenmodes. Fundamentals of ballooning mode structure in toroidal plasma is discussed for the application to MHD instabilities and toroidal drift waves. After an introduction of ballooning transform from real geometry to covering space in section 6.1, the method to satisfy double periodicity flute like perturbation is discussed using the eikonal form in the flux coordinates using Poisson sum and its relation to translational symmetry in 6.2. The 2D hallooning transform and twisted radial Fourier transform are discussed in 6.3. Trapped and passing mode structures in 2D wave equation are discussed using the WKBJ formulation in 6.4. Poisson sum, Bloch theorem and WKBJ solutions are given as Columns.

Further Readings:
Books: White (2006) [818] includes ballooning mode theory in the flux coordinates. Kikuchi (2011) [443] gives formulation using flux coordinates on ballooning transform and its relation to translational symmetry. For WKBJ method, you may read Jeffreys (1962) [397], Heading (2013) [311], White (2010) [819].

Key Papers: Connor-Hastie-Taylor (1978, 1979) [141], [142] is fundamental papers on ballooning transform. Translational symmetry is discussed by Lee-Van Dam (1977) [496] and Zakharov (1979) [851]. Dewar (1979, 1981) [156], [157] gives transparent ballooning formulation in the covering space using the flux coordinates and trapped and passing modes in 2D wave equations.

M. Kikuchi, M. Azumi, *Frontiers in Fusion Research II*,
DOI 10.1007/978-3-319-18905-5_6

6.1 Double Periodicity and Ballooning Mode

In this section, we gives an analytical theory of high-n wave equation using ballooning formalism originally developed for ideal MHD ballooning mode by Connor et al., (Fig. 6.1) [141, 142]. Here we use the flux coordinates (straight field line coordinates: (r, θ, ζ)) where $\boldsymbol{B} = \nabla\alpha \times \nabla\psi$, $\alpha = \zeta - q\theta$, and the radius r is defined using the toroidal flux ϕ as $r = a(\phi/\phi_a)^{1/2}$. Since tokamak has toroidal symmetry, the toroidal mode number n is separable ($\varphi(r, \theta, \zeta, t) = e^{-i\omega t - in\zeta}\varphi(r, \theta)$) while the poloidal mode number m is not. Thus we have to solve following 2-D eigenmode equation in (r, θ) plane where $\theta \in (-\pi, +\pi)$.

$$L(r, \theta)\varphi(r, \theta) = \lambda\varphi(r, \theta) \tag{6.1}$$

where r is the flux label and θ is the poloidal angle to make magnetic field line to be a straight line. The differential operator L and the eigen function φ are periodic in θ so that we express φ in the following form called the "Ballooning Transform".

$$\varphi(r, \theta) = \sum_{m=-\infty}^{+\infty} e^{im\theta} \int_{-\infty}^{+\infty} e^{-im\eta}\hat{\varphi}(r, \eta)d\eta \tag{6.2}$$

Here $a_m \equiv \int e^{-im\eta}\hat{\varphi}(r, \eta)d\eta$ is Fourier coefficient represented by the Fourier integral and $\hat{\varphi}$ is a solution of the following equation in the domain $\eta \in (-\infty, \infty)$

Fig. 6.1 (**a**) Joseph Fourier (1768–1830) is a French mathematician and physicist who invented Fourier analysis. (**b**) Dr. Jack W. Connor. (**c**) Dr. John Bryan Taylor in 1965. JBT and JWC are outstanding theoreticians in plasma physics with many discoveries. JBT is recipient of Alfven and Maxwell Prizes and is well-known by the Taylor relaxation state of the reversed field pinch (RFP). JWC is recipient of Alfven Prize and is a lead author of Ballooning transformation and invariant principle. They have many publications with Dr. R.J. Hastie (Alfven Prize winner as well)

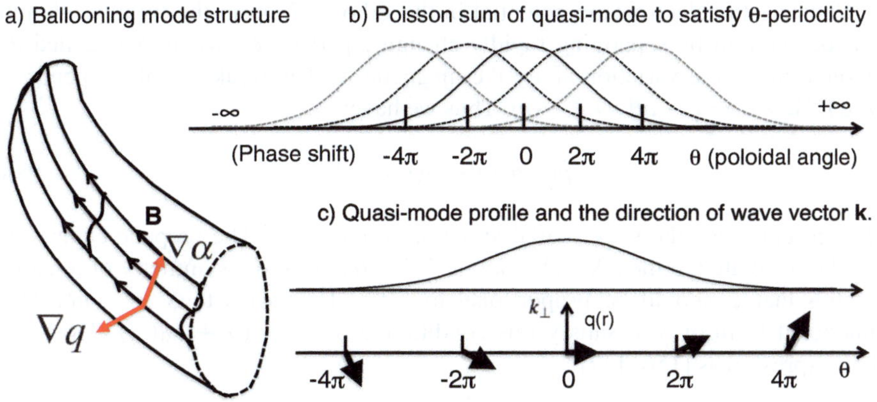

Fig. 6.2 (**a**) Ballooning mode structure of low frequency toroidal drift wave. (**b**) Quasi mode in the covering space $(-\infty, +\infty)$. (**c**) Tilting of wave number vector $\boldsymbol{k}_\perp$ in the covering space for $\theta_0 = 0$

called the "covering space" and the η is an angle variable in the covering space.

$$L(r,\eta)\hat{\varphi}(r,\eta) = \lambda\hat{\varphi}(r,\eta) \tag{6.3}$$

By using the delta function formula for $\sum e^{im(\theta-\eta)}$ (Column 6-1), we can show $\varphi(r,\theta) = \sum_{j=-\infty}^{+\infty}\hat{\varphi}(r,\theta+2\pi j)$. The ballooning transformation is a mapping $(-\pi,\pi) \to (-\infty,+\infty)$.

6.2 1D Ballooning Transform

In this section, we introduce a special case of ballooning transform, here we call "1D Ballooning transform", for the high n flute like perturbation given by the Eikonal form e^{iS}, which has been formulated by Lee-Van Dam [496] and Zakharov [851] including the arbitrary phase shift.

6.2.1 Eikonal Formulation

So-called high-n toroidal flute-like mode is characterized by a long wave length along the magnetic field $\lambda_\parallel \sim qR$ and short perpendicular wave length $\lambda_\perp \sim a/n$ where n is the toroidal mode number (Fig. 6.2a). Since the magnetic field line winds infinitely around the torus except rational surfaces (whose measure is zero), the perturbation also winds infinitely in the Riemann sheets and does not satisfy periodic condition in θ. This implied that the mode is better given in the covering space $\eta \in (-\infty,+\infty)$ for the poloidal angle.

Consider the electrostatic potential in the Eikonal form in the weakly inhomogeneous system by separating rapidly changing part by e^{iS} where S is called the eikonal. Since the variation along the magnetic field is weak, S only depends on (r, α). Here $\alpha = \zeta - q\theta \to \zeta - q\eta$. Thus we have:

$$\hat{\varphi}(r, \eta, \zeta) \equiv u(r, \eta)e^{iS(r,\alpha)} \tag{6.4}$$

Here u represents the slow spatial variation of the wave. The perpendicular wave number is related to the eikonal as $\boldsymbol{k}_\perp = \nabla S(r, \alpha)$. Toroidal symmetry of tokamak implies that $\hat{\varphi}$ should be proportional to $e^{-in\zeta}$. This leads to $S \sim -n\zeta$. Thus functional form of S to satisfy this condition is $S = -n(\alpha + \alpha_0(r))$. The $\boldsymbol{k}_\perp$ is then expressed as [156, 157]:

$$\boldsymbol{k}_\perp = -n\left(\nabla\alpha + \theta_k\nabla q\right) = n\left[-\nabla\zeta + q\nabla\eta + (\eta - \theta_k)\nabla q\right] \tag{6.5}$$

Here, $\theta_k \equiv \alpha_0'(r)/q'(r)$ is a measure of radial wave number and integration gives $\alpha_0(r) = \int \theta_k dq$. The radial wave number $k_r = n(\eta - \theta_k)q'(r)$ goes to $\pm\infty$ as $\eta \to \pm\infty$ for finite magnetic shear $(dq/dr \neq 0)$ (Fig. 6.2c) [158]. The eikonal solution is:

$$\hat{\varphi}(r, \eta, \zeta) = \hat{\varphi}_0(r, \eta, \zeta) \equiv u(r, \eta)exp[-in(\zeta - q\eta + \int_0^q \theta_k dq)] \tag{6.6}$$

From this eigenfunction, we can construct an infinite set of solutions called the "quasi modes" by shifting $\eta \to \eta + 2\pi j$; $\hat{\varphi}_j(r, \eta, \zeta) = \hat{\varphi}_0(r, \eta + 2\pi j, \zeta)$ if the governing linear differential equation for φ is periodic by 2π (Fig. 6.2b).

$$\hat{\varphi}_j(r, \eta, \zeta) = u(r, \eta + 2\pi j)exp[-in(\zeta - q\eta + \int_0^q (\theta_k - 2\pi j)dq)] \tag{6.7}$$

The periodic solution in θ can be constructed by the summation of the quasi-modes.

$$\varphi(r, \theta, \zeta) = e^{-in\zeta} \sum_{j=-\infty}^{+\infty} u(r, \theta + 2\pi j)exp[in(q\theta - \int_0^q (\theta_k - 2\pi j)dq)] \tag{6.8}$$

This expression is a fundamental form of flute-like mode structure in the toroidal sheared magnetic geometry. We may call this the **1D Ballooning transform**.

We define a phase shift $\theta_0(r)$ [308] as:

$$q(r)\theta_0(r) \equiv \int \theta_k dq \tag{6.9}$$

Then, Eq. (6.8) is rewritten as:

$$\varphi(r, \theta, \zeta) = e^{-in\zeta} \sum_{j=-\infty}^{+\infty} u(r, \theta + 2\pi j)e^{inq(\theta - \theta_0 + 2\pi j)} \tag{6.10}$$

Column 6-1: S.D. Poisson and Poisson Summation Formula
Simeon Denis Poisson (1781–1840, Fig. 6.4) is a French Mathematician known by Poisson distribution and Poisson equation. He also found that infinite sum of function $f(n)$ is related to its Fourier transform $\hat{f}(k)$ (Appendix A.8). Poisson summation formula is equivalent to following delta function formula.

$$\sum_{m=-\infty}^{+\infty} e^{imx} = 2\pi \sum_{j=-\infty}^{+\infty} \delta(x - 2\pi j) \tag{6.11}$$

6.2.2 *Translational Symmetry in Ballooning Mode*

In this subsection, we analyze $\varphi(r, \theta, \zeta) \equiv \varphi(r, \theta)e^{-in\zeta}$ from the view point of the translational symmetry by writing (6.10) using the delta function as:

$$\begin{aligned}\varphi(r, \theta) &= \sum_{j=-\infty}^{+\infty} u(r, \theta + 2\pi j)e^{inq(\theta-\theta_0+2\pi j)} \\ &= \sum_{j=-\infty}^{+\infty} \int_{-\infty}^{+\infty} d\eta u(r, \eta + \theta_0)\delta(\eta - (\theta - \theta_0 + 2\pi j))e^{inq\eta}\end{aligned} \tag{6.12}$$

We apply the delta function formula, $2\pi \sum_{j=-\infty}^{+\infty} \delta(x - 2\pi j) = \sum_{m=-\infty}^{+\infty} e^{-imx}$.

$$\varphi(r, \theta) = \sum_{m=-\infty}^{+\infty} \int_{-\infty}^{+\infty} \frac{d\eta}{2\pi} u(r, \eta + \theta_0)e^{i(nq-m)\eta}e^{im(\theta-\theta_0)} \tag{6.13}$$

Defining $\eta' = \eta + \theta_0$, we have:

$$\varphi(r, \theta) = \sum_{m=-\infty}^{+\infty} \int_{-\infty}^{+\infty} \frac{d\eta'}{2\pi} u(r, \eta')e^{i(nq-m)\eta'}e^{-inq\theta_0}e^{im\theta} \tag{6.14}$$

We define $\varphi_0(r, nq - m)$ by the following radial Fourier integral.

$$\varphi_0(r, nq - m) = \int_{-\infty}^{+\infty} \frac{d\eta}{2\pi} u(r, \eta)e^{i(nq-m)\eta} \tag{6.15}$$

Here, we have the radial scale separation r as slow variable and $nq - m$ as fast variable. Then, we have following Fourier expansion in θ from (6.14).

$$\varphi(r,\theta) = \sum_{m=-\infty}^{+\infty} \varphi_m(r,nq)e^{im\theta}, \tag{6.16}$$

$$\text{where } \varphi_m(r,nq) = \varphi_0(r,nq-m)exp\left(-i\int_0^{nq}\theta_k d(nq)\right) \tag{6.17}$$

In this Fourier series expansion in θ, the φ_m resonates at different rational surfaces $q(r_m) = m/n$ as m varies. Namely $\varphi_m(r,nq)$ is a resonant mode at the rational surface r_m. The function $\varphi_0(r,nq-m)$ represents a localized eigenfunction near the rational surface $q(r_m) = m/n$ having translational symmetry as shown schematically in Fig. 6.3a. For a different m, the radial location of the peak given by nq moves as $m-1, m, m+1, --$. This translational symmetry includes a phase factor expected from the "Bloch theorem" (Column 6-2) given by $exp[-in\int\theta_k d(nq)]$. Since this integral at the rational surface $q(r_m) = m/n$ can be approximated as:

$$\int_0^{nq}\theta_k d(nq) \sim \sum_{j=1}^{m}\theta_{k,j}\Delta(nq) = \sum_{j=1}^{m}\theta_{k,j} \tag{6.18}$$

We find the phase factor between m and $m-1$ is $e^{-i\theta_{k,m}}$. Thus the θ_k corresponds to the "Bloch angle" in the crystal lattice. From (6.17) and (6.16),

$$\varphi(r,\theta) \sim \sum_{m=-\infty}^{+\infty} \varphi_0(r,nq-m)e^{im(\theta-\theta_0)} \tag{6.19}$$

since $m^{-1}\int_0^{nq}\theta_k d(nq)(r_m) = \theta_0$. The $\theta_0 = q^{-1}\int\theta_k dq$ is the phase aligned angle of Fourier harmonics so that the phase of the Fourier modes becomes zero at $\theta = \theta_0$ (6.10) schematically shown in Fig. 6.3b. We should be careful that constant θ line is not a straight line in the actual flux coordinates (q,θ,ζ) (Fig. 6.3c).

Column 6-2: Bloch Theorem in Solid State Physics

Felix Bloch (1905–1983) is Swiss physicist who received 1952 Nobel Prize in physics with E.M. Purcell for their development of nuclear magnetic resonance (NMR). He derived the **Bloch Theorem** in 1928.

We consider the electron dynamics in crystal lattice with a lattice constant a. Since the crystal lattice has translational symmetry, the potential energy of the electron satisfy periodicity $V(\boldsymbol{r}) = V(\boldsymbol{r}+\boldsymbol{R})$. The solution of the Schrödinger equation $\psi(\boldsymbol{r})$ in the periodic potential $V(\boldsymbol{r}) = V(\boldsymbol{r}+\boldsymbol{R})$ will

(continued)

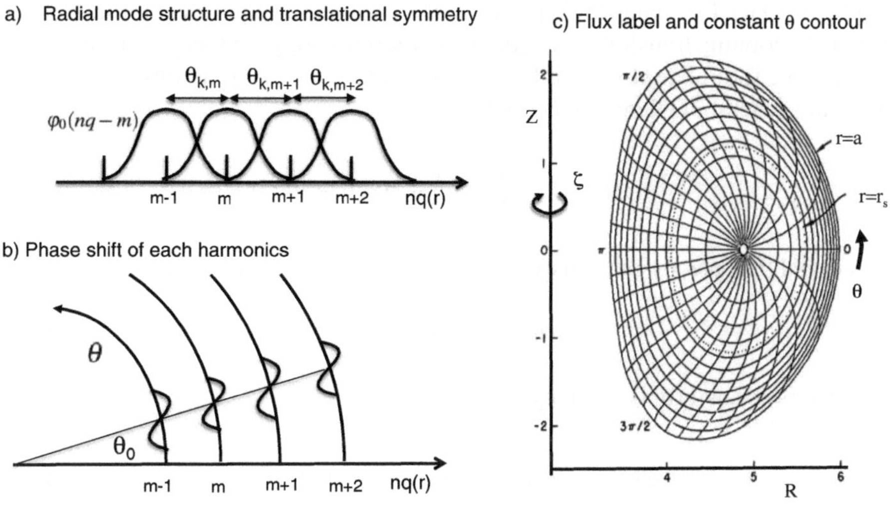

Fig. 6.3 (**a**) Schematics of radial mode structure and translational symmetry. (**b**) Geometrical meaning of phase shift θ_0. (**c**) Schematics of flux coordinates (r, θ, ζ). Modified from Green [263]

satisfy $\psi(\boldsymbol{r}+\boldsymbol{R}) = e^{i\boldsymbol{k}\cdot\boldsymbol{R}}\psi(\boldsymbol{r})$ (see Kittel [459] for proof in one dimensional problem). Here, $\boldsymbol{R}$ is called the lattice vector and $\boldsymbol{k}$ is the crystal wave vector.

The solution to satisfy (6.15) is called the **Bloch wave** and is given by,

$$\psi_k(\boldsymbol{r}) = e^{i\boldsymbol{k}\cdot\boldsymbol{r}}u_k(\boldsymbol{r}) \quad \text{where, } u_k(\boldsymbol{r}+\boldsymbol{R}) = u_k(\boldsymbol{r})$$

If we define the operator $\boldsymbol{T}$ to make a translation $\boldsymbol{r} \rightarrow \boldsymbol{r}+\boldsymbol{R}$, the phase factor $e^{i\boldsymbol{k}\cdot\boldsymbol{R}}$ is the eigenvalue of $\boldsymbol{T}$. Namely $\boldsymbol{T}\psi_k(\boldsymbol{r}) = \psi_k(\boldsymbol{r}+\boldsymbol{R}) = e^{i\boldsymbol{k}\cdot\boldsymbol{R}}\psi_k(\boldsymbol{r})$. This phase factor is determined sometimes in solid state physics by using a boundary condition that $\boldsymbol{T}^N = \boldsymbol{I}$ so that $\boldsymbol{T} = e^{i2\pi/N}\boldsymbol{I}$.

6.3 2D Ballooning Transform

In this section, we introduce different formulation of Ballooning transform, specifically called "2D Ballooning transform". The periodicity problem associated with the flute-like perturbation in the toroidal geometry comes from the $e^{inq(\theta-\theta_0)}$ dependence of $\hat{\varphi}$.

Dewar (Fig. 6.4) [160] showed general method to satisfy θ periodicity by changing Poisson sum in (6.10) to Fourier transform called the twisted radial Fourier transform (TRFT). Zhang-Mahajan [857] also obtained an equivalent formula called the 2D ballooning transform. We adopt latter terminology for this transform.

Here we formulate slightly different way using $\eta \equiv \theta - \theta_0$ as variable. We may write the general 2D flute-like perturbation as:

$$\varphi(r, \theta) = \int_{-\infty}^{+\infty} \frac{d\eta}{2\pi} \hat{\varphi}(\eta, \theta) e^{inq(r)\eta} \tag{6.20}$$

It is trivial that $\varphi(r, \theta)$ can satisfy θ periodicity if $\hat{\varphi}(\eta, \theta)$ is periodic in θ.

$$\hat{\varphi}(\eta, \theta + 2\pi) = \hat{\varphi}(\eta, \theta) \tag{6.21}$$

The inverse transform is given by:

$$\hat{\varphi}(\eta, \theta) = \int_{-\infty}^{+\infty} d(nq)\varphi(r(nq), \theta) e^{-inq\eta} \tag{6.22}$$

We expand $\varphi(r, \theta)$ in Fourier series using nq as a variable rather than r:

$$\varphi(r, \theta) = \sum_{m=-\infty}^{+\infty} \varphi_m(nq) e^{im\theta} \tag{6.23}$$

Using the orthogonality of Fourier harmonics, $\varphi_m(nq)$ is obtained as:

$$\varphi_m(nq) = \int_0^{2\pi} \frac{d\theta}{2\pi} \varphi(r, \theta) e^{-im\theta} \tag{6.24}$$

Substitution of (6.20) into (6.24) gives:

$$\varphi_m(nq) = \int_0^{2\pi} \frac{d\theta}{2\pi} \int_{-\infty}^{+\infty} \frac{d\eta}{2\pi} \hat{\varphi}(\eta, \theta) e^{inq\eta} e^{-im\theta} \tag{6.25}$$

The 1D ballooning transform in Sect. 6.2 corresponds to a special $\hat{\varphi}(\eta, \theta)$ using the delta function, in which the spectrum of Fourier transform is localized at some extended poloidal angle $\eta = \theta - \theta_0 + 2\pi j$.

$$\hat{\varphi}(\eta, \theta) = 2\pi u(\eta + \theta_0) \sum_{j=-\infty}^{+\infty} \delta(\eta - \theta + \theta_0 - 2\pi j) \tag{6.26}$$

We note that this form satisfies the periodicity condition in θ. Substituting this into (6.20), we obtain an equation identical to (6.10).

If we substitute $\eta \to \theta - \theta_0$ and $d\eta \to -d\theta_0$ into (6.20), we have transform by Dewar (twisted radial Fourier transform [160]).

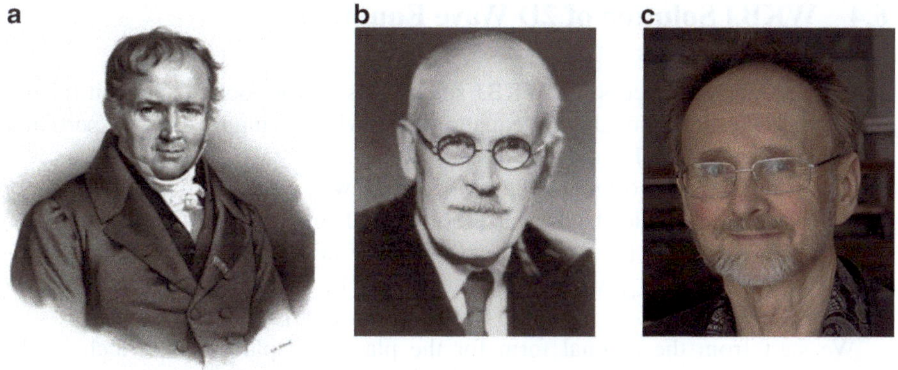

Fig. 6.4 (**a**) S.D. Poisson (1781–1840), (**b**) Harold Jeffreys (1891–1989) is Plumian Professor of Astronomy at University of Cambridge and made significant contribution to the approximate solution of linear second order differential equation now called WKBJ method. He also made important work on Bayesian approach to probability. (**c**) Dr. Robert Dewar. Prof. Em. of Australian National University who made a significant contribution to the mathematical plasma physics including 2-D ballooning transformation and WKB-J solution

$$\varphi(r,\theta) = \int_{-\infty}^{+\infty} \frac{d\theta_0}{2\pi} \varphi_D(\theta,\theta_0) e^{inq(r)(\theta-\theta_0)} \tag{6.27}$$

Here $\varphi_D(\theta,\theta_0) \equiv \hat{\varphi}(\theta-\theta_0,\theta)$, which satisfy $\varphi_D(\theta+2\pi,\theta_0+2\pi) = \varphi_D(\theta,\theta_0)$.

If we replaced θ to θ_0 using $\theta = \eta + \theta_0$ in (6.25), we have an expression by Zhang-Mahajan [857] (Fig. 6.4).

$$\varphi_m(nq) = \int_0^{2\pi} \frac{d\theta_0}{2\pi} \int_{-\infty}^{+\infty} \frac{d\eta}{2\pi} \varphi_Z(\eta,\theta_0) e^{i(nq-m)\eta} e^{-im\theta_0} \tag{6.28}$$

Here $\varphi_Z(\eta,\theta_0) = \hat{\varphi}(\eta,\eta+\theta_0)$, which satisfies $\varphi_Z(\eta,\theta_0+2\pi) = \varphi_Z(\eta\theta_0)$.

Salon 6-1: Jeffreys-Rayleigh-Liouville and WKB [311]
The WKB (Wentzel-Kramers-Brillouin) method (1926) [81, 476, 812] or "WKBJ (Wentzel-Kramers-Brillouin-Jeffreys) method" is the well-known method to solve approximate solution in the quantum mechanics. Historically, "first order WKB solution" is already found by J. Liouville [516] and Rayleigh [622] for a heat conduction problem and the wave solutions and Jeffreys [396] gives the connection formula associated with the turning point. Actually, Mott-Massey (Theory of Atomic Collision) [563] cite this as "Jeffrey method" instead of "WKB method".

6.4 WKBJ Solution of 2D Wave Equation

In this section, we discuss the WKBJ solution using the local dispersion relation of general two dimensional wave equation in slowly varying media with particular application to ideal MHD and toroidal drift waves.

Since this method is originally developed by Dewar, we start from the application to the ideal MHD modes by Dewar [156, 157]. The equation of motion for $\xi \equiv \boldsymbol{\xi} \cdot \nabla\psi$ is expressed as $(\rho\omega^2 + F)\xi = 0$, where F is an operator $F = F(q, \theta, \partial_\theta - inq, -in^{-1}\partial_\theta, -in^{-1}q'\partial_q)$, whose precise form can be found in [157]. Here we use q as radial variable and $\partial_\theta - inq$ comes from $k_\parallel = (nq + i\partial_\theta)/qR$.

We start from the eikonal form for the plasma displacement, which is also called WKBJ form given by $\xi = \hat{\xi}e^{-i\omega t - in(\zeta - q\theta + \int \theta_k dq)}$. The equation of motion is converted to $(\rho\omega^2 + \hat{F})\hat{\xi} = 0$ where $\hat{F} = F(q, \theta, \partial_\theta, q - in^{-1}\partial_\theta, q'(\theta - \theta_k - in^{-1}q'\partial_q))$. Here $' = d/dr$ and we use $\partial_r^{(0)} = inq'(\theta - \theta_k)$ from eikonal part and $\partial_r^{(1)} = q'\partial_q$ from slowly varying part.

The lowest order operator $\hat{F}$ in $1/n$ expansion is $\hat{F}^{(0)} \equiv F(q, \theta, \partial_\theta, q, q'(\theta - \theta_k))$ only including ∂_θ. So the lowest order equation of motion $(\rho\omega^2 + \hat{F}^{(0)})\hat{\xi}^{(0)} = 0$ becomes an ordinary differential equation in θ defined at the covering space $\theta \in (-\infty, +\infty)$. Here $\hat{\xi}^{(0)}$ is lowest order $\hat{\xi}$. Solution with appropriate boundary condition ($\hat{\xi}$ decays faster than $|\theta|^{-1/2}$ as $\theta \to \pm\infty$) will give a local dispersion relation.

$$\omega^2 = \lambda(q, \theta_k) \tag{6.29}$$

Since θ_k appears only in the combination $\theta - \theta_k$, λ is a periodic function of θ_k with 2π period. This leads to infinite solutions $\theta_k = \theta_k^{\pm} + 2\pi j$, where $j = 0, \pm1, \pm2, --$. Depending on parameters, we find topologically different two types of $\lambda(q, \theta_k)$ contour called the "trapped mode" and "passing mode".

6.4.1 Trapped Mode

The $\lambda(q, \theta_k)$ contour can have an extreme so that $\partial\lambda/\partial q = \partial\lambda/\partial\theta_k = 0$ at some point (q_0, θ_{k0}). Usually $\theta_{k0} = 0$ in the ideal MHD and typical contour is shown in Fig. 6.5a where dotted areas are unstable regions ($\lambda < 0$). The $\lambda(q, \theta_k)$ contour becomes up-down symmetric for up-down symmetric tokamak equilibrium.

The local dispersion relation implies $\theta_k = \theta_k(\omega, q)$ where ω is an eigenvalue of the global mode. We consider the case of $\lambda = -\gamma^2$ as shown in Fig. 6.5a. The contour is bounded by two turning points $(q_1, 0)$ and $(q_2, 0)$.

The trajectory in the $\omega^2 = \lambda(q, \theta_k)$ contour (arrow in Fig. 6.5a) is obtained by the following Hamilton (or Ray) equation by introducing a "time-like" variable t_k.

$$\frac{dq}{dt_k} = \frac{\partial \lambda}{\partial \theta_k} \quad , \frac{d\theta_k}{dt_k} = -\frac{\partial \lambda}{\partial q} \tag{6.30}$$

Integration of above Hamilton equation gives the phase parameter $\alpha_0(r) = \int \theta_k \frac{dq}{dt_k} dt_k$. For fixed q value in (q_1, q_2), we have two solutions $\theta_k = \theta_k^+$ and $\theta_k = \theta_k^-$ in $(-\pi, +\pi)$ so that $\xi^+ \sim e^{-in\alpha_0^+(r)} = e^{-in\int \theta_k^+ dq}$ and $\xi^- \sim e^{-in\alpha_0^-(r)} = e^{-in\int \theta_k^- dq}$. Since the radial phase velocity at $\theta = 0$ is given by $v_r^{ph} = k_r/\omega = -n\theta_k q'(r)/\omega$, θ_k^+ solution propagates radially inward and θ_k^- outward. In $q < q_1$ and $q_2 < q$ regions, we have damping solutions as shown in Fig. 6.5b.

At the turning points $(dq/dt_k = \partial\lambda/\partial\theta_k = 0)$, eikonal approximation break down and we have a situation similar to Jeffrey connection rule in Column 6-3. If the mode is localized near $\theta = 0$ region, we have a standing wave if the following Bohr-Sommerfeld condition is satisfied.

$$n\int_{q_1}^{q_2} (\theta_k^+ - \theta_k^-) dq = 2\pi(N + \frac{1}{2}) \quad \text{or} \quad n\oint \theta_k dq = 2\pi(N + \frac{1}{2}) \tag{6.31}$$

While the radial wave number of eikonal form (WKBJ solution) $k_r(\theta) = n(\theta - \theta_k)q'(r)$ (6.5) changes with the extended poloidal angle θ in the covering space $(-\infty, +\infty)$, the 2π shift in each extended cell $(-\pi + 2\pi j, +\pi + 2\pi j)$ cancels out and this condition holds exactly. Here we note that $\oint \theta_k dq$ is the area enclosed by the $\omega^2 = \lambda$ contour. Similar to the quantum mechanics, $N = 0$ gives the ground state which corresponds to the most unstable mode in ideal MHD. Thus Dewar [157] obtained an approximate formula of most unstable mode n as,

$$n_c = \pi / \oint \theta_k dq \tag{6.32}$$

In the β scan for a particular equilibrium series, Dewar found good agreement between n_c variation with β consistent with the numerical calculation $n = n(\beta)$ by the PEST-II code (see Fig. 8.16a) demonstrating the validity of this WKB theory. This WKBJ theory is extended to non-axisymmetric system by Dewar-Glasser [158].

6.4.2 Passing Mode

When the magnetic shear is low, Dewar find a separatrix is introduced inside the unstable region (shown in dotted line in Fig. 6.6a) [156]. Outside the separatrix, equi-λ contour do not close in $\theta_k \in (-\pi, \pi)$ and extends to $\theta_k \in (-\infty, +\infty)$, which implies that $\xi_0 \equiv \hat{\xi} e^{-in\int \theta_k dq}$ is an infinitely multi-valued function.

If we consider the case shown in Fig. 6.7b where $d\theta_k/dt_k < 0$,we can construct the solution $\xi_0(q, \theta)$ as an infinite summation of quasi-mode similar to (6.8). Let $\xi_{0,j}(q, \theta)$ is a quasi mode solution in $(-\pi + 2\pi j, +\pi + 2\pi j)$, we have basic solution in $(-\pi, +\pi)$ as:

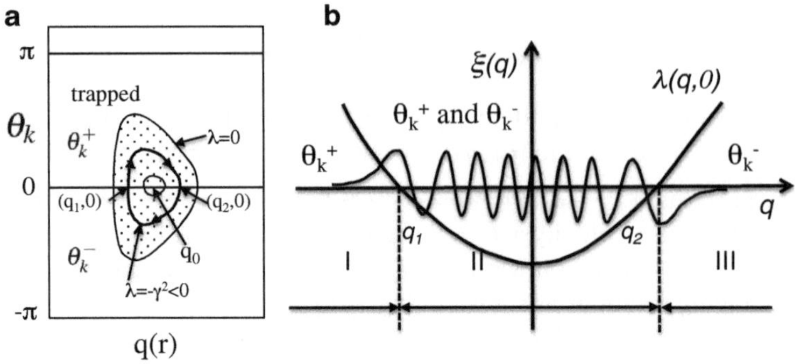

Fig. 6.5 (**a**) Contour plot of λ for unstable up-down symmetric tokamak equilibrium with closed contour. (**b**) Schematics of radial wave structure bounded by two turning points. The wave solutions outside the turning points (region I and III) are decaying solutions

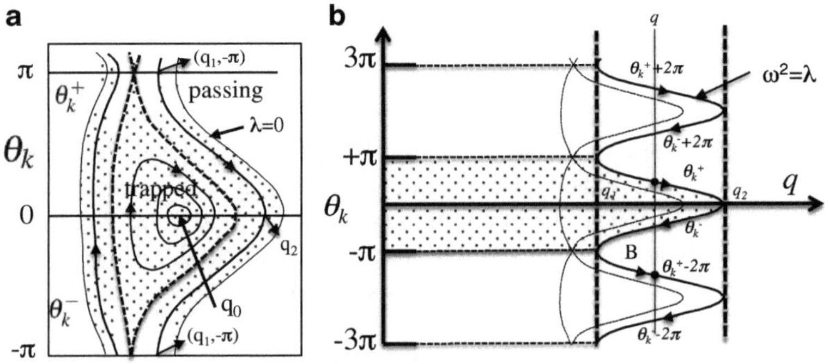

Fig. 6.6 (**a**) $\lambda(\omega, q, \theta_k) = 0$ contour in (q, θ_k) plane including trapped and passing modes. Modified from Dewar [156]. (**b**) Passing mode trajectory in (q, θ_k) plane. *Thin dotted line* corresponds to separatrix in Fig. 6.7a. Area of the *dotted region* is A. Modified from Taylor [757]

$$\xi_{0,0}(q,\theta) = \hat{\xi}_+(q,\theta)exp[-in\int_{q1}^{q}\theta_k^+ dq] + \hat{\xi}_-(q,\theta)exp[-in\int_{q2}^{q}\theta_k^- dq] \tag{6.33}$$

The quasi-mode $\xi_{0,-1}(q,\theta)$ in $(-2\pi, -\pi)$ can be constructed using $\xi_{0,0}(q,\theta)$ as:

$$\xi_{0,-1}(q,\theta) = \xi_{0,0}(q,\theta - 2\pi)exp(-inq\Delta\theta_0) \tag{6.34}$$

where $q\Delta\theta_0$ is given by the integration between two closed dots at q in Fig. 6.6b:

$$q\Delta\theta_0 = \int_{(q,\theta_k^+)}^{(q,\theta_k^+ - 2\pi)} \theta_k \frac{dq}{dt_k} dt_k \tag{6.35}$$

We note that $q\Delta\theta_0$ is the area of the dotted region in (q, q_2) minus area B in Fig. 6.6b. Thus the total solution is given by,

$$\xi(q,\theta,\zeta) = e^{-i\omega t - in\zeta} \sum_{j=-\infty}^{+\infty} \xi_{0,0}(q,\theta + 2\pi j) e^{inq(\theta + \Delta\theta_0 j)} \tag{6.36}$$

Since $\xi(q,\theta+2\pi) = e^{-i\omega t - in\zeta} \sum \xi_{0,0}(q,\theta + 2\pi(j+1)) e^{inq(\theta+\Delta\theta_0(j+1)) + inq(2\pi - \Delta\theta_0)}$, the periodicity $\xi(q,\theta+2\pi) = \xi(q,\theta)$ gives $nq(2\pi - \Delta\theta_0) = 2\pi N$. Let A is the area bounded by $\omega^2 = \lambda$, $\theta_k = \pm\pi$, and $q = 0$, $A = q(2\pi - \Delta\theta_0)$ which corresponds to a dotted area in Fig. 6.6b. The periodicity yields the condition $nA = 2\pi N$. If we solve $\omega^2 = \lambda(q,\theta_k)$ as $q = q(\omega,\theta_k)$, this condition can be expressed as [161]:

$$n \int_{-\pi}^{+\pi} q(\omega,\theta_k) d\theta_k = 2\pi N \tag{6.37}$$

Column 6-3: WKBJ Solution and Jeffrey Connection Rule
We consider the following standard second order ordinary differential equation ($U(x) = 2m(V(x) - E)/\hbar^2$ in quantum mechanics).

$$d^2\psi/dx^2 - U(x)\psi = 0$$

Consider a potential $U(x)$ having two zero points (x_1 and x_2) as shown in Fig. 6.7. Under the condition $|(dU/dx)/U^{3/2}| \ll 1$, approximate solutions at $x > x_1$ and $x < x_1$ can be expressed as follows by using $\kappa = U(x)^{1/2}$ for $U(x) > 0$ and $k = (-U(x))^{1/2}$ for $U(x) < 0$.

$$\psi_I(x) \sim \kappa^{-1/2} exp(-\int_x^{x_1} \kappa dx), \quad \psi_{II}(x) \sim k^{-1/2} cos(\int_{x_1}^{x} k dx + \theta_1)$$

The condition for these solutions breaks at $x = x_1$ since $U(x_1) = 0$. Jeffrey [396] gives a connection formula across $U = 0$ that $\theta_1 = -\pi/4$. This condition assures the exponential damping in region I. Similar solution for $x = x_2$ are:

$$\psi_{II}(x) \sim k^{-1/2} cos(\int_x^{x_2} k dx + \theta_2), \quad \psi_{III}(x) \sim \kappa^{-1/2} exp(-\int_{x_2}^{x} \kappa dx)$$

Apparently, $\theta_2 = -\pi/4$. Since the solution ψ_{II} of (8.43) can be written as $\psi_{II}(x) \sim k^{-1/2} cos(\int_{x_1}^{x} k dx - \int_{x_1}^{x_2} k dx + \pi/4)$ and should be the same with ψ_{II} in (8.34). Therefore, phase difference between two solutions should be $n\pi$.

$$\int_{x_1}^{x_2} k dx = \left(n + \frac{1}{2}\right)\pi$$

In quantum mechanics, it is called the Bohr-Sommerfeld's quantization rule.

Fig. 6.7 Behavior of WKBJ solution in a potential well with two turning points

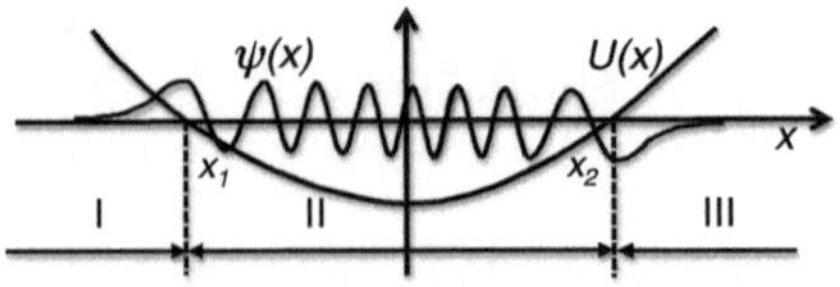

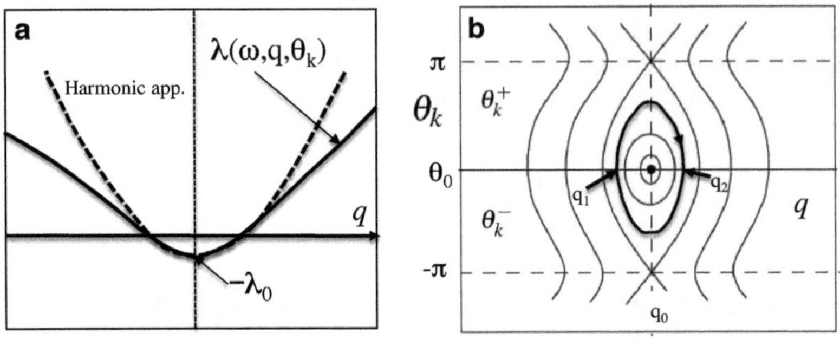

Fig. 6.8 (**a**) Radial variation of the local dispersion relation $\lambda(\omega, q, \theta_0)$ shown by *solid line* and its harmonic approximation by *dotted line*. (**b**) The contour plot of λ for various ω in the phase space (q, θ_k). Modified from Zonca [867]

6.5 Local Dispersion Relation

Following the Dewar formulation, Zonca-Chen [866] developed a WKBJ theory of high n Toroidal Alfven eigenmode (TAE). Connor-Taylor-Wilson [145] and Romanelli-Zonca [636] developed a WKBJ theory of high n Ion Temperature Gradient (ITG) mode. Let the local dispersion relation of WKBJ (eikonal) solution for TAE or ITG (or TEM or ETG) is given by a following form similar to the local dispersion relation of ideal MHD equation (6.29).

$$\lambda(\omega, q, \theta_k) = 0 \tag{6.38}$$

Here ω is the eigenvalue of the global mode. If we solve this equation for θ_k or q, we have $\theta_k = \theta_k(\omega, q)$ or $q = q(\omega, \theta_k)$, respectively. The equip contour plot of $\lambda(\omega, q, \theta_k) = 0$ for different ω is illustrated in Fig. 6.8.

We have trapped and passing modes both of which has discrete spectrum determined by (6.39) and (6.43), respectively.

6.5.1 *Trapped Mode*

If $\lambda(\omega, q, \theta_k)$ has an extreme at (q_0, θ_{k0}), we may expand in a quadratic form as:

$$-\lambda_0 + \frac{1}{2}[\lambda_{qq}(q - q_0)^2 + \lambda_{\theta_k\theta_k}(\theta_k - \theta_{k0})^2] = 0 \tag{6.39}$$

Here, $\lambda_0 = -\lambda(\omega, q_0, \theta_{k0})$. The equip contour surface becomes elliptic as shown in thick line in Fig. 6.8b. This gives two branches $\theta_k^+(\omega, q)$ and $\theta_k^-(\omega, q)$ as:

$$\theta_k^\pm(\omega, q) = \theta_{k0} \pm \sqrt{\lambda_{qq}/\lambda_{\theta_k\theta_k}}\sqrt{a^2 - (q - q_0)^2} \tag{6.40}$$

Here $a^2 = 2\lambda_0/\lambda_{qq} = (q_2 - q_1)^2/4$ since the difference between two turning points (q_1, θ_{k0}) and (q_2, θ_{k0}) is given by $q_2 - q_1 = 2\sqrt{2\lambda_0/\lambda_{qq}}$. The WKBJ solutions are:

$$exp(-in\int^q \theta_k^\pm dq) \tag{6.41}$$

Similar to MHD WKBJ theory, right and left traveling waves produces a standing wave if the following Bohr-Sommerfeld condition is met [145, 636].

$$n\int_{q_1}^{q_2} (\theta_k^+ - \theta_k^-)dq = 2\pi(N + \frac{1}{2}) \tag{6.42}$$

Substituting (6.40) into (6.42) with $N = 0$ and defining $q - q_0 = asiny$, the Bohr-Sommerfeld condition reads:

$$q_2 - q_1 = \frac{2}{\sqrt{n}}\left(\frac{\lambda_{\theta_k\theta_k}}{\lambda_{qq}}\right)^{1/4} \tag{6.43}$$

Or, we have following equation for the eigenvalue ω.

$$\lambda_{\theta_k\theta_k}(\omega, q_0, \theta_{k0}) = \frac{n^2(q_2 - q_1)^4}{16}\lambda_{qq}(\omega, q_0, \theta_{k0}) \tag{6.44}$$

The radial extent of this "trapped mode" is essentially $q_1 \le q \le q_2$.

6.5.2 Passing Mode

If $\lambda(\omega, q, \theta_k) = 0$ do not have extreme at (q_0, θ_{k0}), we may expand λ in the radial direction as follows.

$$\lambda(\omega, q, \theta_k) = \lambda(\omega, q_0, \theta_k) + \lambda_q(\omega, q_0, \theta_k)(q - q_0) = 0 \tag{6.45}$$

Similar to MHD spectrum for passing mode (6.37), we have following Bohr-Sommerfeld quantization condition.

$$n\int_{-\pi}^{+\pi} q(\omega, \theta_k)d\theta_k = 2\pi N \tag{6.46}$$

Using Eq. (6.45), we have following expression [758]:

$$n\int_{-\pi}^{+\pi}\left[q_0-\frac{\lambda(\omega,q_0,\theta_k)}{\lambda_q(\omega,q_0,\theta_k)}\right]d\theta_k=2\pi N \tag{6.47}$$

Radial extent of this passing mode has been given by Kim [448]. We write the local dispersion relation as:

$$\omega=\Omega(r,\theta_k) \tag{6.48}$$

We assume mode amplitude is peak at $q_m=m/n$ and we express Eq. (6.16) in a following form by replacing $m\rightarrow m+j$.

$$\begin{aligned}\varphi(q,\theta)&=e^{im\theta}\sum_{j=-\infty}^{+\infty}\varphi_j(nq-m)e^{ij\theta}\\&=e^{im\theta}\sum_{j=-\infty}^{+\infty}\varphi_0(nq-m-j)e^{-in\int\theta_k dq}e^{ij\theta}\end{aligned} \tag{6.49}$$

We expand θ_k in Taylor expansion as $\theta_k(r-r_m)=\theta_k(0)+\theta_k'(0)(r-r_m)+--$. We assume $\theta_k(0)$ is real and $\theta_k'(0)$ may have real and imaginary part. The real part will contribute tilting of Bloch angle and the imaginary part will characterize envelope of the amplitude. The amplitude modulation can be given by:

$$A(r-r_m)=exp\left[n\int Im[\theta_k]dq\right]=e^{-\alpha(r-r_m)^2} \tag{6.50}$$

Here we define α by $Im[\theta_k'(0)]nq'(r_m)=-2\alpha$.

The α can be determined by the local dispersion relation. If we take the radial derivative of the local dispersion relation (6.48) considering ω is the eigenvalue of the global mode and θ_k can be expressed as $\theta_k(r,\omega)$, we have:

$$(\partial_r\Omega)+(\partial_{\theta_k}\Omega)\theta_k'(0)=0 \tag{6.51}$$

Taking the real part of the above equation, we have:

$$Im[\theta_k'(0)]=\frac{Re[\partial_r\Omega]}{Im[\partial_{\theta_k}\Omega]} \tag{6.52}$$

Here we assumed $Re[\partial_{\theta_k}\Omega]=0$. We have following expression for α.

$$\alpha=-\frac{nq'(r_m)Im[\theta_k'(0)]}{2}=-\frac{nq'(r_m)Re[\partial_r\Omega]}{2Im[\partial_{\theta_k}\Omega]} \tag{6.53}$$

Exercise 6.1. Show that Eq. (6.2) is an infinite summation of the quasi-modes $\varphi(r,\theta) = \sum_{j=-\infty}^{+\infty} \hat{\varphi}(r, \theta + 2\pi j)$.

Answer. We rewrite Eq. (6.2) as:

$$\varphi(r,\theta) = \int_{-\infty}^{+\infty} \frac{d\eta}{2\pi} \sum_{m=-\infty}^{+\infty} e^{im(\theta-\eta)} \hat{\varphi}(r,\eta) \tag{6.54}$$

We apply delta function formula (or Poisson sum rule) in the Column 6-1, we have:

$$\varphi(r,\theta) = \sum_{j=-\infty}^{+\infty} \hat{\varphi}(r, \theta + 2\pi j) \tag{6.55}$$

Chapter 7
Turbulent Transport in Tokamak

Abstract Drift waves in Chap. 5 is most likely candidates of turbulent transport in tokamak and flow shear is key to reduction of turbulent transport. The turbulence itself can produce flow and flow shear through zonal flow and mean flow, which plays essential role for realizing fusion burn in ITER. MHD stability physics of advanced tokamak tells us plasma pressure profile must be controlled to avoid ideal and resistive MHD modes to realize tokamak fusion power. Control of temperature and density profiles requires basic understanding of transport processes across flux surface and to identify key parameters. Turbulent transport is discussed related to its critical temperature gradient transport and the self-organized criticality in Sect. 7.1, flow shear suppression of turbulence in Sect. 7.2, Hasegawa-Mima equation and zonal flow in Sect. 7.3, transport bifurcation to ITB (Internal Transport Barrier) and ETB (Edge Transport Barrier) in Sect. 7.4 and the status of the electromagnetic turbulence in Sect. 7.5 and turbulent momentum transport in Sect. 7.6 in brief.

Further Reading:
Textbooks: Diamond (2010) [168] shows their original approaches to turbulent transport. Balescu (2005) [41] provides comprehensive framework of anomalous transport theories. Textbooks on the fluid turbulence are quite useful for plasma physics as well such as Landau (1987) [495], Kida (1999) [428] and Frisch (1995) [215]. Stringer (1993) [721] is a nice short introduction to basic concepts. Kadomtsev (1965) [407] includes some of fundamentals still useful for tokamak plasma turbulence. Kadomtsev (1976) [409] is well-written book on collective motions behind plasma turbulence.

Review papers: Diamond (2005) [166] is well-written review for zonal flow dynamics. Fujisawa [222] is good review paper on zonal flow experiments. Kadomsev-Pogutse [408] is good introduction to trapped particle instabilities in tokamak. Terry (2000) [764] is useful to understand turbulent decorrelation. Dupree (1972) [181] is a pioneering paper on two point correlation theory which is essential for turbulent decorrelation theory.

M. Kikuchi, M. Azumi, *Frontiers in Fusion Research II*,
DOI 10.1007/978-3-319-18905-5_7

7.1 Critical Temperature Gradient Transport

7.1.1 Structure Formation in Non-equilibrium Open System

Plasma confinement system is a non-equilibrium open system. In the open system, the system can maintain a dynamical state far from thermodynamical equilibrium if the system has thermodynamic forces.

S. Chandrasekhar (Fig. 7.4a) [103] studied the **Benard cell** problem extensively as a typical example of the hydrodynamic thermal stability due to **critical temperature gradient**. If we place viscous fluid such as water or oil between two parallel plates and heat the lower plate to produce a temperature difference ΔT, heat is transferred by the thermal conduction if the temperature difference is small. But a convective cell called the Benard cell appears when the dimensionless quantity the **Rayleigh number** $R = (g\alpha d^4/\kappa\nu)|\,(T_1 - T_2)/d|$ exceeds some critical value, where g is the gravitational constant, α is the thermal expansion coefficient, κ is the thermal conductivity, ν is the kinematic viscosity, d is the distance between two plates, T_1 and T_2 are temperatures at lower and upper plates. Appearance of this convective cell from uniform fluid is the manifestation of the **symmetry breaking**. The quantity $(T_1 - T_2)/d$ is the temperature gradient and there exists a critical temperature gradient for the structure formation.

Ilya Prigogine [582] called states and motions maintained by the thermodynamic forces in the open as **Dissipative Structure**. They shows Benard cell (Fig. 7.1a) as a typical example of the dissipative structure whose driving force is gravity and the dissipation plays an essential role in the structure formation. This dissipative structure can be seen on the surface of the Sun.

Per Bak [39] discussed the sand avalanche in the sand hill as a typical example of critical phenomena and find the system tends to maintain the critical state as shown in Fig. 7.1b. Because of this characteristics, he named it as **Self-Organized Criticality** (SOC). The probability distribution of the size of the sand collapse

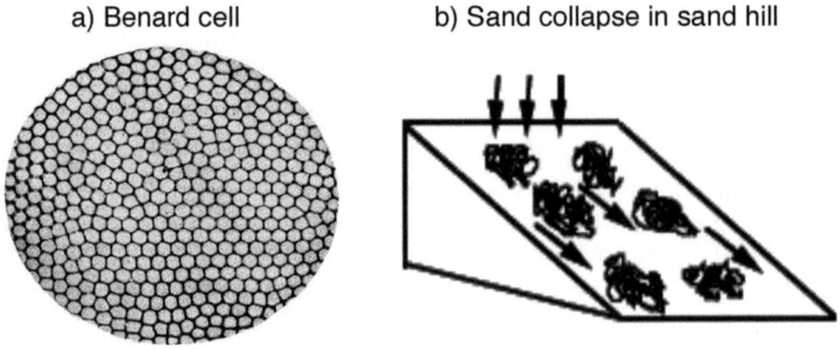

Fig. 7.1 Examples of self-organization. (**a**) Benard cell as a typical dissipative structure. (**b**) Sand collapse in the sand hill as a typical self-organized criticality

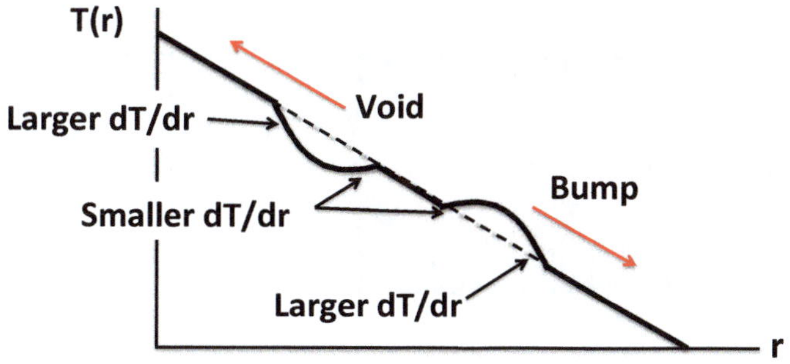

Fig. 7.2 Bump and void avalanches near marginal stability predicted by Diamond and Hahm [163]. Near critical gradient, a void perturbation produces larger dT/dr exceeding instability threshold at upper front leading to uphill propagation of void. Similarly bump gives downhill propagation

follows power law. Such a system can be seen also in earthquake dynamics, called Gutenberg-Richter scaling.

7.1.2 Self-organized Criticality in Tokamak Transport

Heat and particle transports in tokamaks are governed by turbulent transport due to drift waves. Stiffness of temperature profile is observed in the L-mode, which is closely related to the self-organized criticality. The free energy is coming from the temperature and density gradients. The existence of critical temperature gradient in the toroidal ITG/TEM and ETG modes above which drift waves are strongly destabilized and large turbulent heat transport is expected as a form of avalanche which is analogous to the avalanche dynamics in the sand hill.

Diamond-Hahm (Fig. 7.4c) [163] clarified dynamics of turbulent plasma transport near marginal stability. If initial dT_0/dr profile (shown by dotted line in Fig. 7.2) is close to critical temperature gradient $dT/dr|_c$, superposition of void ($\delta T < 0$) gives strong destabilization of the critical temperature gradient instabilities in the inner front leading to the inward propagation of void. In case bump ($\delta T > 0$) is superposed, dT/dr in the outer front becomes larger leading to outward propagation of bump. They noticed the heat flux q must be invariant under the dual transformations $x \to -x$ and $\delta T \to -\delta T$, called **joint reflection symmetry** by Hwa [357]. Existence of avalanche is first shown by the flux-driven gyrofluid simulation [239].

The first gyrokinetic full-f Vlasov simulation of heat flux driven toroidal ITG turbulence by Idomura gives features of self-organized criticality [369]. As will be discussed in Sect. 7.2, the avalanche characteristics in the critical temperature gradient transport in tokamaks is strongly influenced by the flow component $E_r(n=0)$. Therefore, it is Important to see the case without $E_r(n=0)$.

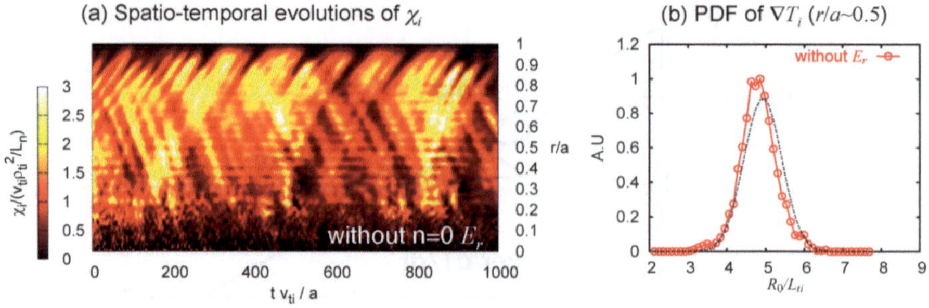

Fig. 7.3 (**a**) Spatio-temporal evolution of χ_i without $E_r(n = 0)$. (**b**) Probability distribution function (PDF) of dT/dr. After Idomura (private communication)

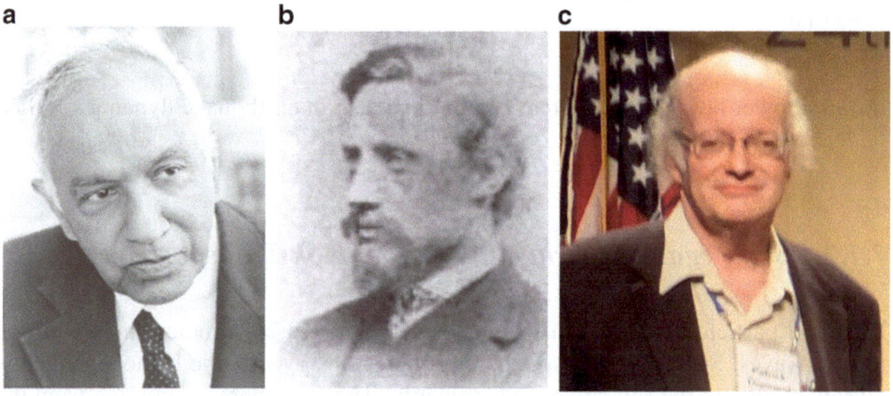

Fig. 7.4 (**a**) S. Chandrasekhar (1910–1995) is Nobel Prize winner in physics for his theoretical prediction of black hole. He is also famous for his works on hydrodynamic and hydromagnetic stability, in which the Benard cell is discussed in detail. (**b**) O. Reynolds (1842–1912) is Professor of Engineering at University of Manchester and made significant contribution to fluid mechanics including Reynolds stress to produce mean flow, similarity law in fluid dynamics. (**c**) Patrick Diamond. Fusion plasma is an object of typical non-equilibrium thermodynamics. PD is a distinguished professor in the field of turbulent transport who pinned up key physics processes in the complex plasma systems such as avalanche dynamics of turbulent transport, Predator-Prey model using Reynolds stress for turbulence-flow interaction, who received the Alfven prize

Spatio-temporal evolution of the thermal diffusivity χ_i in Fig. 7.3a (Y. Idomura, private communication) shows the inward and outward propagating avalanche structures. These avalanches follows the Hwa's joint reflection symmetry, namely the voids propagate in the up-hill, while bumps propagate down-hill. The probability distribution function (PDF) is symmetric around the critical temperature gradient.

7.1.3 Observations of Critical Gradients

7.1.3.1 Critical Electron Temperature Gradient

Significant departure of the electron heat transport from the neoclassical theory has been observed since its early experiments in Russia led by Artsimovich [28]. Possible candidates of turbulent electron transport are Trapped Electron Mode (TEM) by Kadomtsev [408] ($k_\perp \rho_i \sim 1$ in TEM) and toroidal ETG (Electron Temperature Gradient) mode, ($k_\perp \rho_e \sim 1$ in ETG). While ITG turbulence produces turbulent electron heat transport as well as ion heat transport, it is too small to explain measured electron heat transport. Turbulent transport by ETG is thought to be much smaller than that by ITG due to its small scale $k_\theta \rho_e \sim 1$ until FY2000. Jenko [398] and Dorland [173] showed large level of ETG turbulent transport (60× Gyro-Bohm) associated with radially elongated streamer.

First report on existence of critical temperature gradient $dT_e/dr|_c$ for electron transport is reported in 2001 by Hoang [333] in Tore Supra as shown in Fig. 7.5a and also showed dependence of $dT_e/dr|_c$ on magnetic shear as $R/L_T = 5 + 10|s|/q$, where $s = rdq/dr/q$ is magnetic shear.

Jenko showed that following theoretical critical temperature gradient dT_{ec}/dr in toroidal ETG is consistent with this measurement as shown in Fig. 7.5b [400].

$$\left(\frac{R}{L_{Te}}\right)_{crit} = (1 + \tau_e)(1.33 + 1.91s/q), \tag{7.1}$$

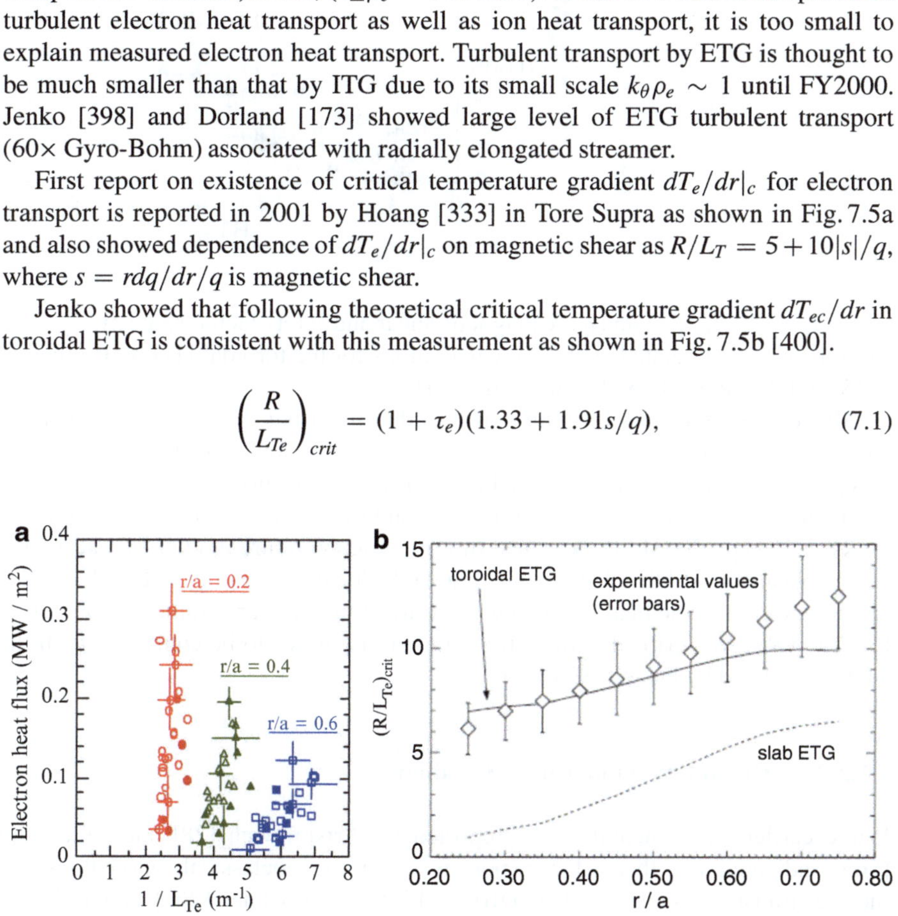

Fig. 7.5 (**a**) Electron heat flux as a function of R/L_T showing the existence of critical electron temperature gradient. Symbols *open circle*, *open triangle*, correspond to $r/a = 0.2, 0.4, 0.6$, respectively, and *full symbols* correspond to another series of discharges [333]. (**b**) Comparison of radial profile of $(R/L_{T_e})_{crit}$ between experimental value and theoretical value [400]. Reproduced with permission from Hoang [333] and Jenko [400]. Copyright American Physical Society

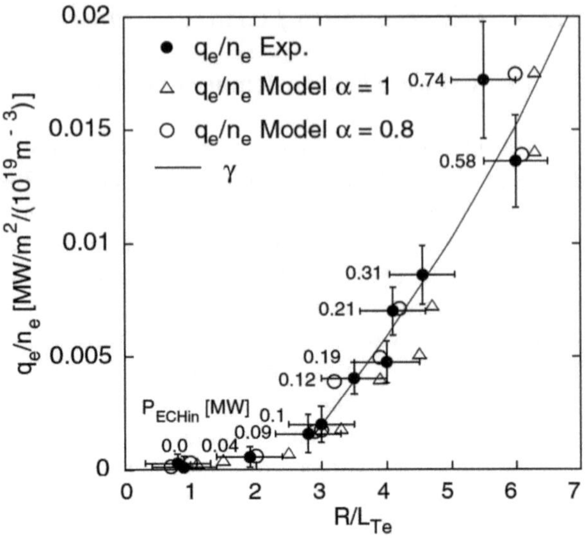

Fig. 7.6 Normalized heat flux q_e/n_e as a function of R/L_{T_e}. Model calculations assumes $\chi_e = q^{3/2}(T_e/eB)(\rho_s/R)[\chi_s(R/L_{T_e} - R/L_{T_{ec}})^\alpha + \chi_0]$. Reproduced with permission from Ryter [644]. Copyright American Physical Society

where $\tau_e = Z_{eff}T_e/T_i$, s and q are magnetic shear and safety factor, respectively. This formula is closely related analytical calculations for the toroidal ITG by Romanelli [635] and the slab ITG by Hahm-Tang [277].

Above ETG threshold formula indicates that for usual tokamak plasma parameters, the ETG becomes unstable if T_e is close to or lower than T_i. In collision less plasmas with dominant electron heating provided by electron cyclotron heating (ECH) having $T_e > T_i$, the ETG modes are stable while the TEM can be dominant.

Ryter [644] showed the existence of critical temperature gradient in the ECRF heated plasma in ASDEX-U tokamak. Use of the localized heating by ECRF enables the observation of the heat flux below the critical temperature gradient as shown in Fig. 7.6. Observed critical temperature gradient is shown to be consistent with the TEM calculated by using GS2 [471].

7.1.3.2 Critical Ion Temperature Gradient

While earlier experimental works reported by Artsimovich [28] and Murakami [565] indicated ion thermal transport is close to neoclassical, anomalous ion thermal transport is observed in DIII in 1986 by Groebner [267] and followed by large tokamak experiments such as in TFTR supershot by Zarnstorff [854]. JET experiments by Mantica in 2009 [539, 540] showed clear evidence of critical temperature gradient in ion transport as shown in Fig. 7.7a.

Theoretical works to determine dT_i/dr threshold starts from earlier work by Horton in 1972 [340] in slab geometry. Toroidal effect completely changes the nature of the ITG as discussed in Chap. 5. Rigorously speaking, stability to drift wave must be analyzed using the Gyrokinetic equation (5.174) to accurately include

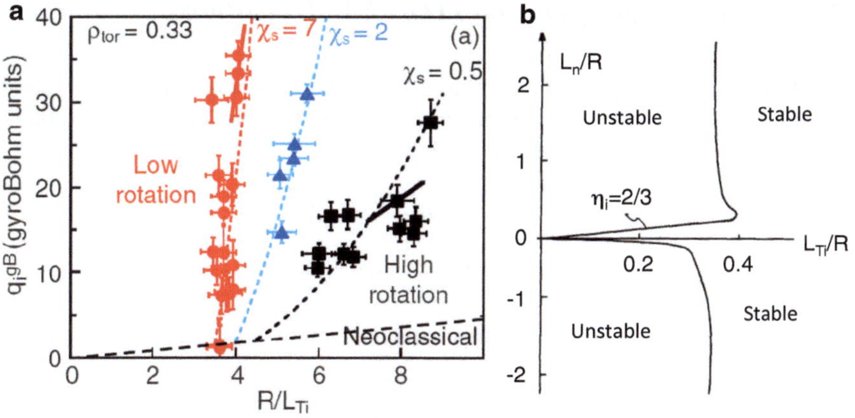

Fig. 7.7 (**a**) Experimental evidence of critical ion temperature gradient in tokamak and dependence of the stiffness on toroidal rotation. Normalized ion heat flux as a function of R/L_{Ti} for different toroidal rotations. Recent EM simulation [128] shows that fast ion pressure gradient may be a cause of de-stiffening instead of rotation. Reproduced with permission from Mantica [540]. Copyright American Physical Society. (**b**) Linear stability of toroidal ITG mode to show existence of two different stability boundaries. Reproduced with permission from Biglari [66]. Copyright AIP Publishing LLC

the kinetic effect. Biglari-Diamond-Rosenbluth in 1989 [66] gives a comprehensive analysis of ITG stability threshold using Gyrokinetic equation (5.108) showing that instability condition is given by $\eta_i = dlnn/dlnT_i = 2/3$ for peaked density profile and $L_{T_i}/R = 0.35$ for flat density shown in Fig. 7.7a. Romanelli in 1989 solved ion Gyrokinetic equation using ballooning transformation and derived an approximate expression of dT_{ic}/dr in terms of $L_{T_i} = T_i/(dT_{ic}/dr)$ for flat density regime $R/L_n < 2(1 + T_i/T_e)$ $(L_n = n/(dn/dr))$ in circular plasma and showed that observed dT_i/dr in DIII-D [267] is close to the theoretical dT_{ic}/dr given by $R/L_{T_i} = \frac{4}{3}(1 + T_i/T_e)$[635].

Guo-Romanelli [270] proposed combined formula including Hahm-Tang slab ITG threshold $R/L_{T_i} = (3/2)(\pi/2)^{0.5}(1 + T_i/T_e)s/q$ [277], which does not quite reproduce slab ITG threshold. More refined formula for critical temperature gradient can be expressed by just adding two formula including impurity effect implied from ETG formula (7.1) as

$$R/L_{T_i} = (1 + \tau_i)(1.33 + 1.91s/q), \tag{7.2}$$

where $\tau_i = T_i/T_e Z_{eff}$, s and q are magnetic shear and safety factor, respectively. While the theoretical critical temperature gradient do not have any mass dependence, Urano [784] shows interesting difference in the heat diffusivity χ_i as a function of $\nabla T_i/T_i$ between hydrogen and deuterium, while mass dependence of $R/L_{T_i}|_c$ is still not clear. Some of earlier historical notes can be seen in a review paper by Connor-Wilson in 1994 [146].

7.1.4 Particle Transport and ITG/TEM Transition

At first, we discuss importance of phase difference between density and potential fluctuation by an elementary procedure. Let $\tilde{n}$ and $\tilde{\phi}$ are density and potential fluctuation and has a phase difference $\Delta\varphi$.

$$\tilde{n}_e = n_1 sin(\omega t + k_y y) \tag{7.3}$$

$$\tilde{\phi} = \phi_1 sin(\omega t + k_y y + \Delta\varphi) \tag{7.4}$$

The particle flux Γ_e due to $E \times B$ convection is given by $\Gamma_e = \langle \tilde{n}_e \tilde{E}_y \rangle / B$ is given by,

$$\Gamma_e = \frac{k_y sin\Delta\varphi}{2} \frac{n_1 \phi_1}{B} \tag{7.5}$$

This means that the phase difference is a key parameter for the turbulent convection.

The gyrokinetic equation for the non-adiabatic part of the perturbed distribution function $\hat{g}_a$ (5.108) with a simple Krook collision operator in Fourier space is given by,

$$\left[\omega - k_\parallel v_\parallel - \boldsymbol{k} \cdot \boldsymbol{v}_d + i\nu_{ei}(v)\right] \hat{g}_a = \frac{e_a}{T_a} f_{aM}(\omega - \omega_*) J_0(k_\perp \rho_a) \boldsymbol{\Phi}(\boldsymbol{k}) \tag{7.6}$$

$$\omega_* = \omega_n \left[1 + \eta_a \left(\frac{K}{T_a} - \frac{3}{2}\right)\right], \omega_n = \frac{k_s T_a}{e_a B} \boldsymbol{n} \cdot \nabla ln n_a, \nu_{ei}(v) = \nu_{ei} \frac{v_{te}^3}{v^3} \tag{7.7}$$

$$\eta_a = \frac{\boldsymbol{n} \cdot \nabla ln T_a}{\boldsymbol{n} \cdot \nabla ln n_a}, \boldsymbol{n} = \frac{\nabla \psi}{|\nabla \psi|}, k_s = \boldsymbol{k} \cdot \boldsymbol{s}, \boldsymbol{s} = \boldsymbol{b} \times \boldsymbol{n}, K = \frac{1}{2} m_a v^2, \tag{7.8}$$

where ν_{ei} is the electron-ion thermal collision frequency (4.73).

For the particle transport, we use the equation for the electron, which is essentially drift kinetic equation but is written in the circular shifted plasma as [19]:

$$(\omega - \omega_G + i\nu_{ei}(v))\hat{g}_a = \left[\omega_D\left(\frac{R}{L_n} + \left(\frac{K}{T_e} - \frac{3}{2}\right)\frac{R}{L_{T_e}}\right) - \omega\right] f_{eM} J_0(k_\perp \rho_s) \hat{\phi}_k, \tag{7.9}$$

where $\omega_G = k_\parallel v_\parallel + \omega_d$, $\omega_d = \omega_D[(v_\parallel^2 + v_\perp^2/2)/v_{Te}^2][cos\theta + (s\theta - \alpha sin\theta)sin\theta]$, $\hat{\phi}_k = e\Phi(\boldsymbol{k})/T_e$, $\omega_D = k_y \rho_s c_s / R$, $c_s = \sqrt{T_e/m_i}$, $\rho_s = c_s/\Omega_i$, $\alpha = -2\mu_0 q^2 R(dP/dr)/B^2$, $s = r(dq/dr)/q$, $v_{Te} = \sqrt{2T_e/m_e}$, and θ is the extended ballooning angle.

The quasi-linear particle flux Γ_{QL} produced by the fluctuating $\boldsymbol{E} \times \boldsymbol{B}$ drift $\tilde{v}_E$ is $\langle \tilde{n}_e \tilde{v}_E \rangle_\psi$, where $\langle\rangle_\psi$ is the flux surface average. The quasi-linear cross field particle flux is closely related to the phase difference between $\tilde{n}_e$ and $\hat{\phi}_k$ and is zero if the phase difference is zero. The Γ_{QL} is given by:

$$\Gamma_{QL} = \sum_k \left\langle \frac{k_y c_s^2}{\Omega_i} \int d^3v C_k f_M J_0(k_\perp \rho_s)^2 |\hat{\phi}_k|^2 \right\rangle_\psi$$

$$\text{where, } C_k = \frac{(\hat{\gamma}_k + \hat{\nu}_k)[R/L_n + (K/T_e - 3/2)R/L_{T_e}] - \hat{\gamma}_k \hat{\omega}_{Gk} + \hat{\omega}_{rk}\hat{\nu}_k}{(\hat{\omega}_{rk} + \hat{\omega}_{Gk})^2 + (\hat{\gamma}_k + \hat{\nu}_k)^2}, \tag{7.10}$$

where $\hat{\gamma}_k = \gamma_k/\omega_D$ and $\hat{\omega}_{rk} = \omega_{rk}/\omega_D$ are the linear growth rate and the real frequency of the unstable mode normalized to $\omega_D = k_y \rho_s c_s/R$ and $\hat{\omega}_{Gk} = \omega_{Gk}/\omega_D$ and $\hat{\nu}_k = \nu_{ei}(v_{te}/v)^3/\omega_D$. Equation (7.10) implies following form of the quasi linear particle flux.

$$\frac{R\Gamma_{QL}}{n_e} = D\frac{R}{L_n} + D_T\frac{R}{L_{T_e}} + RV_p \tag{7.11}$$

Here three terms in RHS are the diffusive flux, the thermo-diffusion and the pure convective term, respectively. In the collision less plasma, the pure convective term is proportional to the toroidal drift ω_d, which is called the "curvature pinch".

From (7.11), stationary density scale length parameter R/L_n^{stat} is given by:

$$R/L_n^{stat} = -C_T R/L_T - C_p, \tag{7.12}$$

where $C_T = D_T/D$ and $C_p = RV_p/D$. There are complicated parametric dependences on the effective collisionality ν_{eff}, the temperature ratio T_e/T_i, the magnetic shear $s = rq'(r)/q$, the scale length ratio L_{T_i}/L_{T_e}, and R/L_{T_i}.

The most important to explain the density peaking is the trapped electron contribution to C_T and C_p called the trapped electron pinch. As seen from the numerator $K/T_e - 3/2$ of C_T (7.10), the low energy trapped electrons diffuse inward and the high energy trapped electrons diffuse outward. Which one is dominant, depends on the denominator, which has resonance at $\omega \sim \omega_d$. For $\omega_d > 0$, the low energy electron has resonance for the ITG mode ($\omega < 0$) leading to the inward convection ($C_T < 0$). The curvature pinch term has two terms, one proportional to ω_d and one proportional to ν. Curvature pinch is inward ($C_p < 0$) at collision less limit but outward flux increases with collision as will be seen in Fig. 7.10.

Dominance of the trapped electron pinch is seen from the collision less GK calculation by Fable-Angioni-Sauter in 2008 [194], which solved the linearized electrostatic collisionless gyrokinetic equation to obtain analytical form of C_T and C_p.

$$C_T = \frac{\int_0^{+\infty} d\epsilon \sqrt{\epsilon} e^{-\epsilon} \frac{f_t \gamma (\epsilon - 3/2)}{\gamma^2 + (\omega - \epsilon\omega_d/2)^2} + \frac{1-f_t}{4k_\parallel v_{te}}}{\int_0^{+\infty} d\epsilon \sqrt{\epsilon} e^{-\epsilon} \frac{f_t \gamma}{\gamma^2 + (\omega - \epsilon\omega_d/2)^2} - \frac{1-f_t}{2k_\parallel v_{te}}} \tag{7.13}$$

$$C_p = -\frac{\omega_d}{k_y \rho_i} \frac{\int_0^{+\infty} d\epsilon \sqrt{\epsilon} e^{-\epsilon} \frac{f_t \gamma \epsilon/2}{\gamma^2 + (\omega - \epsilon\omega_d/2)^2} - \frac{(1-f_t)\omega}{\pi}}{\int_0^{+\infty} d\epsilon \sqrt{\epsilon} e^{-\epsilon} \frac{f_t \gamma}{\gamma^2 + (\omega - \epsilon\omega_d/2)^2} - \frac{(1-f_t)\omega_d}{\pi}}, \tag{7.14}$$

where $\epsilon = K/T_e$.

In case of ETG regime ($k_y\rho_i \gg 1$, ω_d goes up as well), the trapped particle pinch goes to zero and the passing particle contribution to the pinch dominates to have $C_T = -1/2$ and $C_p = \omega/k_y\rho_i$ as seen from above formula.

In case of ITG/TEM regime ($k_y\rho_i \sim 1$), the trapped particle contribution dominates over the passing particle contribution ($\sim 1/k_\parallel v_{te}$) for C_T and C_p since $k_\parallel v_{te} \gg \omega$ except $k_\parallel \sim 0$ in slab like situation at $s \sim 0$.

An important transition in the ion scale drift wave turbulence is the ITG to TEM transition. While ITG rotates in the ion diamagnetic drift direction, TEM rotates in the electron diamagnetic drift direction. Particle transport property is quite different between ITG and TEM, which can be used to identify the mode transition.

In tokamaks, the particle convection by the ITG is inward and that by the TEM is outward due to thermo-diffusion. Recent modulation experiments in Tore Supra tokamak show change of the direction of the convective speed across the stability boundary between ITG and ITG+TEM, which is in good agreement with the numerical calculation of the quasi-linear gyrokinetic code as shown in Fig. 7.8. The stability diagram is close to Fig. 5.2 of Weiland fluid mode. The ITG - (ITG+TEM) boundary is a stable solution against the perturbation.

Thermo-diffusion is the off-diagonal turbulent particle flux driven by dT/dr originating from Coppi [147]. Thermo-diffusion is first reported by Nagashima [569] using ECRH in 1995. There is also an off-diagonal turbulent particle flux proportional to ∇B, called "curvature pinch", also called "turbulent equipartition" (TEP) by Yankov [842]. Garbet [242] showed existence of both particle pinches in gyro fluid simulation of ITG/TEM turbulence. Hoang [334] reported existence of strong particle pinch in full CD plasma in Tore Supra, where Ware pinch [810] vanishes.

Angioni made a systematic comparison of density profile shape between theory and experiments [13, 15, 19]. This density peaking is found to be related to low

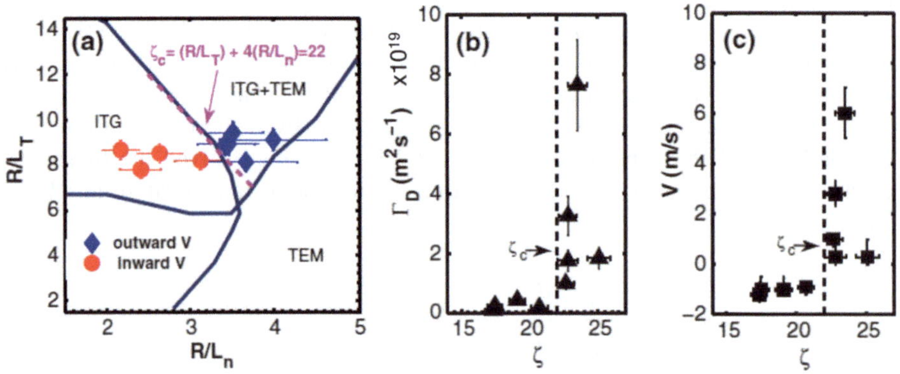

Fig. 7.8 (**a**) Stability diagram of ITG/TEM with experimental data of convection direction. (**b**) Diffusive particle flux v.s. ITG/TEM boundary parameter ζ. (**c**) Convective speed v.s. ITG/TEM boundary parameter ζ. Reproduced with permission from Zhong [859]. Copyright American Physical Society

collisionality [13] and are confirmed in many tokamaks as shown in Fig. 7.9a. Maslov [542] shows a good agreement of R/L_n between theory and experiments as shown in Fig. 7.9b. Fable-Angioni-Sauter [195] made a systematic analysis to identify key physics behind the density profile using quasi-linear gyrokinetic simulation.

Figure 7.10 shows a typical ν_{eff} dependence of ω_R^{QL}, R/L_n^{stat}, C_T, C_p. The real frequency ω_R^{QL} is negative, which means the mode is ITG. In this case, the thermo-diffusion parameter C_T does not change much but the curvature pinch parameter C_p changes from minus to a large positive value. Hence, the density peaking parameter R/L_n^{stat} strongly depends on ν_{eff}.

Density profile control is one of important subjects for steady state tokamak operation since dn/dr drive bootstrap current more strongly and also fusion power density $P_f \propto n^2$. With the increased experimental evidence for the ITG/TEM transition, it become more important to investigate possibility of density peaking by changing TEM stability through shaping control.

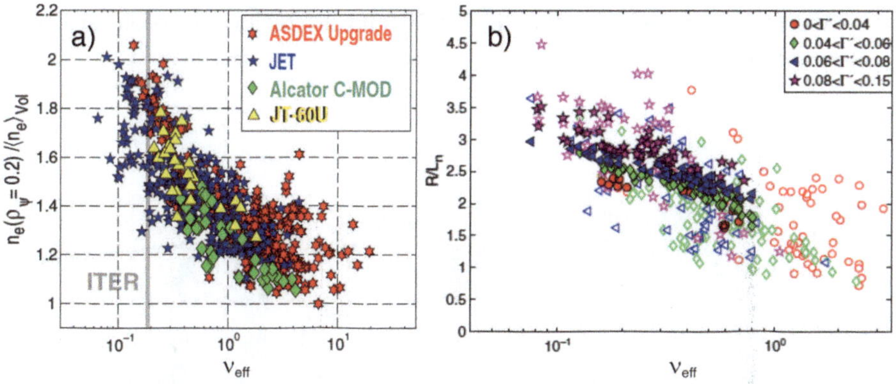

Fig. 7.9 (**a**) Experimental density peaking factor as a function of effective collisionality[19]. (**b**) A comparison of R/L_n between simulation (*un-filled symbol*) and measurement (*filled symbol*) in JET[542]. Reproduced with permission from Angioni [19] and Maslov [542]. Copyright IAEA Vienna

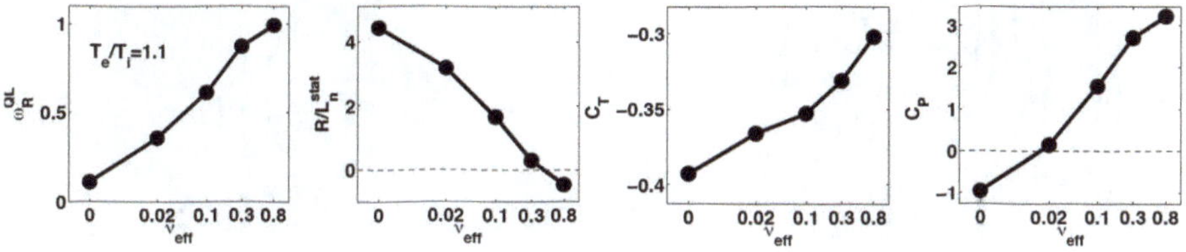

Fig. 7.10 The real frequency ω_R^{QL}, the stationary density scale length parameter R/L_n^{stat}, the thermo-diffusion parameter C_T and the curvature pinch parameter C_p as a function of effective collisionality under the parameter set of $s = 0.8$, $q = 1.4$, $\epsilon = 0.12$, $R/L_{T_e} = R/L_{T_i} = 9$, $T_e/T_i = 1.1$. Reproduced with permission from Fable [195]. Copyright IOP Publishing

Salon 7-1 : Nuclear Fusion Prize for Turbulent Transport
Nuclear Fusion journal is the acknowledged world-leading journal which publishes significant original works in the field of fusion research. IAEA awarding excellent 10 papers as Nuclear Fusion Prize nominees and among which one paper is selected to be a winner of Nuclear Fusion prize. Paper led by T. Luce (2006) received the first Nuclear Fusion prize by the enhanced confinement regime relevant for ITER hybrid mode [527]. The top author of Nuclear Fusion Prize papers in turbulent transport are C. Angioni [15] on particle transport, J.E. Rice [628] on intrinsic rotation, H. Urano [783] for pedestal transport, P. Diamond [167] (see Fig. 7.4) for momentum transport theory, D.G. Whyte [822] for I-mode discovery.

In 2007, the paper led by Clemento Angioni concerning the particle transport studies is selected. He continued to produce many more excellent works even after. Below from left, Prof. W. Burkart (IAEA Deputy Director General), Dr. C. Angioni (Max Planck Institute for Plasma Physics, Garching), and M. Kikuchi (Chairman of Nuclear Fusion board of editors). He wrote a series of original papers [13–18, 20] and also review papers on particle transport [19, 21].

7.2 Flow Shear Suppression of Turbulence

7.2.1 Turbulence-Flow Paradigm

The turbulence can be affected by the flow shear. Flow shear suppression of turbulent transport is uncovered by Biglari-Diamond-Terry [67]. The theory shows that the turbulent suppression by the poloidal flow shear can be given by $\tilde{\xi}_{\omega_s}/\tilde{\xi}_{\omega_s=0} \propto (\Delta\omega_T/\omega_s)^{2/3}$, where $\omega_s = (\Delta r/\Delta\theta)\partial(V_\theta/r)/\partial r$ is the shearing rate, $\Delta\omega_T$ is the random scattering rate of the ambient turbulence, and V_θ is the poloidal $E \times B$ flow.

The theory stems from the pioneering work on two-point correlation theory of Dupree [75, 181]. Key element of his theory is that if we define $x_\pm = (x_1 \pm x_2)/2$, we can neglect (x_+, v_+) dependence for the relative evolution operator T_{12} to $\langle\delta f(1)\delta f(2)\rangle$ since $\partial/\partial x_- \gg \partial/\partial x_+$ and $\partial/\partial v_- \gg \partial/\partial v_+$. Here the two-point equation can be written as $(\partial_t + T_{12})\langle\delta f(1)\delta f(2)\rangle = S$ and the T_{12} can be approximated as $T_{12} \sim v_- \frac{\partial}{\partial x_-} - \frac{\partial}{\partial v_-} D_- \frac{\partial}{\partial v_-}$.

The physical picture of the effect of the flow shear on the turbulent convective cell is shown in Fig. 7.11a. There are two sources of flow shear (dv_E/dr) in tokamaks. First one is the mean flow shear determined by the radial force balance. Second one is the zonal flow shear driven by turbulence itself [290] as shown in Fig. 7.11b.

Hahm [278], Hahm-Burrel [279] gives a condition for flow shear induced fluctuation suppression in the toroidal plasma as follows,

$$\Delta\omega_T \le \omega_E = \left|\frac{\Delta\psi}{\Delta\zeta}\frac{d^2\Phi(\psi)}{d\psi^2}\right| = \frac{\Delta r}{\Delta L_\perp}\frac{R^2 B_\theta^2}{B}\frac{d^2\Phi}{d\psi^2} \tag{7.15}$$

where $\Delta\omega_T$ is the de-correlation rate of the ambient turbulence and ω_E is the shearing rate due to flow shear, $\Delta r = \Delta\psi/RB_\theta$ is the radial correlation length of turbulence, and $R\Delta\zeta$ is the toroidal correlation length which is related to the perpendicular correlation length, $\Delta L_\perp$ as $\Delta L_\perp = RB_\theta\Delta\zeta/B$.

We sketch here how to arrive at Eq. (7.15). Consider a standard nonlinear fluid equation $(\partial_t + \boldsymbol{u}_0 \cdot \nabla + \nu)\delta H = -\tilde{\boldsymbol{u}}_E \cdot \delta H + S$, where δH is a fluctuating field, $\boldsymbol{u}_0 = \boldsymbol{u}_E + u_\parallel \boldsymbol{b}$, and $\boldsymbol{u}_E = \boldsymbol{E}_r \times \boldsymbol{B}/B^2$, ν is the dissipation term to ensure finiteness of the two-point correlation function at zero separation, and the right hand side consists

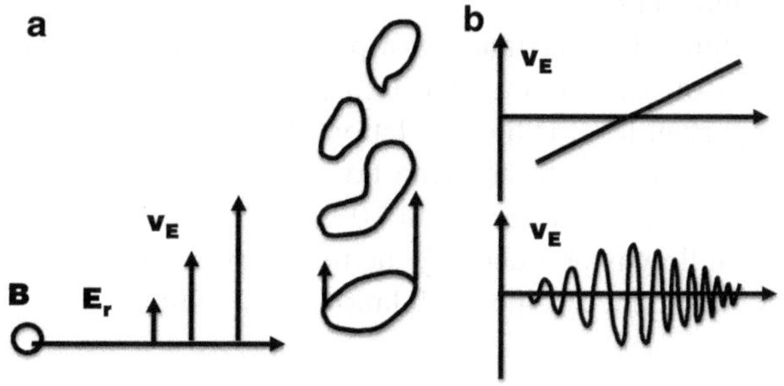

Fig. 7.11 (**a**) Schematics of effect of $E_r \times B$ drift shear on turbulent convective cell. Flow shear produces elongation and tearing of convective cell. (**b**) Two types of v_E shear. Upper v_E shear comes from global force balance and lower v_E shear is Zonal flow produced by the turbulence

of $\boldsymbol{E}\times\boldsymbol{B}$ nonlinearity and the driving force of the turbulence, S. In toroidal geometry, high m modes are decomposed by the ballooning formalism as $\delta H(\theta,\zeta,r) = \sum_n e^{-in\zeta}\sum_m e^{im\theta}\int d\eta e^{-i(m-nq)\eta}\delta H_n(\eta,r)$ where ζ is the toroidal angle and η is the ballooning coordinate. The equilibrium flow $\boldsymbol{u}_0$ does not include the diamagnetic flow and the convective derivative is given by, $\boldsymbol{u}_0\cdot\nabla = -(u_E q/r)\partial_\zeta - (u_\parallel/qR)\partial_\eta$.

Two-point correlation equation is derived by multiplying the fluid equation for $\delta H(\eta_1,\zeta_1,r_1)$ by $\delta H(\eta_2,\zeta_2,r_2)$ as $(\partial_t + T_{12})\langle\delta H(1)\delta H(2)\rangle = S'$, where $\langle\rangle$ is ensemble average and $T_{12} = -r_-(\omega'_\perp\partial/\partial\zeta_- + \omega'_\parallel\partial/\partial\eta_-) - (q/r)^2 D_-\partial^2/\partial\zeta_-^2$. Here $\omega'_\perp = \partial(u_E q/r)/\partial r$ and $\omega'_\parallel = \partial(u_\parallel/qR)/\partial r$. If we take the diffusion approximation for the $\boldsymbol{E}\times\boldsymbol{B}$ nonlinearity[763], the relative diffusion is given by $D_- = 2D_\zeta[(\eta_-/\Delta\eta)^2 + (\zeta_-/\Delta\zeta)^2 + (r_-/\Delta r_0)^2]$. Here $Rq\Delta\eta$, $R\Delta\zeta$, and Δr_0 are the correlation lengths of the ambient turbulence in the parallel, toroidal, and radial directions, respectively, and D_ζ is the diffusion coefficient at large separation. Taking various moments, Hahm obtained the square ratio of radial correlation length with and without flow shear is given by $(\Delta r_t/\Delta r_0)^2 = \Delta\omega_T^2/(\Delta\omega_T^2 + \omega_s^2)$, where $\Delta\omega_T = 4D_\zeta/(r\Delta\zeta/q)^2$ is the random scattering rate of the ambient turbulence defined with respect to the poloidal correlation length $\Delta L_\theta = r\Delta\zeta/q$. The shearing rate is given as $\omega_s^2 = \Delta r_0^2[(\omega'_\perp/\Delta\zeta)^2 + (\omega'_\parallel/\Delta\eta)^2]$ in which the first term of the right hand side is dominant for flute like fluctuations.

Generalization to arbitrary shaped finite aspect ratio tokamak has been described in Hahm [279] to arrive at Eq. (7.15).

Perpendicular correlation length $\Delta L_\perp$ may be estimated as $(\Delta L_\perp)^{-1} \sim m/r = nq/R \sim B_\zeta/B_\theta\Delta\zeta R$. If we use $\Delta r \approx \Delta L_\perp$ [797] and the linear growth rate γ_L for the de-correlation rate, we have following crude flow shear stabilization condition:

$$\gamma_L \leq \omega_E = \frac{R^2 B_\theta^2}{B}\frac{d^2\Phi}{d\psi^2} \tag{7.16}$$

The discovery of flow shear suppression of turbulence gives rise to a paradigm shift in the turbulent transport physics.

Early day's view of the profile regulation was relatively simple. Gradients of plasma profiles such as pressure, temperature and beta produces the free energy to drive drift wave micro fluctuations such as ITG/TEM, ETG and micro-tearing as shown in Fig. 7.12a. While these micro fluctuations have their nonlinear regulation processes, the enhanced transport due to micro fluctuation reduces gradient to reach regulated profiles.

Refined understanding of profile regulation shown in Fig. 7.12b consists of two level of regulations. One is regulation by transport same in Fig. 7.12a. Another important level is the regulation of the micro fluctuations by the flow shear, which comes from both the mean flow and the zonal flow. Here, the mean flow shear is strongly constrained by the 0-th order radial ion force balance equation and the zonal flow is produced by the turbulence itself, either by the inverse cascade of the 2D turbulence or the modulational instability.

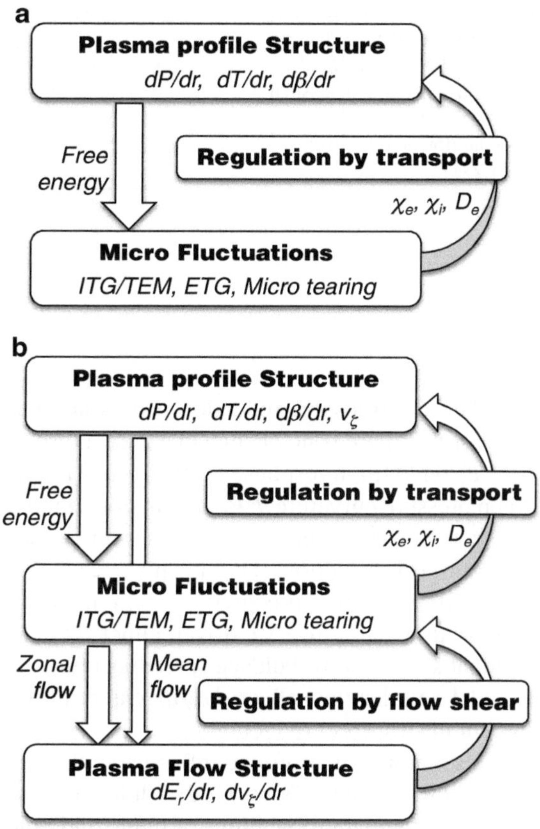

Fig. 7.12 (**a**) Early day's understanding of the profile regulation by the turbulent transport, where free energy from gradients produces turbulence, which regulates gradients due to enhanced turbulent transport. (**b**) Refined understanding of the profile regulation, where the flow structure plays another important role in the regulation of turbulence and flow shear comes from both the mean flow from the profile structure and the zonal flow from the turbulence itself

7.2.2 *Mean Flow and Avalanche Dynamics*

First, we discuss the mean flow effect on turbulent regulation and its effect on the avalanche dynamics. The 0th order radial force balance equation (4.222) combined with neoclassical residual poloidal flow (4.225) is robust in turbulent plasma [369].

$$\boldsymbol{u}_i \cdot \nabla\zeta = -\left[\frac{d\Phi}{d\psi} + \frac{1}{eZ_i n_i}\frac{dP_i}{d\psi}\right] - \frac{qK_1 F(\boldsymbol{B}\cdot\nabla\theta)}{eZ_i\langle B^2\rangle}\frac{dT_i}{d\psi}$$

This constraint plays an important role in the shearing rate ω_E in tokamaks as positive feedback effect of turbulent transport suppression by flow shear (Fig. 7.13). In the cylindrical coordinates, this may be written as:

$$E_r = \frac{1}{eZ_i n_i}\frac{dP_i}{dr} + \frac{K_1}{eZ_i}\frac{dT_i}{dr} + u_{\zeta i}B_\theta \tag{7.17}$$

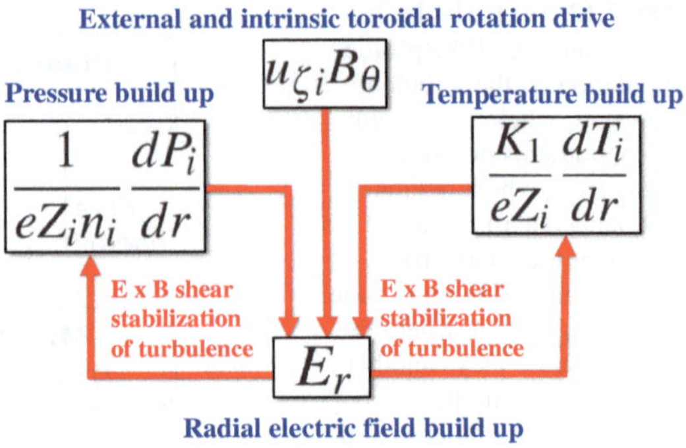

Fig. 7.13 Schematic diagram of positive feedback effect on E_r from dP_i/dr and dT_i/dr, and $u_{\zeta i}$. K_1 is the neoclassical coefficient of residual poloidal flow and is given by (4.221)

When the transport reduction occurs at some region, both dP_i/dr and dT_i/dr increases. If there is no change in external toroidal rotation drive, the radial electric field E_r build up in response to dP_i/dr and dT_i/dr. Therefore, E_r shear is enhanced when pressure curvature d^2P_i/dr^2 and/or temperature curvature d^2T_i/dr^2 is formed.

Once E_r shear is enhanced, further turbulence de-correlation occurs to enhance the transport reduction. This positive feedback effect may lead to formation of localized steep pressure gradient region unstable to MHD instability. Toroidal angular flow $\omega_t = \boldsymbol{u}_i \cdot \nabla\zeta$ driven either by external torque or intrinsic torque (by residual stress and turbulence spreading) further produces radial electric field. The toroidal angular flow shear $d\omega_t/dr$ enters into the dynamics of positive feedback loop in the transport reduction.

This feature has profound influence in the dynamics of the sand collapses in the sand hill discussed in Sect. 7.1.2. While we may expect that sand collapses in sand pile exhibit equal probabilities of bump and void propagations, some of bump and void propagations are prohibited due to this radial force balance constraint.

The radial force balance means that E_r will be balance with other two terms in the right hand side as long as there is no toroidal rotation drive ($u_{\zeta i}$ is determined by the toroidal momentum balance equation). Therefore, important additional constraint for plasma turbulence is that E_r shear build up in proportion to temperature curvature during Void and Bump propagation, namely $dE_r/dr \sim cd^2T_i/dr^2$, where $c > 0$ is constant. Since void has positive temperature curvature ($d^2T_i/dr^2 > 0$) and bump has negative temperature curvature ($d^2T_i/dr^2 < 0$), E_r shear structure with void and bump for positive and negative background dE_{r0}/dr cases becomes as shown schematically in Fig. 7.14. Selective avalanche occurs in cases weakening $|dE_r/dr|$ by void and bump.

This symmetry breaking is observed in the Gyrokinetic full f simulation of the toroidal ITG turbulence [369]. First, two kinds of heat flux propagation are identified, which is a manifestation of joint reflection symmetry. But, this symmetry breaks down when there is dE_r/dr in the plasma as described above and selective avalanches occur. When the background $dE_r/dr > 0$, simulation observes only

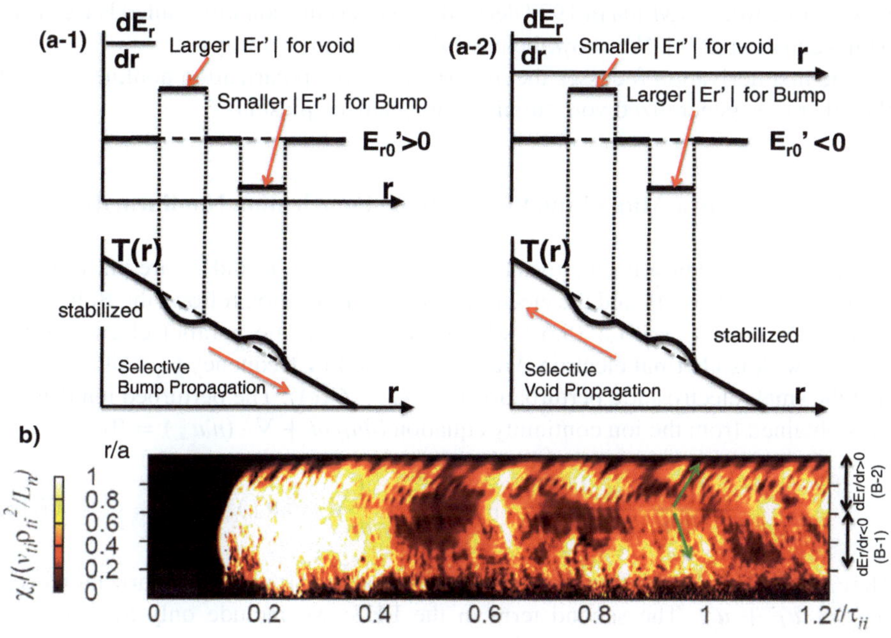

Fig. 7.14 (**a**) Symmetry breaking of void/bump avalanche by dE_r shear. (**a-1**) $E'_{r0} > 0$ case where bump reduces $|E'_r|$. (**a-2**) $E'_{r0} < 0$ case where void reduces $|E'_r|$. After Kikuchi [444]. (**b**) Gyrokinetic full-f simulation of toroidal ITG turbulence, showing outward propagation of bump where $dE_r/dr > 0$ and inward propagation of void where $dE_r/dr < 0$. Modified from Idomura [369]

downhill bump propagation and the uphill void propagation is prohibited. When the background $dE_r/dr < 0$, simulation observes only uphill void propagation and the downhill bump propagation is prohibited as shown in Fig. 7.14.

7.3 Turbulence and Zonal Flow

In this section, we only introduce so-called Hasegawa-Mima equation and some features of zonal flow and do not go into the details of 3D turbulence. Hasegawa-Mima equation is fundamental equation for dissipation-less 2D turbulence which has marked difference with the 3D turbulence with respect to the direction of cascades.

7.3.1 Hasegawa-Mima Equation

Turbulence in strongly magnetized plasma may exhibit two dimensional structure perpendicular to the magnetic field. This is especially true in uniform magnetic field. Hasegawa and Mima in 1977 derived the governing equation called Hasegawa-Mima equation [288]. We show the basic derivation in two ways.

The first derivation focusses the importance of the polarization nonlinearity and the other uses generalized vorticity in the magnetized plasma.

7.3.1.1 Hasegawa-Mima Equation from the Polarization Nonlinearity

We assume turbulence is electrostatic ($\boldsymbol{E} = -\nabla\phi$), and $\boldsymbol{B}$ and T_e are uniform and the ion is cold ($T_i \sim 0$), and the electron density perturbation follows the Boltzmann relation ($\tilde{n}_e/n_e = e\phi/T_e$: (5.6)) which comes from the parallel electron force balance with isothermal electron. We consider the low frequency ($\partial/\Omega_i\partial t = O(\epsilon)$) and the small electrostatic perturbation ($e\phi/T_e = O(\epsilon)$). The perturbed ion density $\tilde{n}_i$ is obtained from the ion continuity equation ($\partial n_i/\partial t + \nabla\cdot(n_i\boldsymbol{u}_\perp) = 0$).

$$\frac{\partial}{\partial t}\frac{\tilde{n}_i}{n_i} + \boldsymbol{u}_\perp\cdot\frac{\nabla n_i}{n_i} + \nabla\cdot\boldsymbol{u}_\perp + \frac{1}{n_i}\nabla\cdot(\tilde{n}_i\boldsymbol{u}_\perp) = 0 \tag{7.18}$$

Here, the perpendicular flow is composed of $\boldsymbol{E}\times\boldsymbol{B}$ drift and the polarization drift, $\boldsymbol{u}_\perp = \boldsymbol{u}_E + \boldsymbol{u}_{pi}$. The second term in the LHS: we include only $\boldsymbol{u}_E$ since the polarization drift $\boldsymbol{u}_{pi}$ is much smaller than $\boldsymbol{u}_E$. The third term: the polarization term is kept since $\nabla\cdot\boldsymbol{u}_{pi} \neq 0$ while $\nabla\cdot\boldsymbol{u}_E = 0$ in the uniform magnetic field. The last term: the $\boldsymbol{E}\times\boldsymbol{B}$ nonlinearity $\nabla_\perp\cdot(\tilde{n}\boldsymbol{u}_E) = 0$ since $\nabla_\perp\cdot(\phi\nabla_\perp\phi\times z) = 0$ and second order polarization term $\nabla_\perp\cdot(\tilde{n}\boldsymbol{u}_{pi}) = O(\epsilon^3)$ is negligible.

Using the leading order expression $d/dt = \partial/\partial t + \boldsymbol{u}_E\cdot\nabla_\perp$, we express $\boldsymbol{u}_{pi}$ as:

$$\boldsymbol{u}_{pi} = -\frac{m_i}{eB^2}\left[\frac{\partial}{\partial t}\nabla_\perp\phi + (\boldsymbol{u}_E\cdot\nabla_\perp)\nabla_\perp\phi\right] \tag{7.19}$$

Using the charge neutrality condition $\tilde{n}_e/n_e = \tilde{n}_i/n_i$, we obtain following equation for the electrostatic potential.

$$\frac{\partial}{\partial t}\frac{e\phi}{T_e} - \frac{\nabla_\perp\phi\times\boldsymbol{b}}{B}\cdot\frac{\nabla n_i}{n_i}$$
$$+\nabla_\perp\cdot\left[-\frac{\nabla_\perp\phi\times\boldsymbol{b}}{B} - \frac{m_i}{eB^2}\left(\frac{\partial}{\partial t}\nabla_\perp\phi + \left(\frac{\nabla_\perp\phi\times\boldsymbol{b}}{B}\cdot\nabla_\perp\right)\nabla_\perp\phi\right)\right] = 0$$

The last term in the $\nabla\cdot\boldsymbol{u}_\perp$ is the $\boldsymbol{E}\times\boldsymbol{B}$ convection of the polarization drift, which is the essential nonlinearity in the Hasegawa-Mima equation. We define $c_s = \sqrt{T_e/m_i}$, $\rho_s = c_s/\Omega_i$, $u_{de} = -T_e n_e'(x)/eBn_e$, and $\hat{\phi} = e\phi/T_e$, above equation reads:

$$\left[1 - \rho_s^2\nabla_\perp^2\right]\frac{\partial\hat{\phi}}{\partial t} + u_{de}\frac{\partial\hat{\phi}}{\partial y} + \rho_s^3 c_s\nabla_\perp\cdot\left[(\nabla_\perp\hat{\phi}\times\boldsymbol{b}\cdot\nabla_\perp)\nabla_\perp\hat{\phi}\right] = 0 \tag{7.20}$$

Since $\nabla_\perp \cdot \left[(\nabla_\perp \hat{\phi} \times \boldsymbol{b} \cdot \nabla_\perp)\nabla_\perp \hat{\phi}\right] = \nabla_\perp \hat{\phi} \times \boldsymbol{b} \cdot \nabla_\perp \nabla_\perp^2 \hat{\phi}$ in the two dimensional system, this equation is reduced to following Hasegawa-Mima (HM) equation:

$$\left[1 - \rho_s^2 \nabla_\perp^2\right] \frac{\partial \hat{\phi}}{\partial t} + u_{de} \frac{\partial \hat{\phi}}{\partial y} + \rho_s^3 c_s \nabla_\perp \hat{\phi} \times \boldsymbol{b} \cdot \nabla_\perp \nabla_\perp^2 \hat{\phi} = 0 \tag{7.21}$$

If we normalize $x/\rho_s \to x$, $\Omega_i t \to t$, $n_i/n_0 \to n_i$, we obtain following equation.

$$\frac{\partial}{\partial t}(\nabla_\perp^2 \hat{\phi} - \hat{\phi}) + \boldsymbol{b} \times \nabla_\perp \phi \cdot \nabla_\perp (\nabla_\perp^2 \hat{\phi} + \ln n_i) = 0 \tag{7.22}$$

Salon 7-1: Vorticity

The vorticity of the neutral fluid is defined by $\boldsymbol{\omega} = \nabla \times \boldsymbol{u}$. Vortex motion is a visualization of vorticity and are seen in the river side. In the rest frame shown in the below figure (a), flow lines may be seen in the mid river, while vortex motion may be seen near the edge. In the moving frame with u of mid river, vortex motion may be seen in the mid river as shown in the below figure (b). To see the net vortex motion, it is important to eliminate directional flow.

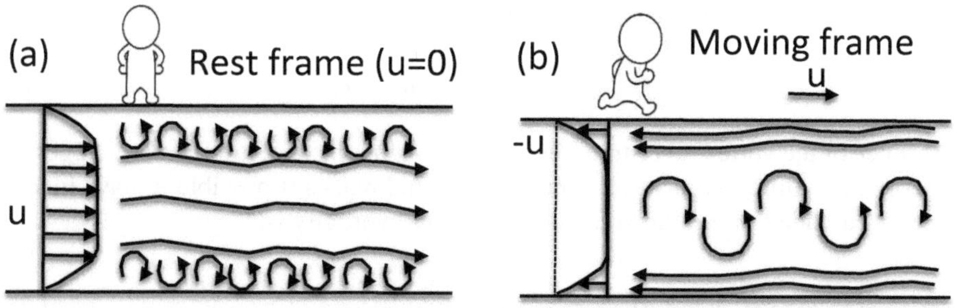

7.3.1.2 Hasegawa-Mima Equation from Generalized Vorticity

Above derivation of the Hasegawa-Mima equation is based on the ion polarization drift. This Hasegawa-Mima equation can be derived using the generalized vorticity. The canonical momentum per unit mass of the magnetized fluid is defined as $\boldsymbol{p} = \boldsymbol{u} + (e_i/m_i)\boldsymbol{A}$. We define the generalized vorticity $\boldsymbol{\Omega}$ as:

$$\boldsymbol{\Omega} = \nabla \times \boldsymbol{p} = \nabla \times \boldsymbol{u} + \frac{e_i}{m_i} \nabla \times \boldsymbol{A} = \boldsymbol{\omega} + \frac{e_i}{m_i} \boldsymbol{B} = \boldsymbol{\omega} + \boldsymbol{\Omega}_i, \tag{7.23}$$

where $\boldsymbol{\omega} = \nabla \times \boldsymbol{u}$, $\boldsymbol{\Omega}_i = e_i \boldsymbol{B}/m_i = \Omega_i \boldsymbol{z}$. The generalized vorticity is a summation of the fluid vorticity and the cyclotron frequency. We note $\boldsymbol{\Omega}$ is divergence free,

$\nabla \cdot \boldsymbol{\Omega} = 0$. Using the relation $\boldsymbol{u} \cdot \nabla \boldsymbol{u} = \frac{1}{2}\nabla u^2 - \boldsymbol{u} \times (\nabla \times \boldsymbol{u})$, the equation of motion, $d\boldsymbol{u}/dt = (e_i/m_i)(\boldsymbol{E} + \boldsymbol{u} \times \boldsymbol{B}) - (1/m_i n_i)\nabla p$ can be given as:

$$\frac{\partial \boldsymbol{u}}{\partial t} = \frac{e_i}{m_i}\boldsymbol{E} + \boldsymbol{u} \times \boldsymbol{\Omega} - \frac{1}{2}\nabla u^2 - \frac{1}{m_i n_i}\nabla P \tag{7.24}$$

We can eliminate above potential terms by taking the rotation ($\nabla \times$).

$$\frac{\partial \boldsymbol{\omega}}{\partial t} = -\frac{e_i}{m_i}\frac{\partial \boldsymbol{B}}{\partial t} + \nabla \times (\boldsymbol{u} \times \boldsymbol{\Omega}) + \frac{1}{m_i n_i^2}\nabla n_i \times \nabla P \tag{7.25}$$

Considering $\nabla \times (\boldsymbol{u} \times \boldsymbol{\Omega}) = \boldsymbol{u}\nabla \cdot \boldsymbol{\Omega} - \boldsymbol{\Omega}\nabla \cdot \boldsymbol{u} + (\boldsymbol{\Omega} \cdot \nabla)\boldsymbol{u} - (\boldsymbol{u} \cdot \nabla)\boldsymbol{\Omega}$ and $\nabla \cdot \boldsymbol{\Omega} = 0$, we obtain following generalized vorticity equation.

$$d\boldsymbol{\Omega}/dt + \boldsymbol{\Omega}\nabla \cdot \boldsymbol{u} = (\boldsymbol{\Omega} \cdot \nabla)\boldsymbol{u} \tag{7.26}$$

Here, we used $\nabla n_i \times \nabla P \sim 0$. The term $(\boldsymbol{\Omega} \cdot \nabla)\boldsymbol{u}$ is called the generalized vortex tube stretching, which is an extension of vertex tube stretching in the neutral fluid as shown in the Column 7-1.

In the two dimensional turbulence, the generalized vorticity has only z-component, $\boldsymbol{\Omega} = \Omega \boldsymbol{z} = (\omega + \Omega_i)\boldsymbol{z}$ and the generalized vortex tube stretching is absent since all physical quantities are uniform in the direction of the magnetic field. Therefore, time evolution of the generalized vorticity $\boldsymbol{\Omega}$ in two dimensional magnetized plasma turbulence is given by,

$$d\boldsymbol{\Omega}/dt + \boldsymbol{\Omega}\nabla \cdot \boldsymbol{u} = 0 \tag{7.27}$$

Absence of the vortex tube stretching in the two dimensional turbulence changes the energy flow direction in wave number space and makes it possible to have inverse energy cascade to form the zonal flow. The fluid vorticity is:

$$\omega = \boldsymbol{\omega} \cdot \boldsymbol{z} \sim \nabla_\perp \times \boldsymbol{u}_E \cdot \boldsymbol{z} = -\left[\nabla_\perp \times \frac{\nabla_\perp \phi \times \boldsymbol{z}}{B}\right] \cdot \boldsymbol{z} = \frac{1}{B}\nabla_\perp^2 \phi \tag{7.28}$$

Therefore, the time derivative of the fluid vorticity is given by:

$$\frac{d\omega}{dt} = \frac{\partial \omega}{\partial t} + \frac{\nabla_\perp \phi \times \boldsymbol{z}}{B} \cdot \nabla_\perp \omega = \rho_s^2 \Omega_i \left[\frac{\partial}{\partial t}\nabla_\perp^2 \hat{\phi} + \rho_s c_s (\nabla_\perp \hat{\phi} \times \boldsymbol{z} \cdot \nabla_\perp)\nabla_\perp^2 \hat{\phi}\right] \tag{7.29}$$

Using the continuity equation ($\nabla \cdot \boldsymbol{u} = -dln n_i/dt$), we have

$$\Omega \nabla \cdot \boldsymbol{u} = -(\omega + \Omega_i) dln n_i/dt \sim -\Omega_i dln n_i/dt \sim -\Omega_i \left[\partial \hat{\phi}/\partial t + \boldsymbol{u}_E \cdot \nabla_\perp ln n_i\right]$$

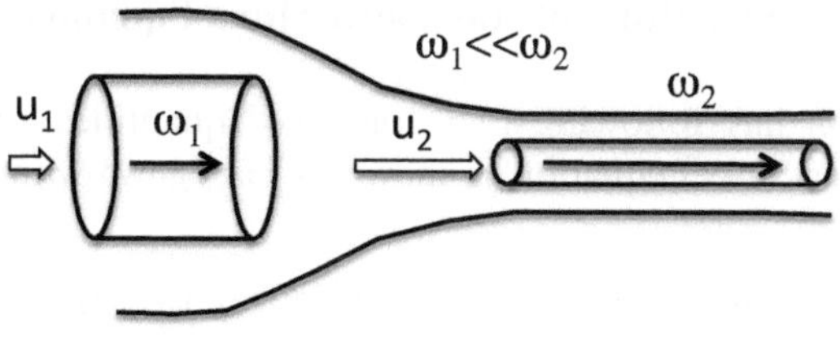

Fig. 7.15 Vortex tube stretching in the wind tunnel. As the flow moves from large diameter to small diameter region in the wind tunnel, the speed u and the vorticity ω increases. This increase in vorticity is related to the conservation of the angular momentum of the vortex and is called "vortex tube stretching"

Thus, we obtain

$$\Omega \nabla \cdot \boldsymbol{u} = -\Omega_i \left[\partial \hat{\phi} / \partial t + u_{de} (\partial \hat{\phi} / \partial y) \right] \tag{7.30}$$

Substituting (7.29) and (7.30) into (7.27) and using $d\Omega_i/dt = 0$, we have Hasegawa-Mima equation (7.21).

Column 7-1 : Vortex Tube Stretching in 3-D Fluid System
In the fluid turbulence, the vorticity $\boldsymbol{\omega} = \nabla \times \boldsymbol{u}$ plays an important role. If ρ is constant in space, the Navier-Stokes equation can be expressed as,

$$\frac{\partial \boldsymbol{u}}{\partial t} = -\nabla \left(\frac{p}{\rho} + \frac{1}{2} u^2 \right) + \boldsymbol{u} \times \boldsymbol{\omega} - \nu \nabla \times \boldsymbol{\omega} \tag{7.31}$$

Taking $\nabla \times$ (7.31), we obtain following vorticity equation.

$$\frac{\partial \boldsymbol{\omega}}{\partial t} + \boldsymbol{u} \cdot \nabla \boldsymbol{\omega} = (\boldsymbol{\omega} \cdot \nabla) \boldsymbol{u} + \nu \nabla^2 \boldsymbol{\omega} \tag{7.32}$$

The appearance of $(\boldsymbol{\omega} \cdot \nabla)\boldsymbol{u}$ term above implies that the vorticity is produced if the flow $\boldsymbol{u}$ changes in the direction of $\boldsymbol{\omega}$, by which the vortex tube is stretched or squeezed. This term is called the "vortex tube stretching" (Fig. 7.15).

If we write $\partial u_i / \partial x_k = s_{ki} + r_{ki}$, where $s_{ki} = (\partial u_k / \partial x_i + \partial u_i / \partial x_k)/2$ (rate of strain tensor, see Column 4-1) and $\omega_{ki} = (\partial u_k / \partial x_i - \partial u_i / \partial x_k)/2$ (vorticity tensor), we obtain following relation for this vortex tube stretching.

$$(\boldsymbol{\omega} \cdot \nabla)\boldsymbol{u} = \boldsymbol{\omega} \cdot \boldsymbol{s}, \tag{7.33}$$

where $\boldsymbol{s}$ is a symmetric tensor having (k, i) component, s_{ki}.

7.3.1.3 Conserved Quantities of Hasegawa-Mima Equation

This HM equation has following two conserved quantities similar to the two dimensional Navier-Stokes equation as shown by Hasegawa-Mima [289].

$$E = \frac{1}{2}\langle(\nabla_\perp\phi)^2 + \phi^2\rangle \ , W = \frac{1}{2}\langle(\nabla_\perp^2\phi)^2 + (\nabla_\perp\phi)^2\rangle, \tag{7.34}$$

where $\langle A\rangle$ is the volume average of A. Here E corresponds to the total energy and W corresponds to the potential enstrophy. Here the enstrophy is the volume integral of half of square vorticity, $\omega^2/2$ and is a conserved quantity (Column 7-2).

Zonal flow in drift wave turbulence is first predicted by Hasegawa in 1979 from energy cascade of Hasegawa-Mima equation [288] and its similarity to geostrophic vortex equation (Charney equation) of Jobian atmosphere is discussed in [290].

Column 7-2 : Enstrophy Conservation in 2-D Turbulence
In case of ideal two dimensional fluid turbulence ($\nu = 0$), the vorticity is conserved along the fluid motion,

$$\frac{d\omega}{dt} = \frac{\partial\omega}{\partial t} + \boldsymbol{u}\cdot\nabla\omega = 0 \tag{7.35}$$

Considering the vector formula $\boldsymbol{\omega}\cdot(\boldsymbol{u}\cdot\nabla)\boldsymbol{\omega} = -\boldsymbol{\omega}\cdot(\boldsymbol{u}\times(\nabla\times\boldsymbol{\omega})) = \boldsymbol{u}\cdot(\boldsymbol{\omega}\times(\nabla\times\boldsymbol{\omega})) = \boldsymbol{u}\cdot\nabla(u^2/2)$, multiplication of $\boldsymbol{\omega}\cdot$ to $d\boldsymbol{\omega}/dt = 0$ gives:

$$\frac{\partial}{\partial t}\left(\frac{\omega^2}{2}\right) + \nabla\cdot\left(\frac{\omega^2}{2}\boldsymbol{u}\right) = 0 \tag{7.36}$$

If we take an integral inside the volume bounded by a rigid body, where $\boldsymbol{u}\cdot d\boldsymbol{S} = 0$, we arrive at a law of enstrophy conservation, U.

$$\frac{\partial U}{\partial t} \equiv \frac{\partial}{\partial t}\int \frac{\omega^2}{2} dV = 0 \tag{7.37}$$

We note that enstrophy conservation holds only for two dimensional ideal fluid turbulence and is closely related to the absence of vortex tube stretching, $\boldsymbol{\omega}\cdot\nabla\boldsymbol{u} = 0$ [472].

Conservation of enstrophy is one of important properties of Hasegawa-Mima equation in the two dimensional turbulence in magnetized plasma.

For more detail, see Hasegawa [291] and Sect. 2.3.1 of Diamond's textbook [168].

7.3.1.4 Dual Cascades of Turbulent Spectrum

Hasegawa [289] clarified that the turbulent cascade in two dimensional drift wave turbulence (HM equation) is a dual cascade similar to the two dimensional fluid turbulence [472]. Namely, the energy flux in k space is directed to lower k (inverse energy cascade) to form a zonal flow and the enstrophy flux in k space is directed to higher k (forward enstrophy cascade). Energy spectrum in the inertial range of the wave number space can be obtained assuming the stationary flow in k space. If k_s is the wave number of the source, energy spectrum in $k < k_s$ and $k > k_s$ are given by:

$$W(k) = C_1 k^{-5/3} \text{ for } k < k_s \tag{7.38}$$

$$W(k) = C_2 k^{-3} \text{ for } k > k_s \tag{7.39}$$

In the atmospheric dynamics, the Rossby wave [541] plays similar behavior, where Lorentz force is replaced to Coriolis force and the $\boldsymbol{E} \times \boldsymbol{B}$ flow $\boldsymbol{u}_E$ is replaced to geostrophic flow $\boldsymbol{u}_g = \boldsymbol{e}_z \times \nabla p/(f\rho)$, where f is the Coriolis parameter. This similarity corresponds to the equation $\boldsymbol{E}+\boldsymbol{u}\times\boldsymbol{B} = 0$ in plasma and $\nabla p + f\rho \boldsymbol{e}_z \times \boldsymbol{u} = 0$ in Westerlies (Jet Stream) dynamics. The reason why Rossby wave behave as two dimensional system is that atmospheric layer is so thin compared with the radius of the Earth.

7.3.1.5 Electron Hasegawa-Mima Equation

The Hasegawa-Mima (HM) equation is derived for the ion scale turbulence based on the ion polarization drift. But it is also important for the electron scale turbulence as discussed by Idomura in 2006 [368]. The electrostatic gyrokinetic Vlasov-Poisson system with gyrokinetic electron and adiabatic ion in the limit of $k_\parallel \to 0$ gives Hasegawa-Mima type equation for electron dynamics [368].

$$\frac{\partial}{\partial t}(\rho_s^2 \nabla_\perp^2 \phi - \tau\phi) + \boldsymbol{b} \times \nabla_\perp \phi \cdot \nabla_\perp (\rho_s^2 \nabla_\perp^2 \phi + \ln n_0) = 0 \tag{7.40}$$

where normalization $x/\rho_{te} \to x$, $t\Omega_e \to t$, $e\phi/T_e \to \phi$, $n_0/\langle n_0 \rangle \to n_0$ are taken and $\rho_s^2 = 1 + \lambda_{De}^2/\rho_{te}^2$ and $\tau = T_e/T_i$, respectively. Here, it is important to note that main nonlinearity in original HM equation comes from $E \times B$ convection of ion polarization drift, nonlinearity for this electron HM (e-HM) equation comes from Debye shielding effect and electron polarization/ finite electron Larmor radius effect.

This e-HM equation also conserves the energy $E = \frac{1}{2}\langle \rho_s^2 (\nabla_\perp \phi)^2 + \tau\phi^2 \rangle$, and the potential enstrophy $W = \frac{1}{2}\langle \rho_s^2 (\nabla_\perp^2 \phi)^2 + \tau(\nabla_\perp \phi)^2 \rangle$. The nonlinear term $\boldsymbol{b} \times \nabla_\perp \phi \cdot \rho_s^2 \nabla_\perp^2 \phi$ leads to inverse energy cascade. Balancing with linear dn_0/dx term gives critical radial wave number for energy condensation $k_x \sim k_\beta$ ($k_\beta = \rho_s^{-1}\sqrt{\beta/2U}$, $\beta = |d\ln n_0/dx|$, $U = \sqrt{2\epsilon}$, $\epsilon = \frac{1}{2}\langle(\nabla_\perp \phi)^2\rangle$), equivalent to Rhines scale length

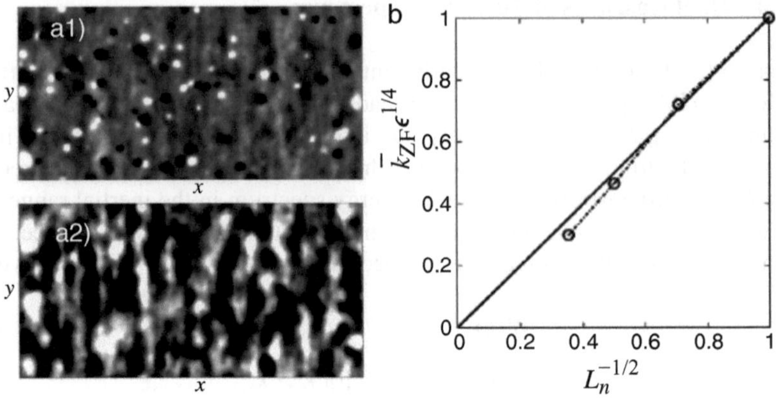

Fig. 7.16 (**a**) Contour plot of ϕ in the electron turbulence. (**a1**) Uniform density. (**a2**) $L_n = 1462\rho_{te}$ to excite zonal flow. (**b**) Normalized zonal flow wave length $k_{ZF}\epsilon^{1/4}$ v.s. $L_n^{-1/2}$ from the gyrokinetic electron turbulence simulation. Reproduced with permission from Idomura [368]. Copyright AIP Publishing LLC

[625] in the β-plane turbulence. The inverse energy cascade occurs from high k_x to k_β to form zonal flow in shear-less ETG turbulence with radial wave number $\sim k_\beta$.

Figure 7.16a shows the contour plot of the electrostatic potential ϕ for uniform density case (a1) and for density gradient case (a2). In case of uniform density, turbulence has isolated coherent vortices. In case of non-zero density gradient (a2), zonal flows are produced and the turbulence becomes anisotropic. The normalized average wave number of zonal flow $k_{ZF}\epsilon^{1/4}$ is inversely proportional to the square root of density scale length, $k_{ZF}\epsilon^{1/4} \sim L_n^{-1/2}$ as shown in Fig. 7.16b.

7.3.2 *Zonal Flow by Modulational Instability*

The zonal flow is $\boldsymbol{E} \times \boldsymbol{B}$ flow symmetric toroidally ($n = 0$) and poloidally ($m = 0$) but is a radially varying electric potential fluctuation with nearly zero frequency.

The zonal flow in the sheared magnetic field is excited through the modulational instability first discussed by Diamond [164]. Chen-Lin-White [112] refine this idea in the toroidal geometry using the ballooning formulation and obtain the quantitative agreement with the numerical simulation. In the axisymmetric system, the electrostatic potential of the pump drift wave $\tilde{\Phi}^p$ can be expressed as a summation of poloidal harmonics with harmonic amplitude $\tilde{\Phi}_l(q)$, where safety factor q is used as a radial coordinate instead of r. The harmonic amplitude $\tilde{\Phi}_l(q)$ can be constructed by the eigenfunction for $l = 0$, $\tilde{\Phi}_0(nq - m)$ peaked at $q = m/n$ and assuming translational symmetry similar to Bloch function in crystal lattice [459, 851].

$$\tilde{\Phi}^p(r,\theta,\zeta) = \sum_{l=-\infty}^{+\infty} \tilde{\Phi}_l(q) e^{i(m+l)\theta} e^{-in\zeta} \tag{7.41}$$

$$\tilde{\Phi}_l(q) = a(\Delta q)\tilde{\Phi}_0(nq - m - l) \tag{7.42}$$

where, each harmonics is located in the region where $q = (m+l)/n$. Here, $\Delta q = l/n$ and $a(\Delta q)$ is slowly varying amplitude corresponding to envelope of modes. This is called Ballooning eigenfunction, originally developed for ideal MHD Ballooning mode by Connor-Hastie-Taylor [141]. For high n modes, this translational symmetry holds and can be assumed a constant ($a(\Delta q) = 1$). This translational symmetry comes from the dense overlap of poloidal harmonics, where all physical quantities are the same for all harmonics.

Three wave parametric excitation of the zonal flow does not work due to wave number and frequency matching constraints. The zonal flow is excited by the modulational instability by four wave interaction, i.e. pump wave, zonal flow, and two sideband waves. The pump drift wave ($\tilde{\Phi}^p$) is the toroidal ITG with single n-value consists of a set of poloidal harmonics oscillating together with a fixed phase relation given by Eq. (7.42). This pump drift wave with wave number $\boldsymbol{k}_d$ couples to the modulating zonal flow with the radial wave number $\boldsymbol{k}_r$ and frequency Ω and induces two side-band drift waves (denoted by $\tilde{\Phi}^{\pm}$) having wave numbers:

$$\boldsymbol{k}_+ = \boldsymbol{k}_d + \boldsymbol{k}_r \text{ , and , } \boldsymbol{k}_- = \boldsymbol{k}_d - \boldsymbol{k}_r \tag{7.43}$$

The zonal flow, the pump wave and sidebands can be expressed using the ballooning eigen function as follows,

$$\tilde{\Phi}^Z = \tilde{\Phi}_0^Z(nq - m)e^{ik_r r - i\Omega t} + c.c. \tag{7.44}$$

$$\tilde{\Phi}^p = e^{-i(n\zeta+\omega_0 t)} \sum_{m=-\infty}^{+\infty} \tilde{\Phi}_0^p(nq - m)e^{im\theta} + c.c. \tag{7.45}$$

$$\tilde{\Phi}^{\pm} = e^{i(\mp(n\zeta+\omega_0 t)+k_r r-\Omega t)} \sum_{m=-\infty}^{+\infty} \tilde{\Phi}_0^{\pm}(nq - m)e^{im\theta} + c.c. \tag{7.46}$$

Here, $k_r = nq'\theta_k$ and $0 \le \theta_k \le \pi$ is called the Bloch angle. The pump wave is in general a most unstable mode, while side bands are linearly stable modes.

The zonal flow generation by the coupling to pump and sidebands is described by Hasegawa-Mima type equation and Chen [112, 113] gives following dispersion relation.

$$(\Gamma_z + \nu_z)(\Gamma_z + \gamma_d) = \gamma_M^2 \tag{7.47}$$

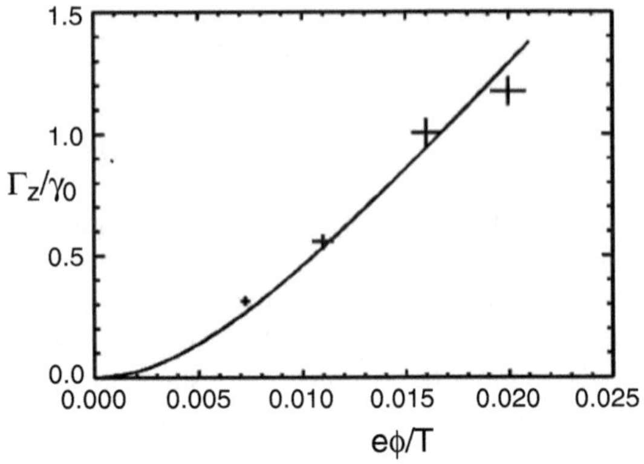

Fig. 7.17 Growth rate of zonal flow normalized to ITG growth rate v.s. normalized ITG amplitude. *Solid line* is (7.48) and + is gyrokinetic simulation. Reproduced with permission from Chen-Lin-White [112]. Copyright AIP Publishing LLC

This dispersion relation gives the growth rate of the zonal flow as:

$$\Gamma_z = -(\gamma_d + \nu_z)/2 + \sqrt{\gamma_M^2 + (\gamma_d - \nu_z)^2/4}, \tag{7.48}$$

where γ_d is the damping rate of sideband, $\nu_z = 1/1.5\epsilon\tau_{ii}$ is the zonal flow damping rate due to ion-ion collision, and γ_M is the characteristic growth rate of modulational instability inversely proportional to the neoclassical polarization (see Diamond [166] for more explanation).

The growth rate of zonal flow normalized to ITG growth rate γ_0 is plotted as a function of the ITG potential amplitude as shown in Fig. 7.17 and the growth rate of zonal flow Γ_z is in good agreement with gyrokinetic simulation.

In the toroidal geometry, the zonal flow couples to the geodesic acoustic mode and may damp through the Landau damping. Rosenbluth-Hinton showed that this zonal flow can have a undamped component due to neoclassical effect, which is found to be an important mechanism to regulate saturation level of turbulence and associated radial heat transport [640]. We show key physics of this residual zonal flow and the dynamics of zonal flow in the following subsections.

7.3.3 Residual Zonal Flow

Polarization plays an important role in the toroidal plasma. In the case of solid state physics, polarization appears even at $\omega = 0$. In the case of plasma, however, polarization drift can only exist transiently as,

$$\boldsymbol{v}_{pa} = \frac{m_a}{e_a B^2}\frac{d\boldsymbol{E}}{dt} \tag{7.49}$$

This polarization drift change sign for ion and electron (negligibly small). This gives rise to the current, called polarization current $\boldsymbol{j}_p^{cl}$.

$$\boldsymbol{j}_p^{cl} = \sum_a e_a n_a \boldsymbol{v}_{pa} = \frac{m_i n_i}{B^2}\frac{d\boldsymbol{E}}{dt} = \epsilon_0 \epsilon_p^{cl}\frac{d\boldsymbol{E}}{dt}, \tag{7.50}$$

where $\epsilon_p^{cl} = m_i n_i/\epsilon_0 B^2 = (\omega_{pi}/\Omega_i)^2 = (k_{Di}\rho_i)^2 \gg 1$, where ω_{pi} and k_{Di} are ion plasma frequency and ion Debye wave number, respectively. This polarization current originates from delayed ion gyro motion from time varying electric field and ϵ_p^{cl} is called classical polarization, which is a low frequency dielectric constant perpendicular to the magnetic field (note: vacuum has extra contribution of 1).

In tokamak, some fraction of charged particles ($f_t \sim \sqrt{\epsilon}, \epsilon \equiv r/R$) are trapped by the magnetic mirror and the radial excursion of trapped particles (Δ_t) is larger than that of passing particles ($\Delta_p = q\rho_i$) given by,

$$\Delta_t = \sqrt{\epsilon}\rho_{pi} = \frac{q\rho_i}{\sqrt{\epsilon}} \tag{7.51}$$

During this trapped particle orbit motion, we also expect similar polarization effect if we have radial electric field E_r,

$$j_{pr}^{nc} = \epsilon_0 \epsilon_p^{nc}\frac{dE_r}{dt} \tag{7.52}$$

$$\epsilon_p^{nc} = \sqrt{\epsilon}k_{Di}^2\Delta_t^2 = \frac{q^2}{\sqrt{\epsilon}}\epsilon_p^{cl}, \tag{7.53}$$

where ϵ_p^{nc} is called neoclassical polarization. Radial excursion of passing particle also have some polarization effect. Summation of classical and neoclassical polarization is total polarization, $\epsilon_p = \epsilon_p^{cl} + \epsilon_p^{nc}$, which is a dielectric constant in the radial direction. The continuity equation of the polarization current is given by,

$$\frac{\partial \rho_p}{\partial t} + \nabla \cdot \boldsymbol{j}_p = S, \tag{7.54}$$

where ρ_p is the polarization charge density, S_k is the external source density. Taking the flux surface average ($\langle\rangle$) and Fourier expansion in space, we have

$$\frac{\partial}{\partial t}\langle \rho_p(\boldsymbol{k})\rangle + \langle i\boldsymbol{k}_\perp \cdot \boldsymbol{j}_p(\boldsymbol{k})\rangle = \langle S(\boldsymbol{k})\rangle \tag{7.55}$$

Polarization current $\boldsymbol{j}_p$ is given by,

$$\boldsymbol{j}_p = \epsilon_0 \epsilon_p \frac{d\boldsymbol{E}}{dt} \tag{7.56}$$

For the electrostatic perturbation ($\boldsymbol{E} = -i\boldsymbol{k}\Phi_k$), above two equations lead to

$$\epsilon_0 \epsilon_p \langle k_\perp^2 \rangle \Phi_k = \langle \rho_p \rangle - \int \langle S(\boldsymbol{k}) \rangle dt \tag{7.57}$$

Hinton-Rosenbluth 1999 (Instead of polarization, they uses the dielectric susceptibility $\tilde{\chi}_k = (\langle k_\perp^2 \rangle \rho_i^2 \Omega_i^2)/\omega_{pi}^2)\epsilon_p$.) gave an expression for this polarization as,

$$\epsilon_p = \frac{\omega_{pi}^2}{\Omega_i^2}\left(1 + \frac{1.6q^2}{\sqrt{\epsilon}}\right), \tag{7.58}$$

where first and second terms of RHS are classical and neoclassical polarization, respectively. The factor "1.6" comes from detailed kinetic calculation including passing particle contribution.

Consider an initial source perturbation $\langle S(\boldsymbol{k}) \rangle = \delta n_k(0)\delta(t)$, build up of the electrostatic potential is determined by the classical polarization shielding for a time scale of a few gyro motion but much shorter than the bounce time of trapped particle. Then, we have,

$$\epsilon_0 \epsilon_p^{cl} \langle k_\perp^2 \rangle \Phi_k(t = +0) = -e_i \delta n_k(0) \tag{7.59}$$

At longer than the bounce time of trapped particles, the electrostatic potential is further shielded by the addition of the neoclassical polarization ϵ_p^{nc} to have,

$$\epsilon_0 (\epsilon_p^{cl} + \epsilon_p^{nc}) \langle k_\perp^2 \rangle \Phi_k(t = +\infty) = -e_i \delta n_k(0) \tag{7.60}$$

Therefore, the ratio of the long term zonal flow potential to the initial zonal flow potential is given by,

$$\frac{\Phi_k(t = \infty)}{\Phi_k(t = 0)} = \frac{\epsilon_p^{cl}}{\epsilon_p^{cl} + \epsilon_p^{nc}} \tag{7.61}$$

Using Hinton formula (7.58), we have

$$\frac{\Phi_k(t = \infty)}{\Phi_k(t = 0)} = \frac{1}{1 + 1.6q^2/\sqrt{\epsilon}} \tag{7.62}$$

The damping of zonal flow to this residual zonal flow has been confirmed in the nonlinear gyrokinetic simulations such as by Idomura [366]. Kinetic equilibrium of tokamak is characterized by canonical Maxwellian distribution, $f_{CM}(P_\phi, \epsilon, \mu) = C_{CM}(P_\phi) exp[-\epsilon/T_a(P_\phi)]$, where P_ϕ, ϵ and μ are canonical angular momentum, kinetic energy, and magnetic moment, respectively. Use of this canonical Maxwellian is important in delta-f simulation to eliminate spurious growth of zonal flow.

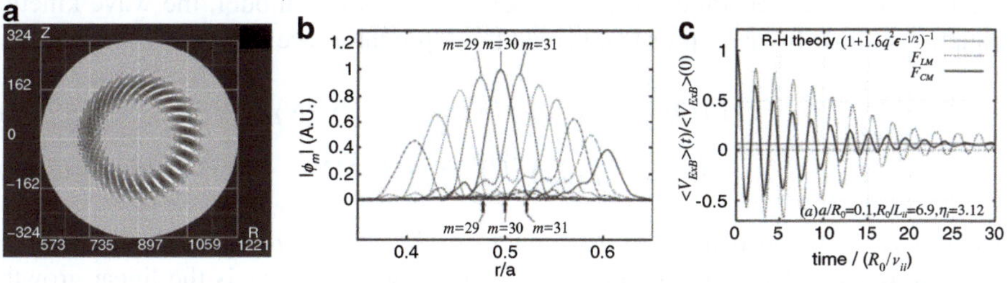

Fig. 7.18 (**a**) Toroidal ITG eigenmode structure in poloidal cross section. (**b**) Radial harmonic structure of toroidal ITG with $n = 15$ and $a/\rho_i = 324$. (**c**) Comparison of the damping of the zonal flow using local and canonical Maxwellian. Reproduced with permission from Idomura [366]. Copyright IAEA Vienna

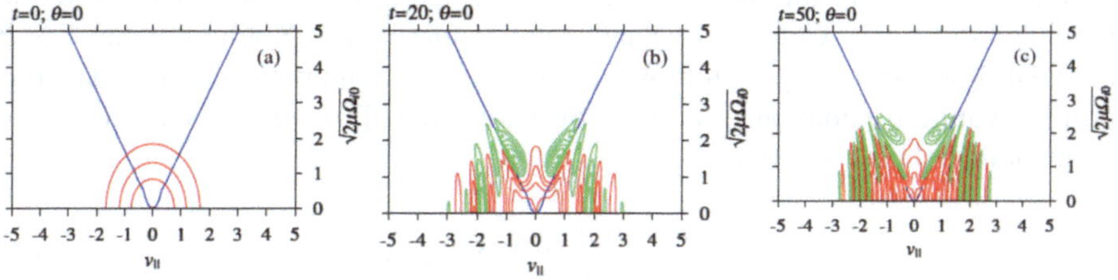

Fig. 7.19 Velocity space profiles of real part of the perturbed distribution function at $\theta = 0$ for different time steps (t=0, 20, 50) of simulation. Positive and negative perturbations are shown in *red* and *green*, respectively. Reproduced with permission from Watanabe [811]. Copyright IAEA Vienna

Figure 7.18 shows the gyrokinetic simulation of the toroidal ITG mode structure and the damping of the zonal flow to the residual zonal flow. The zonal flows are coupled to $m \neq 0$ components through geodesic coupling to the acoustic mode and the zonal flow coupled to $m \neq 0$ is called the Geodesic Acoustic Mode (GAM). This GAM has an oscillating solution, which damps by the Landau damping.

The study of the velocity space structure during the zonal flow (GAM) damping by Watanabe [811] clearly shows that the damping of the electrostatic potential occurs through the phase mixing as time goes and the residual negative perturbed distribution function are seen locally at the trapped particle region as seen in Fig. 7.19.

7.3.4 Zonal Flow Dynamics

7.3.4.1 Diamond's Predator-Prey Model

While the zonal flow is produced by the turbulence, the zonal flow can regulate turbulence through its shear flow. Diamond proposes a self-regulation model of the turbulence, called Predator-Prey model [166]. In this model, the wave kinetic equations for drift wave (prey) and zonal flow (predator) are:

$$(\partial/\partial t - \gamma_L + \gamma_{NL}) \langle N \rangle = -\alpha \langle U \rangle^2 \langle N \rangle \tag{7.63}$$

$$\left(\partial/\partial t + \gamma_{damp}\right) \langle U \rangle^2 = \alpha \langle U \rangle^2 \langle N \rangle, \tag{7.64}$$

where $\langle N \rangle = \sum_k N_k$ is the drift wave intensity, $U = \partial V_{ZF}/\partial r$ is the vorticity of the zonal flow, and $\langle U \rangle^2$ corresponds to the zonal flow energy, γ_L is the linear growth rate of the drift wave, γ_{NL} is the nonlinear damping rate of the drift wave, which may be given in lowest order as $\gamma_{NL} = \gamma_2 \langle N \rangle$. Right hand sides are energy transfer terms between the drift wave and the zonal flow.

Basic idea behind this Predator-Prey model originated from the Reynolds equation in the fluid turbulence [162], the average flow can be produced by the Reynolds stress (see Column 7-3) and stabilize turbulence. Noting that the $E \times B$ poloidal flow shear suppression of the turbulence is important, the poloidal flow profile evolution equation is derived by taking the flux surface average of the momentum balance equation.

$$\frac{\partial \langle V_\theta \rangle}{\partial t} = -\frac{\partial}{\partial r} \langle \tilde{V}_r \tilde{V}_\theta \rangle - \mu \langle V_\theta \rangle \tag{7.65}$$

We use the approximation$\langle V_\theta \rangle \sim \langle V_E \rangle$ noting that the diamagnetic flow do not contribute the suppression of the turbulence. Taking $\partial/\partial r \times$ (7.65) and multiplying by $\langle \partial V_E/\partial r \rangle$ and considering $\partial^2 \langle \tilde{V}_r \tilde{V}_\theta \rangle / \partial r^2 \sim \langle \partial V_E/\partial r \rangle \langle N \rangle$, we have (7.64).

This Predator-Prey system has a stable fixed point in the $(\langle U \rangle^2, \langle N \rangle)$ space if all coefficients $(\gamma_L, \gamma_{damp}, \alpha, \gamma_2)$ are non-negative [166].

This Predator-Prey system gives periodic bursts (limit cycle oscillation: LOC) of the wave and flow if the nonlinear self-stabilization is absent. This LCO rotates clock-wise in $(\langle U \rangle^2, \langle N \rangle)$ plane. Namely, the turbulent fluctuation increases first and then the zonal flow grows and suppress turbulence. Suppression of turbulent fluctuation reduces zonal flow generation and the zonal flow damps to return to the original state.

If we include mean flow shear, the system becomes two Predator- one Prey system given by Kim-Diamond [450]. The system equations for the amplitude of turbulence ϵ, zonal flow shear $V_{ZF} \sim \partial \tilde{V}_E/\partial r$, mean flow shear $V \sim \partial \langle V_E \rangle/\partial r$, and the ion pressure gradient $N \sim \partial p_i/\partial r$:

$$\partial \epsilon/\partial t = \epsilon N - a_1 \epsilon^2 - a_2 V^2 \epsilon - a_3 V_{ZF}^2 \epsilon \tag{7.66}$$

$$\partial V_{ZF}/\partial t = b_1 \frac{\epsilon V_{ZF}}{1 + b_2 V^2} - b_3 V_{ZF} \tag{7.67}$$

$$\partial N/\partial t = -c_1 \epsilon N - c_2 N + Q \tag{7.68}$$

Right hand side terms of (7.66) are linear growth, nonlinear saturation, mean flow and zonal flow suppression. Right hand side terms of (7.67) are zonal flow generation by the Reynolds stress and damping. Here suppression of zonal flow generation by the mean flow is modeled as $1/(1+b_1V^2)$. Right hand terms of (7.68) are turbulent and neoclassical reductions of pressure gradient and the input power. The characteristics of dynamical behavior is discussed in Sect. 7.4.

Column 7-3 : Reynolds Equation
In ideal neutral fluid, the continuity equation and the Euler equation are:

$$\frac{\partial \rho}{\partial t} + \frac{\partial \rho u_i}{\partial x_i} = 0\,, \text{ and } \rho\left(\frac{\partial u_i}{\partial t} + u_j \frac{\partial u_i}{\partial x_j}\right) = -\frac{\partial p}{\partial x_i} \tag{7.69}$$

This Euler equation can be converted to following conservative form.

$$\frac{\partial \rho u_i}{\partial t} = -\frac{\partial \Pi_{ik}}{\partial x_k}\,, \text{ where } \Pi_{ik} = p\delta_{ik} + \rho u_i u_k \tag{7.70}$$

Here, Π_{ki} is called the momentum flux density and is a symmetric tensor (see Landau-Lifschitz [495]). Integrating in volume and using the Gauss's theorem, we have: $\partial/\partial t \int \rho u_i dV = -\oint \Pi_{ik} dS_k$. The $\Pi_{ik} dS_k$ is the i-th component of the momentum flowing out from the volume through the surface dS. We derive so-called Reynolds equation assuming constant ρ for simplicity. This means fluid flow is incompressible as well. Let $\tilde{\boldsymbol{u}}$ is the random velocity fluctuation in the fluid turbulence. This fluctuating velocity field produces the momentum transport. Navier-Stokes equation in this case is given by,

$$\frac{\partial u_i}{\partial t} = -\frac{1}{\rho}\frac{\partial \Pi_{ki}}{\partial x_k}, \text{ where } \Pi_{ki} = p\delta_{ki} + \rho u_k u_i - \mu(\frac{\partial u_i}{\partial x_k} + \frac{\partial u_k}{\partial x_i}) \tag{7.71}$$

We divide velocity to mean and fluctuating flows as $u_i = U_i + \tilde{u}_i$ and substitute into the Navier-Stokes equation, (7.71). The time derivative is divided into slow and fast time variations and take the fast time average of a quantity A denoted by $\langle A \rangle$. We obtain following Reynolds equation for the average flow.

$$\frac{\partial U_i}{\partial t} = -\frac{1}{\rho}\frac{\partial \Pi_{ki}}{\partial x_k} \tag{7.72}$$

$$\Pi_{ki} = p\delta_{ki} + \rho U_k U_i - \mu(\frac{\partial U_i}{\partial x_k} + \frac{\partial U_k}{\partial x_i}) + \rho\langle \tilde{u}_k \tilde{u}_i \rangle \tag{7.73}$$

(continued)

Here, Π_{ki} is the average momentum flux tensor and $R_{ki} \equiv -\rho\langle\tilde{u}_k\tilde{u}_i\rangle$ is called Reynolds stress tensor (sometimes, $R_{ij} = \langle\tilde{u}_k\tilde{u}_i\rangle$ is called Reynolds stress as in Sect. 7.6). This Reynolds stress tensor is a symmetric tensor to satisfy $R_{ki} = R_{ik}$. While its diagonal elements $\rho\langle\tilde{u}_1^2\rangle, \rho\langle\tilde{u}_2^2\rangle, \rho\langle\tilde{u}_2^2\rangle$ are perpendicular stress (i.e. pressure) and are negligible, the off-diagonal elements are shear stress and can be significant for the momentum transport. If $\langle\tilde{u}_k\tilde{u}_i\rangle \neq 0$, $\tilde{u}_k$ and $\tilde{u}_i$ are said to have correlation and if $\langle\tilde{u}_k\tilde{u}_i\rangle = 0$, $\tilde{u}_k$ and $\tilde{u}_i$ are said to have no correlation. Degree of the correlation is characterized by a correlation coefficient, $c_{ki} \equiv \langle\tilde{u}_k\tilde{u}_i\rangle/(\langle\tilde{u}_k^2\rangle^{1/2}\langle\tilde{u}_i^2\rangle^{1/2})$. In Landau-Lifschitz fluid mechanics [495], Reynolds stress are defined as $R_{ij} = \rho\langle\tilde{u}_k\tilde{u}_i\rangle$ and some papers uses this definition. The equation for R_{ij} is called Reynolds stress equation and include $< u_i u_j u_k >$, whose equation also includes $< u_i u_j u_k u_l >$.

7.3.4.2 Dimits Shift

Direct numerical gyrokinetic simulation of ITG turbulence showed that near the linear stability boundary, ITG mode is completely stabilized. In the slightly unstable regime, ITG grows initially but is quenched by the induced zonal flow. This zonal flow is strong enough to reduce ion thermal transport to neoclassical level [169]. This dynamics between zonal flow and turbulence upshifts the critical temperature gradient from critical temperature gradient for linear stability, R/L_{Tclin} to $R/L_{Tclin} + \Delta R/L_{Tc}$ as shown in Fig. 7.20 and $\Delta R/L_{Tc}$ is called Dimits shift [169]. In the Dimits shift regime ($R/L_{Tc,lin} < R/L_T < R/L_{Tc,lin} + \Delta R/L_{Tc}$), free energy from dT/dr is transferred mainly to zonal flow and not to ITG turbulence.

Miki [551] showed interesting transient dynamical interplay between GAM and ambient turbulence in Dimits shift regime by Landau-fluid simulation with fixed profiles, which produces intermittent transport called GAM intermittency as shown in Fig. 7.20b, c. In the Dimits shift regime, zonal flow energy increases with time while $m = 1$ and turbulence energies reduced and quenched. On the other hand, GAM oscillation persists above the Dimits shift regime giving quasi steady intermittent transport driven by GAM. This behavior is consistent with the prediction of Diamond's Predator-Prey model.

7.3.4.3 Geodesic Acoustic Mode

Toroidal effect (especially at high q) sometimes inhibits this zero frequency zonal flow (hereafter, called zonal flow) and produces so-called Geodesic Acoustic Mode (GAM), which is an oscillating zonal flow ($m = n = 0$ electrostatic potential) through geodesic coupling to $m = 1, n = 0$ pressure perturbation, first predicted

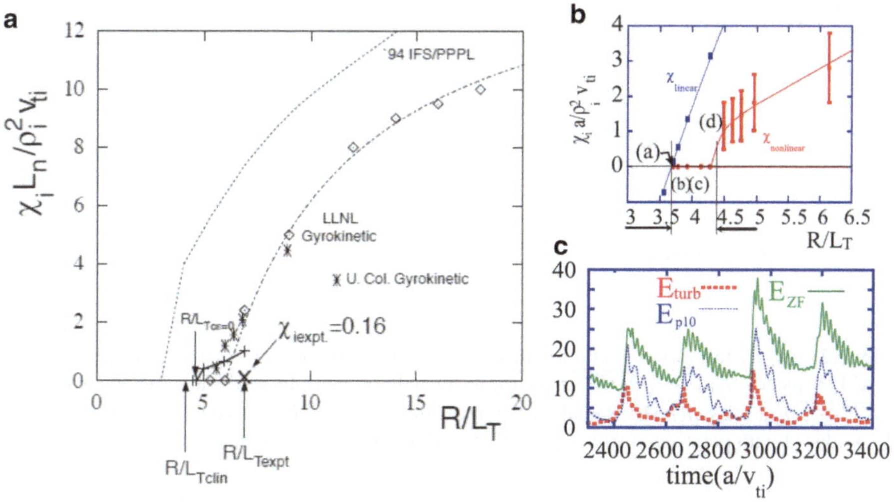

Fig. 7.20 (**a**) Normalized ion thermal diffusivity as a function of R/L_T from gyrokinetic simulation showing Dimits shift as compared with '94 IFS/PPPL gyrofluid simulation. Reproduced with permission from Dimits [169]. Copyright IAEA Vienna. (**b**) Intermittent transport coefficient induced by GAM-turbulence interaction in the Dimits shift regime and temporal evolution of turbulence, zonal and geodesic energies at $R/L_T = 4.05$. (**c**) Temporal evolution of $E_{turb.}$, E_{ZF}, and E_{p10} ($p_{10} =< p sin\theta >$) for $R/L_T = 4.05$. Reproduced with permission from Miki [551]. Copyright American Physical Society

by Winsor [826] and revisited by Hallatschek [281]. Zonal/GAM flow v_E is larger on the outside of the torus and smaller on the inside due to $B \propto 1/R$. This leads to density perturbation $\tilde{n}$ proportional to $-\nabla \cdot \boldsymbol{v}_\perp = (\boldsymbol{E} \times \boldsymbol{B}) \cdot \nabla B^2/B^4$, which produces parallel pressure perturbation, p. Coupled evolution equations for v_E (zonal/GAM flow) and perturbed pressure p in toroidal plasma with circular cross section are give in [554] as,

$$\frac{\partial \langle v_E \rangle}{\partial t} = \frac{1}{r^2}\frac{\partial}{\partial r} r^2 \langle \tilde{v}_{Er} \tilde{v}_{E\theta} \rangle + \frac{\mu_0 \beta}{n_{eq}} \frac{1}{r^2} \frac{\partial}{\partial r} r^2 \langle \tilde{B}_r \tilde{B}_\theta \rangle - \frac{2}{n_{eq}} \frac{a}{R} \langle p \sin\theta \rangle \tag{7.74}$$

$$\frac{\partial}{\partial t} \langle p \sin\theta \rangle = -\langle [\tilde{\phi}, \tilde{p}] \sin\theta \rangle + (\Gamma + \tau) p_{eq} \frac{a}{qR} \langle v \cos\theta \rangle + (\Gamma + \tau) \frac{a}{R} p_{eq} \langle v_E \rangle \tag{7.75}$$

Here, $\langle \cdot \rangle$ is flux surface average, $\langle v_E \rangle = \partial \phi_0 / \partial r$ is GAM flow, $\tilde{v}_{Er} = -(1/r)(\partial \tilde{\phi}/\partial \theta)$ and $\tilde{v}_{E\theta} = (\partial \tilde{\phi}/\partial r)$ are turbulent $E \times B$ drift velocities in radial and poloidal directions, respectively, $\langle p \sin\theta \rangle$ is $(m, n) = (1, 0)$ pressure perturbation, n_{eq} is equilibrium density, β is beta value, $[f, g]$ is Poisson bracket, $\Gamma = 5/3$, $\tau = T_e/T_i$, p_{eq} is equilibrium pressure. Three terms in the right hand side of Eq. (7.74) are Reynolds stress and Maxwell stress, and geodesic transfer term, respectively. Three terms in the right hand side of (7.75) are nonlinear coupling of pressure perturbation to turbulence, sound wave, and zonal flow terms, respectively.

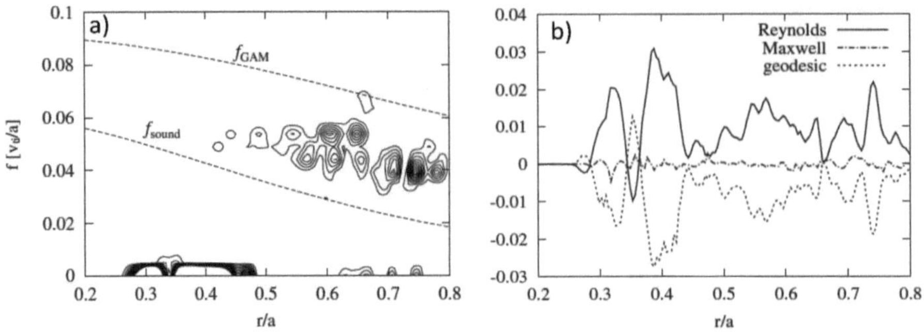

Fig. 7.21 (**a**) Radial variation of normalized zonal flow frequency ($f[v_i/a]$) with normalized pure GAM frequency $f_{GAM} = \sqrt{2(\Gamma + \tau)T_{eq}}(a/2\pi R)$ and normalized sound wave frequency $f_{sound} = \sqrt{(\Gamma + \tau)T_{eq}}(a/2\pi qR)$. (**b**) Radial variation of time averaged $E \times B$ zonal flow energy drives. Reproduced with permission from Miyato [555]. Copyright IAEA Vienna

Global electromagnetic Landau-fluid ITG simulation by Miyato [555] clarified basic properties of GAM. While strong zonal flow dominates in low q regime, zonal flow is weak and GAM is dominant in high q regime (Fig. 7.21a). Zonal flow energy is supplied from Reynolds stress drive $(-\langle\tilde{v}_{Er}\nabla^2\tilde{\phi}\rangle\langle v_E\rangle)$ at low β and zonal flow energy is transferred to (1,0) pressure perturbation through geodesic transfer for both zonal flow and GAM (Fig. 7.21b). The energy supply for $\langle p\sin\theta\rangle$ is from zonal/GAM flows and this energy is transferred mainly to sound wave term $\langle v\cos\theta\rangle$ in case of zonal flow, while it is transferred to turbulence through nonlinear coupling between electrostatic potential and pressure perturbation in case of GAM. The GAM frequency is almost spatially uniform and different from local GAM frequency f_{GAM} (Fig. 7.21a).

Radial uniformity of GAM frequency is also observed experimentally by Ido [363] and they showed that GAM is a radially propagating wave. This is because the GAM has global mode structure and the ω is determined as an eigenvalue. Sasaki [658] derived the radial eigenmode equation of GAM from the dispersion relation $\omega = \omega_{GAM}(1 + ck_r^2\rho_s^2) - i\gamma_{GAM}$ as $\alpha\rho_s^2\omega_{GAM}(r)d^2\phi_{GAM}/dr^2 + (\omega - \omega_{GAM}(r) + i\gamma_{GAM}(r))\phi_{GAM} = 0$ and showed the radial structure of GAM. Here $\omega_{GAM} = 2\pi f_{GAM}$, γ_{GAM} is the local growth rate of GAM and α is a constant. They also show the analytical expression of GAM frequency by solving the radial eigenmode equation via WKBJ method.

Ido [363, 364] showed experimentally in JFT-2M that GAM-turbulence interaction can produce intermittent transport and have high coherence between GAM oscillation and density fluctuation as shown in Fig. 7.22a. This behavior is consistent with theoretical predictions such as by Miki shown in Fig. 7.22b.

B. Scott [671] showed that while turbulent Reynolds stress always transfer energy from small eddies to the larger scale zonal flows, the geodesic curvature couples the zonal flows to finite $k_{\parallel}$ pressure sidebands, by which the nonlinear $\boldsymbol{E} \times \boldsymbol{B}$ pressure advection $(v_E \cdot \nabla\tilde{p}_e)$ quickly delivers the free energy in the zonal flow/side bands

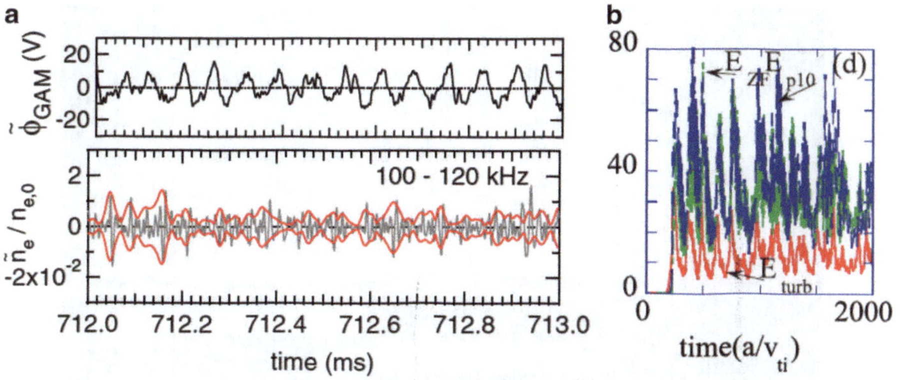

Fig. 7.22 (**a**) Correlation of GAM oscillation and density fluctuation in JFT-2M. Reproduced with permission from Ido [363]. Copyright IOP Publishing. (**b**) GAM-turbulence intermittency just above the Dimits shift regime $R/L_T = 4.47$. Reproduced with permission from Miki [551]. Copyright American Physical Society

back to the turbulence and growth of zonal flow amplitude is limited (geodesic transfer effect). The energy transfer among turbulence, zonal flow and GAM is measured in fusion relevant plasma for the first time by M. Xu [839] in HL-2A.

The existence of Zonal flow/GAM is confirmed experimentally by G.S. Xu [837] in HT-7U, Fujisawa [221] in CHS, Hamada [283] in JIPP T-IIU, and Zhao [858] in HL-2A. The modulation of GAM by low frequency Zonal flow in HL-2A is reported by Liu [512].

7.3.4.4 Streamer and Zonal Flows in ETG Turbulence

Zonal flow is one form of convective cell, which is toroidally symmetric and do not produce cross field transport. There is another form of the convective cell, called the "streamer". The streamer is radially elongated convective cell, which produces significant cross field transport.

In toroidal geometry, both zonal flows and streamers can appear depending on the magnetic shear. We show an example of toroidal ETG turbulence simulation of negative shear plasma which produce the zonal flows near shear less region and streamer in the positive shear region. Gyrokinetic simulation of ETG turbulence using quasi-ballooning formalism and consistent profile evolution (but not flux-driven) by Idomura [367] showed that saturation of ETG turbulence is determined by the quasi-linear modification of dT_e/dr close to $dT_e/dr|_c$.

In the NS configuration, ETG shows two different structures, streamer and zonal flow. Figure 7.23 shows toroidal ETG simulation in NS plasma [367]. In the positive shear region, nonlinear toroidal ETG turbulence produces streamer through coupling among poloidal harmonics, which exhibits 3D character of turbulence. While zonal flow is excited in ITG turbulence by the modulational instability in

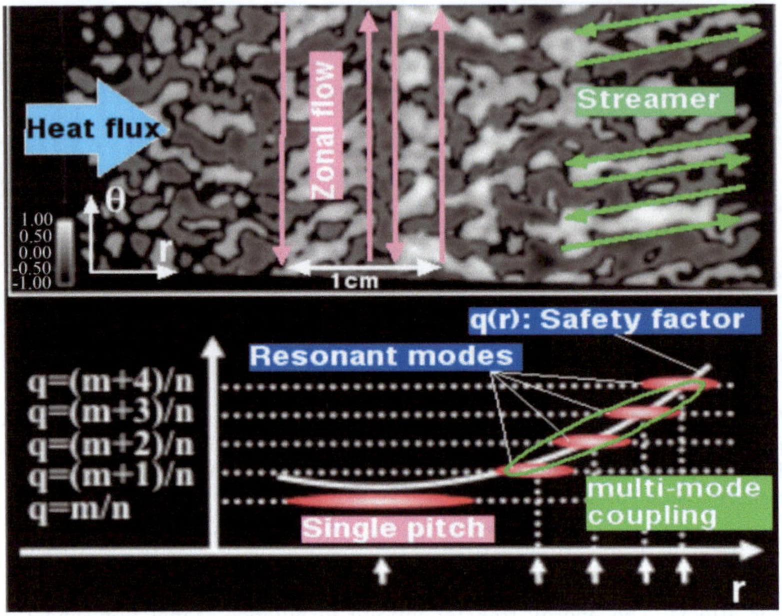

Fig. 7.23 ETG turbulence structures of zonal flow and streamer in toroidal NS plasma. Modified from Idomura [367]

positive shear tokamak, zonal flow generation is weak for ETG turbulence in case of positive shear. However,near the q_{min} in the NS plasma, nonlinear ETG turbulence is dominated by a single poloidal mode, which produces zonal flow through inverse energy cascade in 2D turbulence. When the distance between neighboring mode rational surfaces is much larger than electron gyroradius, $\Delta r \sim \rho_{te}/s \gg \rho_{te}$, ETG turbulent structure becomes 2 dimensional (2D turbulence).

7.4 Edge and Internal Transport Barriers

7.4.1 Edge Transport Barrier

Formation of Edge Transport Barrier (ETB) is associated with H-mode [794]. The H-mode is a standard operation scenario of ITER and the physics basis of H-mode is discussed in detail in the ITER physics basis [175, 387].

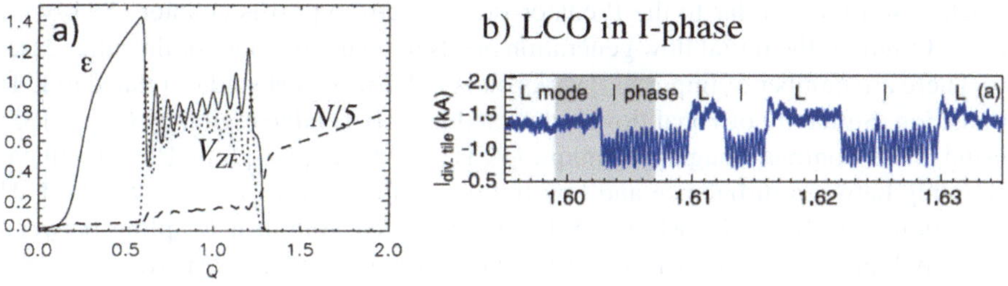

Fig. 7.24 (**a**) Kim-Diamond model of L-I-H transition. Input power $Q = 0.01t$ is used as horizontal axis. *Solid line* is the amplitude of turbulence ϵ, *dotted line* is zonal flow shear V_{ZF}, and *dashed line* is ion pressure gradient $N/5$. A limit cycle oscillation (called I-phase recently) occurs between L-mode and H-mode [450]. (**b**) ASDEX observation of limit cycle oscillation in I-phase [138]. Reproduced with permission from Kim-Diamond [450] and Conway [138]. Copyright American Physical Society

7.4.1.1 L-H Transition

The H-mode is associated with L to H mode transition [795] but the transition can be also transition less [436] or complex multi stage [415]. Measurement of main ion and impurity ions in DIII-D [447] shows that the main ion is rotating in the ion diamagnetic drift direction and the impurity ions are rotating in the electron diamagnetic direction to give $E_r < 0$ near the edge region.

The L to H-mode transition is explained by the flow shear regulation of the micro fluctuation leading to the bifurcation of transport states [322, 323, 713]. The bifurcation theory called Predator-Prey model based on the development of Reynolds stress is first developed by Diamond [162] (Sect. 7.3.4). Diamond [165] reformulates zonal flow excitation via wave kinetic equation and showed that Reynolds stress driven flow can play an important role in the edge region.

Most likely L-H transition physics is Kim-Diamond [450] two Predator-one Prey system described in Sect. 7.3.4. Using the set of Eqs. (7.66)–(7.68) and assuming the mean flow shear relation to the pressure gradient ($V \sim N^2$), dynamics similar to L-H transition is observed by increasing input power with time ($Q = 0.01t$). As the input power Q increases, the amplitude of turbulence ϵ increases while the zonal flow V_{ZF} can not develop due to strong damping until $Q \sim 0.6$. When $\epsilon \geq b_3(1 + b_2V^2)/b_1$, zonal flow generation by Reynolds stress overwhelms the zonal flow damping and the amplitude of turbulence ϵ reduces. In this period of time, the LOC (limit cycle oscillation) can be seen as discussed in Sect. 7.3.4. For sufficiently high input power Q, this self-regulation turns off the turbulence ϵ and the ion pressure gradient N becomes sufficiently large enough to produce strong mean flow $V \sim N^2$ and the quenched state of turbulence can be maintained without the zonal flow (Fig. 7.24a).

Conway [138] showed that the mean and GAM flows and the turbulence interaction is important in the I-phase as an intermediate state between L and H mode (Fig. 7.24b) in the ASDEX-U. Cheng [120] identified two types of limit cycle oscillations (type-Y and -J) during L-I-H transition in HL-2A. While type-Y is the clock-wise LOC similar to the Predator-Prey system, type-J is counter clock-wise LCO. Origin of the initial flow generation needs to be understood for the latter case.

There are number of important works on zonal flow/GAM studies related to L-H transition. Signature of zonal flow is detected in DIII-D edge region by Coda [133] using phase-contrast-imaging technique (PCI) and Moyer [564] found that nonlinear coupling between turbulence and low frequency fluctuations increase during L-H transition in DIII-D. Nagashima [570] also reported nonlinear coupling between small poloidal wave number and turbulent fluctuations in JFT-2M. Role of zonal flow in the L-H transition is first measured by G.S. Xu [838] in the EAST tokamak. Role of zonal flow Pradator-Prey oscillation in triggering H-mode transition is reported by Schmitz [670] in DIII-D, while Kobayshi [461] shows LCO without zonal flow signature.

7.4.1.2 Pedestal Width and Height

Wider ETB width is preferable to support large plasma energy by the pedestal part. During the early-day's JT-60U experiments, pedestal width Δ_{ped} is nearly proportional to $1/B_p$ [436, 438]. Further study in JT-60U showed Δ_{ped} is proportional to $\sqrt{T_i}$ to find $\Delta_{ped} \sim 3.3\epsilon^{0.5}\rho_{pi}$ [297]. Later analysis showed the weaker dependence on ρ_{pi} [298]. In the DIII-D tokamak, $\Delta_{ped} \propto \rho_{pi}^{0.66}$ or $\Delta_{ped} \propto \beta_p^{0.4}$ are reported [592], where $\beta_p \propto nT/I_p^2$ denotes the ratio of the edge kinetic pressure to the local poloidal magnetic pressure. JET tokamak also reported somewhat weaker ρ_{pi} dependence [511]. Urano [783] concluded by comparing hydrogen and deuterium H-mode that the scaling of ETB width follows $\Delta_{ped} \propto a\rho *_{pol}^{0.2} \beta_p^{0.5}$, where $\rho*_{pol} = \rho_{pol}/a$.

Pedestal height is characterized by the pressure at the top of the edge transport barrier (ETB) and is a key factor of the tokamak performance. Snyder made an extensive modeling study based on DIII-D experimental data and also using multi-machine database [706] showing the pedestal height (P_{ped}) and width ($\Delta\psi_N$) are determined by the peeling/ballooning mode stability and the prediction of width scaling based on the onset condition of the kinetic ballooning mode [705].This model is called EPED1 and EPED1.6 [707], which can reproduce experimental results well.

Pedestals in many tokamaks (DIII-D, NSTX, etc.) start from narrow edge transport barrier in which the pressure gradient is limited by the KBM (Kinetic Ballooning Mode) much lower than the ideal ballooning limit and then the pedestal width increases with time keeping the pressure gradient stays nearly that of KBM limit. As pedestal broadens, the drive of low n mode increases and the peeling/ballooning mode becomes unstable, which is ELM.

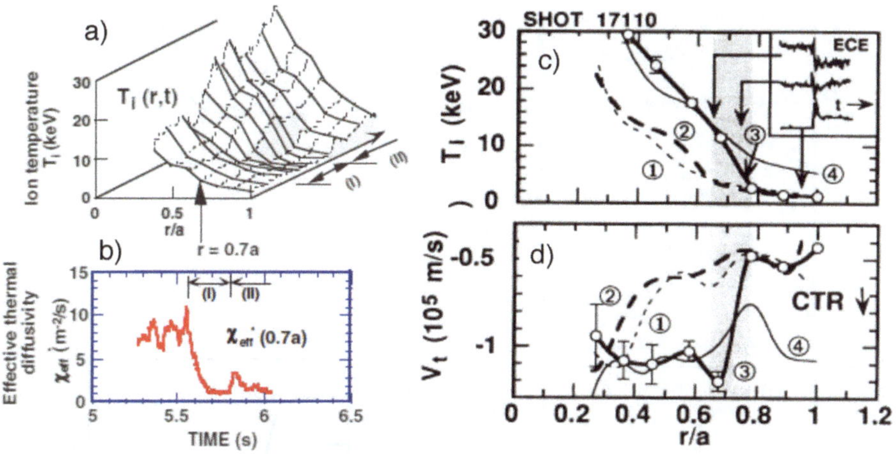

Fig. 7.25 (**a**) Spatio-temporal evolution of the ion temperature during the ITB formation (I) and transition to the H-mode phase (II). (**b**) Time evolution of $\chi_{eff}(07a)$. (**c**, **d**) Ion temperature and toroidal rotation profiles at t = 5.5, 5.6, 5.75, 6.0 s. Reproduced with permission from Koide [463]. Copyright American Physical Society

7.4.2 Internal Transport Barrier

Internal Transport Barrier (ITB) is first discovered by Koide [463] associated with core improved confinement with positive magnetic shear as shown in Fig. 7.25. Intensive central NBI heating produces high central ion temperature above 10 keV and the initial ITB is formed around $r/a \sim 0.5$–0.6 and moves outward $r/a \sim$0.7–0.8. The ITB in the final stage is associated with strong counter toroidal rotation while the counter rotation does not appear for the initial ITB. After the steep ITB formation, MHD burst of m=3 mode occurs near the ITB leading to the formation of the edge transport barrier (H-mode). The characteristics of ITB in WS has been discussed in [464] and compared with NS in [465]. Sometimes improved confinement inside $q = 1$ surface by Kamada [411] and by Hugon [354] is also called ITB.

Figure 7.26a is typical T_i, T_e and n_e profiles with ITB in NS plasma. Figure 7.26b shows both the ion thermal diffusivity χ_e and the electron thermal diffusivity χ_i from transport analysis and E_r profiles using the radial electric field expression (4.203). The $E \times B$ shearing rate is order of $\omega_E \sim 10^5\,\mathrm{s}^{-1}$ as compared with $v_{ti}/R \sim 2.7 \times 10^5\,\mathrm{s}^{-1}$ and χ_i inside the ITB is close to the ion neoclassical one and χ_e is also significantly reduced [691, 692]. Here "neoclassical" means formula obtained from moment equation ((8.134) in [443]). This steep ITB is associated with jump in E_r shear (also [651]) due to temperature curvature formation later discussed as curvature transition by Ida [362]. Koide-Burrell [466] gives a comparison of ITB characteristics between JT-60U and DIII-D. Figure 7.26c shows a comparison of χ_i profiles in co, counter and balanced NBI in NS plasmas by Sakamoto [649].

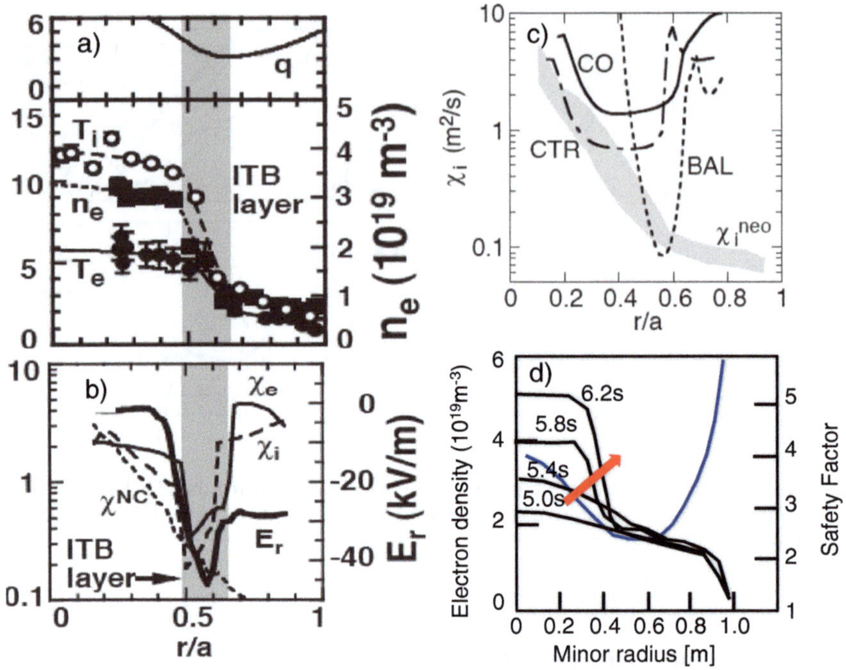

Fig. 7.26 (**a**) Profiles of q, T_i, T_e and n_e for NS plasma [691]. (**b**) Profiles of χ_i, χ_e, E_r and χ_{NC} for (**a**) [691]. (**c**) Difference in χ_i of ITBs in balanced, co and counter NB injection in NS plasma [649]. (**a**)–(**c**) reproduced with permission from Shirai[691] and Sakamoto[649]. Copyright IAEA Vienna. (**d**) Time evolution of the density ITB in NS discharge. Modified from Nazikian [579]

A steeper ITB tends to be formed in balanced neutral injection, while wider and weaker ITB is formed in co- and counter NB injection cases in Fig. 7.26c [649]. Since too strong ITB tends to trigger plasma disruption, control of ITB strength through the toroidal rotation drive has been pursued [652].

Figure 7.26d shows time evolution of the density ITB in NS plasma in JT-60U. Radial correlation measurement using correlation reflectometer during the density ITB formation by Nazikian [579] showed long correlation length $L_r \sim 20$ cm in L-mode phase and $L_r \leq 4$ mm in ITB. These characteristics seems to be consistent with theoretical understanding of avalanche/streamer in L-mode and $E\times B$ flow shear turbulence de-correlation discussed above. Xiao [835] showed that the density ITB can also be formed in Ohmically heated plasma with a coincidence with TEM-ITG transition in HL-2A tokamak.

Magnetic shear dependence of ITG heat transport by Waltz [798] showed no particular role of q_{min} but showed reduced transport with negative shear. Breakup of eigenmode across q_{min} surface is discussed as possible cause of ITB formation in NS plasma by Kishimoto [457] and flux driven gyro fluid simulations of ITG turbulence by Garbet [240, 241] but it may not be relevant due to omission of non-resonant mode [97]. Miyato [556] indicates stronger zonal flow excitation at low q_{min} as alternative candidate for ITB formation.

Formation of ITB starts at central region and propagates outward as shown by Koide [464] for WS plasma and by Fujita [225] for NS plasma. These observations may be consistent with prediction by Lebedev-Diamond [501] with density ITB. At present, there is no quantitative explanation of ITB formation by Gyrofluid and Gyrokinetic simulations.

7.5 Electromagnetic Turbulence

Understanding of electromagnetic turbulence is still not matured and needs further development for the quantitative understanding of β degradation of thermal confinement. Thus, this section is more descriptive on status of development.

The efficient steady state tokamak reactor may operate at high normalized beta. The thermal energy confinement time in the saturated-Ohmic and NB-heated L-mode showed strong degradation with β, $\tau_{E_{th}} \propto \beta^{-7/12}$ [437]. The empirical thermal H-mode confinement also follows strong degradation with beta as shown by Urano [782] in JT-60U and by Vermale [788] in AUG close to IPB98(y,2) scaling of $B\tau_{E_{th}} \propto \beta^{-0.9}$ [387]. Thus it is important to clarify physics of turbulent transport with beta.

7.5.1 ITG/TEM/KBM Turbulence

As plasma beta increases, plasma turbulence exhibits electromagnetic (EM) feature, which modifies especially ITG turbulence. There are two candidates as EM turbulence, kinetic ballooning mode (KBM) by W. Tang [749] and micro tearing mode by J. Drake [176]. In the electromagnetic ITG/TEM and the kinetic ballooning mode, the perturbed vector potential $A_\parallel(z,t)$ has the odd parity along the magnetic field. On the other hand, $A_\parallel(z,t)$ has even parity along the magnetic field in the micro tearing mode and has strong resonant component (Fig. 7.27).

The parallel electron heat flux is given by:

$$\tilde{q}_{e\parallel} = -n_{e0}\chi_{e\parallel}\left(\frac{d\tilde{T}_e}{dz} + \frac{\tilde{B}_x}{B_0}\frac{d\tilde{T}_{e\parallel}}{dx} + \frac{\tilde{B}_x}{B_0}\frac{dT_{e0}}{dx}\right), \tag{7.76}$$

where $\tilde{B}_x$ is the radial component of the magnetic field fluctuation. The odd parity (ITG/TEM,KBM) modes contribute through the first term. The third term is related to the field line diffusivity D_M and gives the heat transport due to streaming along the stochastic field lines which comes from the even parity (micro tearing) modes. The second term of the heat flux is nonlinear term.

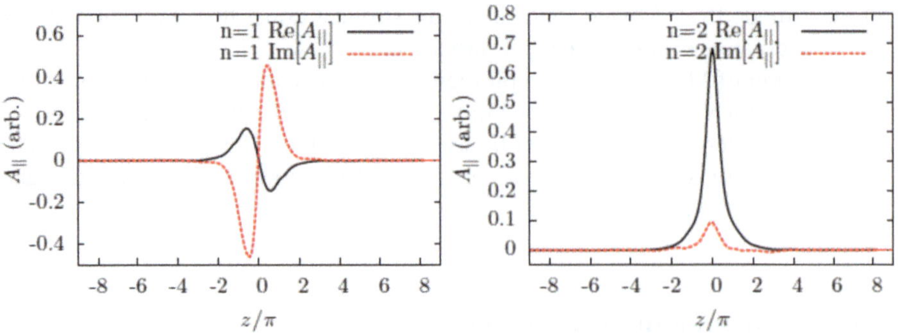

Fig. 7.27 Typical odd and even parity mode structures of $A_{\|}$. Electromagnetic ITG/TEM and KBM has odd parity and the micro tearing has even parity. Reproduced with permission from Hatch [299]. Copyright American Physical Society

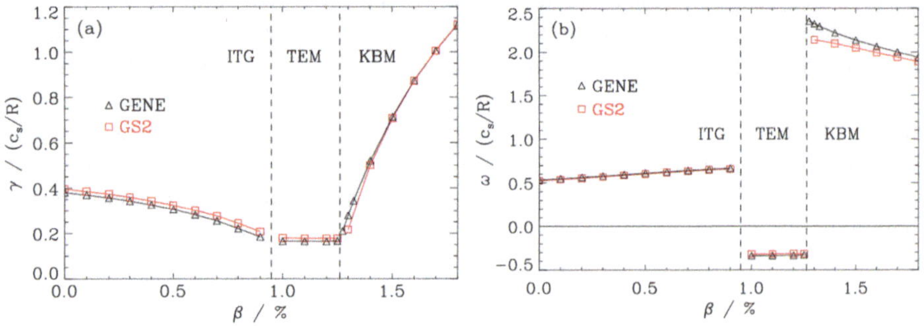

Fig. 7.28 Linear stability analysis of the electromagnetic drift wave turbulence using GS2 and GENE codes. (**a**) Growth rate. (**b**) Real frequency, as a function of β. Reproduced with permission from Pueschel [617]. Copyright AIP Publishing LLC

Linear stability of KBM with ideal ballooning mode is analyzed by Hong [339], and its gyro fluid turbulence simulation is given by Snyder [702]. Pueschel [617] showed that depending on parameters, growth rate of ITG is reduced as plasma beta increases and ITG-TEM transition occurs (say, at $\beta \sim 1\,\%$) and finally transition to KBM occurs (say, at $\beta \sim 1.3\,\%$) as shown in Fig. 7.28.

Nonlinear turbulent simulation of finite β ITG turbulence is started by Pueschel-Jenko [618], showing that Dimits shift due to zonal flow becomes larger with β and stabilize ITG mode over a wide range of β. This tendency is opposite to the experimental observation of strong transport enhancement with β.

Waltz in 2010 [799] showed that saturation level in the electromagnetic ITG/TEM turbulence takes long time in the simulation and can excite non-linear sub-critical drift Alfven wave to enhance turbulent transport rate, which is closely related to the non-linear sub-critical excitation of the kinetic ballooning mode by Waltz in 1985 [796] (Column 7-4). Recent publication by Pueschel [619] addressed this issue and shows that such a runaway phenomena occurs in electromagnetic

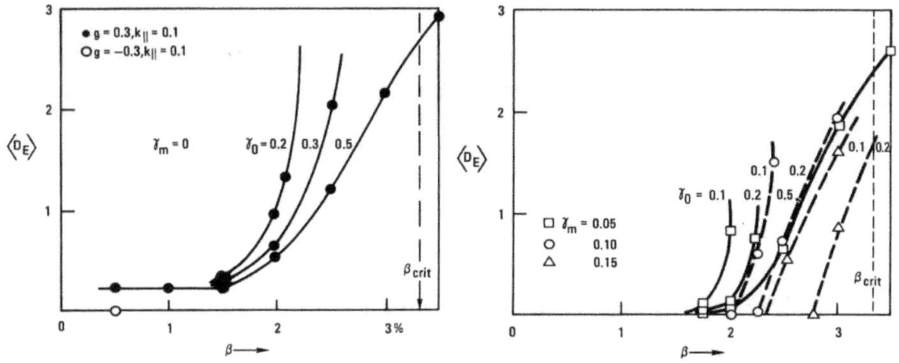

Fig. 7.29 Transport coefficients $\langle D_E \rangle = \langle \sum \tilde{\phi}_k^* i k_y \tilde{p}_k \rangle$ due to subcritical turbulence for marginal stability $\gamma_m = 0$ and damping case $\gamma_m > 0$. Here γ_m and γ_0 are parameters characterize γ_k^d. Reproduced with permission from Waltz [796]. Copyright American Physical Society

ITG/TEM turbulence. The saturation mechanism of EM ITG/TEM turbulence is still an open issue.

Column 7-4 : Electromagnetic Subcritical Turbulence

Self-sustained turbulent motion having a parameter below the onset of the linear instability is said to be a "subcritical turbulence". Subcritical turbulence is a well-known feature of hydrodynamic flows such as Poiseuille pipe flow shown by Orszag-Kells [591].

Waltz in 1985 [796] showed occurrence of the subcritical turbulence in the ideal 2D MHD turbulence using the model normalized equations for the perturbed electrostatic potential $\tilde{\phi}_k$, the parallel vector potential $\tilde{A}_k$, and the perturbed pressure $\tilde{p}_k$.

$$\frac{d}{dt} k_\perp^2 \tilde{\phi}_k = i g \omega_* \tilde{p}_k - \tilde{\nabla}_\parallel (k_\perp^2 \tilde{A}_k) - \gamma_k^\phi k_\perp^2 \tilde{\phi}_k \tag{7.77}$$

$$\frac{\partial}{\partial t} \tilde{A}_k = -\frac{1}{\beta} \tilde{\nabla}_\parallel \tilde{\phi}_k - \gamma_k^A \tilde{A}_k \tag{7.78}$$

$$\frac{d}{dt} \tilde{p}_k = -i \omega_* \tilde{\phi}_k - \gamma_k^p \tilde{p}_k \tag{7.79}$$

Here nonlinear mode couplings are contained in the $\tilde{E} \times B$ convective time derivative $(\partial/\partial t + \tilde{\boldsymbol{u}}_E \cdot \nabla_\perp)$ and the magnetic flutter of the parallel gradient $(\nabla_\parallel + \tilde{\boldsymbol{B}}_\perp / B_0 \cdot \nabla_\perp)$.

For $\beta > \beta_{critical} = k_\parallel^2 / g = L_p / R q^2$, the curvature drive g exceeds the resistance to field-line bending and all modes with $k_x = 0$ becomes unstable.

(continued)

Assuming a model damping rates $\gamma_k^{\phi} = \gamma_k^{A} = \gamma_k^{p} = \gamma_k^{d}$, he obtained a enhance transport rate of the pressure flux $\langle \tilde{u}_E \tilde{p} \rangle$ due to subcritical turbulence at $\frac{1}{2}\beta_{critical} < \beta < \beta_{critical}$ as shown in Fig. 7.29.

7.5.2 *Micro Tearing Mode Turbulence*

Recently, micro tearing mode is also thought to be an important candidate for the electron transport as shown by Guttenfelder [274] for ST and Doerk [171] for standard tokamak. The details of the fundamentals of micro tearing modes will be postponed to the future publication and we only show some historical background [144].

The resistive tearing mode is developed by Furth-Killeen-Rosenbluth [232]. Kinetic theory of tearing mode is first developed by Hazeltine [310], where a dispersion relation for the kinetic tearing mode is derived in the collision less regime and clarified importance of dT_e/dr for the stability of tearing mode. Drake [176] studied micro-tearing mode driven by dT/dr in collisional/semi-collisional regime. Drake in 1980 [177] also shows that nonlinear saturation of the micro tearing mode is given by $\tilde{B}_x/B_0 \sim \rho_e/L_{T_e}$ and the micro tearing mode can be a potential candidate for electron anomalous transport due to Rechester-Rosenbluth mechanism [630]. Here Richester-Rosenbluth heat transport is given as $\chi_e = D_M v_{Te}$, where the D_M is a magnetic field diffusivity given by, $D_M = R\,(\delta B/B)^2$. Using the saturation level $\delta B/B = \rho_{Te}/L_T$, we have:

$$\chi_e = R v_{Te} \left(\frac{\rho_{Te}}{L_T} \right)^2 \tag{7.80}$$

Wong [832, 833] showed experimental electron thermal diffusivity is close to the expected value of the micro tearing mode. Guttenfelder [274] showed nonlinear gyrokinetic micro tearing mode simulation gives transport comparable to experimental results from NSTX as shown in Fig. 7.30a.

Doerk [171] also showed through gyrokinetic micro tearing mode simulation that electron heat transport by micro tearing mode follows Richester-Rosenbluth formula and can be important in standard tokamak. Figure 7.30b shows the Poincare plots of magnetic field structure for $R/L_{Te} = 2.5$ and $R/L_{Te} = 3.5$ to show how the magnetic field becomes stochastic with the increase of R/L_{Te}.

Magnetic field structure under perturbation is still an active field of research as recently discovered by White [820] that the simple perturbation of the ideal MHD equation do not preserve its topology and the Lagrangian approach is necessary to formulate correct magnetic perturbation in ideal MHD similar to the Hamiltonian guiding center theory [821].

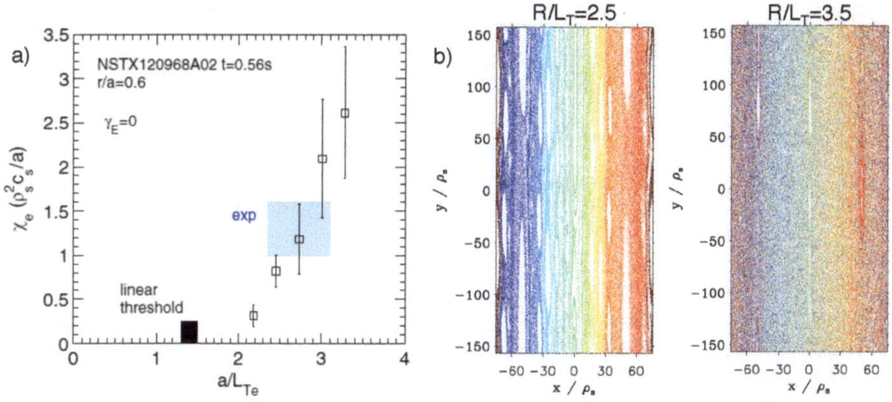

Fig. 7.30 (**a**) Normalized electron thermal diffusivity versus a/L_{T_e} from gyrokinetic simulation compared with experiment. After Guttenfelder [274]. (**b**) Poincare plot of magnetic field under micro-tearing turbulence for $R/L_T = 2.5$ and $R/L_T = 3.5$. Reproduced with permission from Doerk [171]. Copyright American Physical Society

7.6 Turbulent Momentum Transport

Momentum transport has long been a subject of interest. This field is active and not matured yet and we introduce this topics, briefly.

Existence of non-diffusive toroidal momentum flux is first observed by Nagashima [568] using modulation technique in JT-60U. Off-diagonal contribution to the momentum flux is also shown in JFT-2M by Ida [361].

Rice showed that the plasma can spin in the toroidal direction without external torque input and this phenomena is called the **Intrinsic Rotation**.

Observation of intrinsic rotation [627] in Alcator C-Mod, and inter-machine comparison by Rice [628] clarified parametric dependence of intrinsic rotation $\Delta v_\phi \sim \Delta W/I_p$ and has a significant impact for the momentum transport studies to find hidden torque input (see Fig. 7.31a). Scarabosio [668] found slightly different scaling $\Delta v_\phi \sim T_i/I_p$.

Existence of intrinsic torque is shown in DIII-D by Solomon [709]. Parametric dependence of intrinsic rotation has been investigated by Yoshida [845] ($v_\phi \propto dP/dr$), Rice [629] ($v_\phi \propto dT/dr$), and Angioni [20] ($dv_\phi/dr \propto dn/dr$). Shi [685] observed spontaneous co-toroidal rotation during LHCD in EAST tokamak. Yoshida [846] also showed that co-intrinsic rotation inside of the local ECRH deposition layer and counter-intrinsic rotation outside of the local ECRH deposition layer.

On theoretical front, Dominguez [172] is first to calculate residual stress for intrinsic rotation. Garbet [240] gives first numerical Gyrofluid ITG simulation to show toroidal flow generation by the Reynolds stress $\Pi_{r\parallel} = \langle v_{Er} v_{E\parallel} \rangle$ (Fig. 7.31b).

The momentum flux density Π_{kj} and the Reynolds stress are important physical quantities in the turbulent momentum transport originating from the Reynolds

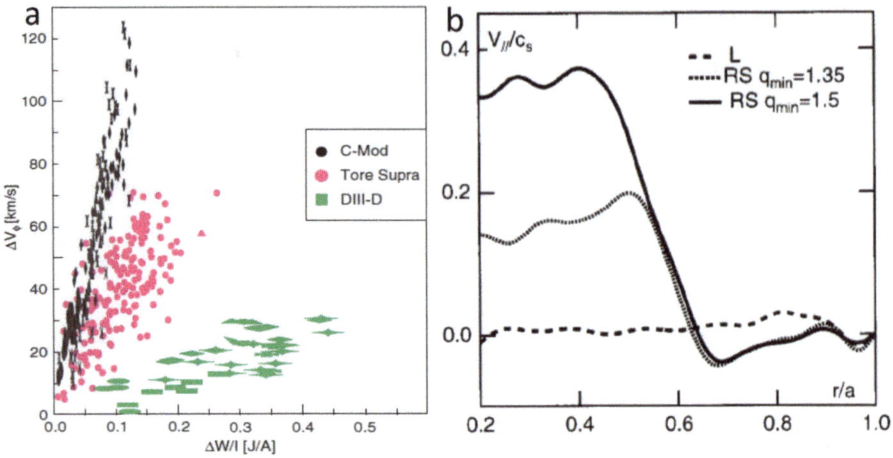

Fig. 7.31 (**a**) Observed intrinsic rotations from 3 tokamaks in proportional to $\Delta W/I_p$. Reproduced with permission from Rice [628]. Copyright IAEA Vienna. (**b**) Intrinsic rotation by Reynolds stress in ITG turbulence. Reproduced with permission from Garbet [240]. Copyright AIP Publishing LLC

equation in fluid mechanics (see Column 7-3). In the fluid mechanics, force is given as a divergence of the momentum flux density (or Reynolds stress).

Diamond [167] clarified basic mechanism of turbulent momentum transport. Mean field momentum flux driven by electrostatic turbulence $\Pi_{r,\phi}$ is given by Reynolds stress $\langle\tilde{v}_r\tilde{v}_\phi\rangle$ as $\Pi_{r,\phi} = \langle n\rangle\langle\tilde{v}_r\tilde{v}_\phi\rangle$ by neglecting convective flux $\langle\tilde{v}_r\tilde{n}\rangle\langle v_\phi\rangle$ and 3rd order flux $\langle\tilde{n}\tilde{v}_r\tilde{v}_\phi\rangle$ and $\langle\tilde{v}_r\tilde{v}_\phi\rangle$ is further decomposed as follows,

$$\langle\tilde{v}_r\tilde{v}_\phi\rangle = -\chi_\phi\frac{\partial\langle v_\phi\rangle}{\partial r} + V\langle v_\phi\rangle + \Pi^R_{r,\phi} \tag{7.81}$$

Here, χ_ϕ is the turbulent viscosity, V is the convective velocity and $\Pi^R_{r,\phi}$ is called the residual stress. The turbulent viscosity is closely related to turbulent ion thermal diffusivity and the ratio is called Prandtl number ($P_r = \chi_\phi/\chi_i$). The residual stress $\Pi^R_{r,\phi}$ is defined as momentum flux other than diffusive and convective fluxes [272] and is proportional to $\partial\langle n\rangle/\partial r$ and/or $\partial\langle T\rangle/\partial r$.

Residual stress originating from resonant and non-resonant wave-particle interaction is given by $\Pi^{wave}_{r,\phi} = \int d\boldsymbol{k} v_{gr} k_\parallel \delta N$. Using the solution of wave kinetic equation, quasi particle density is given as $\delta N = \tau_c[k_\theta V'_E\partial\langle N\rangle/\partial k_r - v_{gr}\partial\langle N\rangle/\partial r]$. Here, V'_E and τ_c are electric field shear and correlation time of δN response, respectively. First and second terms of δN are related to $k_\parallel$ symmetry breaking and radial inhomogeneity of turbulent amplitude,respectively as clarified by Gurcan-Diamond [271]. Figure 7.32 shows schematics of two mechanism of intrinsic rotation.

Peeters [608] showed Coriolis drift can drive momentum pinch for long wave length ITG ($k_\perp\rho_i \ll 1$) and Hahm [280] gives more general theory of momentum pinch due to symmetry breaking due to magnetic curvature. Camenen [93] showed

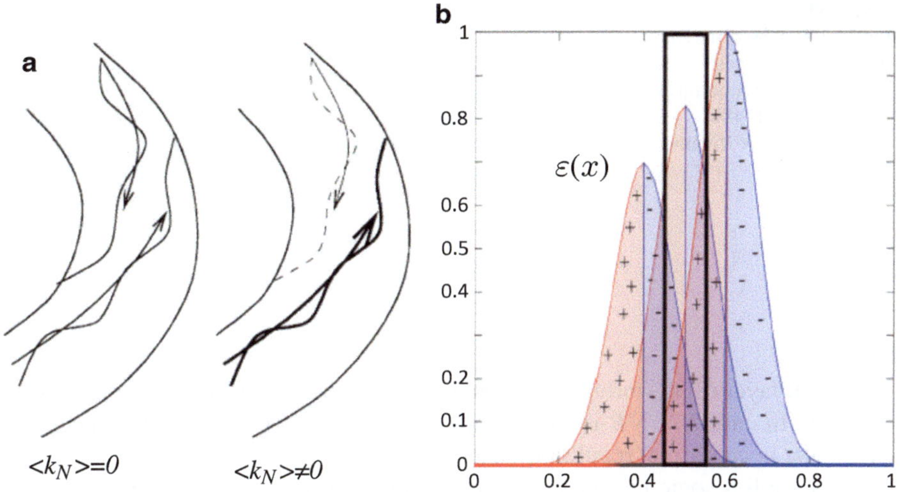

Fig. 7.32 (**a**) Turbulent parallel force by $k_{\parallel}$ symmetry breaking. Sheared flow changes the population density of waves in one direction relative to the other [271]. (**b**) Turbulent parallel force by the fluctuation intensity gradient. Left from the peak of the eigenmode has $k_{\parallel} > 0$ and right has $k_{\parallel} < 0$ as shown in the figure $(+, -)$. Intensity gradient produces symmetry breaking as well [273]. Reproduced with permission from Gurcan-Diamond[271, 273]. Copyright AIP Publishing LLC

existence of momentum pinch due to up-down asymmetry theoretically and [94] also confirmed it experimentally. Wang [805] studies parametric dependence of intrinsic rotation by gyrokinetic simulation of ITG and trapped electron mode (TEM) turbulences and reported that intrinsic rotation by ITG is proportional to dT_i/dr and, while intrinsic rotation by collisionless trapped electron mode (TEM) increases with both dT_e/dr and dn_e/dr [806]. Angioni [20] reported reversal of intrinsic rotation from co to counter correlates with large density gradient at TEM-dominant regime or ITG-TEM transition. Wang also reported that enhanced $k_{\parallel}$ symmetry breaking due to larger q variation at lower plasma current is consistent with I_p part of the Rice scaling, $\Delta v_{\phi} \sim 1/I_p$. Bortolon [74] showed that direction of intrinsic toroidal rotation can be reversed at critical density. Gyrokinetic full f Vlasov simulation of toroidal ITG turbulent momentum transport with consistent momentum conservation properties (see discussion by Scott [672]) by Idomura [369] showed build up of intrinsic co and counter rotations in a time scale of τ_{ii}.

While Reynolds stress plays an essential role in the fluid picture, the gyrokinetic theories by Wang-Diamond [808] and Garbet [244] showed existence of turbulent parallel acceleration mechanism, which can not given by the divergence of momentum flux density. This is purely a kinetic effect and is called the **Turbulent Acceleration**.

Boundary condition at separatrix is important since so-called no-slip condition is usually used for the turbulent simulation studies. Concerning this topics, Bombard [486] showed in C-Mod that strong ballooning character of transport leads to a net

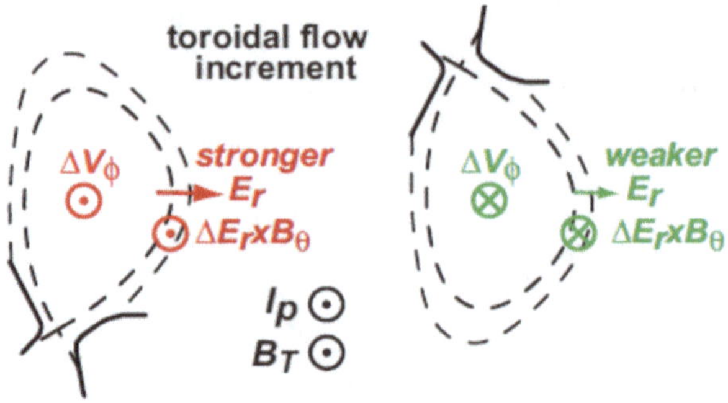

Fig. 7.33 Excitation of toroidal flow by different divertor geometries. Reproduced with permission from Bombard [486]. Copyright IAEA Vienna

scrape off layer (SOL) toroidal momentum whose sign is dependent on the X-point location. In case $\boldsymbol{B} \times \nabla B$ is toward (away from) the X-point, positive (negative) increment in co-current rotation is observed (Fig. 7.31).

This implies some neoclassical effect can play a role to exhaust momentum to the SOL flow. Takizuka [744, 745] shows that neoclassical orbit excursion from pedestal to SOL can be a mechanism to transfer momentum to SOL ion (Fig. 7.33).

One important question is the effect of the intrinsic torque on generalized Ohm's law as well as its relation to experimental observation.

7.7 Plasma Confinement in Tokamak

Plasma confinement study relies on dimensional analysis of empirical data. In this chapter, we introduce dimensional analysis in neutral fluid originated from Reynolds and the invariant principle originated from J. Connor.

7.7.1 Similarity Law in Fluid Mechanics and Invariant Principle

Osborne Reynolds (1842–1912) shown in Fig. 7.4b is an outstanding physicist and also great engineer in fluid mechanics. His most famous work is the experimental study of the transition condition from the laminar to turbulent flow in fluid flow inside the pile. From these experiments, he derived the dimensionless quantity called Reynolds number for the dynamical similarity.

Since then, fluid mechanics made a significant progress in understanding laminar and turbulent flows in the fluid mechanics [491]. There are many dimensionless quantities in the fluid mechanics, such as Prandtl number, Nusselt number, Rayleigh number. Using these numbers, we can normalize dynamics of fluid flow and heat transport or formation of the convective cell. The viscous fluid dynamics is governed by the following Navier-Stokes equation.

$$\frac{\partial \boldsymbol{u}}{\partial t} + \boldsymbol{u} \cdot \nabla \boldsymbol{u} = -\frac{1}{\rho} \nabla p + \nu \nabla^2 \boldsymbol{u} \tag{7.82}$$

Physical quantities here are the velocity $\boldsymbol{u}$, time t, length $\boldsymbol{x}$, mass density ρ, pressure p, and the kinematic viscosity ν.

Since any physical quantities can be normalized using three different physical quantities in mechanics, we normalize this equation using typical scale length L, typical speed U and mass density ρ as $\hat{\nabla} = L\nabla$, $\hat{t} = Ut/L$, $\hat{p} = p/\rho U^2$, and $\hat{\boldsymbol{u}} = \boldsymbol{u}/U$ to obtain,

$$\frac{\partial \hat{\boldsymbol{u}}}{\partial \hat{t}} + \hat{\boldsymbol{u}} \cdot \hat{\nabla} \hat{\boldsymbol{u}} = -\hat{\nabla} \hat{p} + \frac{1}{R_e} \hat{\nabla}^2 \hat{\boldsymbol{u}}, \tag{7.83}$$

where $R_e = UL/\nu$ is called the Reynolds number. The flows in the geometrically similar systems are also mechanically similar if the Reynolds numbers R_e are same. Physically, the Reynolds number is the ratio of the inertial term to the viscous term, $R_e \sim \boldsymbol{u} \cdot \nabla \boldsymbol{u} / \nu \nabla^2 \boldsymbol{u}$.

Consider a drag force F on a body of characteristic size $a[L]$ moving with a velocity $u[L/T]$ through a fluid of mass density $\rho[M/L^3]$ and a kinematic viscosity $\nu[L^2/T]$. Dimensional analysis is able to find powers of only 3 variables. Above drag force may be expressed as $F \sim \rho^i u^j \nu^k a^l$, while F should have a dimension of $[ML/T^2]$. Dimensional analysis gives $F = (\rho u^2 a^2)(ua/\nu)^{-k}$, i.e. $F = (\rho u^2 a^2) f(R_e)$, where $f(x)$ is a function of x and R_e is the Reynolds number. The dimensionless quantity ua/ν is the Reynolds number and the form of $f(R_e)$ can be determined by the experiment.

Dimensional analysis is closely related to the invariant property of the system equation under scale transformation. Connor [143] reformulated Lamb's discussion on the similarity law [491] as 'Invariant Principle'. For the Navier-Stokes equation, there are six physical quantities $\boldsymbol{u}, t, \boldsymbol{x}, \rho, p, \nu$. If we consider scale transformations on these six quantities, invariance conditions for scale transformation gives three $(6 - 3 = 3)$ relations since Navier-Stokes equation has four terms. If we choose $\boldsymbol{u}, \boldsymbol{x}, \rho$ as leading variables, we find following invariant scale transformations.

$$\boldsymbol{u} \to \alpha \boldsymbol{u}\ ,\ t \to \alpha^{-1} t\ ,\ p \to \alpha^2 p\ ,\ \nu \to \alpha \nu \tag{7.84}$$

$$\boldsymbol{x} \to \beta \boldsymbol{x}\ ,\ t \to \beta t\ ,\ \nu \to \beta \nu \tag{7.85}$$

$$\rho \to \gamma \rho\ ,\ p \to \gamma p \tag{7.86}$$

To show (7.84), for example, $\boldsymbol{u} \to \alpha \boldsymbol{u}$ leads to $t \to \alpha^{-1} t$ by a invariant condition between $\partial \boldsymbol{u}/\partial t$ and $\boldsymbol{u} \cdot \nabla \boldsymbol{u}$. Transformation laws $p \to \alpha^2 p$ and $\nu \to \alpha \nu$ are obtained by the invariant condition with two terms in the RHS of the Navier-Stokes equation.

The drag force $F \sim \rho u x^3/t$ on a body of characteristic size $a[L]$ moving with a velocity $u[L/T]$ through a fluid of mass density $\rho[M/L^3]$ and a kinematic viscosity $\nu[L^2/T]$ to be given by a power law $F \sim \rho^i u^j \nu^k a^l$. Substitution of the scale transformation (7.84), we obtain $F \to \alpha^2 \rho u x^3/t \sim \alpha^{j+k} \rho^i u^j \nu^k a^l$, thus we have $2 = j + k$. From (7.85), we have $F \to \beta^2 \rho u x^3/t \sim \beta^{k+l} \rho^i u^j \nu^k a^l$, thus we have $2 = k + l$. From (7.86), we have $F \to \gamma \rho u x^3/t \sim \gamma^i \rho^i u^j \nu^k a^l$, thus we have $1 = i$. In the end, we have $F \sim \rho u^2 a^2 (ua/\nu)^{-k}$.

We can not determine k by the dimensional analysis, i.e. $F = \rho u^2 a^2 f(R_e)$. Thus Connor's invariant principle is equivalent to the dimensional analysis and provides a systematic way to find relevant dimensionless quantities.

7.7.2 Invariant Principle of Plasma Confinement

Plasma transport across the magnetic surface (particle diffusion, thermal conduction and convection) is a complex nonlinear phenomena and to accurately predict the theoretical transport coefficients is a difficult challenge. Therefore, the prediction of the plasma confinement is usually performed by an extrapolation based on the experimental data. The most useful method is a dimensional analysis that has been used in the fluid mechanics.

Most general equations on the plasma confinement are Vlasov equation, Maxwell's equation and the charge neutrality condition.

$$\frac{\partial f_a}{\partial t} + \boldsymbol{v} \cdot \frac{\partial f_a}{\partial \boldsymbol{x}} + \frac{e_a}{m_a} (\boldsymbol{E} + \boldsymbol{v} \times \boldsymbol{B}) \cdot \frac{\partial f_a}{\partial \boldsymbol{v}} = C_a(f_a) \tag{7.87}$$

$$\nabla \times \boldsymbol{E} = -\frac{\partial \boldsymbol{B}}{\partial t} \tag{7.88}$$

$$\nabla \times \boldsymbol{B} = \mu_0 \boldsymbol{J} \tag{7.89}$$

$$\boldsymbol{J} = \sum_a \int e_a \boldsymbol{v} f_a d\boldsymbol{v} \tag{7.90}$$

$$\sum_a e_a \int f_a d\boldsymbol{v} = 0 \tag{7.91}$$

Here, f_a is a distribution function of the species a, $\boldsymbol{v}$ is the velocity variable, $\boldsymbol{x}$ is the position, $\boldsymbol{E}$ is the electric field, $\boldsymbol{B}$ is the magnetic field, and $C_a(f_a)$ is a bi-linear collision term. We follow the Connor's invariant scale transformation method. The above set of equations has seven physical quantities, $f_a, \boldsymbol{v}, \boldsymbol{x}, t, \boldsymbol{E}, \boldsymbol{B}, \boldsymbol{J}$. Invariant condition of Eqs. (7.87)–(7.91) provides seven relations of which one is degenerated and (7.91) does not give any relation. Therefore, only one $(7-6 = 1)$ invariant scale transformation exists. If we take $\boldsymbol{v}$ as leading variable, we obtain following relation.

$$f_a \to \alpha^5 f_a, \boldsymbol{v} \to \alpha \boldsymbol{v}, \boldsymbol{x} \to \alpha^{-4} \boldsymbol{x}, \boldsymbol{B} \to \alpha^5 \boldsymbol{B}, t \to \alpha^{-5} t, \boldsymbol{E} \to \alpha^6 \boldsymbol{E}, \boldsymbol{J} \to \alpha^9 \boldsymbol{J} \tag{7.92}$$

To derive this, we start with scale transformation $\boldsymbol{v} \rightarrow \alpha \boldsymbol{v}, f_a \rightarrow \alpha^i f_a, \boldsymbol{x} \rightarrow \alpha^j \boldsymbol{x}, \boldsymbol{B} \rightarrow \alpha^k \boldsymbol{B}, t \rightarrow \alpha^l t, \boldsymbol{E} \rightarrow \alpha^m \boldsymbol{E}$, and $\boldsymbol{J} \rightarrow \alpha^n \boldsymbol{J}$. Scale invariance of all terms in the Vlasov-FP equation (7.87) gives $i - l = 1 + i - j = i + k = 2i$ and $k + 1 = m$. Other equations (7.88)–(7.90) give $m - j = k - l, k - j = n, n = 4 + i$. From these equations, we obtain $i = 5, j = -4, k = 5, l = -5, m = 6, n = 9$, which leads to the invariant transformation (7.92).

Under this transformation, heat flux q_{ar} is expressed and transformed as.

$$q_{ar} \equiv \int (\boldsymbol{v} - \boldsymbol{u}_a)_r \frac{m_a}{2} |\boldsymbol{v} - \boldsymbol{u}_a|^2 f_a d\boldsymbol{v} \rightarrow \alpha^{11} q_{ar} \tag{7.93}$$

Assuming these heat and particle fluxes are functions of the density (n), the temperature (T), the magnetic field (B) and the plasma minor radius (a) and the inverse aspect ratio (a/R), the safety factor (q), the vertical elongation (κ), and the triangularity(δ), we expand q_{ar} as Taylor series on n_a, T_a, B, and a as follows,

$$q_{ar} = \sum D_{pqrs}(a/R, q, \kappa, \delta) n_a^p T_a^q B^r a^s \tag{7.94}$$

Under the invariant transformation (7.92), q_{ar} are transformed as,

$$q_{ar} = \sum D_{pqrs} \alpha^{8p+2q+5r-4s} n_a^p T_a^q B^r a^s \tag{7.95}$$

This Taylor expansion expression of heat flux must agree with Eq. (7.93), leading to following relations.

$$8p + 2q + 5r - 4s = 11 \tag{7.96}$$

We can eliminate s using (7.96) to obtain,

$$q_{ar} = a^{-11/4} \sum_{p,q,r} D_{pqrs} (n_a a^2)^p (T_a a^{0.5})^q (B a^{1.25})^r \tag{7.97}$$

The energy confinement time τ_E is defined as

$$\tau_E = \frac{1.5 \int n_a T_a dV}{P_{loss}} \sim \frac{n_a T_a V}{q_{ar} S} \sim \frac{(n_a a^2)(T_a a^{0.5}) a^{1.25}}{\sum_{p,q,r} D_{p,q,r} (n_a a^2)^p (T_a a^{0.5})^q (B a^{1.25})^r} \tag{7.98}$$

Therefore, $B\tau_E$ is a function of $n_a a^2, T_a a^{0.5}, B a^{1.25}$, while τ_E is not, as follows,

$$B\tau_E = f(a/R, q, \kappa, \delta, n_a a^2, T_a a^{0.5}, B a^{1.25}) \tag{7.99}$$

Instead of using three dimensional parameters $n_a a^2, T_a a^{0.5}, B a^{1.25}$, we can use three dimensionless quantities, the toroidal beta $\beta \sim n_a T_a / B^2$, the normalized

Larmor radius $\rho_a^* \sim T_a^{0.5}/aB$, and the normalized collision frequency $\nu_a^* \sim n_a a/T_a^2$ as follows,

$$B\tau_E = f(a/R, q, \kappa, \delta, \beta, \rho_a^*, \nu_a^*), \tag{7.100}$$

where we have following formula for the ion.

$$\beta = \frac{nT}{B^2/2\mu_0} = 4.03 \times 10^{-2} \frac{T_i(keV)n(10^{20}m^{-3})}{B(T)^2} \tag{7.101}$$

$$\rho^* = \frac{\rho_i}{a} = \frac{\sqrt{2m_iT_i}}{eZ_iB} = 4.57 \times 10^{-3} \frac{\sqrt{A_iT_i(keV)}}{Z_iB(T)} \tag{7.102}$$

$$\nu_i^* = \frac{Rq}{\varepsilon^{1.5} v_{Ti}\tau_{ii}} = 4.9 \times 10^{-12} \frac{Rqn_iZ_i^4 \ln \Lambda}{\varepsilon^{1.5} T_i(keV)^2} \tag{7.103}$$

If we replace charge neutrality equation (7.91) to the Poisson equation, $B\tau_E$ dependence on λ_D/a recovers.

$$B\tau_E = F(a/R, q, \kappa, \delta, \beta, \rho_a^*, \nu_a^*, \lambda_D/a) \tag{7.104}$$

Here, $\lambda_D = (\varepsilon_0 T/e^2 n)^{0.5}$ is Debye length. The charge neutrality is usually satisfied in the plasma confinement devices and the dependence on λ_D/a disappears in this case. While the coulomb collision plays an essential role in plasma confinement in the neoclassical transport, the collision may not play key role in the turbulent transport in collision less plasmas. In this case, the invariance condition for the scale transformation to collision term disappears and we have five independent relations. Since we have seven variables, we have two independent transformations to keep system equations invariant. We show the results using $\boldsymbol{v}, \boldsymbol{x}$ as leading variables and proof is left for the exercise.

$$f_a \to \alpha^{-3} f_a, \boldsymbol{v} \to \alpha \boldsymbol{v}, \boldsymbol{B} \to \alpha \boldsymbol{B}, t \to \alpha^{-1} t, \boldsymbol{E} \to \alpha^2 \boldsymbol{E}, \boldsymbol{J} \to \alpha \boldsymbol{J} \tag{7.105}$$

$$f_a \to \beta^{-2} f_a, \boldsymbol{x} \to \beta \boldsymbol{x}, \boldsymbol{B} \to \beta^{-1} \boldsymbol{B}, t \to \beta t, \boldsymbol{E} \to \beta^{-1} \boldsymbol{E}, \boldsymbol{J} \to \beta^{-2} \boldsymbol{J} \tag{7.106}$$

In this collision less case, the dependence is reduced to

$$B\tau_E = f(na^2, T/a^2B^2) = F(\beta, \rho_i/a) \tag{7.107}$$

Let consider a power law scaling $B\tau_E = (na^2)^p (T/a^2B^2)^q$ and combine with energy balance equation $nTa^3 \sim \tau_E P_{ext}$, we obtain following scaling.

$$\tau_E \sim n^{\frac{p-q}{1-q}} B^{\frac{1+3q}{1-q}} P_{ext}^{\frac{q}{1-q}} a^{\frac{2p-5q}{1-q}} \tag{7.108}$$

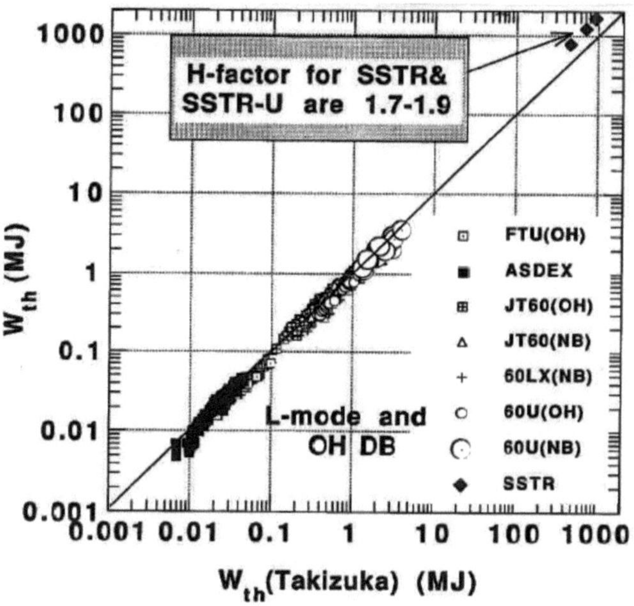

Fig. 7.34 A comparison of experimental data from FTU, ASDEX, JT-60, JT-60LX, JT-60U with the scaling (7.109). Modified from Kikuchi [437]

The confinement database of JT-60 [435] gives $\tau_E \sim (nI_p)^{0.5} P_{ext}^{-2/3}$ [437] to obtain $p = -0.5$, and $q = -2$. An important point here is that density and power (or temperature) dependences in one machine gives a constraints on size scaling (but the ε dependence must be based on inter-machine database). Figure 7.34 shows a scaling of thermal plasma energy W_{th} in relation to the scaling of the thermal energy confinement time $\tau_{Eth} = W_{th}/P_{ext}$. Medium size tokamak data such as ASDEX agrees with this scaling.

$$W_{th} = 0.026 A_i^{0.3} \kappa^{0.875} I_p^{0.5} B_t^{7/18} \bar{n}_e^{0.5} R_p^{1.44} a_p^{0.92} P_{ext}^{1/3}, \tag{7.109}$$

where $W_{th}, I_p, B_t, n_e, R_p, a_p, P_{ext}$ are given in [MJ, MA, T, $10^{19} m^{-3}$, m, m, MW]. If we use $\tau_{th} = W_{th}/P_{ext} = a^2/4\chi_{eff}$, we have

$$\chi_{eff} \sim \frac{T}{eB_t} (\rho_i^*)^{0.5} \beta^{7/12} (\nu^*)^{-1/12} q^{1.5} \tag{7.110}$$

This thermal diffusivity scaling has temperature and density dependence, $\chi \sim T^2 n^{0.5}$ in agreement with Efthimion's TFTR result $\chi \sim T^{2\pm 0.5}$ [182].

The ρ^* power is particularly important and ρ^{*0} is called Bohm transport and ρ^{*1} is called gyro-Bohm transport. Above example $\rho^{*0.5}$ is called weak gyro-Bohm.

Chapter 8
MHD Stability

Abstract Ideal, resistive and kinetic MHD instabilities are described. Since the birth of steady state tokamak research in 1990, there are number of important understandings of MHD modes related to steady state tokamak. In Sect. 8.1, we introduce the spectral property of MHD operator such as continuous spectrum and spectral gap, in brief. Marginal stability is discussed using the Newcomb equation in Sect. 8.2. Section 8.3 deals with flow effect on ideal MHD based on the Frieman-Rotenberg equation briefly. Localized MHD instabilities such as ELM, peeling/ballooning modes, infernal and barrier localized modes are discussed in Sect. 8.4. In Sect. 8.5, we introduce progress of resistive MHD such as the classical tearing mode (TM), the neoclassical tearing mode (NTM), the double tearing mode (DTM). Kinetic MHD equation is introduced in Sect. 8.6. In Sect. 8.7, Alfven eigenmode (AE) is introduced extensively. An important resistive/kinetic instability called the resistive wall mode (RWM) is introduced in Sect. 8.8. Control of ideal, resistive and kinetic MHD is essential element of fusion research. For three types of advanced tokamak operation (WS, NS, CH) to realize efficient steady state operation, ideal MHD modes such as peeling/ballooning modes for edge plasma, infernal modes for core plasma and BLM for ITB, resistive MHD modes such as NTM, DTM, RWM, kinetic MHD modes such as TAE, RSAE are understood well including control knob, while some kinetic MHD are still in progress in understanding. Key issue is that plasma profile does not match optimum profile for high beta advanced tokamak operation since profile is determined by the turbulent transport. Hence, combined understanding of MHD and turbulent transport is essential to optimize operation scenario for steady state tokamak operation.

Further Readings:
Textbooks: Freidberg [211] is an excellent introductory text on ideal MHD theory including original cylindrical resistive wall mode theory. Biskamp [68], [69] are good textbook on nonlinear MHD phenomenas and the magnetic reconnections. Hazeltine-Meiss [308] gives excellent introduction of fluid theory. Miyamoto [566] gives some introduction of Alfven eigenmode theory. Kikuchi-Lackner-Tran [445] Chap. 2 by K. Lackner et al. is useful to understand MHD instabilities in Tokamak.

Reviews: For the RWM, Chu-Okabayashi [127] is useful. As for the Alfven eigenmode, readers are encouraged to read review papers such as by Gorelenkov [256].

M. Kikuchi, M. Azumi, *Frontiers in Fusion Research II*,
DOI 10.1007/978-3-319-18905-5_8

8.1 Spectral Property of MHD Operator

Plasma stability is one of important long-standing subjects in plasma physics. The linear stability relies on functional theory in the infinite dimensional (square integrable) functional space called the Hilbert space [468].

The continuous spectrum of ideal MHD has significant importance not only in ideal MHD stability but also in stabilizing resistive wall mode and as damping mechanism of Alfven eigenmodes. The linearized ideal MHD equation is given by,

$$\left[-\rho\omega^2 - \boldsymbol{F}\right]\boldsymbol{\xi} = 0, \tag{8.1}$$

where $\boldsymbol{F}$ is self-adjoint linear force operator and ρ, ω, $\boldsymbol{\xi}$ are mass density, eigen frequency, and plasma displacement vector, respectively. Actual form of $\boldsymbol{F}$ is obtained by the following linearized ideal single-fluid MHD equations by Bernstein et al [62] and its energy integral is given in the Column 8-1.

$$\rho\frac{\partial^2\boldsymbol{\xi}}{\partial t^2} + \nabla p_1 + \frac{1}{\mu_0}\left[\boldsymbol{B}_1 \times (\nabla \times \boldsymbol{B}) + \boldsymbol{B} \times (\nabla \times \boldsymbol{B}_1)\right] = 0 \tag{8.2}$$

$$p_1 = -\boldsymbol{\xi}\cdot\nabla p - \gamma p \nabla\cdot\boldsymbol{\xi} \tag{8.3}$$

$$\boldsymbol{B}_1 = -\nabla \times (\boldsymbol{B} \times \boldsymbol{\xi}), \tag{8.4}$$

where γ is the ratio of specific heat and subscript 0 is suppressed for simplicity.

Due to this self-adjointness, spectrum of $\boldsymbol{F}$ is either discrete spectrum or continuous spectrum and no resolvent spectrum in the linear MHD equation (8.1) [443] (see Appendix A.5). If some ω^2 gives non trivial $\boldsymbol{\xi}$ to satisfy (8.1), the spectrum is called discrete spectrum (in this case, $\left[-\rho\omega^2 - \boldsymbol{F}\right]^{-1}$ does not exist). The continuous spectrum is characterized by the existence of the inverse operator $\left[-\rho\omega^2 - \boldsymbol{F}\right]^{-1}$ and unbounded non-square integrable eigenfunction.

Well-known linear operator having continuous spectrum is $(x - \lambda)f = 0$, whose unbounded (non-square integrable) eigenfunction is delta function, $f = c\delta(x - \lambda)$, which appears in linearized Vlasov equation to produce Landau damping.

In linear ideal MHD equation, linear force operator has singular local eigenfunction such as logarithmic singularity $\xi \sim ln(r - r_s)(= \int dr/(r - r_s))$ or the large solution $(r - r_s)^{-n}(n > 1)$, which are also non-square integrable. This type of eigenfunction to produce continuous spectrum is found for the first time in electrostatic wave in non-uniform plasma by Barston in 1964 [46] and analyzed by Sedlacek in 1971 [673] and applied to MHD modes by Uberoi in 1972 [780].

In plasma physics, both discrete spectrum and continuous spectrum exists similar to the quantum mechanics (Fig. 8.1a). Continuous spectrum in velocity space produces so-called Landau damping [493] through phase mixing (Fig. 8.1b) and the continuous spectrum in real space produces various damping including shear Alfven continuum damping at the shear Alfven resonance (Fig. 8.1c). In shear

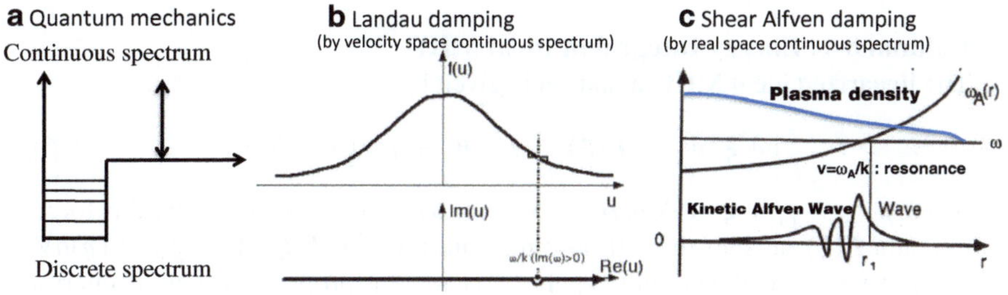

Fig. 8.1 (**a**) Discrete and continuous spectrums in quantum mechanics. (**b**) Landau damping by velocity space continuous spectrum. (**c**) Alfven continuum damping by real space continuous spectrum

Alfven continuum damping, inclusion of the kinetic effect implies damping occurs due to mode conversion to the kinetic Alfven wave (KAW) introduced in Sect. 5.2.2, which is discovered by Hasegawa and Chen in 1975 [293].

The shear Alfven resonance can be seen from the cold wave resonance. The cold wave dispersion relation is given as in Stix [717].

$$n_\perp^2 = \frac{(R - n_\parallel^2)(L - n_\parallel^2)}{S - n_\parallel^2} \tag{8.5}$$

For low frequency $\omega \ll \Omega_i$, $S \sim c^2/v_A^2$ and the low frequency resonance ($n_\perp \to \infty$) is given by $\omega = k_\parallel v_A$. This shear Alfven resonance appears as one of regular singular points of Hain-Lüst equation to be discussed in the next subsection.

As discussed in Sect. 5.2.2, this shear Alfven wave dispersion relation is modified by the finite Larmor radius effect to a kinetic Alfven wave using $\alpha = T_e/T_i + 3/4$ as $\omega^2 = k_\parallel^2 v_A^2 [1 + \alpha k_\perp^2 \rho_i^2]$.

Damping of the KAW occurs mainly through electron Landau damping for low beta plasma [294]. Mode conversion of shear Alfven wave to KAW and its damping was confirmed by Weisen [816] by excitation of shear Alfven wave using external antenna structure in TCA tokamak in 1989.

The spectral property of resistive MHD equation is still under development. Dewar [159] is the first to study spectral property of resistive Aflven wave spectrum showing some important characteristics. When the finite resistivity is included, van der Holst [785] showed that the ideal continua disappear and are replaced by damped global waves located on specific curves in the complex frequency plane. It is also shown that for very small toroidicity the topology of the resistive spectrum is completely different from the cylindrical one. Matsumoto-Tokuda [543] also investigate the spectrum of resistive reduced MHD equation.

Column 8-1: Energy Integral in Ideal MHD

The linearized ideal MHD equation is given by:

$$\rho\partial^2\xi/\partial t^2 = \boldsymbol{F}(\xi) = \boldsymbol{J}\times\boldsymbol{B}_1 + \boldsymbol{J}_1\times\boldsymbol{B} - \nabla p_1, \tag{8.6}$$

where $p_1 = -\gamma p\nabla\cdot\xi - \xi\cdot\nabla p$, $\boldsymbol{B}_1 = \nabla\times(\xi\times\boldsymbol{B})$, $\mu_0\boldsymbol{J}_1 = \nabla\times\boldsymbol{B}_1$. Here this force operator $\boldsymbol{F}(\xi)$ satisfies the self-adjoint condition, $\int\eta\cdot\boldsymbol{F}(\xi)dV = \int\xi\cdot\boldsymbol{F}(\eta)dV$ and $\boldsymbol{F}(\xi)$ is called the Hermite operator. Using this property, following energy conservation relation is obtained from above equation of motion.

$$\frac{1}{2}\int_{P+V}\rho\,(\partial\xi/\partial t)^2\,dV = \frac{1}{2}\int_{P+V}\xi\cdot\boldsymbol{F}(\xi)dV, \tag{8.7}$$

where subscript "P" stands for plasma and "V" stands for vacuum. Thus $\delta W = -(1/2)\int_{P+V}\xi\cdot\boldsymbol{F}(\xi)dV$ is a potential energy for the plasma displacement. Assuming $e^{-i\omega t}$ dependence, we have $\omega^2 = \delta W/K$, $K = (1/2)\int_{P+V}\rho|\xi|^2 dV$. Consider a plasma surrounded by the vacuum region and the ideal wall at $\boldsymbol{r}_w$. The distorted plasma surface is given by $\boldsymbol{r}_s = \boldsymbol{r}_p + \xi$, where $\boldsymbol{r}_p$ is unperturbed surface. Let $\boldsymbol{n}$ is the unit normal vector to ideal wall, the boundary conditions are:

$$\boldsymbol{n}\cdot\boldsymbol{B}_1|_{r_w} = 0,\ \text{where},\ \boldsymbol{n}\cdot\boldsymbol{B}_1|_{r_p} = \boldsymbol{n}\cdot\nabla\times(\xi_\perp\times\boldsymbol{B})|_{r_p} \tag{8.8}$$

$$\left[p_1 + \boldsymbol{B}\cdot\boldsymbol{B}_1/\mu_0 + \xi\cdot\nabla\left(p + B^2/2\mu_0\right)\right]_{r_p} = \left[\boldsymbol{B}\cdot\boldsymbol{B}_1/\mu_0 + \xi\cdot\nabla\left(B^2/2\mu_0\right)\right]_{r_p}, \tag{8.9}$$

The energy integral δW_p is given as:

$$\delta W_p = -\frac{1}{2}\int_P \xi\cdot\left[\frac{(\nabla\times\boldsymbol{B}_1)\times\boldsymbol{B} + (\nabla\times\boldsymbol{B})\times\boldsymbol{B}_1}{\mu_0} + \nabla(\gamma p\nabla\cdot\xi + \xi\cdot\nabla p)\right]dV \tag{8.10}$$

Taking the partial integration of $\xi\cdot\nabla(\gamma p\nabla\cdot\xi + \xi\cdot\nabla p)$, $\xi\cdot(\nabla\times\boldsymbol{B}_1)\times\boldsymbol{B}$ and after some algebra (see Miyamoto [552], for example), we have:

$$\delta W = \delta W_P + \delta W_V \tag{8.11}$$

$$\delta W_P = \frac{1}{2}\int_P\left[\frac{\boldsymbol{B}_1^2}{\mu_0} - \frac{1}{\mu_0}\xi\cdot(\nabla\times\boldsymbol{B})\times\boldsymbol{B}_1 + (\gamma p\nabla\cdot\xi + \xi\cdot\nabla p)\nabla\cdot\xi\right]dV \tag{8.12}$$

$$\delta W_V = (2\mu_0)^{-1}\int_V(\nabla\times\boldsymbol{A}_1)^2 dV = \int_V \boldsymbol{B}_1^2/2\mu_0 dV \tag{8.13}$$

(continued)

Here we assumed no surface current proportional $\sim \delta(\boldsymbol{r}-\boldsymbol{r}_s)$. According to Furth [233], the energy integral δW_p can be rewritten as:

$$\delta W_P = \int_P dV[\delta W_{SA} + \delta W_{MS} + \delta W_{SW} + \delta W_{IC} + \delta W_{KI}] \tag{8.14}$$

Here, $\delta W_{SA} = \boldsymbol{B}_1^2/2\mu_0$, $\delta W_{MS} = \boldsymbol{B}^2(\nabla\cdot\boldsymbol{\xi}_\perp + 2\boldsymbol{\xi}_\perp\cdot\boldsymbol{\kappa})^2/2\mu_0$, $\delta W_{SW} = \gamma p(\nabla\cdot\boldsymbol{\xi})^2/2$, $\delta W_{IC} = (\boldsymbol{\xi}_\perp\cdot\nabla p)(\boldsymbol{\xi}_\perp\cdot\boldsymbol{\kappa})/2$, $\delta W_{KI} = -J_\parallel \boldsymbol{b}\cdot(\boldsymbol{B}_{1\perp}\times\boldsymbol{\xi}_\perp)/2$. Here, $\delta W_{SA}, \delta W_{MS}, \delta\delta W_{SW}, \delta W_{IC}, \delta W_{KI}$ are energies from shear Alfven wave, magneto sonic wave, sound wave, interchange, and kink terms.

8.1.1 MHD Spectrum of the Cylindrical Plasma

8.1.1.1 Shear Alfven and Sound Wave Singularities

In the cylindrical (r,θ,z) plasma, pressure balance equation is given by,

$$\frac{d}{dr}\left(P + \frac{B_z^2}{2\mu_0}\right) + \frac{B_\theta^2}{\mu_0 r} = 0 \tag{8.15}$$

Using Fourier decomposition $(e^{i(kz+m\theta-\omega t)})$ and eliminating ξ_θ, ξ_z, and $\boldsymbol{B}_1$, Appert [24] in 1974 derived following 1-st order ordinary differential equations for radial displacement $X \equiv r\xi_r$ and the perturbed total pressure $P_1 = p_1 + \boldsymbol{B}_1\cdot\boldsymbol{B}/\mu_0$,

$$D\frac{dX}{dr} = C_1 X - rC_2 P_1 \tag{8.16}$$

$$D\frac{dP_1}{dr} = \frac{C_3}{r}X - C_1 P_1 \tag{8.17}$$

Here,

$$D = \left[\rho\omega^2 - \frac{F^2}{\mu_0}\right]\left[\rho\omega^2\left(\gamma p + \frac{B^2}{\mu_0}\right) - \frac{\gamma p F^2}{\mu_0}\right] \tag{8.18}$$

$$C_1 = \frac{2B_\theta}{\mu_0 r}\left[\rho^2\omega^4 B_\theta - \frac{mF}{r}\left\{\rho\omega^2\left(\gamma p + \frac{B^2}{\mu_0}\right) - \gamma p\frac{F^2}{\mu_0}\right\}\right] \tag{8.19}$$

$$C_2 = \rho^2\omega^4 - \left(k^2 + \frac{m^2}{r^2}\right)\left[\rho\omega^2\left(\gamma p + \frac{B^2}{\mu_0}\right) - \gamma p\frac{F^2}{\mu_0}\right] \tag{8.20}$$

$$C_3 = D\left[\rho\omega^2 - \frac{F^2}{\mu_0} + \frac{2B_\theta}{\mu_0}\frac{d}{dr}\left(\frac{B_\theta}{r}\right)\right] + \rho\omega^2\left(\rho\omega^2 - \frac{F^2}{\mu_0}\right)\left(\frac{2B_\theta^2}{\mu_0 r}\right)^2$$
$$- \left[\gamma p\left(\rho\omega^2 - \frac{F^2}{\mu_0}\right) + \rho\omega^2\frac{B_z^2}{\mu_0}\right]\left(\frac{2B_\theta F}{\mu_0 r}\right)^2 \tag{8.21}$$

where, $F = mB_\theta/r + kB_z$ and $k = -n/R$. If $C_2 \neq 0$, we have $P_1 = (C_1 X - DdX/dr)/rC_2$ from (8.16) and we have following second order ordinary differential equation (called Hain-Lüst equation) by substituting this into (8.17).

$$\frac{d}{dr}\left(f\frac{dX}{dr}\right) + gX = 0, \tag{8.22}$$

$$\text{where } f = \frac{D}{rC_2},\ g = \frac{C_2C_3 - C_1^2}{rC_2D} - \frac{d}{dr}\left(\frac{C_1}{rC_2}\right) \tag{8.23}$$

While $C_2 = 0$ looks like a singular point, nothing will brow up for (8.16) and (8.17). So, it is not a real singularity. Radial positions satisfying $D(r) = 0$ however produce singular solution divergent at this positions and are called singular points. Zeros of first and second bracket of D are called shear Alfven and sound wave singularities.

$$\text{Shear Alfven singularity: } \mu_0\rho\omega^2 = F^2 \tag{8.24}$$

$$\text{Sound Wave singularity: } \rho\omega^2(\mu_0\gamma p + B^2) = \gamma p F^2 \tag{8.25}$$

The method of Frobenius series to obtain singular solution of second order ordinary differential equation (see Column 8-2) is well developed (see for example, Ince [372]). These singularities (resonances) produce the continuous spectrum of the ideal MHD force operator $\boldsymbol{F}$ similar to the case of Landau damping. For a low beta plasma, these singularities (resonances) are approximately given by $\omega^2 = k_\parallel^2 V_A^2 \equiv \omega_A^2$ (shear Alfven resonance) and $\omega^2 = k_\parallel^2 C_s^2 \equiv \omega_s^2$ (sound wave resonance), where $V_A = B/\sqrt{\mu_0\rho}$ is the Alfven velocity, $C_s = \sqrt{\gamma p/\rho} \sim \sqrt{\gamma(T_i + T_e)/m_i}$ is the sound velocity, and $k_\parallel = (n - m/q)/R$ is the parallel wave number, where $q = B_z R/B_\theta r$ is the cylindrical safety factor.

Radial profiles of ω_A^2 and ω_s^2 are schematically shown in Fig. 8.2a in the next section, which are well separated and we can discuss for a single singularity, say concentrate on ω_A^2. Different type of singularities appears for $d\omega_A/dr \neq 0$ (everywhere) and $d\omega_A/dr = 0$ at some radius r_0.

(a) Monotonic ω_A^2 profile:
Let ω_A^2 spans $\omega_A^2(r_1) \leq \omega_A^2 \leq \omega_A^2(r_2)$. Any choice of $\omega^2 \in [\omega_A^2(r_1), \omega_A^2(r_2)]$ leads to the singularity. Let $r = r_0$ satisfies $\omega^2 = \omega_A^2(r_0)$, we have $\omega^2 - \omega_A^2 \sim -(\omega_A^2)'_0(r - r_0)$. Then, $f \sim r - r_0 \equiv x$. The eigenmode equation near singular point becomes:

$$0 = \frac{d^2X}{dx^2} + \frac{f'}{f}\frac{dX}{dx} + \frac{g}{f}X \sim \frac{d^2X}{dx^2} + \frac{1 + b_1x + -}{x}\frac{dX}{dx} + \frac{c_1x + c_2x^2 + -}{x^2}X \tag{8.26}$$

In this case, indicial equation becomes $\kappa^2 = 0$ leading to a logarithmic singular solution $X \sim A_1(x)ln|x|$ (see Column 8-2).

The case for a sound wave resonance can be discussed similarly. Due to radial inhomogeneity, eigenvalue ω^2 can be a band of values of ω_A^2 and ω_s^2, which constitutes the continuous spectrum.

(b) Non-monotonic ω_A^2 profile:
If the profile of ω_A^2 has a local minimum at $r = r_0$, we have $d\omega_A(r_0)/dr = 0$ and $\omega^2 - \omega_A^2 \sim -\frac{1}{2}(\omega_A^2)_0''(r - r_0)^2$. Then, $f \sim (r - r_0)^2$. The eigenmode equation near singular point becomes:

$$0 = \frac{d^2X}{dx^2} + \frac{f'}{f}\frac{dX}{dx} + \frac{g}{f}X \sim \frac{d^2X}{dx^2} + \frac{2 + b_1x + -}{x}\frac{dX}{dx} + \frac{c_0 + c_1x + -}{x^2}X \tag{8.27}$$

In this case, indicial equation becomes $\kappa^2 + \kappa + c_0 = 0$ leading to an algebraic singular solution $X \sim A_i(x)x^{\kappa_i}$ $(i = 1, 2)$, where $\kappa_{1,2} = (-1 \pm \sqrt{1 - 4c_0})/2$ if $\kappa_1 - \kappa_2$ is not integer (see Column 8-2).

The value $\omega^2 = \omega_A^2(r_0)$ is a minimum of the continuous spectrum and may also be a cluster point of the discrete spectrum. The case for a sound wave resonance can be discussed similarly except the value is a maximum of the continuous spectrum and the discrete spectrum may appear above the sound wave continuum.

The connection between the singular solution and the continuous spectrum originates from the fact that a jump of the solution becomes possible at the singular point, which effectively increases the freedom of the eigen function to allow ω^2 to take any value.

Column 8-2: Regular Singular Points in 2nd Order ODE Tenenbaum [761], Ince [372]
We introduce the basic property of the ordinary differential equation (ODE). We consider following second order ODE.

$$y'' + F_1(x)y' + F_2(x)y = 0 \tag{8.28}$$

Here $' \equiv d/dx$. A point $x = x_0$ (we set $x_0 = 0$ without loss of generality) is called an **ordinary point** of (8.28) if F_1 and F_2 are analytic. Here, analytic means that each function has a Taylor series expansion $(a_0 + a_1x + a_2x^2 + --)$ near $x = 0$. A point $x = 0$ is called a **singularity** of (8.28) if one or more of F_1 and F_2 are not analytic at $x = 0$.

(continued)

The singularity can be categorized into regular and irregular singularity. If $x = 0$ is a singularity and if one or both $xF_1(x)$ and $x^2F_2(x)$ are not analytic at $x = 0$, the point $x = 0$ is called a **irregular singular point**. For example, $x = 0, 1$ are irregular singular points of $x^2(x-1)^2y'' + y' + 2x^2y = 0$. If $x = 0$ is a singularity and if both $xF_1(x)$ and $x^2F_2(x)$ are analytic at $x = 0$, the point $x = 0$ is called a **regular singular point**. For example, $x = 0, 1$ are regular singular points of $x(x-1)y'' + y' + 2xy = 0$.

J. Fuchs gives a theorem on a necessary and sufficient condition of regular singular point (Sect. 15.3 of [372]). If $x = 0$ is a regular singular point, F_1 and F_2 can be expanded as $F_1(x) = (b_0 + b_1x + b_2x^2 + -)/x$, $F_2(x) = (c_0 + c_1x + c_2x^2 + --)/x^2$. The solution is given by the **method of Frobenius series**.

$$y(x) = x^\kappa A(x) \equiv x^\kappa(a_0 + a_1x + a_2x^2 + --), \ (a_0 \neq 0) \tag{8.29}$$

If $\kappa = 0$ or positive integer, the series becomes the Taylor series. However, for negative values of κ or non-integral positive values of κ, (8.29) is not a Taylor series. Substituting these expressions into (8.28), we find κ should satisfy the **indicial equation**, $\kappa^2 - \kappa + b_0\kappa + c_0 = 0$ from lowest order equation. Let κ_1 and κ_2 are solutions ($\kappa_1 > \kappa_2$). If $\kappa_1 - \kappa_2$ is not integer, we have two solutions:

$$y_1(x) = x^{\kappa_1}A_1(x), \ y_2(x) = x^{\kappa_2}A_2(x), \tag{8.30}$$

where all $a_{1i}, a_{2i} (i = 1, 2, --)$ can be determined from higher order equations. If $\kappa_1 - \kappa_2$ is integer or zero, we have different solutions:

$$y_1(x) = x^{\kappa_1}A_1(x), y_2(x) = x^{\kappa_1}A_1(x)lnx + A_2(x)x^{\kappa_2}, \ x > 0 \tag{8.31}$$

For both cases, the first solution y_1 is called the **Small solution** and the second solution y_2 is called the **Large solution**, both of which appear in the linear MHD theories in both 1 and 2 dimensions.

In this book, the solution (8.30) appears in Sect. 8.2 and the solution (8.31) including the logarithmic solution appears in Sects. 8.5, 8.6 as well.

8.1.1.2 Discrete Spectrum and Cluster Point: GAE

We note that the Alfven velocity is concave and the sound velocity is convex in the radial direction if the density and temperature profiles are centrally peaked. Schematic radial profiles of the ω_A^2 and ω_s^2 are given for fixed n, m in Fig. 8.2a.

Due to radial inhomogeneity, eigenvalue ω^2 can be a band of values shown by ω_A^2 and ω_s^2 and the eigenfunction is divergent at the resonance points. Below this continuum, there are some discrete spectrum, which is called GAE (Global Alfven Eigenmode).

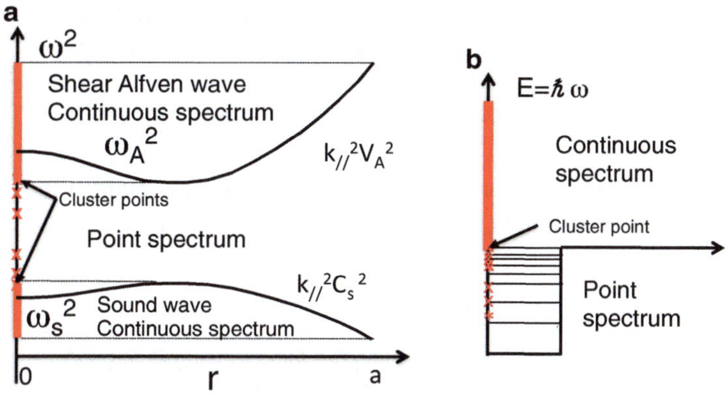

Fig. 8.2 (**a**) Schematic diagram of discrete and continuous spectrum in the cylindrical plasma. *Red lines* are continuous spectrum and *red times symbol* are discrete spectrum. (**b**) Discrete and continuous spectrum in quantum mechanics

In the quantum mechanics, it is well know that the Schrödinger equation has point and continuous spectrum. In this case, situation is much simpler than plasma physics. Figure 8.2b shows a nuclear potential and the energy spectrum $E = \hbar\omega$. The energy states are discrete spectrum inside the nuclear potential and are continuous spectrum in the free space. The transition from discrete to continuous occurs due to the loss of the boundary condition at finite potential boundary.

Conditions for the existence of GAE are investigated by Appert [25], Mahajan [528], and Goedbloed [250]. From the classification of singularities in previous section, the discrete spectrum occurs when following condition is satisfied inside the plasma, by which either the shear Alfven continuum has local minimum or the sound wave continuum has the local maximum.

$$\frac{d\omega_A}{dr} = 0 \quad \text{or} \quad \frac{d\omega_S}{dr} = 0 \tag{8.32}$$

Appert [25] analyzed the characteristics of discrete spectrum including higher order case such as $\omega^2 - \omega_A^2 \sim (r - r_0)^4$. Conditions for the existence of GAE can be discussed using the method of Frobenius series [251]. As discussed in the previous section, the indicial equation becomes $\kappa^2 + \kappa + c_0 = 0$. If $1 - 4c_0 < 0$, these indices become complex and the sequence of discrete shear Alfven modes appears clustering at the lower end of the shear Alfven continuum.

Since $c_0 = (x^2 g/f)_{x\to 0}$ from (8.27), condition of the appearance of the cluster point at the minimum of the shear Alfven continuum is given by Goedbloed [250],

$$0 < \frac{\mu_0 \rho}{8} \frac{d^2}{dr^2} \omega_A^2 < A \tag{8.33}$$

$$A = \left(\frac{2kB_\theta}{r}\right)^2 \left(1 - \frac{\mu_0 \gamma p}{B^2}\right) - \left(\frac{G}{B}\right)^2 r\frac{d}{dr}\left(\frac{B_\theta^2}{r^2} + \frac{2kB_\theta B^2}{r^2 G}\right) \tag{8.34}$$

$$G = (\boldsymbol{k} \times \boldsymbol{B})_r = \frac{mB_z}{r} - kB_\theta \tag{8.35}$$

Similarly, the condition for the appearance of the cluster point at the maximum of the sound wave continuum are given by Goedbloed [250],

$$0 < \frac{\mu_0 \rho}{8}\frac{d^2}{dr^2}\omega_S^2 < S \tag{8.36}$$

$$S = -\frac{(\mu_0\gamma p)^2 F^4}{(\mu_0\gamma p + B^2)^3} - \frac{(\mu_0\gamma p)^2 rF^2 (B_\theta^2/r^2)'}{B^2(\mu_0\gamma p + B^2)^2} + \frac{4k^2B_\theta^2(\mu_0\gamma p)^2}{r^2B^2(\mu_0\gamma p + B^2)} \tag{8.37}$$

Approximate eigen frequency of discrete spectrum is obtained by Appert, Mahajan, and Goedbloed using the WKBJ solution (see Column 6-3) near the local minimum of the shear Alfven continuum and the local maximum of the sound wave continuum.

If we define $\psi = \sqrt{f(r)}X(r)$, Eq. (8.22) can be converted as:

$$\frac{d^2\psi}{dr^2} - U(r)\psi(r) = 0 \tag{8.38}$$

$$\text{where,}\quad U(r) = \frac{1}{\sqrt{f(r)}}\frac{d^2\sqrt{f(r)}}{dr^2} - \frac{g(r)}{f(r)} \tag{8.39}$$

We assume eikonal approximation $k_r^2L^2 \gg 1$ where L is a characteristics scale length of equilibrium quantity. In this case, the first term of RHS of (8.39) can be neglected and $U(r) \sim -g(r)/f(r)$. The WKBJ solution (under $|U'(r)/U(r)^{3/2}| \ll 1$ condition) is given by,

$$X(r) \sim (fg)^{-1/4} exp(i\int k_r dr) \quad k_r = \sqrt{g/f} \tag{8.40}$$

Here f and g are given in (8.23). The equation $k_r = \sqrt{g/f}$ gives an approximate dispersion relation of the discrete spectrum below the continuum and approximately given by Appert[25]: $\mu_0\rho\omega^2 = F^2 - A_1/k_r^2$, where $A_1 = (2kB_\theta/r)^2 - (G/B)^2 rd(B_\theta^2/r^2)/dr$. This GAE is produced by the magnetic field line curvature (or B_θ effect) [250].

Another important explanation of the occurrence of discrete spectrum below the shear Alfven continuum (GAE) is given by Mahajan [529]. He converted MHD equation into the Schrödinger type equation in Fourier space, by which he derived an effective potential. By looking at structure of the effective potential, it is clearly shown that the discrete spectrum appears when the effective potential is convex enough as shown in Fig. 8.3a. He also showed that the inclusion of kinetic Alfven wave further increases the potential well structure as shown in Fig. 8.3b implying

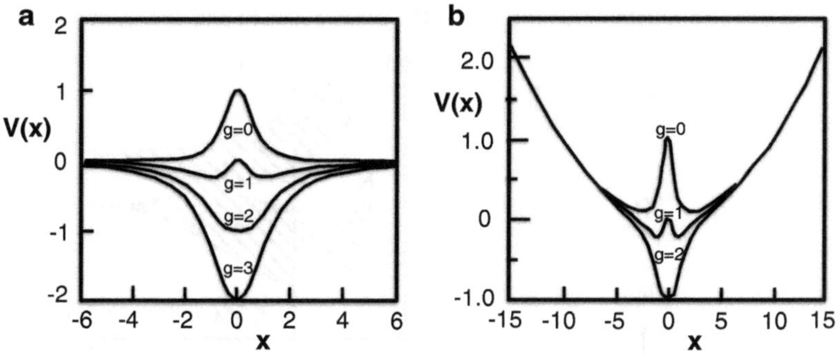

Fig. 8.3 (**a**) Effective potential structure for ideal cylindrical MHD equation in Fourier space. (**b**) Effective potential structure for kinetic cylindrical MHD equation including the kinetic Alfven wave in Fourier space. Modified from Mahajan [529]

that inclusion of kinetic effect changes the continuous spectrum to the discrete spectrum.

TCA tokamak experiment by de Chambrier showed large loading resistance just below the shear Alfven continuum (1982) [153]. The discrete spectrum below the shear Alfven continuum is found by Ross [641] numerically and investigated theoretically by Appert [25], Mahajan [528] and Goedbloed [250]. This mode is named Global Alfven Eigenmodes (GAE). This GAE is confirmed experimentally by Evans [187].

8.1.2 Spectrum Gap in the Periodic Potential

The Schrödinger equation in the periodic crystal lattice takes a form of Mathieu equation (see Column 8-3) and the energy spectrum takes a band structure whose forbidden band width becomes larger when the potential depth ($\sim h$) is bigger.

Figure 8.4a, b shows the structure of crystal lattice and the change of the electron energy spectral characteristics in P-type semi-conductor as a function of the lattice constant d. When d is infinity, the electron energy spectrum has only discrete spectrum bounded by a single potential well. As d reduces, it allows energy band and the electron is shared in the periodic potential. The coupling energy mainly comes from adjacent lattices and the band width is determined by this coupling energy. When the energy bands from different energy states intersect, there appears a forbidden band below which is valence band and above which is conduction band. In case of P-type semi-conductor, excess electron behaves as hydrogen like atom to produce discrete spectrum just below conduction band.

In tokamak, poloidal variation produces periodic potential structure for the shear Alfven (and sound) waves. So, it is natural to see semi-Mathieu-type equation in the MHD eigenmode equation. The wave equation in the periodic potential will have band structure in the continuum and a perturbation can produce discrete spectrum.

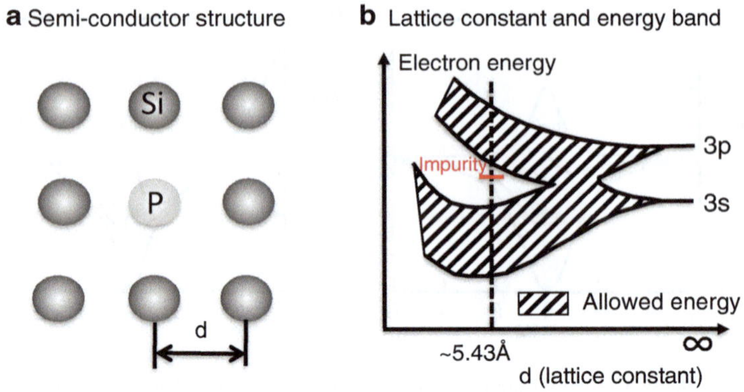

Fig. 8.4 (**a**) Schematic structure of P-type semi conductor. (**b**) Electron energy level in the semi-conductor as a function of lattice constant d. An impurity driven discrete spectrum is possible inside the forbidden band

High n shear Alfven wave is radially localized and the governing equation is given as a semi-Mathieu type equation by Cheng-Chen-Chance as [116]:

$$\frac{d^2\psi}{d\theta^2} + [\Omega^2(1 + 4\epsilon cos\theta) - F(\theta)]\psi = 0 \tag{8.41}$$

where $\psi = \sqrt{1 + s^2\theta^2}\Phi$, and Φ is the electrostatic potential after ballooning transformation in the covering space $[-\infty, +\infty]$ of the poloidal angle θ, $\epsilon = r/R$ and $F = s^2/(1 + s^2\theta^2)$ is a function for a case of low beta and circular plasma. We note that $F \to 0$ as $\theta \to \infty$ and Eq. (8.41) becomes a Mathieu equation. Details of the high-n TAE will be described in Sect. 8.7.

The cylindrical plasma has symmetry in the axial and poloidal directions. This provides a property that toroidal and poloidal mode numbers n and m are good quantum numbers. Tokamak is an axisymmetric system but the poloidal symmetry can easily break by the introduction of the toroidicity, the elongation, and the higher shaping. However, periodicity in the poloidal direction still holds and the resonant reflection of waves in the periodic potential give rise to the spectrum gap.

Historically speaking, the gap structure in the continuous MHD spectrum is first shown for the elongated plasma by Dewar in 1974 [155] as shown in Fig. 8.5a. If we define elliptic tokamak equilibrium by $\psi = \psi_0(r) + \psi_1 cos2\theta$, we define the elliptic deformation $\Delta(r) = -\psi_1(r)/(d\psi_0(r)/dr)$. The shear Alfven wave dispersion relation is given by the coupling of $m(k_{\|m} = (n - m/q)/R)$ and $m+2(k_{\|m+2} = (n-(m+2)/1)/R)$ modes with mode coupling constant $\delta \sim \Delta'(r)$ as,

$$\begin{vmatrix} \frac{\omega^2}{v_A^2} - k_{\|m}^2 & \delta\frac{\omega^2}{v_A^2} \\ \delta\frac{\omega^2}{v_A^2} & \frac{\omega^2}{v_A^2} - k_{\|m+2}^2 \end{vmatrix} = 0 \tag{8.42}$$

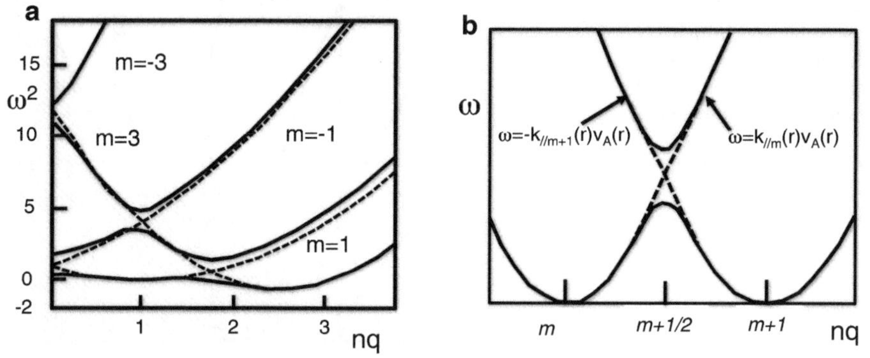

Fig. 8.5 (**a**) Shear Alfven gap spectrum due to elongation coupling. Modified from Dewar [155]. (**b**) Simple explanation of gap formation in toroidal shear Alfven resonance

In toroidal geometry where in-out inhomogeneity $B \sim B_0/(1 + \epsilon cos\theta)$ exists, Shear Alfven Resonance condition is given by a coupling of $m(k_{\|m} = (n-m/q)/R)$ and $m + 1(k_{\|m+1} = (n - (m + 1)/q)/R)$ modes with mode coupling constant ϵ as,

$$\begin{vmatrix} \frac{\omega^2}{v_A^2} - k_{\|m}^2 & \epsilon\frac{\omega^2}{v_A^2} \\ \epsilon\frac{\omega^2}{v_A^2} & \frac{\omega^2}{v_A^2} - k_{\|m+1}^2 \end{vmatrix} = 0 \tag{8.43}$$

This gives the spectrum gap (forbidden band) of ω for the shear Alfven continuum as shown in Fig. 8.5b.

Column 8-3: Mathieu Equation

If the system has periodicity, the relevant ODE may contain trigonometric function. Typical example is the following Mathieu equation.

$$\frac{d^2y}{dx^2} + (\lambda - hcos2x)y = 0 \tag{8.44}$$

Mathieu equation do not have finite singular points.

This Mathieu equation is obtained in the periodic crystal lattice (see Fig. 8.4a). Consider the one-dimensional Schrödinger equation.

$$-\frac{\hbar^2}{2m}\frac{d^2\psi}{dz^2} + V(x)\psi = E\psi \tag{8.45}$$

(continued)

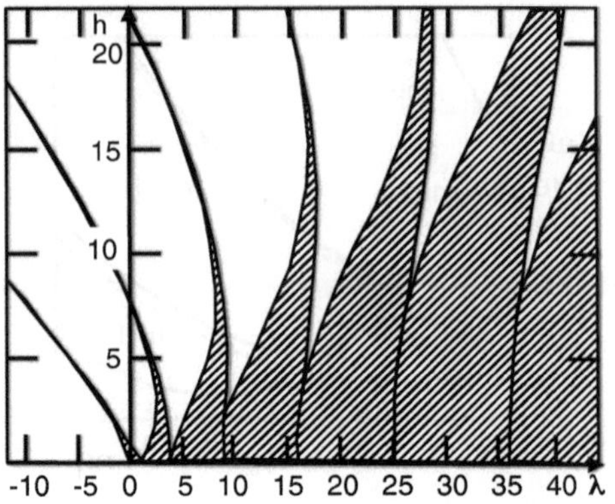

Fig. 8.6 Allowable eigenvalue spectrum of Mathieu function in (λ, h) space. Hatched region is the allowable spectrum and all $\lambda > 0$ region becomes possible for no periodic potential ($h = 0$)

For a lattice constant d, we have $V = V_0 cos(2\pi z/d)$. Defining $x = \pi z/d$, $\lambda = 2md^2E/\pi^2\hbar^2$, and $h = 2md^2/\pi^2\hbar^2$, we have following Mathieu equation.

$$\frac{d^2\psi}{dx^2} + (\lambda - hcos2x)\psi = 0, \tag{8.46}$$

where λ is the normalized eigenvalue and h is the normalized height of the periodic potential. Allowable eigenvalue forms a continuous spectrum as shown in Fig. 8.6. For $h = 0$, allowable λ spectrum is continuous spectrum in $\lambda \geq 0$. For a finite periodic potential height, this continuous spectrum break up to include forbidden bands.

There are four types of solutions in the Mathieu equation, $ce_{2m}(x; h^2), ce_{2m+1}(x; h^2), se_{2m}(x; h^2), se_{2m+1}(x; h^2)$ where $ce_n(x, h^2 \rightarrow 0) = cosnx$ ($n \geq 2$) and $se_n(x, h^2 \rightarrow 0) = sinnx$ ($n \geq 2$) (see Wittaker-Watson [827] or Ince [372].

8.2 Newcomb Equation

It is very important to identify stability boundary (or marginal stability point) of tokamak operation. Existence of continuous spectrum makes it difficult to accurately calculate ω^2 from linear MHD equation in 1D ((8.6), (8.7)) or 2D ((8.105), (8.106) in Sect. 8.7) near marginal stability.

Marginal stability calculation is improved in the PEST-2 code [266], and similarly in KINX code [154] but is limited to positive Mercier index case. So, it is essential to separate out the non-square integrable eigenfunction from the eigenmode equation to evaluate discrete spectrum. Pletzer [612] introduced bell-shaped localization function $H(r)$ ($H = 1$ for $|r - r_0| < \epsilon$ and $H = 0$ for $|r - r_0| > \delta$) to eliminate big solution part of the eigenfunction.

The marginal stability is described by the Newcomb equation to minimize Bernstein's energy integral. It is called 1D Newcomb equation in cylindrical plasma by Newcomb [580] and 2D Newcomb equation for axisymmetric plasma by Tokuda [770].

In Newcomb equation, singular surface is always rational surface for concerned mode, while singular surface changes with ω in the linear MHD equation and is different from rational surface, in general.

8.2.1 1D Newcomb Equation

For marginal stability, the 1D Newcomb equation with single regular singular point in $r \in [0, a]$ is given by Newcomb [580],

$$L(\xi) = \frac{d}{dr}\left(f(r)\frac{d\xi}{dr}\right) - g(r)\xi = 0, \tag{8.47}$$

where $f(r) = f_0(r - r_0)^2 + \cdot\cdot, f_0 > 0$, $g(r) = g_0 + \cdot\cdot$, $g_0 \neq 0$. The $r = r_0$ is a regular singular point (see Column 8-2). The solution comprises "small" (square integrable with power of $r^{-0.5+\mu}$) and "big" (non-square integrable with power of $r^{-0.5-\mu}$) solutions, where $\mu = \sqrt{1/4 + g_0/f_0}$ and $1/4 + g_0/f_0 > 0$ is Suydam stability criterion.

Tokuda [769] converted this marginal stability problem to the eigenvalue problem by keeping the behavior of eigenfunction near the regular singular point as follows,

$$L(\xi) = -\lambda\rho\xi,\ \rho = \rho_0(r - r_0)^2 \tag{8.48}$$

This equation is similar to the Stürm-Liouville equation (see Appendix A.6) for string vibration (f($\sim$ elastic modulus) > 0) [150]. But f can be zero or positive in our case. So, this equation is not Stürm-Liouville equation in exact sense.

Choice of this functional form for ρ does not change leading singularity of eigenfunction near the regular singular point (μ does not change since $\rho(r_0) = 0$) and this equation coincide with Newcomb equation at marginal stability ($\lambda = 0$). This makes it very efficient to find stability boundary to ideal MHD.

8.2.2 2D Newcomb Equation

As in the case of cylindrical plasma, the energy integral inside the plasma is minimized under the incompressibility condition $\nabla \cdot \boldsymbol{\xi} = 0$ in case of axisymmetric torus.

By using $X = \boldsymbol{\xi}\cdot\nabla r$ and $V = r\boldsymbol{\xi}\cdot\nabla(\theta - \zeta/q)$ in the flux coordinates (r, θ, ζ) with $r = [2R_0 \int_0^{\psi} (q/F)d\psi]^{1/2}$, the energy integral W_p can be expressed in a following form [770].

$$W_p = \frac{\pi}{2\mu_0}\int_0^1 dx \int_0^{2\pi} d\theta L\left(X, \frac{\partial X}{\partial\theta}, \frac{\partial X}{\partial x}, V, \frac{\partial V}{\partial\theta}\right) \tag{8.49}$$

Here, $x \equiv r/a = 1$ is the plasma surface and L is Lagrangian as given in detail in [770]. Absence of $\partial V/\partial x$ term in L leads to following Euler-Lagrange equation for V,

$$\frac{\partial}{\partial\theta}\left[\frac{\partial L}{\partial(\partial V/\partial\theta)}\right] - \frac{\partial L}{\partial V} = 0 \tag{8.50}$$

Integration of above equation in $\theta = [0, 2\pi]$ leads to following solvable condition.

$$\int_0^{2\pi} \frac{\partial L}{\partial V} d\theta = 0 \tag{8.51}$$

Fourier expansion of X and V for θ are defined as $X(x,\theta) = \sum X_m(x)e^{im\theta}$, $V(x,\theta) = -i\sum V_m(x)e^{im\theta}$. Substitution of these equations into equations (8.50) and (8.51) gives linear equations for V_m. And the solution is substituted into energy integral. The energy integral is now given by $\boldsymbol{Y} = x\boldsymbol{X}, \boldsymbol{X} = (\cdot\cdot, X_{-2}, X_{-1}, X_0, X_1, X_2, \cdot\cdot)^t$ (t: transposed) in a following form.

$$W_p[\boldsymbol{Y}] = \int_0^1 \hat{L}(\boldsymbol{Y}, \frac{d\boldsymbol{Y}}{dx})dx \tag{8.52}$$

This leads to the following Euler-Lagrange equation to minimize energy integral.

$$\frac{d}{dx}\frac{\partial\hat{L}}{\partial(d\boldsymbol{Y}/dx)} - \frac{\partial\hat{L}}{\partial\boldsymbol{Y}} = 0 \tag{8.53}$$

Since $\hat{L}$ is given by a quadratic form of $\boldsymbol{Y}$ and $d\boldsymbol{Y}/dx$, the Euler-Lagrange equation is reduced to following form of 2nd order ordinary differential equation.

$$\boldsymbol{N}(\boldsymbol{Y}) \equiv \frac{d}{dx}\boldsymbol{f}\frac{d\boldsymbol{Y}}{dx} + \boldsymbol{g}\frac{d\boldsymbol{Y}}{dx} + \boldsymbol{h}\boldsymbol{Y} = 0, \tag{8.54}$$

where $\boldsymbol{f}, \boldsymbol{g}, \boldsymbol{h}$ are constant $(2M + 1) \times (2M + 1)$ matrices. Here M is number of Fourier harmonics. This is called "two-dimensional Newcomb equation". Diagonal elements of $\boldsymbol{f}$ have $(n/m - 1/q)^2$ dependence similar to one-dimensional Newcomb equation and the radius of q=m/n is the regular singular point. Small and big solution exists near the singular point and Mercier condition is derived as local stability condition. Once Mercier condition is met, Kink and Peeling modes can be studied using this two-dimensional Newcomb equation. Tokuda [770] converted this 2D Newcomb equation into eigenvalue equation by adding an artificial kinetic energy integral term similar to the 1D Newcomb equation as follows,

$$W[\boldsymbol{Y}] = W_p - \lambda W_k \tag{8.55}$$

$$W_k = 2\pi^2 \int \sum_m \rho_m(\hat{r}) |Y_l(\hat{r})|^2 dx \tag{8.56}$$

$$\rho_m(x) = \begin{cases} F^2(\frac{m}{q} - n) & m = nq(r_0) \\ F^2 & m \neq nq(r_0) \end{cases} \tag{8.57}$$

This choice of ρ_m conserves leading singularity of eigenfunction at the singular radius. The Euler-Lagrange equation to minimize energy integral $W[\boldsymbol{Y}]$ is given as follows,

$$\boldsymbol{N}_\lambda \boldsymbol{Y} = [\boldsymbol{N} + \lambda \mathrm{Diag}(\rho_m(x))]\boldsymbol{Y} = 0 \tag{8.58}$$

The linear operator $\boldsymbol{N}_\lambda$ is related to 2D Newcomb operator as $\boldsymbol{N}_\lambda = \boldsymbol{N} + \lambda$ Diag$(\rho_m(x))$. Here, Diag$(\rho_m(x))$ is diagonal matrix having $\rho_m(x)$ as (m, m) component. This equation is quasi Sturm-Liouville equation similar to the 1D case.

MARG2D [770] solves Eq. (8.58) to identify stability boundary. The code is extended to free boundary mode by inclusion of vacuum solution by Aiba et al. [4]. Figure 8.7 shows comparison of MARG2D and ERATO-J [33] to identify critical β_p value for stability. While stability boundary is not quite sharp in ERATO-J, MARG2D can determine sharp stability boundary.

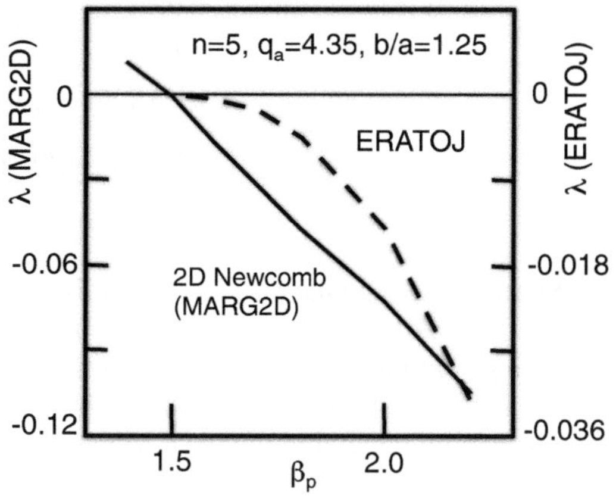

Fig. 8.7 Comparison of ERATO-J and MARG2D to identify stability boundary. Modified from Aiba [4]

8.3 Frieman-Rotenberg Equation

Active use of plasma flow to stabilize global MHD modes (especially RWM in Sect. 8.8) is important for all advanced tokamak regimes (WS, NS and CH). But inclusion of flow bring quite rich physics to ideal MHD stability theory.

The linearized equation of motion with mass flow is given by Frieman and Rotenberg [212] and is called Frieman-Rotenberg equation (see Column 8-4).

$$\rho\frac{\partial^2 \boldsymbol{\xi}}{\partial t^2} + 2\rho(\boldsymbol{u}\cdot\nabla)\frac{\partial \boldsymbol{\xi}}{\partial t} = \boldsymbol{F}(\boldsymbol{\xi}) \tag{8.59}$$

$$\boldsymbol{F} = \boldsymbol{F}_s(\boldsymbol{\xi}) + \boldsymbol{F}_d(\boldsymbol{\xi}) \tag{8.60}$$

$$\boldsymbol{F}_s(\boldsymbol{\xi}) = \nabla[\boldsymbol{\xi}\cdot\nabla P + \gamma P\nabla\cdot\boldsymbol{\xi}] + (\nabla\times\boldsymbol{B}_1)\times\boldsymbol{B}/\mu_0 + \boldsymbol{J}\times\boldsymbol{B}_1$$

$$\boldsymbol{F}_d(\boldsymbol{\xi}) = \nabla\cdot[\rho\boldsymbol{\xi}(\boldsymbol{u}\cdot\nabla)\boldsymbol{u} - \rho\boldsymbol{u}(\boldsymbol{u}\cdot\nabla)\boldsymbol{\xi}]$$

$$\boldsymbol{B}_1 = \nabla\times(\boldsymbol{\xi}\times\boldsymbol{B})$$

Both $\boldsymbol{F}_s$ and $\boldsymbol{F}_d$ are Hermite operators but the convective term $2\rho(\boldsymbol{u}\cdot\nabla)\partial\boldsymbol{\xi}/\partial t$ is anti-Hermitian operator. And the system as a whole is not self-adjoint. While ω spectrum of Hermite MHD operator lies on real or imaginary axis, spectrum of Frieman-Rotenberg (FR) equation spreads in complex plane bringing some difficulty in searching eigenvalue. Moreover, eigenvalue analysis may fail to identify most unstable solution when time evolution of this solution is no longer $exp(\lambda t)$. MHD stability codes treating toroidal flow have been developed during 1990 decade such as MARS by Chu et al. [124] and more recently MINERVA by Aiba et al. [7] and CASTOR-FLOW by Chapman et al. [106]. In case we can assume $\boldsymbol{\xi} = \hat{\boldsymbol{\xi}} exp(\lambda t)$, eigen value $\lambda = \gamma + in\omega$ (γ is growth rate, ω is real frequency, n is toroidal mode number) including toroidal and poloidal flows can be expressed as follows [7, 124],

$$\gamma^2 = -\frac{\delta W_p}{\delta K} - n^2\omega^2,\ in\,\omega = -\frac{\delta W_c}{\delta K} \tag{8.61}$$

$$\delta K = \langle\hat{\boldsymbol{\xi}}|\rho|\hat{\boldsymbol{\xi}}\rangle \tag{8.62}$$

$$\delta W_p = -\langle\hat{\boldsymbol{\xi}}|\boldsymbol{F}(\hat{\boldsymbol{\xi}})\rangle \tag{8.63}$$

$$\delta W_c = \langle\hat{\boldsymbol{\xi}}|\rho(\boldsymbol{u}\cdot\nabla)|\hat{\boldsymbol{\xi}}\rangle \tag{8.64}$$

Spectral and singular structures of this equation has been investigated in depth by Hirota [324]. He extended definition of wave energy in the bounded plasma including continuous spectrum using action-angle formulation, where wave energy is given by $\omega_n\mu_n$ for discrete mode and by $\int \omega\mu(\omega)d\omega$ for continuous mode.

The energy integral in the ideal MHD theory is represented by the sum of these wave energy. In the flowing plasma, these wave energy can be negative and its interaction with positive dissipation at the wall can destabilize the mode. Also resonance causes continuum damping if wave energies of continuous and discrete modes are same sign, while resonant instability occurs if they have different sign.

Column 8-4: Derivation of Frieman-Rotenberg Equation and Eigenvalue
In order to derive the Frieman-Rotenberg equation, we use following Eulerian expression of the perturbed quantities in the flowing ideal MHD plasma.

$$\rho_1 = -\nabla \cdot (\rho_0 \boldsymbol{\xi}) \tag{8.65}$$

$$p_1 = -\gamma p_0 \nabla \cdot \boldsymbol{\xi} - \boldsymbol{\xi} \cdot \nabla p_0 \tag{8.66}$$

$$\boldsymbol{B}_1 = \nabla \times (\boldsymbol{\xi} \times \boldsymbol{B}_0) \tag{8.67}$$

$$\boldsymbol{v}_1 = \partial \boldsymbol{\xi}/\partial t + (\boldsymbol{u} \cdot \nabla)\boldsymbol{\xi} - (\boldsymbol{\xi} \cdot \nabla)\boldsymbol{u} \tag{8.68}$$

$$\mu_0 \boldsymbol{j}_1 = \nabla \times \boldsymbol{B}_1 \tag{8.69}$$

where $\boldsymbol{u}$ is the slow equilibrium flow to which we have $\nabla \cdot (\rho_0 \boldsymbol{u}) = 0$. According to Frieman [212], we can prove these relations using the relation between Lagrangian and Eulerian perturbations. Let ΔA as Lagrangian perturbation of a physical quantity A, and A_1 as Eulerian perturbation, we have following general relation.

$$\Delta A = A_1 + (\boldsymbol{\xi} \cdot \nabla)A \tag{8.70}$$

Let $\boldsymbol{x} = \boldsymbol{x}_0 + \boldsymbol{\xi}$, where $\boldsymbol{x}_0$ is the Eulerian fixed coordinates and $\boldsymbol{x}$ is the Lagrangian moving coordinates. The Jacobian of coordinate transformation $J \equiv \partial(x_1, x_2, x_3)/\partial(x_1^0, x_2^0, x_3^0) = det(\boldsymbol{I} + \nabla^0 \boldsymbol{\xi}) \sim 1 + \nabla^0 \cdot \boldsymbol{\xi}$. Here $\nabla^0 \equiv \partial/\partial \boldsymbol{x}^0$.
Proof of (8.65): Mass conservation law during the perturbation is given by $\rho dV = \rho_0 dV_0$ where dV is the volume element in the moving Lagrangian coordinates and dV_0 is that in fixed Eulerian coordinates. This leads to $\rho = J^{-1}\rho_0 \sim \rho_0(1 - \nabla^0 \cdot \boldsymbol{\xi}_i)$and the Lagrangian perturbation is given by $\Delta\rho = -\rho_0 \nabla^0 \cdot \boldsymbol{\xi}$. Using (8.70), we have $\rho_1 = -\rho_0 \nabla^0 \cdot \boldsymbol{\xi} - (\boldsymbol{\xi} \cdot \nabla^0)\rho_0 = -\nabla^0 \cdot (\rho_0 \boldsymbol{\xi})$ to reach equation (8.65).
Proof of (8.66): By using the thermodynamic equation of state $p\rho^{-\gamma} = C$, we have $p\rho^{-\gamma} = p_0 \rho_0^{-\gamma}$. This leads to $p = p_0/J^\gamma = p_0(1 - \gamma \nabla^0 \cdot \boldsymbol{\xi})$ and the Lagrangian perturbation is given by $\Delta p = -\gamma p_0 \nabla^0 \cdot \boldsymbol{\xi}$. Using (8.70), we have $p_1 = -\gamma p_0 \nabla^0 \cdot \boldsymbol{\xi} - (\boldsymbol{\xi} \cdot \nabla^0)p_0$ to reach equation (8.66).
Proof of (8.67): Using the induction law, $\partial \boldsymbol{B}/\partial t = \nabla \times (\boldsymbol{v} \times \boldsymbol{B})$ and assuming the strong magnetic field and flow is small, we can directly integrate in time at fixed position $\boldsymbol{x}_0$ to reach $\boldsymbol{B}_1 = \nabla^0 \times (\boldsymbol{\xi} \times \boldsymbol{B}_0)$. Proof of (8.69) is trivial.
Proof of (8.68) (Newcomb [581]): Let $\boldsymbol{v}(\boldsymbol{x}, t)$ is the Eulerian velocity expressed in the Lagrangian coordinates $(\boldsymbol{x}, t)$ and $\dot{\boldsymbol{x}}(\boldsymbol{x}_0, t)$ is the Lagrangian velocity expressed in the Eulerian coordinates $(\boldsymbol{x}_0, t)$, we have $\boldsymbol{v}(\boldsymbol{x}, t) = \dot{\boldsymbol{x}}(\boldsymbol{x}_0, t)$. Similarly, the Eulerian displacement $\boldsymbol{\xi}(\boldsymbol{x}, t)$ in the Lagrangian coordinates is the Lagrangian displacement $\boldsymbol{x}_1(\boldsymbol{x}_0, t)$ in the Eulerian coordinates; $\boldsymbol{\xi}(\boldsymbol{x}, t) = \boldsymbol{x}_1(\boldsymbol{x}_0, t)$. We have following equations for *i-th* component.

(continued)

$$v_i(\boldsymbol{x}, t) = \dot{x}_i(\boldsymbol{x}_0, t)$$

$$\xi_i(\boldsymbol{x}, t) = x_{i1}(\boldsymbol{x}_0, t)$$

If we take an Eulerian perturbation of 1st equation and the time derivative of 2nd equation and take a leading order $v_i \to u_i$, we have:

$$v_{i1}(\boldsymbol{x}_0, t) + \xi_j \frac{\partial u_i}{\partial x_j} = \dot{x}_{i1}(\boldsymbol{x}_0, t), \qquad \frac{\partial \xi_i}{\partial t}(\boldsymbol{x}_0, t) + u_j \frac{\partial \xi_i}{\partial x_j} = \dot{x}_{i1}(\boldsymbol{x}_0, t)$$

If we eliminate $\dot{x}_{i1}(\boldsymbol{x}_0, t)$ from above equations, we have equation (8.68).

Having these relations in mind, we linearize the MHD fluid equation of motion $\rho(\partial \boldsymbol{v}/\partial t + \boldsymbol{v} \cdot \nabla)\boldsymbol{v} = \boldsymbol{j} \times \boldsymbol{B} - \nabla p$ as:

$$\rho_1 \boldsymbol{u} \cdot \nabla \boldsymbol{u} + \rho_0 \left[\frac{\partial \boldsymbol{v}_1}{\partial t} + (\boldsymbol{u} \cdot \nabla)\boldsymbol{v}_1 + (\boldsymbol{v}_1 \cdot \nabla)\boldsymbol{u} \right]$$
$$= \boldsymbol{j}_0 \times \boldsymbol{B}_1 + \boldsymbol{j}_1 \times \boldsymbol{B}_0 - \nabla p_1 \tag{8.71}$$

Substituting (8.65) – (8.69) into (8.71), we have:

$$\begin{aligned} &\rho_0 \left(\frac{\partial^2 \boldsymbol{\xi}}{\partial t^2} + (\boldsymbol{u} \cdot \nabla)\frac{\partial \boldsymbol{\xi}}{\partial t} - \left(\frac{\partial \boldsymbol{\xi}}{\partial t} \cdot \nabla\right)\boldsymbol{u}\right)\right) + \rho_0 (\boldsymbol{u} \cdot \nabla)\frac{\partial \boldsymbol{\xi}}{\partial t} + \rho_0 \left(\frac{\partial \boldsymbol{\xi}}{\partial t} \cdot \nabla\right)\boldsymbol{u} \\ &+ \rho_0 (\boldsymbol{u} \cdot \nabla)\left[(\boldsymbol{u} \cdot \nabla)\boldsymbol{\xi} - (\boldsymbol{\xi} \cdot \nabla)\boldsymbol{u}\right] + \rho_0 ([(\boldsymbol{u} \cdot \nabla)\boldsymbol{\xi} - (\boldsymbol{\xi} \cdot \nabla)\boldsymbol{u}] \cdot \nabla)\boldsymbol{u} \\ &= \boldsymbol{j}_0 \times \boldsymbol{B}_1 + \boldsymbol{j}_1 \times \boldsymbol{B}_0 - \nabla p_1 + (\boldsymbol{u} \cdot \nabla \boldsymbol{u}) \nabla \cdot (\rho_0 \boldsymbol{\xi}) \end{aligned} \tag{8.72}$$

If we define $\boldsymbol{R} = (\boldsymbol{u} \cdot \nabla)\boldsymbol{\xi} - (\boldsymbol{\xi} \cdot \nabla)\boldsymbol{u}$, we have:

$$\rho_0 \partial^2 \boldsymbol{\xi}/\partial t^2 + 2\rho_0 (\boldsymbol{u} \cdot \nabla)\partial \boldsymbol{\xi}/\partial t = \boldsymbol{j}_0 \times \boldsymbol{B}_1 + \boldsymbol{j}_1 \times \boldsymbol{B}_0 - \nabla p_1 + \boldsymbol{F}_d(\boldsymbol{\xi}) \tag{8.73}$$

$$\boldsymbol{F}_d(\boldsymbol{\xi}) = [\nabla \cdot (\rho_0 \boldsymbol{\xi})](\boldsymbol{u} \cdot \nabla \boldsymbol{u}) - \rho_0 (\boldsymbol{u} \cdot \nabla)\boldsymbol{R} - \rho_0 (\boldsymbol{R} \cdot \nabla)\boldsymbol{u} \tag{8.74}$$

Noting a following relation (see exercise), we can rewrite $\boldsymbol{F}_d(\boldsymbol{\xi})$.

$$-(\boldsymbol{u} \cdot \nabla)\boldsymbol{R} - (\boldsymbol{R} \cdot \nabla)\boldsymbol{u} = \boldsymbol{\xi} \cdot \nabla[(\boldsymbol{u} \cdot \nabla)\boldsymbol{u}] - \boldsymbol{u} \cdot \nabla[(\boldsymbol{u} \cdot \nabla)\boldsymbol{\xi}]$$

$$\boldsymbol{F}_d(\boldsymbol{\xi}) = [\nabla \cdot (\rho_0 \boldsymbol{\xi})](\boldsymbol{u} \cdot \nabla)\boldsymbol{u} + \rho_0 \boldsymbol{\xi} \cdot \nabla[(\boldsymbol{u} \cdot \nabla)\boldsymbol{u}] - \rho_0 \boldsymbol{u} \cdot \nabla[(\boldsymbol{u} \cdot \nabla)\boldsymbol{\xi}] \tag{8.75}$$

(continued)

Using $\nabla \cdot (\rho_0 \boldsymbol{u}) = 0$ and suppressing 0, we have:

$$\boldsymbol{F}_d(\boldsymbol{\xi}) = \nabla \cdot [(\rho\boldsymbol{\xi})(\boldsymbol{u}\cdot\nabla)\boldsymbol{u} - \rho\boldsymbol{u}(\boldsymbol{u}\cdot\nabla)\boldsymbol{\xi}] \equiv \nabla \cdot [\rho\boldsymbol{\xi}\boldsymbol{u}\cdot\nabla\boldsymbol{u} - \rho\boldsymbol{u}\boldsymbol{u}\cdot\nabla\boldsymbol{\xi}] \tag{8.76}$$

Thus, we obtain following Frieman-Rotenberg equation [212].

$$\rho\frac{\partial^2\boldsymbol{\xi}}{\partial t^2} + 2\rho(\boldsymbol{u}\cdot\nabla)\frac{\partial\boldsymbol{\xi}}{\partial t} = \boldsymbol{F}(\boldsymbol{\xi}) \tag{8.77}$$

$$\boldsymbol{F} = \boldsymbol{F}_s(\boldsymbol{\xi}) + \boldsymbol{F}_u(\boldsymbol{\xi}) \ , \ \boldsymbol{B}_1 = \nabla\times(\boldsymbol{\xi}\times\boldsymbol{B})$$

$$\boldsymbol{F}_s(\boldsymbol{\xi}) = \nabla[\boldsymbol{\xi}\cdot\nabla P + \gamma P\nabla\cdot\boldsymbol{\xi}] + (\nabla\times\boldsymbol{B}_1)\times\boldsymbol{B}/\mu_0 + \boldsymbol{J}\times\boldsymbol{B}_1$$

$$\boldsymbol{F}_d(\boldsymbol{\xi}) = \nabla\cdot[\rho\boldsymbol{\xi}(\boldsymbol{u}\cdot\nabla)\boldsymbol{u} - \rho\boldsymbol{u}(\boldsymbol{u}\cdot\nabla)\boldsymbol{\xi}]$$

Eigenvaue of this Frieman-Rotenberg equation is discussed by Chu [124] and Aiba [6]. If the flow is relatively small, the displacement $\boldsymbol{\xi}$ can be approximated by the eigenmode with the complex eigenvalue $\lambda = \gamma + i\omega$; $\boldsymbol{\xi} = \hat{\boldsymbol{\xi}}e^{\lambda t}$. We define a volume integral $\langle\hat{\boldsymbol{\xi}}|\boldsymbol{A}|\hat{\boldsymbol{\xi}}\rangle \equiv \int \hat{\boldsymbol{\xi}}^* \cdot \boldsymbol{A}\cdot\hat{\boldsymbol{\xi}}dV$. Multiplying $\hat{\boldsymbol{\xi}}^*$ from the left of (8.77) and integrating over the volume, we have following cubic equation for the eigenvalue λ.

$$\delta K\lambda^2 + 2\delta W_c\lambda + \delta W_p = 0 \tag{8.78}$$

$$\delta K = \langle\hat{\boldsymbol{\xi}}|\rho|\hat{\boldsymbol{\xi}}\rangle \ , \ \delta W_c = \langle\hat{\boldsymbol{\xi}}|\rho(\boldsymbol{u}\cdot\nabla)|\hat{\boldsymbol{\xi}}\rangle \ , \ \delta W_p = -\langle\hat{\boldsymbol{\xi}}|\boldsymbol{F}(\hat{\boldsymbol{\xi}})\rangle$$

For known eigenvector $\hat{\boldsymbol{\xi}}$, we can obtain the eigenvalue as:

$$\lambda = \frac{-\delta W_c \pm \sqrt{\delta W_c^2 - \delta K\delta W_p}}{\delta K} \tag{8.79}$$

Since the operator ρ and $\boldsymbol{F}$ are Hermitian and $\rho(\boldsymbol{u}\cdot\nabla)$ is anti-Hermitian, δK and δW_p are real and δW_c is imaginary. We have unstable modes in case $\delta W_c^2 - \delta K\delta W_p \geq 0$, and stable modes in case $\delta W_c^2 - \delta K\delta W_p < 0$.

$$in\omega = -\frac{\delta W_c}{\delta K}\gamma = \pm\frac{\sqrt{\delta W_c^2 - \delta K\delta W_p}}{\delta K} \ (\text{ if } \delta W_c^2 - \delta K\delta W_p \geq 0\ ) \tag{8.80}$$

$$in\omega = \frac{-\delta W_c \pm \sqrt{\delta W_c^2 - \delta K\delta W_p}}{\delta K}, \gamma = 0\ (\text{ if } \delta W_c^2 - \delta K\delta W_p < 0\ ) \tag{8.81}$$

(continued)

Consider the unstable case. From (8.80), we have:

$$in\omega = -\delta W_c/\delta K\,,\ \gamma^2 = -n^2\omega^2 - \delta W_p/\delta K \tag{8.82}$$

Exercise. Show following relation used in deriving Frieman-Rotenberg equation. $-(\boldsymbol{u}\cdot\nabla)\boldsymbol{R}-(\boldsymbol{R}\cdot\nabla)\boldsymbol{u} = \boldsymbol{\xi}\cdot\nabla[(\boldsymbol{u}\cdot\nabla)\boldsymbol{u}]-\boldsymbol{u}\cdot\nabla[(\boldsymbol{u}\cdot\nabla)\boldsymbol{\xi}]$, where $\boldsymbol{R} = (\boldsymbol{u}\cdot\nabla)\boldsymbol{\xi}-(\boldsymbol{\xi}\cdot\nabla)\boldsymbol{u}$

Answer :
$-(\boldsymbol{u}\cdot\nabla)\boldsymbol{R}-(\boldsymbol{R}\cdot\nabla)\boldsymbol{u} = -u_i\partial_i[u_j\partial_j\xi_k - \xi_j\partial_j u_k] - [u_i\partial_i\xi_j - \xi_i\partial_i u_j]\partial_j u_k$
$= \xi_i\partial_i(u_j\partial_j u_k) - u_i\partial_i(u_j\partial_j\xi_k) + u_i\partial_i(\xi_j\partial_j u_k) - u_i(\partial_i\xi_j)(\partial_j u_k) - \xi_i u_j\partial_i\partial_j u_k$
$= \xi_i\partial_i(u_j\partial_j u_k) - u_i\partial_i(u_j\partial_j\xi_k) = \boldsymbol{\xi}\cdot\nabla[(\boldsymbol{u}\cdot\nabla)\boldsymbol{u}] - \boldsymbol{u}\cdot\nabla[(\boldsymbol{u}\cdot\nabla)\boldsymbol{\xi}]$

8.4 Ideal Localized Modes

8.4.1 Edge Localized Modes

The steady state tokamak reactor will operate at high q (safety factor) and high β_p (poloidal beta) regime. Exploration of small or minute ELM in this regime is crucial for the feasibility of divertor power handling. The high pressure gradient at ETB produces Edge Localized Modes (ELM) as shown in Fig 8.8a. Urano [781] showed that the inter ELM transport is close to ion neoclassical transport.

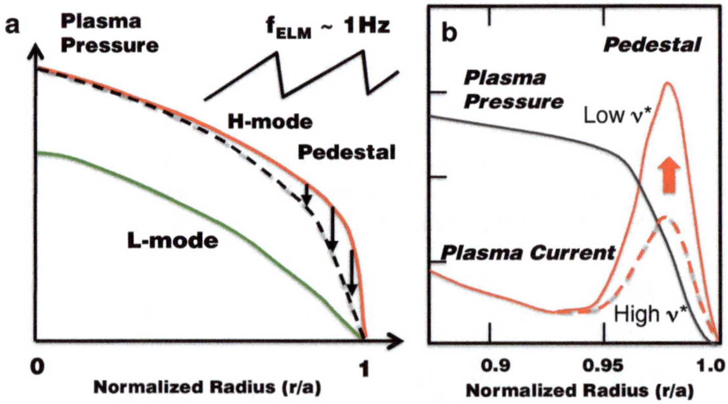

Fig. 8.8 (**a**) Schematics of L-mode and H-mode pressure profiles and the pressure collapse near the edge region due to ELM (Edge Localized Mode). (**b**) Edge pedestal and the build up of the edge bootstrap current with lowering edge collisionality

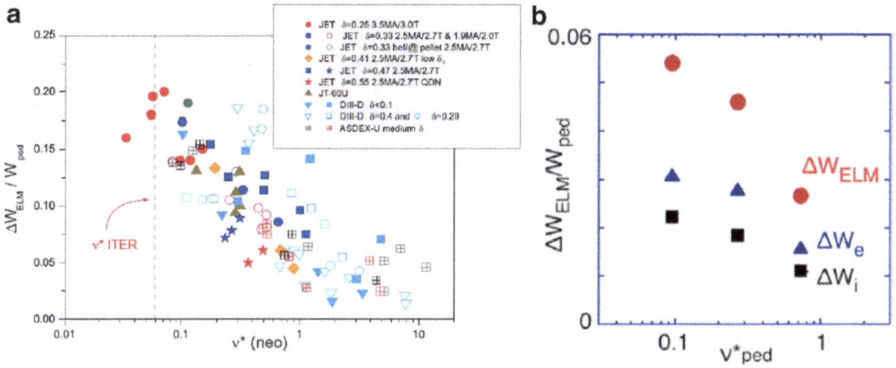

Fig. 8.9 (**a**) ELM energy loss ΔW_{ELM} over the pedestal stored energy W_{ped} as a function of pedestal collisionality from various tokamaks. Reproduced with permission from Loarte [524]. Copyright IOP Publishing. (**b**) Integrated simulation of ELM energy fraction qualitatively in agreement with the experiment. Reproduced with permission from Hayashi [307]. Copyright IAEA Vienna

Experimentally, the ELM energy loss ΔW_{ELM} increases with decreasing edge collisionality as reported by Loarte in 2003 as shown in Fig. 8.9a [524]. The increase of ELM energy loss at low collisionality is attributed to the increase of the edge bootstrap current (Fig. 8.8b). The increase of the surface current (i.e. edge bootstrap current) destabilize lower n (toroidal mode number) peeling mode, which has larger radial extent as shown by Hayashi [307] through the integrated simulation of 1.5D transport equations with MARG2D calculation shown in Fig. 8.9b.

This ELM energy loss ΔW_{ELM} over the pedestal plasma stored energy (energy between green and red line in Fig. 8.8a) depends on the edge collisionality ν^* [524]. Edge plasma of ITER and beyond is collisionless and we may expect $\Delta W_{ELM}/W_{ped} = 0.2$ for the edge collisionality expected in ITER. If $\Delta W_{ELM}/W_{ped} = 0.2$ happens for the ITER, $\Delta W_{ELM} \sim 21$ MJ for $W_{ped} = 0.3 W_{th}$, $W_{th} = 350$ MJ. The effect of this ELM energy loss can be imagined since this energy loss is comparable to the total plasma energy of large tokamaks such as JT-60U [458]. For a ELM interaction surface area 3 m^2, ITER ELM energy density is $\sim$7 MJ/m^2. Such a high transient energy deposition is quite damaging for divertor plate and reduction of ELM energy loss to $\sim$0.5 MJ/m^2 is envisaged to reach $\sim$300 full power discharges having $\sim 10^5$ ELMs within a divertor lifetime. This requires 14 times reduction. Assuming ELM frequency $\sim 1 Hz$, DEMO operating for 300 days has 25,920,000 ELMs. Development of DEMO relevant ELM mitigation techniques is still a horizon after the ITER success.

Since this ELM heat flux may become a serious problem to shorten divertor lifetime of ITER [302], mitigation of ELM becomes an important issue for the reliable operation of ITER. Evans [190], [193] reported successful mitigation of Type I ELM in ITER like discharge with reduced pressure gradient by the Resonant Magnetic Perturbation (RMP) technique as shown in Fig. 8.10. Snyder

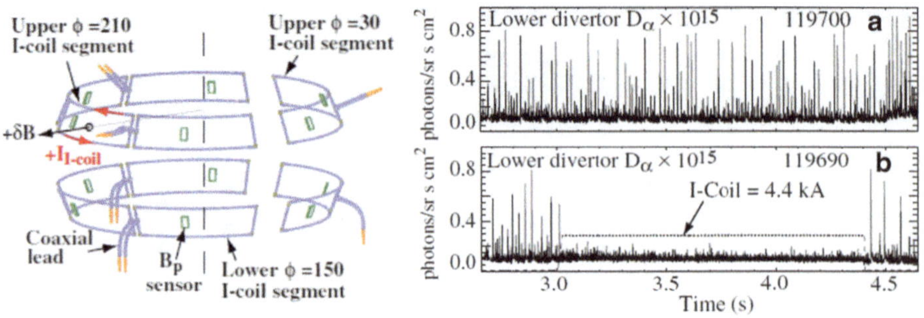

Fig. 8.10 Suppression of ELM by the Magnetic Perturbation. Reproduced with permission from Evans [190]. Copyright IAEA Vienna

[704] explained this suppression of Type-I ELM is due to the lowering the pressure gradient below peeling/ballooning limit through field line stochastization. While Evans [190] uses $n = 3$ magnetic perturbation, ELM suppression using lower n is actively pursued in many tokamaks. Liang [507] showed effectiveness of $n = 1$ in mitigating ELM in JET and Jeon [401] showed the complete suppression of ELM using $n = 1$ MP coils in KSTAR. Suttrop [727] also reported stabilization of ELM by using $n = 2$ in ASDEX-U, respectively. J-TEXT recently rebuilt in Huazhong University of Science and Technology (HUST) [863] also investigates effect of RMP on MHD and also particle transport [352].

The RMP application for H-mode ELM suppression in the divertor tokamak JFT-2M is first presented by Mori [559] and the response to the external rotating RMP is reported by Oasa [583]. Excitation of edge bootstrap current in collisionless plasma (see Fig. 8.8b) is problematic for the edge MHD stability. Active reduction of edge bootstrap current is an important subject for future tokamak research. To improve edge stability, we have to control edge bootstrap current which might be possible by methods proposed by Fisch [203] and Helander [316, 318].

While ELM suppression by RMP is successful, the non axisymmetric perturbation can have enhanced effect on the X-point. The magnetic separatrix in the tokamaks is a relatively fragile structure that is easily split into a pair of intersecting invariant manifolds by small non-axisymmetric magnetic perturbations, called the homoclinic tangles first implied DIII-D VH mode [188] and further discussed by Evans (2004: Homoclinic tangles, bifurcations and edge stochasticity in diverted tokamaks) [189], and the experimental signatures are also discussed by Evans [191]. Wingen (2009) analyzed the existence of laminar flux tubes with short connection lengths numerically. Stable (dashed line) and unstable (solid line) separatrix manifolds are shown in Fig. 8.11b. Similar to the homoclinic tangles in dynamical system, the manifolds start oscillating as they approach the X-point. It is shown that the unstable manifolds intersects with the inner divertor target and the stable manifolds intersect with the outer divertor target. Kirk [456] observed lobes of the homoclinic tangles near the X-point in the RMP experiments in MAST.

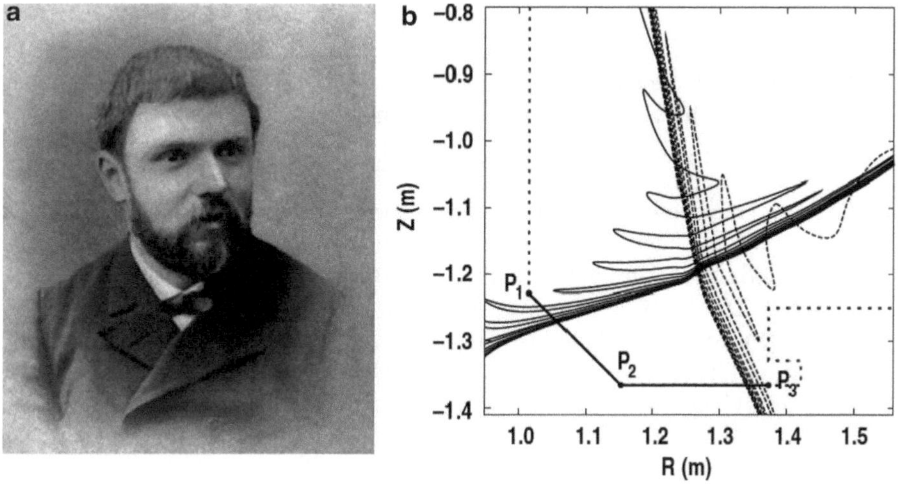

Fig. 8.11 (**a**) French Mathematician H. Poincare (1854–1912). (**b**) Homoclinic tangles near the separatrix in DIII-D by the application of the non-axisymmetric field. Reproduced with permission from Wingen [825]. Copyright IAEA Vienna

Salon 8-1: H. Poincare and Homoclinic Tangles
From the magnetic field line Hamilton equation (2.4), we understand that separatrix line for a single null is a homoclinic orbit. If we start from X-point, the time ("time" is integrated toroidal angle for the magnetic field line Hamilton mechanics) required to return to X-point is infinity and will stop there since the poloidal field at the X-point is zero. The concept of homoclinic orbit and homoclinic tangle originates from H. Poincare's textbook on the Celestial Mechanics [613]. See Lichtenberg-Lieberman [508] for details.

Salon 8-2: Nuclear Fusion Prize for MHD Stability
Nuclear Fusion journal is the acknowledged world-leading journal which publishes significant original works in the field of fusion research. IAEA awarding excellent 10 papers as Nuclear Fusion Prize nominees and among which one paper is selected to be a winner of Nuclear Fusion prize. The top author of Nuclear Fusion Prize papers in MHD stability are T. Evavs [190] on ELM suppression by resonant magnetic perturbation, S. Sabbagh [645] on the resistive wall stabilization in rotating high β plasma, P.B. Snyder [707] for pedestal theory. Steve Sabbagh (see Salon 1-1) is an adjunct professor of

(continued)

Columbia university and senior research scientist at PPPL. He is an expert in MHD and confinement but also NTV studies [860] as well. Phillip Benjamin Snyder started his career from the study of the kinetic ballooning mode based on the gyrofluid theory and simulation [701, 702] and later concentrated on ELM and peeling/ballooning mode [703, 704, 706, 707]. Combination of KBM and the peeling/ballooning mode theory gives rise to the EPED model.

In 2008, the paper led by Todd Evans concerning the ELM suppression is selected. Below from left is Prof. W. Burkart (IAEA Deputy Director General), Dr. T. Evans (General Atomics), and M. Kikuchi (Chairman of Nuclear Fusion board of editors). He has made the experimental confirmation of GAE [187], and series of RMP studies [188, 189, 193]. He is also an expert on Chaos, Complexity and Transport including the Homoclinic tangle [192].

8.4.2 Ballooning and Peeling Modes

Ballooning mode is a pressure driven local ideal MHD mode with long wave length along $\boldsymbol{B}$ and short wave length perpendicular to $\boldsymbol{B}$ [141] (see Column 8-5 for details if not familiar).

Peeling mode is an external mode localized near the plasma edge driven by the finite edge current [536]. This mode can be coupled to the ballooning mode and thought to be a cause of ELM in tokamak.

The growth rate of medium n peeling mode is rather low compared with violent low n kink instability and care must be taken to identify stability boundary as discussed in Sect. 5.1.2 and special codes are developed such as MARG2D [4, 770] and ELITE [824]. While ELITE code is specifically developed for $n \geq 5$ using $1/n$ expansion, MARG2D code can analyze stability boundary for any toroidal mode number n efficiently, due to its special eigenvalue treatment. Effect of toroidal flow is also implemented in MARG2D code using the Frieman-Rotenberg equation with self-consistent flow equilibrium solution.

The stability of peeling mode is well characterized by the pedestal current density j_{ped} and the normalized pressure gradient $\alpha = -2\mu_0 q^2 R(dP/dr)/B^2$. Figure 8.12 shows stability of ideal MHD modes ($n = 1, 2, 3 - -, 18$) for low q ($q_{95} = 3.7$) and high q ($q_{95} = 7.3$) operation in (α, j_{ped}) diagram including the effect of toroidal plasma rotation [5]. The stable region against ideal MHD modes in high q regime is much wider than in low q ($q_{95} \sim 3$) regime. The edge ballooning mode becomes stabilized with small $j_{ped} > 0.1$ in high q operation, while not in low q operation. Finite n ($n = 3 \sim 18$) peeling-ballooning modes become most unstable when α reaches critical value ($\alpha_c \sim 6 - 8$) for $j_{ped} \sim 0.4 - 0.7$. When the pedestal current density $j_{ped} > 0.8$, plasma becomes unstable to $n = 1$ mode. Excitation of large edge bootstrap current in collisionless edge plasma with $j_{ped} > 0.8$ will lead to $n = 1$ or low n global mode (depending on the proximity of stabilizing wall) and control of edge bootstrap current to the stable region is important.

Kamada [412] identified ELM regimes of Type I and grassy ELM (or Type II ELM) by varying plasma triangularity and safety factor as shown in Fig. 8.13a. Grassy ELM has been found at high $q_{95} = 6$ and high triangularity $\delta = 0.5$ with

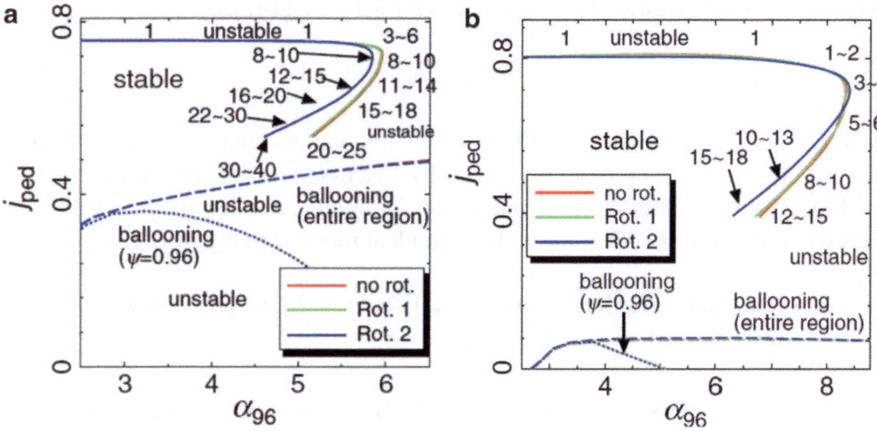

Fig. 8.12 (**a**) Stability diagram of the low-q ($q_{95} = 3.7$) equilibrium in (j_{ped}, α_{96}) plane. (**b**) Stability diagram of the high-q ($q_{95} = 7.3$) equilibria in (j_{ped}, α_{96}) plane. Both calculation include effect of toroidal rotation by solving the Frieman-Rotenberg equation, in which toroidal rotation weakly destabilize the peeling-ballooning modes. Reproduced with permission from Aiba [5]. Copyright IAEA Vienna

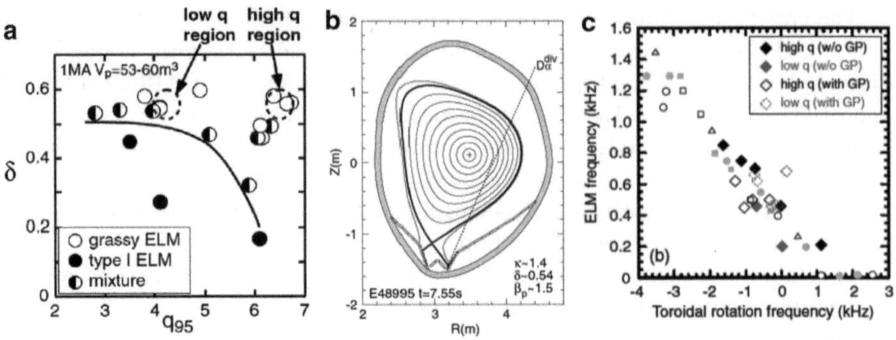

Fig. 8.13 (**a**) Grassy and Type I ELM regimes in (q, δ) plane. (**b**) Typical equilibrium shape with grassy ELM in JT-60U. (**c**) Effect of toroidal rotation on ELM frequency. Reproduced with permission from Oyama [595]. Copyright IAEA Vienna

low vertical elongation as shown in Fig. 8.13b. Significant modification of ELM characteristics by co and counter toroidal rotation is also reported by Oyama [593] as shown in Fig. 8.13c. With more counter toroidal rotation, discharges exhibit high frequency grassy ELMs with comparable pedestal pressure with Type I ELMs.

It seems that kinetic effect is important to explain these grassy ELM characteristics since the growth rate is small in such a localized mode to reach diamagnetic drift frequency. It is also true that co and counter asymmetry will not appear in the ideal MHD. Azumi [33] in 1980 shows that diamagnetic drift or finite Larmor radius effect becomes important for the ballooning modes and $-\gamma^2$ has to be replaced to $\omega(\omega + \omega^*)$. The ω_{*i} effect implies $\gamma < \omega_{*i}/2$ may be stabilized. He also included WKBJ solution by Dewar [156] to give the growth rate $\gamma(n)$ as:

$$\gamma^2(n) = \gamma_\infty^2 - \gamma_{fn}^2 - \omega_{*i}^2/4 \tag{8.83}$$

Here $\gamma_{fn}^2 \sim n^{-1}$ and $\omega_{*i}^2/4 \sim n^2$ and the growth rate is maximum at intermediate n. The ω_{*i} effect, i.e. $\omega^2 \to \omega(\omega + \omega_{*i})$, is also discussed in 1982 by Tang [750].

Hastie [296] includes the sound wave for high n modes to give following dispersion relation when it is unstable to an ideal mode with a growth rate γ_I.

$$-\omega(\omega - \omega_{*i})[(1 + 2q^2)\omega_s^2 - \omega(\omega - \omega_{*e})] = (1 + 2q^2)\gamma_I^2[\omega_s^2 - \omega(\omega - \omega_{*e})] \tag{8.84}$$

where q is the safety factor, $\omega_s = c_s/Rq$ is a sound wave frequency for the connection length of ballooning mode, $c_s^2 = (T_e + T_i)/m_i$, $\omega_{*e} = (n/en_e)(dp_e/d\psi)$ while n is the toroidal mode number.

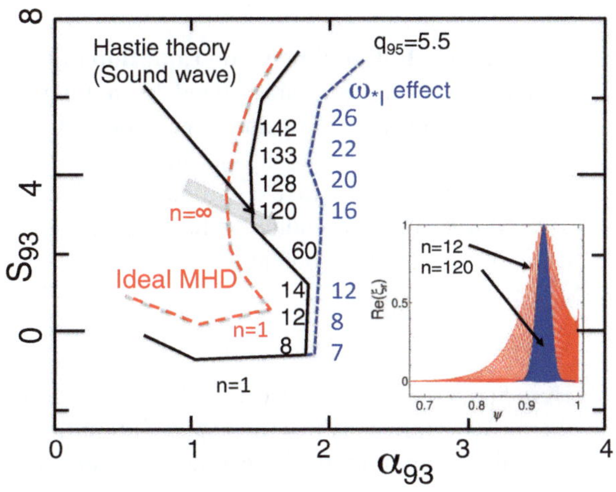

Fig. 8.14 Stability diagram in (α_{93}, s_{93}) plane for ideal (*red broken line*), ω_* corrected (*blue dashed line*), Hastie theory (*black line*). *Arrow* shows trace of the experiment. Modified from Aiba [8]

Aiba [8] discussed the physics behind the grassy ELM of JT-60U [412] using this Hastie formulation. Figure 8.14 shows the stability diagram for the peeling/ballooning modes corresponding to the JT-60U grassy ELM plasma condition. Experimental trace is given by the transparent arrow. Red broken line gives the stability boundary of ideal MHD modes, the blue broken line gives that with ω_{*i} effect, and the black line gives that for Hastie theory. Sound wave modification on the stability boundary is important for higher n modes but has small effect on low n modes. He showed grassy ELM may be due to the high n ballooning modes with $n \sim 120$. While $n \sim 12$ mode has peeling character (eigenmode is not zero at the surface), $n = 120$ mode has ballooning character (eigenmode is zero at the surface). Understanding of the physics behind the different ELM behavior for co and counter toroidal rotations is still an open question.

While we can obtain much higher beta (or α) by increasing the elongation, this leads to the situation that beta is limited by the low n ideal modes through the second stability access with high edge bootstrap current. In such case, ELM becomes a violent instability in reactor relevant plasmas with low toroidal mode number n.

There are another type of H-mode called Quiescent H-mode (QH mode) observed in DIII-D tokamak [86–88], [89]. This mode is ELM free but has small MHD activity called the "Edge Harmonic Oscillation" (EHO). New QH mode with high frequency EHO is observed in EAST tokamak [809]. SMBI (supersonic molecular beam injection) can also mitigate ELM [836].

The nonlinear behavior of ELM is an active field of research. Kirk [455] observed filament structures persisting ~200 μs during ELM in the MAST. Zhu [861] identified three major nonlinear effects (radial convection, line vending, and magneto sonic coupling) involved in the ballooning filament. Yun [849] observed detailed growth and bursts of ELM filaments using 2 dimensional ECE imaging system in the KSTAR tokamak. Nonlinear NIMROD calculation of ELM filament shows stabilizing influence of edge bootstrap current by Zhu [862].

Column 8-5: Ballooning Mode Equation [443]

Ballooning mode is characterized by a long wave length parallel to the magnetic field $\lambda_{\parallel} \sim Rq$ and very short wave length perpendicular to the magnetic field as discussed in Chap. 6. The expression of the equilibrium magnetic field in the flux coordinates is given by $\boldsymbol{B} = \nabla\alpha \times \nabla\psi$, where $\alpha = \zeta - q\theta$. The perpendicular displacement $\boldsymbol{\xi}_{\perp}$ is given by the eikonal form using the electrostatic potential φ as:

$$\boldsymbol{\xi}_{\perp} = \frac{\boldsymbol{B} \times \nabla\varphi}{B^2}, \; \varphi(r,\theta,\zeta) \propto e^{iS(r,\alpha)} \tag{8.85}$$

Here $S(r,\alpha) = -n(\alpha + \alpha_0(r))$ is the eikonal and the radius r is defined using the toroidal flux ϕ as $r = a(\phi/\phi_a)^{1/2}$. Solution to satisfy periodicity in θ is the infinite summation of quasi-modes given by Eq. (6.10).

$$\varphi(r,\theta,\zeta) = \sum_{j=-\infty}^{\infty} u(r,\theta + 2\pi j)e^{-in(\alpha+q\theta_0+2\pi qj)} \equiv F(r,\theta)e^{-in\alpha} \tag{8.86}$$

Here, $F(r,\theta) = \sum u(r,\theta + 2\pi j)e^{inq(\theta+2\pi j-\theta_0)} = \sum F_1(\theta + 2\pi j)$.

For the Ballooning type localized perturbation, the energy integral δW is dominated by the shear Alfven term δW_{SA} and the interchange term δW_{IC} defined in Column 8-1. Those energy integrals can be given in terms of a slowly varying function $F(r,\theta)$ and the eikonal α as:

$$\delta W = \frac{1}{2\mu_0}\int\left[\frac{|\nabla\alpha|^2}{B^2}(\boldsymbol{b}\cdot\nabla F)^2 - 2\mu_0 p'(\psi)\kappa_w F^2\right]dV \tag{8.87}$$

Here, $\kappa_w = (\boldsymbol{b} \times \nabla\alpha)\cdot\boldsymbol{\kappa}/B$ is weird component of the curvature. Using the formula $\boldsymbol{B}\cdot\nabla = J^{-1}\partial/\partial\theta$ in the Clebsch coordinates (ψ,θ,α), the Euler-Lagrange equation to minimize this energy integral is given by,

$$\frac{1}{J}\frac{\partial}{\partial\theta}\left[\frac{|\nabla\alpha|^2}{JB^2}\frac{\partial F}{\partial\theta}\right] + \mu_0 p'(\psi)\kappa_w F = 0 \tag{8.88}$$

This is the Ballooning mode equation for marginal stability. Since (8.88) is linear equation, the solution can be constructed from the solution of following equation in the covering space $(-\infty, +\infty)$.

$$\frac{1}{J}\frac{\partial}{\partial\theta}\left[\frac{|\nabla\alpha|^2}{JB^2}\frac{\partial F_1}{\partial\theta}\right] + \mu_0 p'(\psi)\kappa_w F_1 = 0 \tag{8.89}$$

8.4.3 Infernal Mode and Barrier Localized Mode

8.4.3.1 Infernal Modes

Advanced tokamak operation in WS and NS plasmas are associated with core improved confinement. Large pressure gradients in core weak shear regime in WS and pitch minimum regime in NS leads to the loss of magnetic shear stabilization of pressure driven ideal MHD modes. In these regimes, both $n = 1$ and intermediate n modes become most unstable in some cases.

Ballooning mode theory including finite n correction has been developed using WKBJ formalism by Dewar [156] (equation (6.32)) and agreed with PEST calculation down to $n = 5$ as shown by Dewar-Manickam [157]. This theory predicts that the largest n modes are most unstable within the ballooning formalism. But it also showed interesting oscillatory behavior of beta limit in the intermediate n regime as a function of $1/n$ if it is treated as continuous variable as shown in Fig. 8.15a.

Hastie-Taylor [295] showed this oscillatory behavior can be expected by the breakdown of radial dense-coupled ballooning mode structure. If the magnetic shear is finite, radial coupling of various resonant MHD modes $(m, m \pm 1, m \pm 2, m \pm 3, --)$ becomes strong for high n modes since radial separation between modes is small. However, if the magnetic shear is very low, $s \sim 0$, radial mode separation becomes larger and the Connor's standard ballooning mode theory [141] based on dense radial mode coupling breaks down.

Manickam [535] showed that when the magnetic shear is sufficiently weak, this oscillation can result in bands of unstable n values, where the ballooning mode theory predicts complete stability. This mode is called the **infernal mode**.

Improved core confinement regimes in TFTR [301] and JT-60U [458] are associated with fast internal disruption or major disruption, where the beta limit

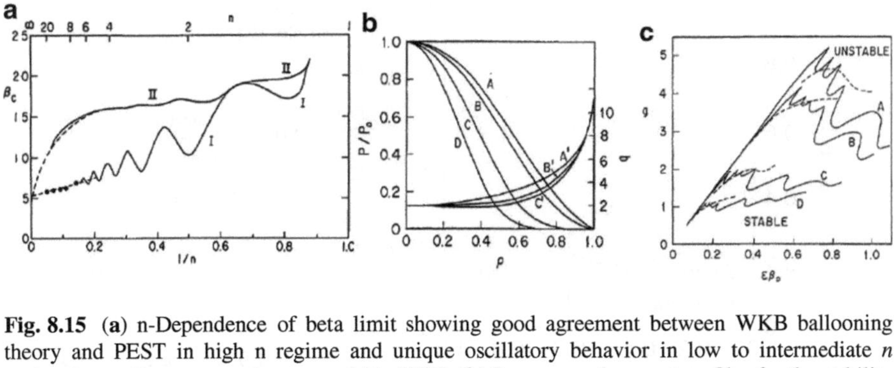

Fig. 8.15 (**a**) n-Dependence of beta limit showing good agreement between WKB ballooning theory and PEST in high n regime and unique oscillatory behavior in low to intermediate n regime by treating n as continuous variable [157]. (**b**) Pressure and current profiles for the stability calculation. A: $dp/d\psi \sim 0.3(1-\psi)^{0.5} + 0.7$, to C: $dp/d\psi \sim (1-\psi)^2$, D: $dp/d\psi \sim (1-\psi)^5$. (**c**) Stability boundaries in ($\epsilon\beta_p$, g), where g is normalized beta, β_N. Oscillating region in c shows region of observed β_p collapses in high β_p experiments in JT-60U [598]. Reproduced with permission from Dewar [157] and Ozeki [598]. Copyright IAEA Vienna

deviates from Troyon scaling and limited by $\epsilon\beta_p$ [546] called β_p collapse [376] . Ideal MHD stability of core improved confinement has been analyzed by Ozeki-Azumi et al. [597, 598]. For the current and pressure profiles in Fig. 8.15b, c shows stability diagram in ($\epsilon\beta_p$, β_N) for various pressure profiles (A-D) in WS plasma (magnetic shear $\sim$ 0 in the central region) without wall stabilization and slightly high internal inductance $l_i = 1.2$ with $q_s/q_0 = 4$ given in Fig. 8.15b [598]. Intensive central heating produces peaked pressure profile and observed β_p collapses is consistent with region of $\epsilon\beta_p = 0.2 - 0.6$ for pressure profile C. Region of $\epsilon\beta_p = 0.2 - 0.4$ is characterized with low q_0 and intermediate $n = 2 - 5$ infernal modes becomes most unstable. While broader pressure profile give rise to much higher stable $\beta_N \sim 5(l_i \sim 1.2)$ without wall stabilization [598], or $\beta_N \sim 5.8(l_i \sim 1)$ with moderately far wall $r_{wall}/a = 1.5$ by Howl [350], resultant broader bootstrap current seems not consistent with moderately high internal inductance for the Steady State Tokamak operation.

Ozeki-Azumi et al. [597] analyzed ideal MHD stability of weakly negative shear (WNS) plasma in circular plasma with $R/a = 3$ showing that significant improvement in $\epsilon\beta_p$ limit can be obtained by modification of pressure profile as shown in Fig. 8.16. Pressure profile (B) has much higher $\epsilon\beta_p$ limit than that for pressure profile (A). While experimentally achievable $\epsilon\beta_p$ is $\sim$ 0.6, tayloring of pressure profile in negative shear regime and reduction of pressure gradient near q_{min} improves stability against low n = 1–5 internal modes. Again, importance of relative location of ITB and q_{min} radius is stressed.

Improved confinement regimes in NS plasma has been explored since their discoveries by Strait in DIII-D [720], by Levinton in TFTR [502] and by Fujita in JT-60U [224] as well as early experiments in JET [354]. The NS plasma terminated mostly with ideal $n = 1$ kink-ballooning mode and sometimes with infernal mode

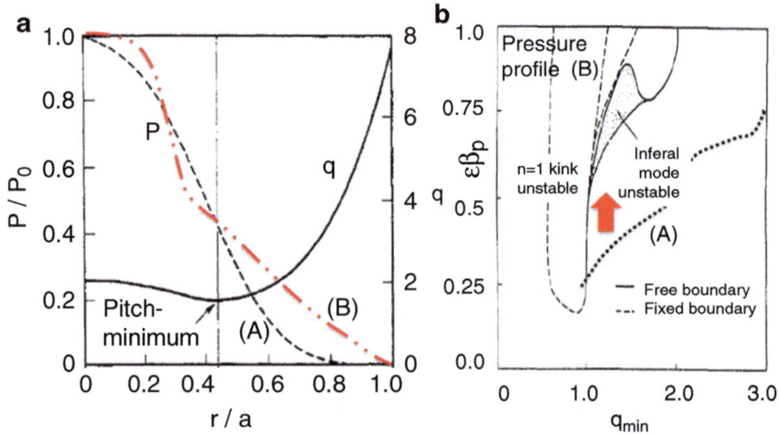

Fig. 8.16 (**a**) Pressure profiles A and B in weak negative shear plasma and (**b**) $\epsilon\beta_p$ stability improvement from A to B in (q_{min}, $\epsilon\beta_p$). Modified from Ozeki [597]

near the ideal beta limit as discussed by Taylor [759] or Ishii [380]. But it sometimes terminated with precursor oscillation possibly due to Rutherford regime in DTM as discussed in Sect. 8.5.3.

8.4.3.2 Barrier Localized Mode

Barrier localized mode (BLM) is an ideal MHD instability driven by the large local pressure gradient at the ITB, in both WS [463, 739] and NS [538] plasmas and leads to relaxation of ITB without major disruption. Figure 8.17 shows time slices of WS discharge profile in JT-60U [463]. After the formation of ITB, BLM becomes unstable and the ITB relaxation led to the formation of H-mode edge. This BLM is estimated as $m/n = 3/1$ associated with steep ITB near $q = 3$ surface. In case ITB is recoverable after BLM, quasi steady-state ITB is possible but may end with termination of ITB as shown by Koide [464].

On the appearance of $m/n = 3/1$ mode, Takeji showed that $n = 1$ mode can be most unstable under the situation that the bootstrap current driven by the steep pressure gradient reduces the local magnetic shear ($s \sim 0$), while $n = 2$ mode is most unstable if the local shear is not weak [739]. While intermediate n peeling modes are responsible to ELM, $n = 1$ or $n = 2$ semi global modes are responsible for the BLM. BLM in NS plasma is analyzed by Manickam [538] for both JT-60U and TFTR. JT-60U achieved wider ITB radius and steeper ITB is formed for balanced or perpendicular neutral beam injection and more frequently observed in JT-60U than other tokamaks such as TFTR. Softening of BLM and long sustainment of quasi steady-state improved confinement with ITB requires further investigation.

Integrated demonstration of fully non-inductive operation with large bootstrap current fraction (f_{BS}) is primary importance to realize steady state tokamak reactor. The pulse length of f_{BS} discharge is still limited [652] as shown in Fig. 8.18a. Sustainment at reactor relevant q_{95} regime (Fig. 8.20b) is faced by ITB induced disruption [654] and further understanding of ITB transport physics is required.

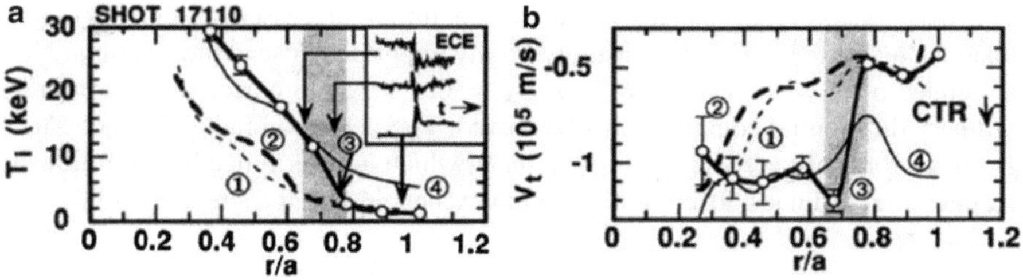

Fig. 8.17 (**a**) Relaxation of the ITB associated with the barrier localized mode (BLM) shown by the ECE temperature perturbation. (**b**) Relaxation of toroidal rotation dip associated with BLM. Reproduced with permission from Koide [463]. Copyright American Physical Society

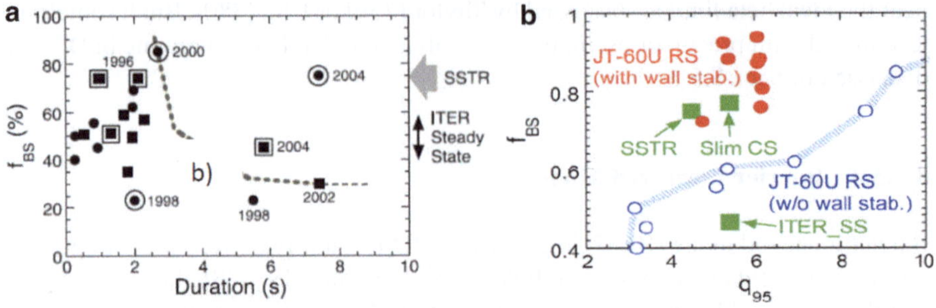

Fig. 8.18 (**a**) Extended high f_{BS} discharge duration by rotation control [652]. (**b**) Attainment of reactor relevant moderate q_{95} high f_{BS} discharge with wall stabilization shown in *red symbol*, which is terminated by disruption [654]. Reproduced with permission from Sakamoto [652, 654]. Copyright IAEA Vienna

8.5 Resistive MHD Modes

8.5.1 Classical Tearing Mode

The tearing mode (TM) is an important resistive instability associated with the reconnection of magnetic field at the resonant rational surface $r = r_s$. This mode is de-stabilized by changing the topology of the magnetic field, while it is stable within the ideal MHD context. Furth-Killeen-Rosenbluth [232] analyzes the resistive instabilities in a sheet pinch and gives the growth rate of the tearing mode as $\gamma \sim \eta^{3/5}$ (or $\gamma \sim S^{-3/5}$, where $S = \mu_0 a V_A/\eta = \tau_R/\tau_A$ is Lundquist number, $\tau_R = \mu_0 a^2/\eta$ resistive diffusion time, $\tau_A = a/V_A$ Alfven transit time, and $V_A = B/\sqrt{\mu_0 n_i m_i}$ is Alfven velocity).

8.5.1.1 Linear Tearing Modes

The linear stability of the tearing mode is further discussed in the cylindrical tokamak by Furth [234]. The resistivity is only important close to the singular radius r_s defined by $F = \boldsymbol{k} \cdot \boldsymbol{B} = k_z B_z + k_\theta B_\theta = 0$, where more accurate fourth order ordinary differential equation must be solved. The perturbed radial magnetic field B_{r1} has a form of $i\psi(r)exp(\gamma t + ik_z + im\theta)$, which is determined by the following second order ordinary differential equation except close to the singular surface.

$$\frac{d}{dr}\left(H\frac{d\psi}{dr}\right) - \left[\frac{g}{F^2} + F^{-1}\frac{d}{dr}\left(H\frac{dF}{dr}\right)\right]\psi = 0 \tag{8.90}$$

$$g = \frac{(m^2-1)F^2}{r(k_z^2+k_\theta^2)} + \frac{k_z^2}{k_z^2+k_\theta^2}\left(2\mu_0\frac{dP}{dr} + rF^2 + \frac{2F(k_zB_z-k_\theta B_\theta)}{r(k_z^2+k_\theta^2)}\right), \tag{8.91}$$

where $k_\theta = m/r$, $k_z = -n/R$, $H = r/(k_z^2 + k_\theta^2)$. If the r_s falls within the plasma ($r_s < a$), the marginally stable kink mode with a boundary condition solution $\psi(r_s) = 0$ is a solution of the equation (8.90). Since $B_{r1} \sim F(r)\xi_r$, ξ_r is finite at the singular surface. In the case of the tearing mode, on the other hand, ξ_{r1} tends to diverge as $r \to r_s$ without including more exact fourth order finite resistivity equation. while $\xi_r \to \infty$, $\psi_{r_s} \sim B_{r1}$ remain finite. We consider the current driven tearing instability for the simplest case of zero pressure gradient, $dP/dr \equiv 0$. To determine tearing-mode stability, we solve (8.90) to obtain $\psi_1(r)$ at $0 < r < r_s$ with $\psi_1 \to r^{m-1}$ at $r = 0$, and $\psi_2(r)$ at $r_s < r < b$ with $\psi_2(r_s) = \psi_1(r_s)$ and $\psi_2(b) = 0$ at ideal wall at $r = b$. While ψ is continuous across the singular radius r_s, $d\psi/dr$ has jump across the r_s. An important quantity Δ' is defined for this jump.

$$\Delta' \equiv d(\psi_2 - \psi_1)/dr|_{r_s}/\psi(r_s) \tag{8.92}$$

Furth-Killeen-Rosenbluth [232] showed that the condition for the tearing mode instability is $\Delta' > 0$ by the finite resistivity analysis near the singular radius.

Near the singular radius r_s, (8.90) can be approximated as:

$$\frac{d^2\psi}{dx^2} - \frac{\kappa}{x}\psi = 0, \text{ where } \kappa = \left[\frac{g}{F} + \frac{d}{dr}\left(H\frac{dF}{dr}\right)\right]_{r_s} \frac{a}{HdF/dr|_{r_s}}, \tag{8.93}$$

where $x = (r - r_s)/a$. The point $x = 0$ is a regular singular point and its indicial equation is $\alpha^2 - \alpha = 0$. Thus the solution has analytic and logarithmic function as discussed in the section Column 8-2. Let $\psi_1(x)$ is solution in $x < 0$ ($r < r_s$) and $\psi_2(x)$ is a solution in $x > 0$ ($r > r_s$). They are given as follows.

$$\psi_1(x) = \psi(r_s)\phi_s(x) + A_1\phi_n(x), \psi_2(x) = \psi(r_s)\phi_s(x) + A_2\phi_n(x) \tag{8.94}$$

$$\phi_s(x) = 1 + \kappa x ln|x| + \frac{1}{2}\kappa^2 x^2 ln|x| - \frac{3}{4}\kappa^2 x^2 + -- \tag{8.95}$$

$$\phi_n(x) = x + \frac{1}{2}\kappa x^2 + \frac{1}{12}\kappa^2 x^3 + -- \tag{8.96}$$

Then, stability parameter for the tearing mode is given as $\Delta' = (A_2 - A_1)/a\psi(r_s)$.

8.5.1.2 Rutherford Regime

The linear phase of the tearing mode is very short for strongly magnetized high temperature plasma and the mode goes into the nonlinear region, called Rutherford regime. The growth of the perturbed magnetic field in the Rutherford regime follows $B_{r1} \sim \psi_1 \sim t^2$. This means magnetic island width w evolves as $w \sim t$. This nonlinear regime is called the Rutherford regime according to Rutherford's pioneering work [643] under constant ψ approximation (i.e. 1st order perturbed flux function $\psi_1(r_s) \neq 0$). The evolution equation is improved by Biskamp [68] as follows,

$$\frac{dw}{dt} = \frac{1.22\eta}{\mu_0}\Delta'(w), \text{ where } w = 4\sqrt{\psi_1(r_s)/\psi_0''(r_s)} \tag{8.97}$$

with $\Delta'(w) = [\psi_1'(r_s + w/2) - \psi_1'(r_s - w/2)]/\psi_1(r_s)$. For the derivation of above equation and the details of nonlinear tearing mode theory, see Biskamp [68].

Island structure of tearing mode is observed using soft X-ray tomography by Sauthoff [666] as shown in Fig. 8.19a and radial phase inversion of tearing mode B_θ is measured by Robinson-McGuire [633] as shown in Fig. 8.19b. This mode is important for all tokamak regimes including advanced tokamak operation.

Suppression of $m/n = 2/1$ tearing mode using direct island heating is shown for the first time in 1992 by Hoshino [342] in JFT-2M. Since island formation produces temperature flattening by the short-circuit effect of magnetic island, heating near the magnetic island, especially O-point, has a stabilizing effect as shown by Yoshioka [844] and shown in the reduced resistive MHD simulation by Kurita [478]. The heating effect on Rutherford equation is expressed by adding additional term to equation (8.97) as follows,

$$\frac{dw}{dt} = \frac{1.22\eta}{\mu_0}\Delta'(w) + C_2\frac{J_0}{\tilde{\psi}_m}\int_{-w/2}^{w/2}\tilde{\eta}_m dr, \tag{8.98}$$

where $C_2, J_0, \tilde{\eta}_m, \tilde{\psi}_m$ are constant of order 1, equilibrium current density, perturbed resistivity, and perturbed poloidal flux function, respectively. Since $\tilde{\eta}_m < 0$ with ECH, second term in RHS is stabilizing.

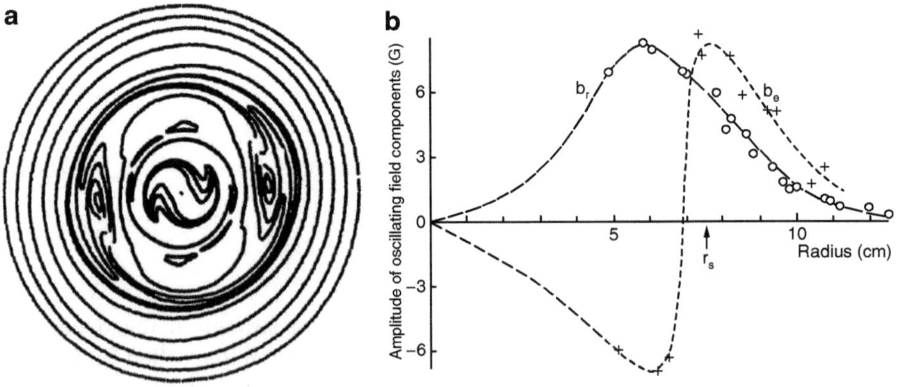

Fig. 8.19 (**a**) Tomographic reconstruction of $m/n = 2/1$ tearing mode [666]. (**b**) Eigenmode structures of tearing mode B_{r1} and $B_{\theta 1}$ [633]. Reproduced with permission from Sauthoff [666] and Robinson [633]. Copyright IAEA Vienna

8.5.2 Neoclassical Tearing Mode

Efficient steady-state operation of tokamak relies on the maximum utilization of the bootstrap current. But, this leads to an important new instability, called neoclassical tearing mode (NTM) [98].

Since the bootstrap current is proportional to pressure gradient ($J_{bs} \sim dP/dr$), local fattening of the pressure gradient near the magnetic island produced negative perturbed helical current, which enhances magnetic island and destabilize tearing mode in positive shear regime, while this negative perturbed helical current reduce magnetic island size and is stabilizing in negative shear regime [443]. Therefore, this NTM is important for all advanced tokamak operating regimes (WS, NS, CH) except regions of negative shear. While the destabilization of NTM by loss of bootstrap current is inversely proportional to island width ($1/w$), there is another important mechanism to stabilize NTM by the excitation of parallel current $J_{\|}$ due to NC polarization current $J_{\perp p}$ ($\nabla_{\|} J_{\|} + \nabla_{\perp} \cdot J_{\perp p} = 0$)) inversely proportional to cubic of island width ($1/w^3$) in both collisional regime [699] and collisionless regime [823]. This gives rise to the NTM is linearly stable but has some threshold island width above which NTM becomes unstable in case of positive shear. In case of negative shear, the parallel current $J_{\|}$ induced by the polarization current changes sign and stays stabilizing.

This NTM is observed for the first time by Chang-Callen in TFTR in 1995 [104] as shown in Fig. 8.20 and explained well by the Rutherford equation including bootsrap destabilization term but can not be explained without including bootstrap current destabilization term. The loss of bootstrap current inside the magnetic island is essential for the NTM and this loss of bootstrap current inside the magnetic island is confirmed using MSE diagnostics in JT-60U by Oikawa [587, 588].

Localized ECCD inside magnetic island can compensate lost bootstrap current to stabilize NTM. First complete NTM stabilization is demonstrated in ASDEX-U using 2nd harmonic X-mode ECCD by Gantenbein [238] and in JT-60U using fundamental O-mode ECCD by Isayama [374] as shown in Fig. 8.21a–c.

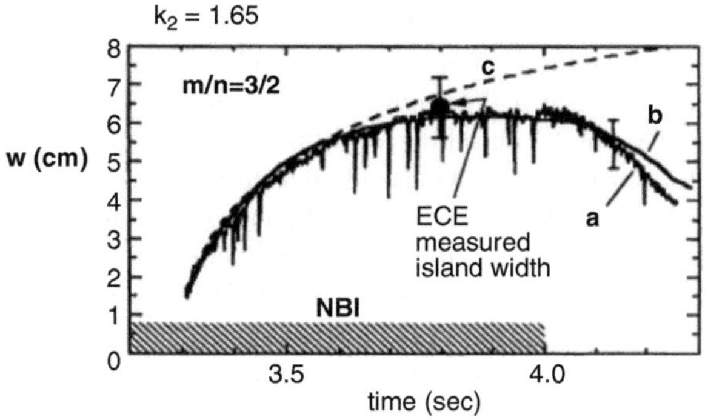

Fig. 8.20 Observation of $m/n = 3/2$ NTM in TFTR. *a*: Measured evolution of island width, *b*: calculated island width using modified Rutherford equation, *c*: calculated with fixed parameters ($\eta, w_c \equiv \epsilon^{1/2} L_q \beta_p / L_p, \Delta'$). Reproduced with permission from Chang [104]. Copyright American Physical Society

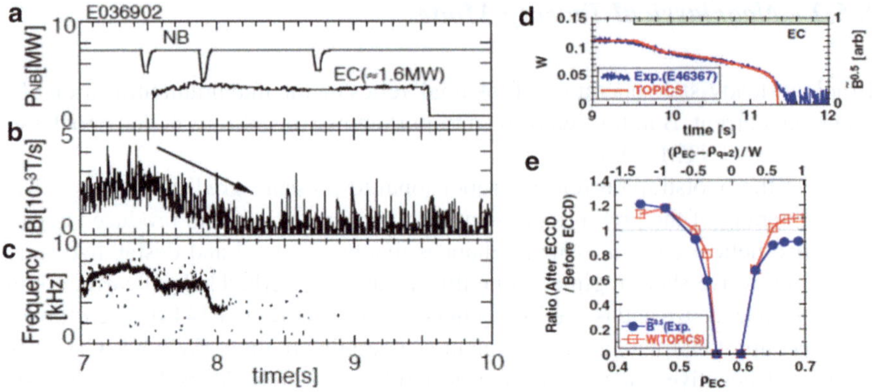

Fig. 8.21 (**a**)–(**c**) Stabilization of NTM for fundamental O-mode ECCD. (**a**) NB and EC injection power, (**b**) $\dot{B}$ for $n = 2$ mode. (**c**) Frequency of $n = 2$ mode. Reproduced with permission from Isayama [374]. Copyright IOP Publishing. (**d**)–(**e**) Comparison of measurement and numerical calculation based on modified Rutherford equation in JT-60U. Reproduced with permission from Isayama [375]. Copyright IAEA Vienna

The Rutherford equation including this bootstrap current destabilization effect and polarization current stabilizing effect (or neoclassical polarization effect $w_{pol} \sim \rho_{pi}$) [793] and finite perpendicular transport ($w_d \sim (\chi_\perp/\chi_\parallel)^{1/4}$) effect [205] is given as [487],

$$\frac{dw}{dt} = \frac{1.22\eta}{\mu_0}\left[\Delta'(w) + \epsilon^{1/2}\frac{L_q}{L_p}\frac{\beta_p}{w}\left(\frac{w^2}{w^2 + w_d^2} - \frac{L_q}{L_p}\frac{\epsilon\rho_{\theta_i}^2}{w^2}\right)\right], \tag{8.99}$$

where $\epsilon = r/R$, $L_q = q/(dq/dr)$, $L_p = p/(dp/dr)$, and β_p is poloidal beta value.

The characteristic transport effect island width is given by Fitzpatrick [205] as $w_d = 5.1(qL_q/\epsilon k_\theta)^{1/2}(\chi_\perp/\chi_\parallel)^{1/4}$, where poloidal wave number $k_\theta = m/r$, $\chi_\perp$ and $\chi_\parallel$ are perpendicular and parallel heat diffusivities, respectively. The characteristics threshold island width for nonlinear excitation of NTM is given as, $w_{pol} = |L_q/L_p|^{1/2}\epsilon^{1/2}\rho_{\theta_i}$. More sophisticated formula with non-divergent polarization term at $w = 0$ and other terms is summarized for example by Sauter [665]. Even if the equilibrium is stable to classical tearing mode $\Delta'(0) \leq 0$, the NTM can be destabilized at finite island width $w \geq w_{crit}$. So, we need seed island to destabilize NTM. Benchmarking of modified Rutherford equation for ECCD stabilization of NTM is important for its extrapolation to ITER control [662]. Efforts are successful in calibrating coefficients [303] and the result is successfully applied to radial scanning experiments [375] as shown in Fig. 8.21d–e and provide power requirement [304] in ITER.

Magnetic island can be distorted by the viscous drag of shear flow as shown in Fig. 8.22a and has a stabilizing effect on classical tearing mode and NTM [700].

Magnetic island is identified by the 180 degree phase shift of the temperature fluctuations across the magnetic island. Ren [624] and Chu [124] showed that this

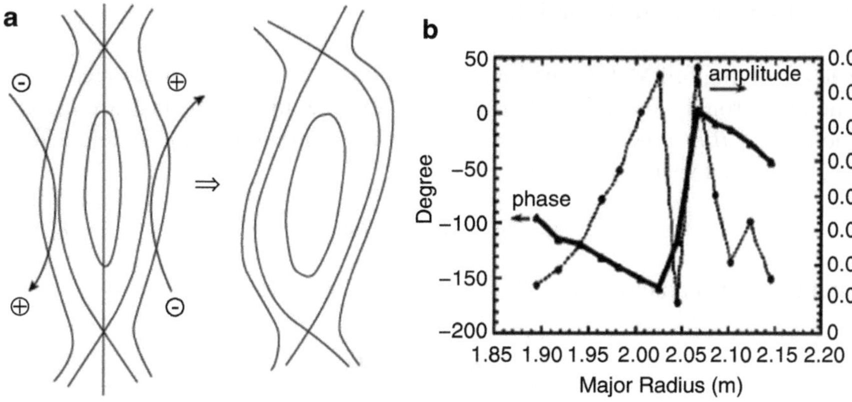

Fig. 8.22 (**a**) Viscous force on the magnetic island and distortion of magnetic island due to viscous drag. After Ren [624]. (**b**) Phase shift across island showing lower than 180°. Reproduced with permission from Ren [624]. Copyright AIP Publishing LLC

phase shift can be changed due to the viscous force on the magnetic island as illustrated in Fig. 8.22a. In the two dimensional geometry, the magnetic field is given by $\boldsymbol{B} = \boldsymbol{e}_z \times \nabla\psi + B_z\boldsymbol{e}_z$ and the flow velocity is given by $\boldsymbol{u} = \boldsymbol{e}_z \times \nabla U$. The two dimensional plasma equilibrium equation for the flux function is given by Ren et al. [624],

$$\Delta\psi + F(\psi) - \mu_0\rho U'\Delta U = \mu_0\nu\frac{\nabla\psi\cdot\boldsymbol{e}_z\times\nabla\Delta U}{(\nabla\psi)^2} \tag{8.100}$$

Here $F(\psi) = B_zB_z' + \mu_0H'$ with H being the enthalpy function and ν is the kinematic viscosity. An analytical estimate of the phase shift is give as $\Delta\phi = 8k\nu\frac{dV}{dr}\frac{\omega\mu_0G}{8B^2}$, where k is the wave length of the magnetic perturbation along the lengthwise direction of the magnetic island, w is the width of the magnetic island, dV/dr is the radiant of the flow on the island separatrix and $G \sim 1$ is a constant. This equation gives quantitative agreement of the phase shift reduction shown in Fig. 8.22b.

Except JT-60U, most medium to large tokamaks uses tangential neutral beam injection, which drives co-toroidal rotation. After the reorientation of tangential NBI, Buttery [90] and La Haye [488] showed that NTM is 'destabilized' with reduced toroidal rotation using Co/Counter NB injection capability. La Haye [489] showed that NTM onset β_N reduces with reduced co rotation associated with weaker absolute negative rotation shear ($d\Omega_\zeta/dr < 0$), where $\Omega_\zeta = \boldsymbol{u}_{i0}\cdot\nabla\zeta$ and correlates with normalized flow shear $n_{FS} = -L_s\tau_A d\Omega_\zeta/dr$, where $L_s = qL_q/\epsilon$.

8.5.3 Double Tearing Mode

Two rational surfaces can exist for hollow current profile in tokamak, which may be unstable to double tearing mode (DTM).

8.5.3.1 Reduced MHD Equation

Instead of solving full MHD equation, it is useful to solve so-called Reduced MHD equation for strongly magnetized plasma [68]. Consider the plasma in a strong axial magnetic field $B_z \boldsymbol{e}_z$ (and hence B_z is constant in lowest order) and poloidal field given by the axial vector potential $A_z = -\psi$ as,

$$\boldsymbol{B} = \boldsymbol{e}_z \times \nabla\psi + B_z \boldsymbol{e}_z \tag{8.101}$$

Integration of the Faraday's law $\partial \boldsymbol{B}/\partial t = \nabla \times (\boldsymbol{u} \times \boldsymbol{B})$ gives,

$$\frac{\partial \boldsymbol{A}}{\partial t} = \boldsymbol{u} \times \boldsymbol{B} - \nabla G, \tag{8.102}$$

where G is a scalar potential. Neglecting $\boldsymbol{A}_\perp$, the perpendicular flow $\boldsymbol{u}_\perp$ can be given by using stream function $\phi = G/B_z$ as follows.

$$(\boldsymbol{u} \times \boldsymbol{B})_\perp = \nabla_\perp G \text{ , or, } \boldsymbol{u}_\perp = \boldsymbol{e}_z \times \nabla\phi \tag{8.103}$$

From the z component of the (8.102), we have $\partial\psi/\partial t + \boldsymbol{u} \cdot \nabla\psi = B_z(\partial\phi/\partial z)$, or:

$$\frac{\partial\psi}{\partial t} - \boldsymbol{B} \cdot \nabla\phi = 0 \tag{8.104}$$

Assuming the uniform density ρ = constant and taking the axial component of the vorticity equation $\nabla \times [\rho(\partial \boldsymbol{u}/\partial t + \boldsymbol{u} \cdot \nabla \boldsymbol{u}) = -\nabla p + \boldsymbol{j} \times \boldsymbol{B}]$, the equation for the stream function ϕ is obtained using the vorticity $\omega = \nabla_\perp^2 \phi$:

$$\frac{\partial\omega}{\partial t} + \boldsymbol{u} \cdot \nabla\omega = \frac{1}{\rho} \boldsymbol{B} \cdot \nabla j_z \tag{8.105}$$

Equations (8.104) and (8.105) for two stream functions are called the lowest order reduced MHD equations.

If we include dissipation mechanisms such as the resistivity η and the kinematic viscosity ν, we have following set of reduced MHD equation.

$$\frac{\partial\psi}{\partial t} - \boldsymbol{B} \cdot \nabla\phi = \eta j_z - E_z \tag{8.106}$$

$$\frac{\partial\omega}{\partial t} + \boldsymbol{u} \cdot \nabla\omega - \frac{1}{\rho} \boldsymbol{B} \cdot \nabla j_z = \nu\nabla^2\omega \tag{8.107}$$

$$J_z = \mu_0^{-1} \nabla_\perp^2 \psi, \omega = \nabla_\perp^2 \phi, \boldsymbol{B} = \boldsymbol{e}_z \times \nabla\psi + \boldsymbol{e}_z B_z, \boldsymbol{u} = \boldsymbol{e}_z \times \nabla\phi \tag{8.108}$$

Column 8-6: Reduced MHD Equation

A set of the reduced MHD equations in a low β cylindrical tokamak geometry (r, θ, ζ) with periodicity of $2\pi R_0$ ($\zeta = z/2\pi R_0$) assuming $\rho = 1$ is given as follows [381],

(continued)

$$\frac{\partial \omega}{\partial t} = \frac{1}{r}[\omega, \phi] + \frac{1}{r}[\psi, j] + \frac{B_0}{R_0}\frac{\partial j}{\partial \zeta} + \nu \nabla_\perp^2 \omega$$

$$\frac{\partial \psi}{\partial t} = \frac{1}{r}[\psi, \phi] + \frac{B_0}{R_0}\frac{\partial \phi}{\partial \zeta} + \eta j - E$$

$$\mu_0 j = \frac{\partial^2 \psi}{\partial r^2} + \frac{1}{r}\frac{\partial \psi}{\partial r} + \frac{1}{r^2}\frac{\partial^2 \psi}{\partial \theta^2}$$

$$\omega = \frac{\partial^2 \phi}{\partial r^2} + \frac{1}{r}\frac{\partial \phi}{\partial r} + \frac{1}{r^2}\frac{\partial^2 \phi}{\partial \theta^2}$$

$$[a, b] = \frac{\partial a}{\partial r}\frac{\partial b}{\partial \theta} - \frac{\partial b}{\partial r}\frac{\partial a}{\partial \theta}$$

Here ψ is the poloidal flux function, ϕ is the stream function, η is the resistivity, ν is the kinematic viscosity, j is the toroidal current density, ω is the vorticity, E is the electric field. B_0 is the toroidal magnetic field, R_0 is the major radius, time is normalized to the poloidal Alfven transit time $\tau_{pa} = \sqrt{\mu_0 \rho} a / B_\theta(a)$ and the uniform density is assumed. The resistivity η is normalized to $\eta = \tau_{pa}/\tau_\eta$, where $\tau_\eta = \mu_0 a^2/\eta$. The poloidal flux ψ is related to the magnetic field as $\boldsymbol{B} = B_0 \boldsymbol{e}_\zeta + \nabla \psi \times \boldsymbol{e}_\zeta$ and the stream function ϕ is related to the velocity field as $\boldsymbol{u} = \nabla \phi \times \boldsymbol{e}_\zeta$.

The reduced MHD equation in toroidal geometry is given as follows [383],

$$\frac{\partial \omega}{\partial t} = \frac{1}{r}[\omega, \phi] + \frac{1}{r}[\psi, j] + \frac{B_0}{R_0}\frac{\partial j}{\partial \zeta} + \frac{1}{r}\left[p, h^2\right] + \nu \nabla_\perp^2 \omega$$

$$\frac{\partial \psi}{\partial t} = \frac{1}{r}[\psi, \phi] + \frac{B_0}{R_0}\frac{\partial \phi}{\partial \zeta} + \eta j - E$$

$$\frac{\partial p}{\partial t} = \frac{1}{r}[p, \phi] + \kappa_\perp \nabla_\perp^2 p$$

$$\mu_0 j = \frac{1}{r}\frac{\partial}{\partial r}\left(r g^{rr}\frac{\partial \psi}{\partial r} + r g^{r\theta}\frac{\partial \psi}{\partial \theta}\right) + \frac{\partial}{\partial \theta}\left(g^{r\theta}\frac{\partial \psi}{\partial r} + g^{\theta\theta}\frac{\partial \psi}{\partial \theta}\right)$$

$$\omega = \frac{1}{r}\frac{\partial}{\partial r}\left[r h^2\left(g^{rr}\frac{\partial \phi}{\partial r} + g^{r\theta}\frac{\partial \phi}{\partial \theta}\right)\right] + \frac{\partial}{\partial \theta}\left[h^2\left(g^{r\theta}\frac{\partial \phi}{\partial r} + g^{\theta\theta}\frac{\partial \phi}{\partial \theta}\right)\right]$$

$$h = \frac{R}{R_0}$$

Here, $g^{rr}, g^{r\theta}, g^{\theta\theta}$ are the contravariant components of the metrics of the flux coordinates (straight field line coordinates) system (r, θ, ζ).

8.5.3.2 Linear Stability

Stability of DTM is analyzed in 1973 for cylindrical tokamak by Furth et al. [234]. DTM is a possible cause of rapid current penetration and disruptive instability during current ramp [718]. Formation of hollow current profile and its rapid relaxation associated with $m/n = 4/1$ mode is observed by Hutchinson [355].

While negative shear (NS) operation showed excellent plasma confinement to exceed $Q_{DT}^{eq} = 1$ in JT-60U [378], plasma is disrupted by ideal and resistive MHD modes. While NS can be stable to ideal MHD if pressure profile is optimized, strong ITB (Internal Transport Barrier) near q_{min} may destabilize ideal MHD instabilities. This mode of operation can also be unstable to the DTM.

Linear eigenmode structure of DTM can be obtained using (8.90) as well in the cylindrical tokamak. The radial proximity of two rational surface $\Delta r_s/2a$ is an important parameter for DTM, where eigenfunctions of inner and outer tearing mode are coupled to form DTM if $\Delta r_s/2a$ is sufficiently small.

Figure 8.23 shows linear eigenmode structures for three values of $\Delta r_s/2a$ given by Ishii-Azumi [381]. In case $\Delta r < 0.15$, strong mode coupling occurs and the growth rate follows $m = 1$ internal resistive tearing scaling $\gamma \sim \eta^{1/3}$. If the radial separation $\Delta r_s/2a$ is large, modes in two rational surfaces are decoupled and the linear growth rate of the mode follows tearing mode scaling $\gamma \sim \eta^{3/5}$. The regime $0.15 < \Delta r < 0.32$ is called intermediate regime.

Pritchett [616] derived the dispersion relation for the DTM in sheet pinch:

$$\hat{\lambda}^{5/4}\hat{\lambda}_h \Gamma[(\hat{\lambda}^{3/2} - 1)/4]/\Gamma[(\hat{\lambda}^{3/2} + 5)/4] = 8 \tag{8.109}$$

$$\hat{\lambda} = \gamma\tau_h(S/\alpha^2(dB_0/dx)^2)^{1/3}, \hat{\lambda}_h = \gamma_h\tau_h(S/\alpha^2(dB_0/dx)^2)^{1/3}, \tag{8.110}$$

where the scaled resistive growth rate $\hat{\lambda}$ and the ideal MHD driving energy $\hat{\lambda}_h$. Pritchett [616] also derived a coupling condition as a DTM given by $\Delta r_s/2a < (ma/r_s)^{-7/9}S^{-1/9}$, where S is Lundquist number. Flow and flow shear effect on linear DTM is studied by Wei [813] and Wang [807] showing strong stabilizing effect through decoupling between two rational surfaces.

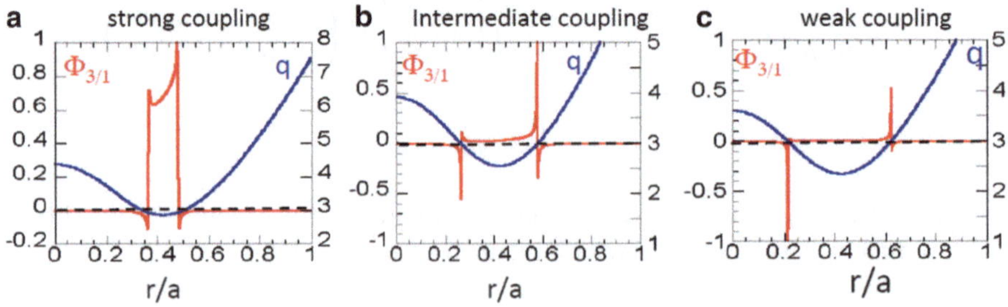

Fig. 8.23 Comparison of linear radial eigenmode structures among (**a**) small (strong coupling), (**b**) intermediate and (**c**) large (weak coupling) Δr. Modified from Ishii [381]

8.5.3.3 Nonlinear Stability

Nonlinear behavior of DTM can be analyzed using the reduced MHD equations introduced in Sect. 8.5.3.1. More refined reduced equations in the cylindrical tokamak and with toroidal correction are given in Column 8-6. Figure 8.24 shows nonlinear DTM time evolution of kinetic energies of various m/n modes. In case of small Δr, mode grows exponentially until the collapse. In case of large Δr, mode just goes into Rutherford regime and saturates. On the other hand, non-linear destabilization of DTM is found in the intermediate regime, which is attributed to new type of reconnection discussed later.

Takeji [740] identified two types of resistive instabilities in JT-60U NS plasma. One is non-disruptive resistive interchange mode and other is disruptive tearing mode shown in Fig. 8.25. In the former case, temperature fluctuation was observed only near the inner rational surface (typically q=3) showing no phase inversion across the rational surface and the stability criterion of the resistive interchange mode D_R [249] is violated ($D_R > 0$). These observations are consistent with resistive interchange mode. The observation of resistive interchange mode is first reported in DIII-D by Chu [123] at $\beta_N = 1.5$ resulted in disruption showing importance of pressure peaking factor, while this resistive interchange mode in JT-60U at $\beta_N < 1$ does not lead to disruption.

For the latter case, precursor oscillation is observed before thermal collapse. The temperature perturbation $\widetilde{T_e}$ just before the 1st thermal collapse show clear phase inversion near the outer $q = 3$ surface with a growth time of $\gamma^{-1} \sim 0.5ms$ indicating tearing mode and do not have any fluctuation near the inner $q = 3$ surface [740].

Ishii-Azumi [383] investigated resistive instability in toroidal geometry under the condition of $D_R > 0$ near the inner $q = 3$ surface, showing that mode structure is not interchange but is DTM and they concluded existence of Rutherford regime may cause precursor oscillation and nonlinearly destabilized DTM [381] later go to thermal quench by the explosive instability. Ishii-Smolyakov further investigated effect of toroidal flow and external magnetic perturbation [384] showing that inner magnetic island may disappears. Other important effect is loss of bootstrap current inside the magnetic island [98] on DTM since loss of bootstrap current has stabilizing effect in negative shear ($s < 0$) regime and destabilizing effect in positive shear ($s > 0$) regime [443], which may lead to single tearing mode at rational surface in positive shear region. This is still left for future study.

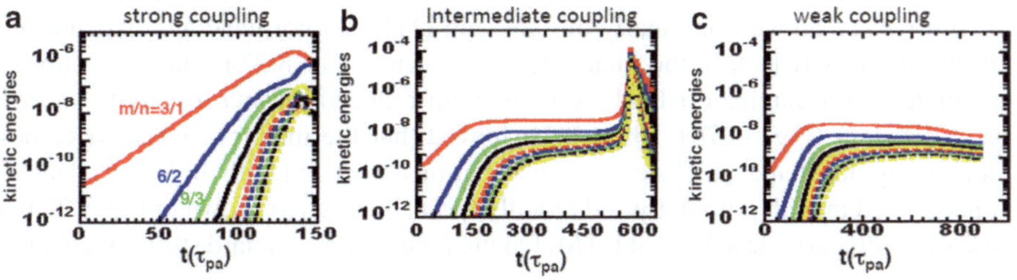

Fig. 8.24 Comparison of nonlinear evolutions of DTMs among (**a**) small (strong coupling), (**b**) intermediate and (**c**) large (weak coupling) Δr. Modified from Ishii [381]

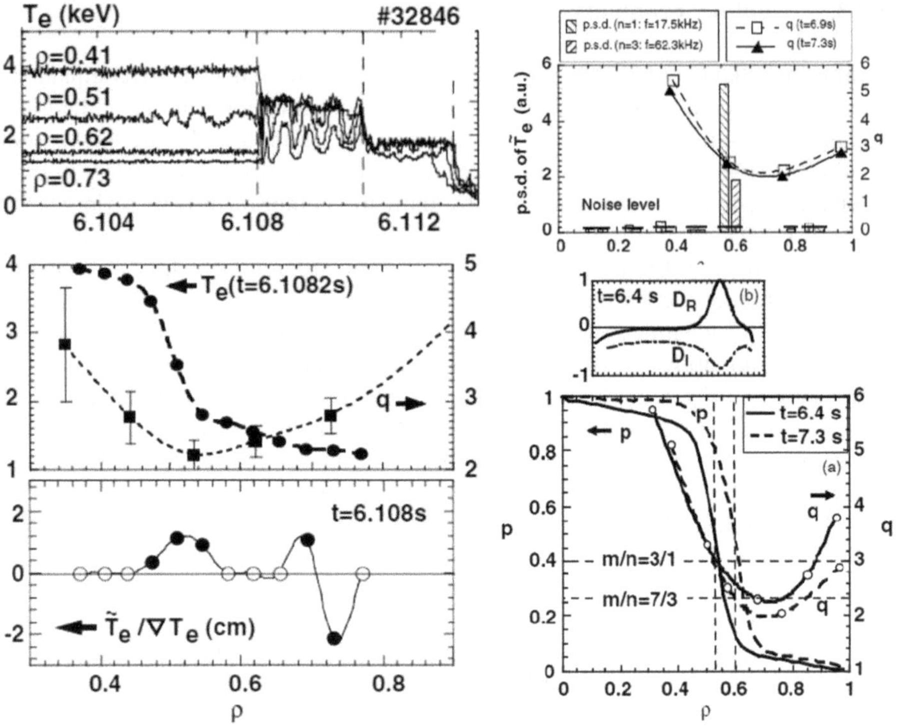

Fig. 8.25 Time evolution of disruptive tearing mode in NS discharge. Reproduced with permission from Takeji [740]. Copyright IAEA Vienna

8.5.3.4 Fast Reconnection

Magnetic reconnection is one of most important physics subjects in plasma physics. Fast reconnection observed in solar flare is one of mysteries in plasma physics since usual Sweet-Parker sheet reconnection do not have fast reconnection time [69].

As shown in Fig. 8.26, for sufficiently low resistivity η, a slowly growing Rutherford regime-like phase is seen after the initial linear growth and it changes to an explosive growth phase. The period of Rutherford regime increases with lower resistivity η. The magnetic reconnection process in this nonlinearly destabilized DTM has marked differences with usual Sweet-Parker sheet reconnection. The growth rate of above nonlinearly destabilized DTM do not have any dependence on plasma resistivity η in the final stage as shown in Fig. 8.27a. The peak current associated with magnetic island increases with reduced plasma resistivity so that ηj = constant (Fig. 8.27b). Ishii [382] showed that the magnetic island becomes more triangular shape and the current distribution associated with magnetic island is concentrated in both poloidally and radially (point current) as shown in Fig. 8.28a in case of nonlinearly destabilized DTM. On the other hand, standard strongly coupled

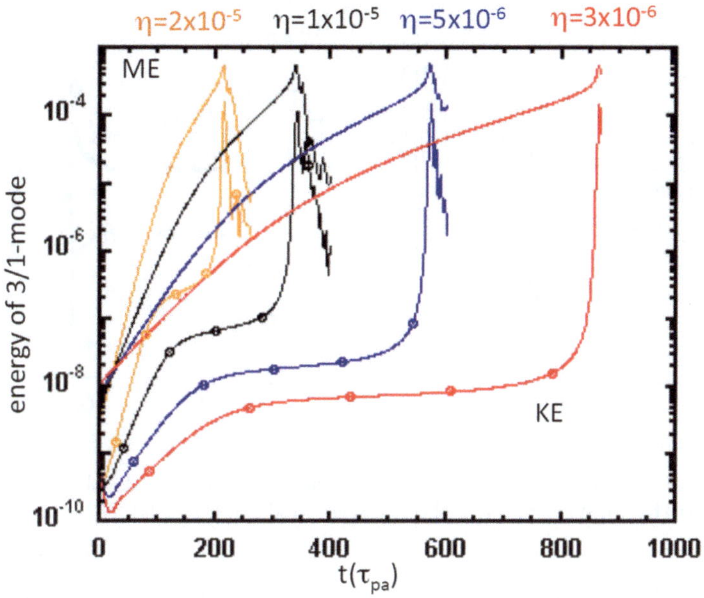

Fig. 8.26 Time evolution of non-linear DTM for intermediate Δr for different resistivity η. ME and KE are magnetic energy and kinetic energy, respectively. Modified from Ishii [381]

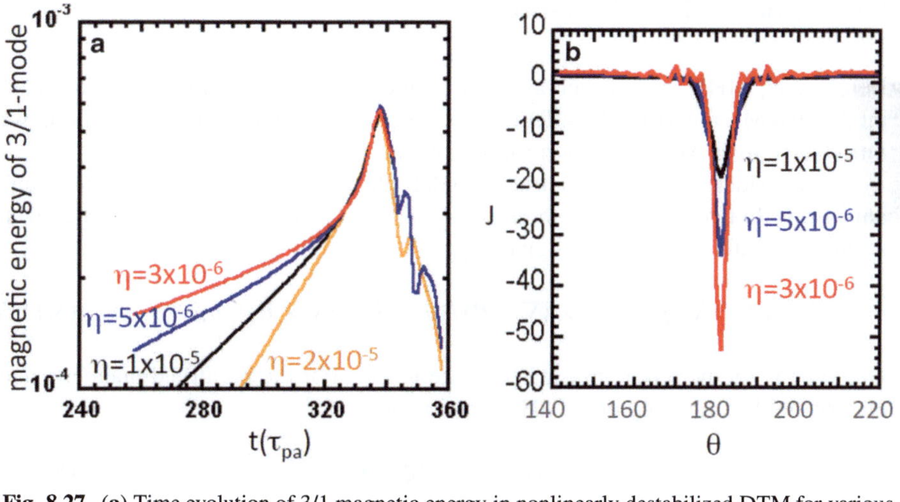

Fig. 8.27 (a) Time evolution of 3/1 magnetic energy in nonlinearly destabilized DTM for various plasma resistivity η. (b) Current distributions for various η. Modified from Ishii [382]

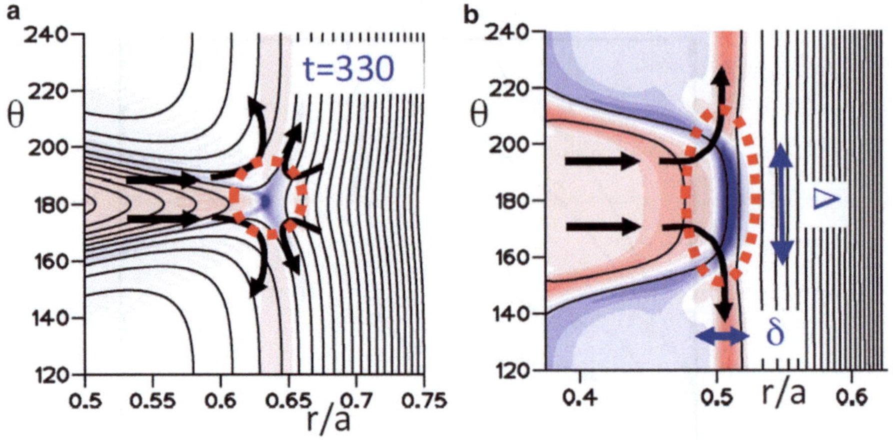

Fig. 8.28 (**a**) Contour plot of current distribution for nonlinearly destabilized DTM (point current). (**b**) That for standard DTM (sheet current). Modified from Ishii [382]

DTM (Fig. 8.28b) has sheet current distribution during explosive growth similar to Sweet-Parker reconnection.

8.6 Kinetic MHD Equation

In tokamak, MHD modes such as Alfven Eigenmode (AE) and the Resistive Wall mode (RWM) are strongly affected by the kinetic effects. In this section, we introduce kinetic MHD equation following Cheng [119].

We consider an axisymmetric toroidal plasma with isotropic thermal and anisotropic hot components with $n_h \ll n_{th}$, $T_h \gg T_{th}$ and $\beta_h \ll \beta_{th}$. The linearized momentum balance equation is given by,

$$\omega^2 \rho \boldsymbol{\xi} = \nabla \cdot \boldsymbol{p}_1 + \boldsymbol{B}_1 \times (\nabla \times \boldsymbol{B})/\mu_0 + \boldsymbol{B} \times (\nabla \times \boldsymbol{B}_1)/\mu_0, \tag{8.111}$$

where $\boldsymbol{p}_1$ is the perturbed plasma pressure tensor. For the perturbed magnetic and electric fields, following ideal MHD relation holds.

$$\boldsymbol{B}_1 = \nabla \times (\boldsymbol{\xi} \times \boldsymbol{B}) \tag{8.112}$$

$$\boldsymbol{E}_1 = i\omega \boldsymbol{\xi} \times \boldsymbol{B} \tag{8.113}$$

Also the thermal plasma pressure may follow adiabatic response.

$$p_{1th} + \boldsymbol{\xi} \cdot \nabla p_{th} + \gamma p_{th} \nabla \cdot \boldsymbol{\xi} = 0 \tag{8.114}$$

The perturbed plasma pressure tensor $\boldsymbol{p}_1$ can be given by following CGL form.

$$\boldsymbol{p}_1 = p_{1\perp}\boldsymbol{I} + (p_{1\parallel} - p_{1\perp})\boldsymbol{bb} \tag{8.115}$$

The parallel and perpendicular pressures $p_{1\parallel}$ and $p_{1\perp}$ are given using the perturbed velocity distribution function f_{1a} by,

$$\begin{bmatrix} p_{1\parallel} \\ p_{1\perp} \end{bmatrix} = \sum_a \int d^3v f_{1a} \begin{bmatrix} 2(\varepsilon - \mu B) \\ \mu B \end{bmatrix}, \tag{8.116}$$

where $\varepsilon = m_a v^2/2$ is the particle kinetic energy and $\mu = m_a v_\perp^2/2B$ is the magnetic moment.

The velocity space integral can be expressed in terms of the velocity coordinates $(\varepsilon, \lambda, \sigma)$ as follows,

$$\int dv^3 = \sum_\sigma \frac{\sqrt{2}\pi}{m_a^{3/2}} \int_0^\infty d\varepsilon \sqrt{\varepsilon} \int_0^{h(\psi,\theta)} \frac{d\lambda}{h\sqrt{1-\lambda/h}}, \tag{8.117}$$

where $\lambda = \mu B_0/\varepsilon$ is the pitch angle, B_0 is the vacuum magnetic field at the tokamak major radius R_p, $h = B_0/B(\psi, \theta)$, and $\sigma = \pm 1$ is the direction of particle parallel velocity. On a flux surface, the circulating particles correspond to $0 \le \lambda \le h_{min}$, and the trapped particles correspond to $h_{min} \le \lambda \le h$ at a given θ, where $h_{min} = Min[h(\psi, \theta)]$ on the ψ surface.

The perturbed distribution function f_{1a} can be given by a summation of the adiabatic and non-adiabatic parts as follows,

$$f_{1a} = e\Phi_1 \frac{\partial F_a}{\partial \varepsilon} - \frac{\mu B_{1\parallel}}{B}\frac{\partial F_a}{\partial \mu} + g_a, \tag{8.118}$$

where Φ_1 is the perturbed electrostatic potential, $B_{1\parallel}$ is the perturbed parallel magnetic field, F_a is the unperturbed distribution function, and the non-adiabatic distribution function g_a is governed by the following drift kinetic equation,

$$\frac{\partial g_a}{\partial t} + (v_\parallel \boldsymbol{b} + \boldsymbol{v}_d)\cdot\nabla g_a = -\frac{iF_a}{T_a}\left(\tilde{\omega} - \omega_*^T\right)\left(e\Phi_1 - ev_\parallel A_{1\parallel} + \mu B_{1\parallel}\right), \tag{8.119}$$

where $\tilde{\omega} = -T_h\omega\partial lnF_a/\partial K$, and $\omega_*^T = -i(T_h/\Omega_h)\boldsymbol{b}\times(\nabla F_a)\cdot\nabla$ only operates on the perturbed quantities, $\boldsymbol{v}_d = (\boldsymbol{b}/\Omega_h)\times(\nabla(\mu B + \kappa v_\parallel^2)$ is the toroidal drift velocity, T_h is the average temperature of the hot particle, Ω_h is the cyclotron frequency of the hot particle, and $A_{1\parallel}$ is the perturbed parallel vector potential.

Since we are interested in MHD-type modes, we have $E_{1\parallel} = 0$ from Eq. (8.113) to find following relation.

$$i\omega A_{1\parallel} = \boldsymbol{b}\cdot\nabla\Phi_1 \tag{8.120}$$

For low beta toroidal plasma, we may neglect perpendicular component of the perturbed vector potential $A_{1\perp}$ to the $\boldsymbol{E}_{1\perp}$ to find $\boldsymbol{E}_{1\perp} \sim -\nabla_\perp \Phi_1$. Then,

$$\Phi_1 - v_\parallel A_{1\parallel} = \frac{i}{\omega}\left(\frac{d\Phi_1}{dt} + \boldsymbol{v}_d \cdot \boldsymbol{E}_{1\perp}\right) \tag{8.121}$$

The drift kinetic equation (8.119) can be solved the method of characteristics through the integration along the unperturbed guiding center trajectory.

$$g_a = \frac{eF_a}{T_a}\frac{\tilde{\omega} - \omega_*^T}{\omega}\Phi_1 + g_{1a} \tag{8.122}$$

Assuming $d[F_a(\tilde{\omega} - \omega_*^T)]/dt \sim 0$, we have following expression for g_{1a}.

$$g_{1a} = \int_{-\infty}^{t} dt' \frac{-ieF_a}{T_a}(\tilde{\omega} - \omega_*^T)\left[\frac{i\boldsymbol{v}_d \cdot \boldsymbol{E}_{1\perp}}{\omega} + \frac{m_h \mu B_{1\parallel}}{e}\right] \tag{8.123}$$

Substituting (8.122) into (8.118) and noting the following relation

$$\frac{eF_a}{T_a}\frac{\tilde{\omega} - \omega_*^T}{\omega}\Phi_1 = -\boldsymbol{\xi}_\perp \cdot \nabla F_a - e\Phi_1 \frac{\partial F_a}{\partial \varepsilon}, \tag{8.124}$$

the perturbed distribution function f_{1a} can be given by

$$f_{1a} = -\boldsymbol{\xi}_\perp \cdot \nabla F_a - \frac{\mu B_{1\parallel}}{B}\frac{\partial F_a}{\partial \mu} + g_{1a} \tag{8.125}$$

The expressions for g_{1a} are obtained for trapped and passing particles to obtain perturbed parallel and perpendicular pressures, $p_{1\parallel}, p_{1\perp}$. Linearized kinetic MHD equations can be obtained from the momentum balance equation (8.111) and the induction equation (8.113). But the system are now non-Hermitian integro-differential eigenmode equations.

Taking an inner product of equation (8.111) with $\boldsymbol{\xi}^*$ and integrating over the plasma volume with an assumption of a fixed conducting boundary, we obtain following quadratic form.

$$D(\omega) = \delta W_f + \delta W_k - \delta K = 0, \tag{8.126}$$

where δK is the fluid kinetic energy, δW_k is the potential energy of energetic particles, and δW_f is the potential energy of MHD fluid as follows,

$$\delta K = \omega^2 \int dV \rho |\boldsymbol{\xi}|^2 \tag{8.127}$$

$$\delta W_k = -\int dV[\nabla \cdot \boldsymbol{\xi}^* p'_{1\perp} + (p'_{1\perp} - p'_{1\parallel})(\boldsymbol{\kappa} \cdot \boldsymbol{\xi}_\perp^* - \boldsymbol{b} \cdot \nabla \boldsymbol{\xi}_\parallel^*)] \tag{8.128}$$

$$\delta W_f = \int dV[\gamma p_{th}|\nabla \cdot \boldsymbol{\xi}|^2 + |\boldsymbol{B}_1|^2/\mu_0 + |\nabla \cdot \boldsymbol{\xi}_\perp + 2\boldsymbol{\kappa} \cdot \boldsymbol{\xi}_\perp|^2 B^2/\mu_0$$
$$+ \frac{\boldsymbol{J} \cdot \boldsymbol{B}}{B^2}(\boldsymbol{B}_1 \times \boldsymbol{B}) \cdot \boldsymbol{\xi}_\perp^* - 2(\boldsymbol{\kappa} \cdot \boldsymbol{\xi}_\perp^*)(\boldsymbol{J} \times \boldsymbol{B}) \cdot \boldsymbol{\xi}_\perp] \quad (8.129)$$

Here $p'_{1\parallel}$ and $p'_{1\perp}$ are given by,

$$\begin{bmatrix} p'_{1\parallel} \\ p'_{1\perp} \end{bmatrix} = \sum_a \int d^3 v g_{1a} \begin{bmatrix} 2(\varepsilon - \mu B) \\ \mu B \end{bmatrix} \quad (8.130)$$

8.7 Alfven Eigen Modes

8.7.1 Global MHD Spectrum in Axisymmetric Tokamak

The linearized equations of ideal single-fluid MHD are given by (8.2)–(8.4). We decompose the plasma displacement $\boldsymbol{\xi}$ and the perturbed magnetic field $\boldsymbol{B}_1$ in three orthogonal directions $\nabla\psi/|\nabla\psi|$, $\boldsymbol{b} \times \nabla\psi/|\boldsymbol{b} \times \nabla\psi|$ and $\boldsymbol{b}$ as follows.

$$\boldsymbol{\xi} = \frac{\xi_\psi}{|\nabla\psi|^2}\nabla\psi + \frac{\xi_s}{B^2}(\boldsymbol{B} \times \nabla\psi) + \frac{\xi_b}{B^2}\boldsymbol{B} \quad (8.131)$$

$$\boldsymbol{B}_1 = \frac{Q_\psi}{|\nabla\psi|^2}\nabla\psi + \frac{Q_s}{|\nabla\psi|^2}(\boldsymbol{B} \times \nabla\psi) + \frac{Q_b}{B^2}\boldsymbol{B} \quad (8.132)$$

so that,

$$\xi_\psi = \boldsymbol{\xi} \cdot \nabla\psi,\ \xi_s = \boldsymbol{\xi} \cdot \frac{\boldsymbol{B} \times \nabla\psi}{|\nabla\psi|^2}, \xi_b = \boldsymbol{\xi} \cdot \boldsymbol{B} \quad (8.133)$$

$$Q_\psi = \boldsymbol{B}_1 \cdot \nabla\psi, Q_s = \boldsymbol{B}_1 \cdot \frac{\boldsymbol{B} \times \nabla\psi}{|\nabla\psi|^2}, Q_b = \boldsymbol{B}_1 \cdot \boldsymbol{B} \quad (8.134)$$

The three components of the induction equation (8.4) can be written as,

$$Q_\psi = \boldsymbol{B} \cdot \nabla\xi_\psi, Q_s = \frac{|\nabla\psi|^2}{B^2}(\boldsymbol{B} \cdot \nabla\xi_s - S\xi_\psi) \quad (8.135)$$

$$Q_b = B^2\boldsymbol{B} \cdot \nabla\frac{\xi_b}{B^2} - B^2\nabla \cdot \boldsymbol{\xi} - 2\boldsymbol{\kappa} \cdot (\boldsymbol{B} \times \nabla\psi)\xi_s - \frac{2(\boldsymbol{\kappa} \cdot \nabla\psi)B^2}{|\nabla\psi|^2}\xi_s + \mu_0 P'\xi_\psi, \quad (8.136)$$

where $\boldsymbol{\kappa} = \boldsymbol{b} \cdot \nabla\boldsymbol{b}$ is the magnetic curvature, $P' = dP/d\psi$, and S is the negative local magnetic shear defined as follows,

$$S = \frac{\boldsymbol{B} \times \nabla\psi}{|\nabla\psi|^2} \cdot \nabla \times \left[\frac{\boldsymbol{B} \times \nabla\psi}{|\nabla\psi|^2}\right] \quad (8.137)$$

Similarly, the momentum equation (8.2) can be decomposed as,

$$\mu_0\omega^2\rho\xi_\psi = \mu_0\nabla\psi\cdot\nabla P_1 - |\nabla\psi|^2\boldsymbol{B}\cdot\nabla\frac{\boldsymbol{B}\cdot\xi_\psi}{|\nabla\psi|^2} - (|\nabla\psi|^2 S - \mu_0\boldsymbol{B}\cdot\boldsymbol{J})Q_s - 2(\boldsymbol{\kappa}\cdot\nabla\psi)Q_b \tag{8.138}$$

$$\begin{aligned}\mu_0\omega^2\rho|\nabla\psi|^2\xi_s &= \mu_0(\boldsymbol{B}\times\nabla\psi)\cdot\nabla P_1 - \mu_0(\boldsymbol{B}\cdot\boldsymbol{J})(\boldsymbol{B}\cdot\nabla\xi_\psi) - B^2\boldsymbol{B}\cdot\nabla Q_s \\ &\quad -2\boldsymbol{\kappa}\cdot(\boldsymbol{B}\times\nabla\psi)Q_b\end{aligned} \tag{8.139}$$

$$\omega^2\rho\xi_b = \boldsymbol{B}\cdot\nabla(p_1 + P'\xi_\psi), \tag{8.140}$$

where $P_1 = p_1 + \boldsymbol{B}_1\cdot\boldsymbol{B}/\mu_0$ is the perturbed total pressure. The $\nabla\cdot\boldsymbol{\xi}$ can express as:

$$\nabla\cdot\boldsymbol{\xi} = \frac{\nabla\psi\cdot\nabla\xi_\psi}{|\nabla\psi|^2} + \xi_\psi\nabla\cdot\frac{\nabla\psi}{|\nabla\psi|^2} + \frac{(\boldsymbol{B}\times\nabla\psi)\cdot\nabla\xi_s}{B^2} - 2\boldsymbol{\kappa}\cdot(\boldsymbol{B}\times\nabla\psi)\xi_s + \boldsymbol{B}\cdot\nabla\frac{\xi_b}{B^2} \tag{8.141}$$

Using Eqs. (8.135)–(8.136) and (8.140), we can eliminate Q_ψ, Q_s, Q_b, and ξ_b. From (8.3), (8.138), (8.139) and (8.141), we obtain following linearized ideal MHD eigenmode equation.

$$\nabla\psi\cdot\nabla\begin{bmatrix}P_1\\ \xi_\psi\end{bmatrix} = C\begin{bmatrix}P_1\\ \xi_\psi\end{bmatrix} + D\begin{bmatrix}\xi_s\\ \nabla\cdot\boldsymbol{\xi}\end{bmatrix}, E\begin{bmatrix}\xi_s\\ \nabla\cdot\boldsymbol{\xi}\end{bmatrix} = F\begin{bmatrix}P_1\\ \xi_\psi\end{bmatrix} \tag{8.142}$$

where, C, D, E, F are (2×2) matrix operators involving surface derivatives $\boldsymbol{B}\cdot\nabla$ and $(\boldsymbol{B}\times\nabla\psi)\cdot\nabla$ as follows,

$$C = \begin{bmatrix}2\kappa_\psi & C_{12}\\ 0 & -|\nabla\psi|^2\nabla\cdot\left(\frac{\nabla\psi}{|\nabla\psi|^2}\right)\end{bmatrix}, \quad F = \begin{bmatrix}-\kappa_s + \frac{(\boldsymbol{B}\times\nabla\psi)}{B^2}\cdot\nabla & F_{12}\\ -\frac{\mu_0}{B^2} & -\frac{\kappa_\psi}{|\nabla\psi|^2}\end{bmatrix} \tag{8.143}$$

$$D = \begin{bmatrix}(|\nabla\psi|^2 S/\mu_0 - \boldsymbol{B}\cdot\boldsymbol{J})\frac{|\nabla\psi|^2}{B^2}\boldsymbol{B}\cdot\nabla & \gamma_s P\kappa_\psi\\ |\nabla\psi|^2\left[\kappa_s - \frac{\boldsymbol{B}\times\nabla\psi}{B^2}\cdot\nabla\right] & |\nabla\psi|^2\left[1 + \frac{\gamma_s P\boldsymbol{B}\cdot\nabla}{\omega^2\rho}\left(\frac{\boldsymbol{B}\cdot\nabla}{B^2}\right)\right]\end{bmatrix} \tag{8.144}$$

$$E = \begin{bmatrix}\frac{\omega^2\rho|\nabla\psi|^2}{B^2} + \boldsymbol{B}\cdot\nabla\left(\frac{|\nabla\psi|^2\boldsymbol{B}\cdot\nabla}{\mu_0 B^2}\right) & 2\gamma P\kappa_s\\ 2\kappa_s & \frac{\gamma\mu_0 P + B^2}{B^2} + \frac{\gamma P}{\omega^2\rho}\boldsymbol{B}\cdot\nabla\left(\frac{\boldsymbol{B}\cdot\nabla}{B^2}\right)\end{bmatrix}, \tag{8.145}$$

where

$$C_{12} = \omega^2\rho + P'\kappa_\psi + |\nabla\psi|^2\boldsymbol{B}\cdot\nabla\frac{\boldsymbol{B}\cdot\nabla}{\mu_0|\nabla\psi|^2} + (\boldsymbol{B}\cdot\boldsymbol{J} - S|\nabla\psi|^2/\mu_0)\frac{S|\nabla\psi|^2}{B^2} \tag{8.146}$$

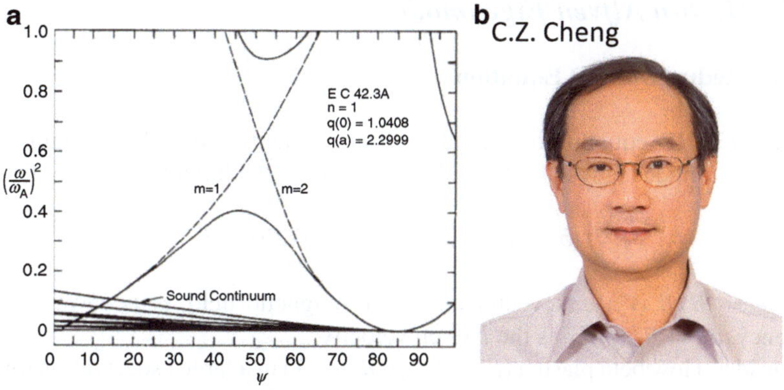

Fig. 8.29 (**a**) Shear Alfven gap spectrum due to toroidal coupling of $m = 1$ and $m = 2$ modes. Reproduced with permission from Cheng [116]. Copyright AIP Publishing LLC. (**b**) C.Z. (Frank) Cheng, who made a significant contribution to TAE theories and experiments

$$F_{12} = \boldsymbol{B} \cdot \nabla \frac{|\nabla \psi|^2}{\mu_0 B^2} S - \frac{\boldsymbol{J} \cdot \boldsymbol{B}}{B^2} \boldsymbol{B} \cdot \nabla - P' \kappa_s \tag{8.147}$$

$$\kappa_\psi = 2\boldsymbol{\kappa} \cdot \nabla \psi, \kappa_s = 2\boldsymbol{\kappa} \cdot (\boldsymbol{B} \times \nabla \psi)/B^2 \tag{8.148}$$

$$\nabla \psi \cdot \nabla = |\nabla \psi|^2 \frac{\partial}{\partial \psi} + (\nabla \psi \cdot \nabla \theta) \frac{\partial}{\partial \theta} + (\nabla \psi \cdot \nabla \zeta) \frac{\partial}{\partial \zeta} \tag{8.149}$$

For a given plasma equilibrium, we may solve ξ_s and $\nabla \cdot \boldsymbol{\xi}$ in terms of P_1 and ξ_ψ from the 2nd equation of (8.142). If E^{-1} exists, solution of ξ_s and $\nabla \cdot \boldsymbol{\xi}$ is substituted into the 1st equation of (8.142) and P_1 and ξ_ψ can be obtained. This procedures fails if the inverse of E does not exist for a given ω at some surface ψ. If the inverse operator E^{-1} does not exist, only non square integrable solution is possible as eigenfunction of (8.142). Such situation can be found by,

$$E \begin{bmatrix} \xi_s \\ \nabla \cdot \boldsymbol{\xi} \end{bmatrix} = 0 \tag{8.150}$$

Then corresponding set of eigenvalue ω^2 forms continuous spectrum, which is the toroidal generalization of $D = 0$ in Eq. (8.18). Typical solution of (8.150) is shown in Fig. 8.29a. The shear Alfven and sound wave continuum $(\omega/\omega_A)^2$ is given as a function of the flux label ψ, where we can see spectrum gap.

Exercise: Derive (8.140) from (8.111).
Hint: Use $\boldsymbol{J} \times \boldsymbol{B} = \nabla P$ and the vector formula $\boldsymbol{b} \cdot (\nabla \times \boldsymbol{a}) = \nabla \cdot (\boldsymbol{a} \times \boldsymbol{b} + \boldsymbol{a} \cdot (\nabla \times \boldsymbol{b})$.

8.7.2 High-n Alfven Eigenmodes

8.7.2.1 Reduced MHD Equation

A reduced MHD equation for the electrostatic potential Φ for a medium n shear Alfven eigenmode is given by Rosenbluth [639] (see Column 8-7 for proof):

$$(\boldsymbol{b}\cdot\nabla)[\nabla^2(\boldsymbol{b}\cdot\nabla)\Phi] + \nabla\cdot\left[(\omega^2/v_A^2)\nabla\Phi\right] = 0 \tag{8.151}$$

Here, $\boldsymbol{b} = \boldsymbol{B}/B$ is a unit vector along the magnetic field, ω is a wave angular frequency, $v_A = B/\sqrt{\mu_0\rho}$ is the Alfven velocity.

Consider a low beta plasma ($\beta = 2\mu_0 p/B^2 \ll 1$) in a large aspect ration tokamak ($a \ll R$) with a circular cross section. We expand Φ in the Fourier decomposition in poloidal and toroidal harmonics as:

$$\Phi(r,\theta,\zeta,t) = \sum_m \Phi_m(r)exp(-im\theta + in\zeta - i\omega t) \tag{8.152}$$

In this cylindrical coordinates, $\boldsymbol{b}\cdot\nabla\Phi \to ik_{\|m}\Phi_m$, where

$$k_{\|m}(r) = \frac{1}{R}\left(n - \frac{m}{q(r)}\right) \tag{8.153}$$

Here, $q(r) = rB_\zeta/RB_\theta$ is a cylindrical safety factor. In the toroidal geometry, the magnetic field strength is $B \sim B_0[1-\epsilon cos\theta]$ where $\epsilon = r/R$. This leads to a coupling among poloidal harmonics. We assume the ordering $k_{\|m}r \ll 1$ and $m \gg 1$.

The lowest order form of equation (8.151) can be obtained neglecting the toroidal effect. We obtain following uncoupled equation for the poloidal harmonic Φ_m for the shear Alfven mode in the cylindrical plasma.

$$\frac{d}{dr}\left[\left(\frac{\omega^2}{v_A^2} - k_{\|m}^2\right)\frac{d\Phi_m}{dr}\right] - \frac{m^2}{r^2}\left(\frac{\omega^2}{v_A^2} - k_{\|m}^2\right)\Phi_m = 0 \tag{8.154}$$

This equation exhibits singular structure at $r = r_s$ defined by the shear Alfven resonance condition $\omega = k_{\|m}v_A(r_s)$ as given by (8.24). Apparently, sound wave resonance is neglected in this reduced MHD treatment. This singularity is a regular singular point, where non square-integrable solution becomes possible leading to the appearance of continuous spectrum. At the singular point, collisionless wave absorption occurs due to phase mixing process of continuous spectrum. If the shear Alfven continuum has off-axis minimum in ω^2, discrete spectrum called GAE appears just below the continuum as discussed in Sect. 8.1.

To a lowest order in the toroidal effect, Φ_m couples only to the neighboring side bands, $\Phi_{m\pm1}$. Using (8.151), we obtain following system of coupled equations.

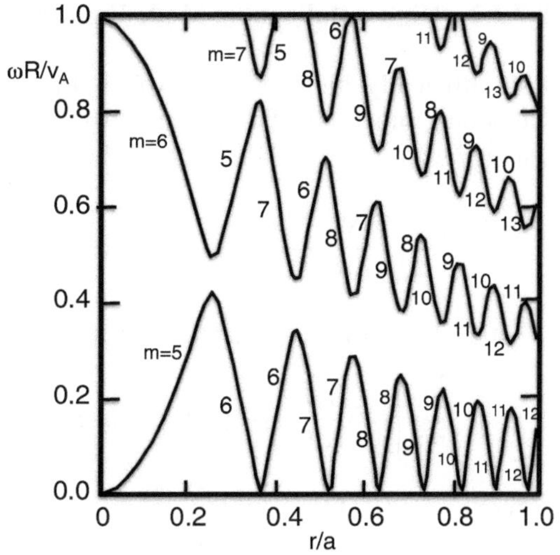

Fig. 8.30 The Alfven continuum resonance curves for the normalized frequency $\omega R/v_A$ for $n = 5$ in the circular tokamak with $q(0) = 1.05$, $q(a) = 2.5$, and $a/R = 0.25$. Modified from Rosenbluth [639]

$$\frac{d}{dr}\left[\left(\frac{\omega^2}{v_A^2} - k_{\parallel m}^2\right)\frac{d\Phi_m}{dr}\right] - \frac{m^2}{r^2}\left(\frac{\omega^2}{v_A^2} - k_{\parallel m}^2\right)\Phi_m + \epsilon_c \frac{\omega^2}{v_A^2}\frac{d^2}{dr^2}(\Phi_{m-1} + \Phi_{m+1}) = 0 \tag{8.155}$$

Here ϵ_c is a toroidicity coupling constant originally obtained by Fu [216] and refined by Berk [55] as $\epsilon_c = 2(\Delta'(r) + \epsilon) \sim 5\epsilon/2$.

The precise condition of singularity in this coupled equations is the vanishing of the determinant whose elements are the coefficients of the terms with the highest derivatives of the ϕ_m as follows.

$$\begin{vmatrix} 0\ldots & \ddots & \epsilon_c\frac{\omega^2}{v_A^2} & 0 & 0 & 0 & \ldots 0 \\ 0\ldots & \epsilon_c\frac{\omega^2}{v_A^2} & \frac{\omega^2}{v_A^2} - k_{\parallel m-1}^2 & \epsilon_c\frac{\omega^2}{v_A^2} & 0 & 0 & \ldots 0 \\ 0\ldots & 0 & \epsilon_c\frac{\omega^2}{v_A^2} & \frac{\omega^2}{v_A^2} - k_{\parallel m}^2 & \epsilon_c\frac{\omega^2}{v_A^2} & 0 & \ldots 0 \\ 0\ldots & 0 & 0 & \epsilon_c\frac{\omega^2}{v_A^2} & \frac{\omega^2}{v_A^2} - k_{\parallel m+1}^2 & \epsilon_c\frac{\omega^2}{v_A^2} & \ldots 0 \\ 0\ldots & 0 & 0 & 0 & \ldots & \ddots & \end{vmatrix} = 0 \tag{8.156}$$

Solving Eq. (8.156), the Alfven continuum resonance curves for the normalized frequency $\omega R/v_A$ can be obtained as shown in Fig. 8.30.

As shown in Fig. 8.30, spectrum gap naturally produces minimum and maximum in shear Alfven continuum in ω^2 at around $nq = m + 1/2$, where $k_{\parallel m} = -k_{\parallel m+1}$ is satisfied. Two cylindrical shear Alfven waves with poloidal mode numbers m and $m + 1$ having same frequency $\omega = v_A(nq = m)/(2q_m R)$ $(q_m = (m - 1/2)/n)$ propagate in opposite directions along the magnetic field since $k_{\parallel m} = -k_{\parallel m+1}$. Their beating forms a standing wave which is localized into a bound state to form a discrete spectrum.

8.7.2.2 High n Ballooning Mode Equation

For sufficiently high n mode Alfven Eigenmode, we can start from the ideal ballooning mode equation given in (8.89) by adding inertial term from kinetic energy integral $\int \frac{1}{2}\rho\omega^2|\xi|^2 dV$ under the approximation of $\nabla \cdot \boldsymbol{\xi} = 0$.

$$\frac{1}{J}\frac{d}{d\theta}\left[\frac{|\nabla\alpha|^2}{JB^2}\frac{d\Phi}{d\theta}\right] + \frac{\rho\omega^2}{B^2}|\nabla\alpha|^2\Phi + 2\mu_0 p'(\psi)\kappa_w \Phi = 0 \tag{8.157}$$

In the large aspect ratio and low beta circular tokamak, this equation becomes:

$$\frac{d}{d\theta}\left[f(\theta)\frac{d\Phi}{d\theta}\right] + g(\theta)\Phi + \Omega^2 f(\theta)(1 + 4\epsilon cos\theta)\Phi = 0 \tag{8.158}$$

where $\epsilon = r/R$, $\Omega^2 = \omega^2/\omega_A^2$, $\omega_A = v_A/qR, f(\theta) = (1 + 2\Delta' cos\theta)(1 + h^2(\theta))$, $h(\theta) = s\theta - \alpha_p sin\theta$, $s = rq'(r)/q$ is the magnetic shear, $\alpha_p = -2\mu_0 q^2 R p'(r)/B^2$, Δ' is a radial derivative of Shafranov shift, $g(\theta) = 2\mu_0 p'(r) rq\kappa_w = \alpha_p[cos\theta + h(\theta) sin\theta]$ [116, 218]. If we define $\psi = \sqrt{f(\theta)}\Phi$, this ballooning mode equation leads to the Mathieu like equation in Sect. 8.1, (8.41).

$$\frac{d^2\psi}{d\theta^2} + [\Omega^2(1 + 4\epsilon cos\theta) - F(\theta)]\psi = 0 \tag{8.159}$$

$$\text{where, } F(\theta) = \frac{1}{\sqrt{f(\theta)}}\frac{d^2\sqrt{f(\theta)}}{d\theta^2} - \frac{g(\theta)}{f(\theta)} \tag{8.160}$$

If we neglect toroidal effect $\epsilon \to 0$, pressure gradient and Shafranov shift, the ballooning mode equation (8.158) becomes:

$$\frac{d}{d\theta}\left[(1 + s^2\theta^2)\frac{d\Phi}{d\theta}\right] + \Omega^2(1 + s^2\theta^2)\Phi = 0 \tag{8.161}$$

This equation is equivalent to the cylindrical high n reduced MHD equation (8.154) in the $n \to \infty$ limit. Consider the quasi-mode expansion of Φ_m with the eigenmode expressed in the Fourier transform in the radial direction:

$$\Phi_m = \frac{1}{2\pi}\sum_{k=-\infty}^{\infty} e^{i(k\theta - n\zeta)} \int_{-\infty}^{\infty} \tilde{\Phi}(nq - k)(\eta) e^{i\eta(nq-k)} d\eta \tag{8.162}$$

This leads to the following transformations to produce (8.161) from (8.154).

$$\frac{d\Phi}{dr} = \frac{dq}{dr}\frac{d\Phi}{dq} \to \frac{inqs\eta}{r}\tilde{\Phi}, k_\parallel \Phi \to \frac{i}{Rq}\frac{d\tilde{\Phi}}{d\eta} \tag{8.163}$$

Here $s = r(dq/dr)/q$ is magnetic shear.

If we retain the toroidal effect ϵ and neglect pressure gradient and Shafranov shift, the ballooning mode equation (8.158) becomes:

$$\frac{d}{d\theta}\left[(1+s^2\theta^2)\frac{d\Phi}{d\theta}\right]+\Omega^2(1+s^2\theta^2)(1+4\epsilon cos\theta)\Phi=0 \tag{8.164}$$

If we define $\psi \equiv \sqrt{1+s^2\theta^2}\Phi$ and $F = s^2/(1+s^2\theta^2)$, we obtain semi-Mathieu equation (8.41): $d^2\psi/d\theta^2 + [\Omega^2(1+4\epsilon cos\theta) - F(\theta)]\psi = 0$. We can solve this equation by the two-scale asymptotic matching method using $\epsilon \ll 1$ as a small parameter [115]. Since the lowest eigenvalue λ of Mathieu equation (8.44) for small h is $\lambda \sim 1$, we obtain a lowest eigenvalue given as $\Omega^2 = 1/4$. Thus the lowest order solution becomes $\psi \sim \psi_c cos\Omega\theta + \psi_s sin\Omega\theta \sim \psi_c cos(\theta/2) + \psi_s sin(\theta/2)$. Including slow θ dependence for ψ_c and ψ_s, we will get equation for $\psi_c(\theta)$ and $\psi_s(\theta)$. Cheng-Chen-Chance solved this equation analytically to obtain TAE discrete spectrum inside the gap for even parity and no discrete spectrum for the odd parity. As is clear from the following equations, high n TAE frequency lies lower region of the gap for low magnetic shear ($s \ll 1$) and it goes up to the upper region of the gap as it increases ($s \gg 1$).

$$\Omega^2 \sim \frac{1}{4}\frac{1}{1+4\epsilon(1-s^2\pi^2/8)} \text{ (for } s \ll 1) \tag{8.165}$$

$$\Omega^2 \sim \frac{1}{4}\frac{1}{1-4\epsilon(1-\pi^2/72s^4)} \text{ (for } s \gg 1) \tag{8.166}$$

Column 8-7: Derivation of High-n Reduced MHD Equation
In the ideal MHD theory, the perturbed parallel electric field vanishes, i.e. $E_\parallel = -\boldsymbol{b}\cdot\nabla\phi - \partial A_\parallel/\partial t = 0$. This leads to a relation between the electrostatic potential ϕ and the parallel vector potential $A_\parallel$.

$$-\boldsymbol{b}\cdot\nabla\phi + i\omega A_\parallel = 0 \tag{8.167}$$

The charge neutrality condition is given by the divergence free condition for the current density, i.e. $\nabla\cdot\boldsymbol{j} = 0$.

$$\boldsymbol{b}\cdot\nabla j_\parallel + \nabla\cdot\boldsymbol{j}_\perp = 0, \tag{8.168}$$

where $j_\parallel$ is the parallel current density governed by Faraday's law.

$$\mu_0 j_\parallel = -\nabla^2 A_\parallel = -\nabla^2(\boldsymbol{b}\cdot\nabla\phi)/i\omega \tag{8.169}$$

(continued)

The perpendicular current is given by the polarization current, $\boldsymbol{j}_\perp = e_i n_i \boldsymbol{u}_{pi}$, where $\boldsymbol{u}_{pi}$ is the ion polarization drift given by (5.25). Using the fluid velocity $\boldsymbol{u}_i = \boldsymbol{E} \times \boldsymbol{B}/B^2 = -\nabla\phi \times \boldsymbol{b}/B$ and $\rho = e_i n_i$, we have

$$\boldsymbol{j}_\perp = \frac{\boldsymbol{b} \times \rho(\partial \boldsymbol{u}_i/\partial t)}{B} = \frac{i\omega\rho\nabla_\perp\phi}{B^2} \tag{8.170}$$

Substituting expressions for $j_\parallel$ and $\boldsymbol{j}_\perp$ into the charge neutrality condition, we have (8.151).

8.7.3 Categories of AE Modes

Large number of Alfven eigenmodes (AEs) are identified theoretically and experimentally such as GAE (Global Alfven Eigenmode), TAE (Toroidicity induced Alfven Eigenmode), EAE (Ellipticity induced Alfven Eigenmodes), NAE (Non-circular triangularity induced Alfven Eigenmodes), RSAE (Reversed Shear Alfven Eigenmode), BAE (Beta-induced Alfven Eigenmode), BAAE (Beta-induced Alfven Acoustic Eigenmode), KTAE (Kinetic Toroidicity-induced Alfven Eigenmode), and CAE (Compressional Alfven Eigenmode). The radial location and frequency range of these AEs for standard ITER operation are shown in Fig. 8.31a.

The TAE is formed in the TAE gap by the coupling of m and $m+1$ modes, where m and $m+1$ harmonics propagate in opposite directions along the magnetic field (note: $k_{\parallel m+1} = -k_{\parallel m}$ and see Fig. 8.31b) and their beat generates a standing wave and the equilibrium inhomogeneity localizes the standing wave into a bound state [866]. The nominal frequency of the TAE is $\omega_{TAE} = \omega_A/2$, where $\omega_A = V_A/qR$. TAE is first observed in TFTR by K.L. Wong [829].

Shear Alfven gap is also possible through other coupling mechanisms such as elongation of the plasma shape or higher shaping both found by Betti-Freidberg, which are called Ellipticity induced Alfven Eigenmodes (EAE) [63] ($\omega_{EAE} \sim \omega_A$) and Noncircular triangularity induced Alfven Eigenmodes (NAE) [64] ($\omega_{NAE} \sim 3\omega_A/2$), respectively. EAE is first observed in JET by Fasoli [199] and NAE is first observed in JT-60U by Kramer [474].

The shear Alfven eigenmode structures of continuum, GAE and gaps excited by the external means are shown in Fig. 8.32. In the shear Alfven resonance, the eigenmode structure is singular at the resonance (Fig. 8.32a) while it does not show any singularity for the discrete modes such as GAE (Fig. 8.32b) and TAE/ EAE (Fig. 8.32c).

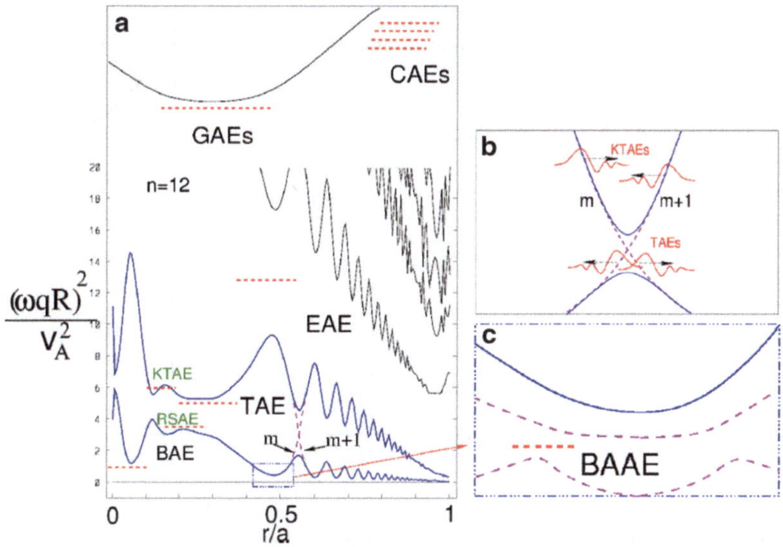

Fig. 8.31 Alfven continuum gap structure in ITER and AEs such as CAE, GAE, EAE, TAE, KTAE, RSAE, BAE, BAAE calculated by NOVA [117]. *Dashed curves* correspond to the coupling of Alfven and sound waves. *Dashed horizontal lines* correspond to the eigenmode solutions. Reproduced with permission from Gorerenkov [256]. Copyright IAEA Vienna

The effect of finite pressure gradient (α_p) to high-n TAE is analyzed by Fu-Cheng [217]. With the increase in α_p, TAE frequency is reduced to reach lower boundary of the frequency gap where strong absorption by continuous spectrum occurs. Therefore there exists a critical value of α_p above which TAE is damped. The condition is given as $\alpha_c = s^2/(1+s)$ for $s < 1$ and $\alpha_c = s + 1 - \sqrt{2s+1}$ for $s \gg 1$. This condition is broken if he magnetic shear is smaller than the inverse aspect ratio, $s < \epsilon$. Near the magnetic axis, special two kinds of TAE can be observed called cTAE (core-localized TAE) if the magnetic shear is small enough to satisfy $s < \epsilon$. One of the cTAE is first found by Fu [219] giving the $\alpha_c = 3\epsilon + 2s^2$, where $\epsilon = r/R$ and $s = rdq/qdr$ is the magnetic shear. Another cTAE is found by Berk [56]. These cTAEs are found to induce insignificant α particle loss in the reactor condition [96].

The kinetic effect such as finite ion Larmor radius and finite electron mass effects has important effects. Important kinetic theory of toroidicity-induced Alfven eigenmode is developed by Mett-Mahajan [550]. Inclusion of the kinetic Alfven wave (KAW, see Sect. 5.2.2) changes the mode structure of TAE localized at frequency gap to propagate in the outward direction with enhanced parallel electric field. Outward propagating KAW will be damped more strongly by the electron Landau damping. This mechanism is called the "radiative damping".

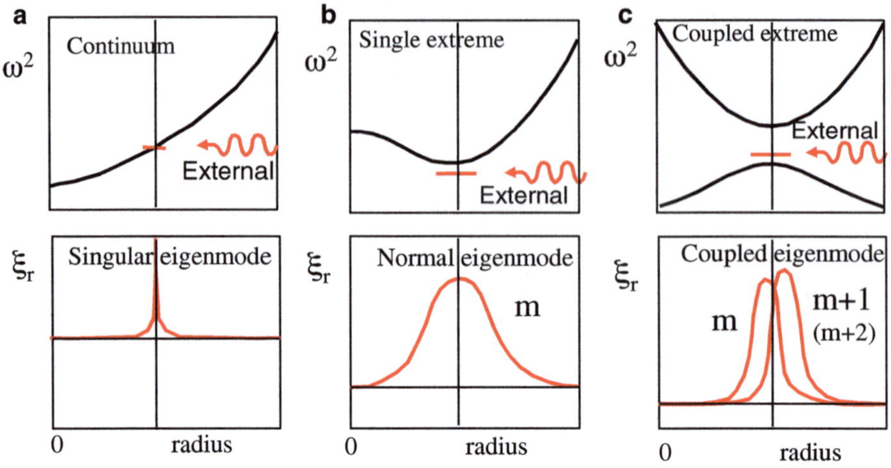

Fig. 8.32 Frequency spectrum and eigenmode structures of the plasma displacement in (**a**) continuum, (**b**) single extreme (GAE), and (**c**) coupled extreme (TAE/EAE). Modified from Heidbrink [314]

Candy-Rosenbluth [95] clarified kinetic effect to TAE produces new AE mode called KTAE (Kinetic Toroidicity-induced Alfven Eigenmode), whose frequency lies just above the TAE gap and propagates to the region between Alfven resonances and survives locally as shown in Fig. 8.31b. While KTAE is also possible below the gap, it propagates to outward subject to strong damping.

The eigenmode equation is obtained by adding finite Larmor radius term to Eq. (8.154).

$$\bar{\rho}^2 \frac{d^4}{dr^4}\Phi_m + \frac{d}{dr}\left[\left(\frac{\omega^2}{v_A^2} - k_{\|m}^2\right)\frac{d\Phi_m}{dr}\right] - \frac{m^2}{r^2}\left(\frac{\omega^2}{v_A^2} - k_{\|m}^2\right)\Phi_m$$
$$+ \epsilon_c \frac{\omega^2}{v_A^2}\frac{d^2}{dr^2}(\Phi_{m-1} + \Phi_{m+1}) = 0 \qquad (8.171)$$

where $\bar{\rho}^2 = \rho_i^2(3\omega^2/4v_A^2 + \tau k_{\|m}^2)$ and $\tau = T_e/T_i$. This KTAE is observed in JET by Fasoli [200].

GAE (see Sect. 8.1) is characterized by $d\omega_A/dr = 0$ can also exist in tokamak which is dominated by a single mode and localized near the extremum of the shear Alfven continuum.

A special case of $d\omega_A/dr = 0$ occurs with hollow q profile (NS: negative shear) and so-called RSAE (Reversed Shear Alfven Eigenmode) can exist. The mode structure of the RSAE is close to a cylindrical mode with toroidal mode number n and single poloidal mode number m. RSAE is associated with a minimum of the safety factor q profile as seen in Fig. 8.33a and produces local extremum of the shear Alfven continuum similar to the GAE. The mode frequency of RSAE is:

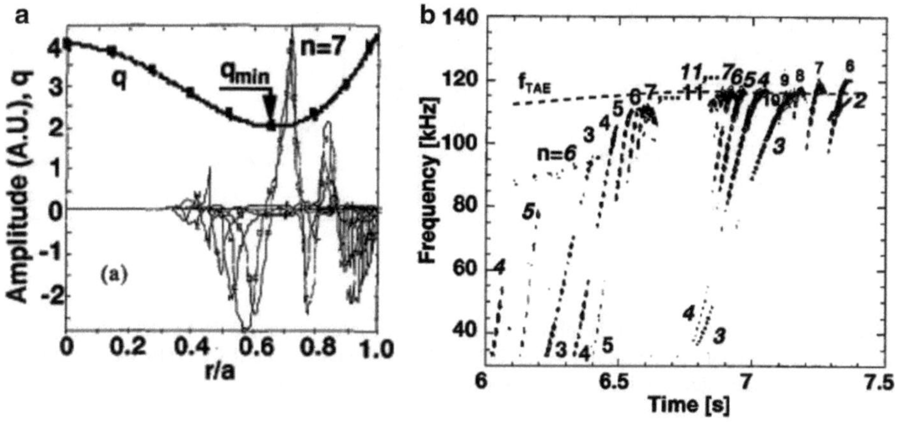

Fig. 8.33 (**a**) Eigenmode structure of RSAE [452]. (**b**) Typical frequency up chirping of RSAE in JT-60U. Reproduced with permission from Kimura [452] and Kusama [481]. Copyright IAEA Vienna

$$\omega \sim \omega_A \equiv k_{\parallel} V_A = V_A(m/q_{min} - n)/R, \tag{8.172}$$

where V_A is the Alfven velocity and R is major plasma radius. When q_{min} is rational number ($q_{min} = m/n$), the mode frequency is zero. The TAE gap near q_{min} is $q_{TAE} = m/n \pm 1/2n$ and $\omega_{TAE} = V_A/2q_{TAE}R$. Equation (8.176) implies RSAE may start from zero frequency at $q_{min} = m/n$ and chirp up to TAE frequency.

Kramer [475] showed finite pressure effect sets lowest RSAE frequency to a fraction (0.2–0.4) of TAE frequency according to the beta value (0.25–1 %) in agreement with JT-60U observations. When q_{min} decreases from m/n, the mode frequency chirps up to the TAE frequency as seen in Fig. 8.33b, whose time evolution is given by $d\omega_A/dt \sim mV_A dq_{min}^{-1}/Rdt$. RSAE is sometimes called Alfven Cascade Mode (ACM) due to this frequency chirping.

Breizman [77] showed that RSAE can exist due to second order in inverse aspect ratio within the ideal MHD theory as well as other higher order effects such as finite pressure effect [79, 475].

This RSAE is first observed in JT-60U (1996) by Kimura [452, 453] and is explained as a special case of GAE mode by Fukuyama [230, 231]. Kusama [481] showed clear chirping characteristics of RSAE, whose frequency approaches to TAE frequency as shown in Fig. 8.33b. Berk [59] developed a theory of RSAE to explain up chirping in RSAE. Nazikian also showed α particle driven AE in WNS TFTR plasma [577] is RSAE [578]. RSAE mode has been used to determine q_{min} [684].

Takechi [736] further studied time evolution and structure of RSAE. As q_{min} decreases in the range of $(m+1/2+\epsilon)/n < q_{min} < (m+1)/n$, there are two RSAEs, HRSAE (Higher frequency RSAE) and LRSAE (Lower frequency RSAE), whose frequencies are given by $f_{HRSAE} \sim (n-m/q_{min})v_A/2\pi R$ and $f_{LRSAE} \sim ((m+1)/q_{min} - n)v_A/2\pi R$. When q_{min} decreases further in the range of $m/n < q_{min} < (m + 1/2 +$

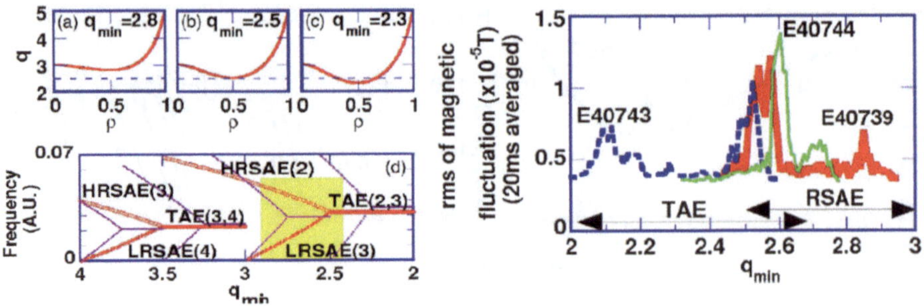

Fig. 8.34 Amplification of AE amplitude by RSAE-TAE interaction and RSAE-TAE diagram. As q_{min} reduces, there are up chirping RASE (LRSAE) and down chirping RSAE (HRSAE) to collide to form TAE. Reproduced with permission from Takechi [736]. Copyright IAEA Vienna

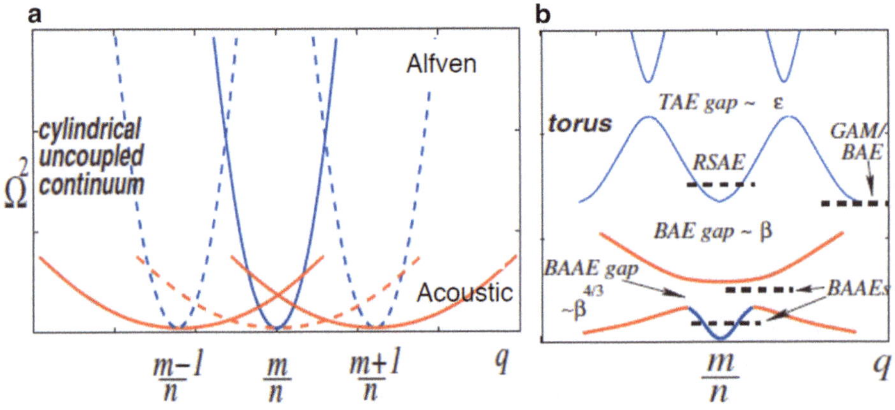

Fig. 8.35 (**a**) Alfven and sound wave continuum in cylindrical tokamak. (**b**) BAE and BAAE gap structure by the interaction between Alfven and sound wave resonances. Reproduced with permission from Gorelenkov [255]. Copyright AIP Publishing LLC

$\epsilon)/n$, TAE gap is formed and TAE frequency is given by $f_{TAE} \sim v_A/4\pi q_{TAE}R$, where $q_{TAE} = (m - 1/2)/n$. Schematic diagram of AE evolution is shown in Fig. 8.34. Importantly, AE mode amplitude becomes bigger when the transition from RSAE to TAE mode occurs.

At finite beta, inclusion of compressibility ($\nabla \cdot \boldsymbol{\xi} \neq 0$) in the MHD equation produces coupling of acoustic wave and the shear Alfven wave. An additional beta induced gap structure appears below the Alfven continua as seen in Fig. 8.35a. Inside this beta induced gap, Beta induced Alfven eigenmode (BAE) exists found by M.S. Chu [122] and A.D. Turnbull [778] numerically and experimentally by Heidbrink [313]. The BAE mode frequency is roughly half the TAE mode frequency. The BAE can be destabilized by the energetic electrons as shown by Cheng [120] in HL-2A. Coupling between shear Alfven continuum and sound wave continuum also produces another important gap structure BAAE gap as seen in Fig. 8.35b and also Beta induced Alfven Acoustic Eigenmode (BAAE) by Gorelenkov [254].

A unique $n = 0$ chirping mode is observed in JET ICRF experiment, where energetically inverted ion distribution function is formed in the high field side. Berk explained it as Global Geodesic Acoustic Mode (GGAM) [60]. The dispersion relation of GGAM is given by $\omega^2 = (\gamma p/\rho R^2)(2 + 1/q^2)$.

There are several damping mechanisms, such as electron Landau damping by Fu-Van Dam [216], ion Landau damping by Betti-Freidberg [64], collisional damping of trapped electrons by Fu-Cheng [218] continuum damping by Zonca-Chen [865] and Berk [55], and radiative damping by Mett-Mahajan [550]. Direct measurement of the damping rate was made by Fasoli [198]. The balance between instability drive and damping gives the threshold $d\beta_{fast}/dr$, discussed for burning plasma experiments in [253]. Not only α particles but also MeV-class N-NBI fast ion drive Alfven eigenmodes and beam deposition profile should be carefully considered for the steady state operation scenario. So the off-axis neutral beam injection current drive (NBCD) scenario is favorable to reduce central $d\beta_{fast}/dr$. Actually tilting of N-NBI is considered to avoid TAE in ITER [614].

An interim review of Alfven Eigenmodes can be found in Heidbrink [312], Mahajan [530], Wong [830], Zonca [868], Chen-Zonca [114], Heidbrink [314], Breizman-Sharapov [78], and Gorelenkov-Pinches-Toi [256].

8.7.4 Energetic Particle Modes (EPM)

The AEs in previous section are normal modes of the background thermal plasma (a kind of cavity modes). In these modes, the energetic particles (EPs) only contribute to the imaginary part of the dielectric constant.

L. Chen [111] showed that completely new class of instabilities called Energetic Particle Modes (EPM) appear when the EP pressure is comparable to the thermal pressure. Since EPs modify real part of the dielectric constant, mode can be destabilized where strong continuum damping is expected. The EPM is a type of beam mode with the background plasma supplying neutralizing charge [314].

Important EPMs are the fishbone instability [109] as an interaction between internal kink mode and EPs observed in PDX, EGAM (EP driven GAM), where both the mode frequency and structure are strongly modified by the EP dynamics [220]. The EWM given in Sect. 8.8.5 may be a kind of EPM.

Following the historical back ground, we start from the Fishbone instability" and then describe energetic particle mode (EPM).

8.7.4.1 Fishbone Instabilities

The fishbone instability is first observed in PDX by McGuire [545]. L. Chen (1984) explained this instability as EP driven internal kink mode [109].

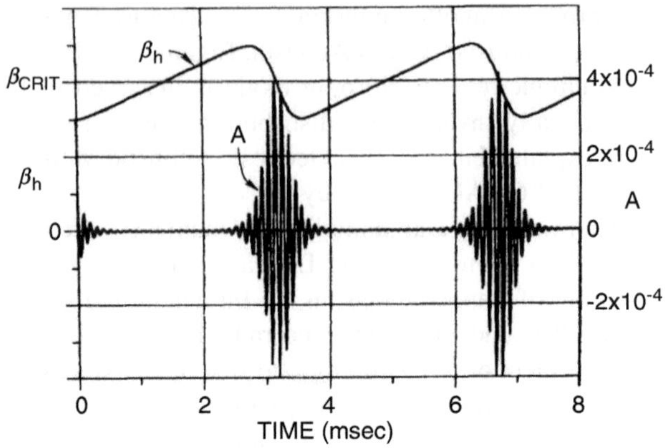

Fig. 8.36 Time evolution of energetic trapped particle beta β_h and mode amplitude of the internal kink component in the fishbone instability. Reproduced with permission from L. Chen [109]. Copyright American Physical Society

It occur when the plasma is marginally stable to the internal kink mode ($\delta W_{MHD} \sim 0$) and is destabilized by the energetic particle contribution to the energy integral ($\delta W_k > 0$). They derived a system of equations for mode amplitude A and energetic trapped particle beta β_h to show similar behavior with the experiments as shown in Fig. 8.36.

The linear gyro kinetic equation for hot particle under electromagnetic perturbation can be derived as shown by Antonsen-Lane [22] similar to the electrostatic one (5.100) and (5.108).

$$f_{h1} = \frac{e_h}{m_h}\left[\Phi_1 \frac{\partial}{\partial K} - \frac{\mu}{\Omega_h} B_{1\parallel} \frac{\partial}{\partial \mu}\right] f_{h0} + g_h \tag{8.173}$$

$$\left[v_\parallel \frac{\partial}{\partial l} - i(\omega - \omega_{dh})\right] g_h = i\frac{e_h}{m_h} Q\delta\psi \tag{8.174}$$

where finite Larmor radius effect is neglected. Here, $K = v^2/2$ and $\mu = v_\perp^2/2B$ are the kinetic energy and magnetic moment per unit mass, respectively and $\Omega_h = e_h B/m_h$ is the cyclotron frequency of hot ion, $\partial/\partial l \equiv \boldsymbol{b} \cdot \nabla$ is the parallel derivative, $\delta\psi = \tilde{\Phi} - v_\parallel A_{1\parallel} + v_\perp^2 B_{1\parallel}/2\Omega_h$, $Q = (\omega\partial/\partial K + \hat{\omega}_{*h}) f_{0h}$: $\hat{\omega}_{*h} = -(i/\Omega_h)(\boldsymbol{b} \times \nabla ln f_{h0}) \cdot \nabla$, $\omega_{dh} = -i\boldsymbol{v}_{dh} \cdot \nabla$, $\boldsymbol{v}_{dh}$ is the toroidal drift velocity of the hot ion, and perturbed quantities Φ_1 and $A_{1\parallel}$ are related to $\boldsymbol{\xi}$ as $\nabla\Phi_1 = -i\omega\boldsymbol{\xi} \times \boldsymbol{B}$, and $\omega A_{1\parallel} = -i\partial\Phi_1/\partial l$ from vanishing parallel electric field condition. Here the bounce average of f is defined as $\bar{f} = (\oint f dl/|v_\parallel|)/(\oint dl/|v_\parallel|)$ and $J = (\mu B/2K)\nabla\cdot\boldsymbol{\xi}_\perp - (1 - 3\mu B/2K)(\boldsymbol{\xi}_\perp \cdot \boldsymbol{\kappa})$ and $\kappa = \boldsymbol{b} \cdot \nabla \boldsymbol{b} = \partial \boldsymbol{b}/\partial l$ is the curvature of the magnetic field.

For a frequency ω much smaller than the hot ion transit and bounce frequencies, Eq. (8.182) is solved for trapped (t) and untapped (u) particles as $g_{h,u} = -e_h Q\Phi_1/m_h\omega$ and $g_{h,t} = -e_h Q\Phi_1/m_h\omega + \delta g_{h,t}$, where $\delta g_{h,t} = 2QK\bar{J}/(\omega - \bar{\omega}_{dh})$.

Using these perturbed velocity distribution function for trapped and untrapped particles, we can evaluate parallel and perpendicular pressure perturbations of fast particle, $p_{1f\perp}$ and $p_{1f\parallel}$.

$$\begin{bmatrix} p_{1f\perp} \\ p_{1f\parallel} \end{bmatrix} = 2^{7/2}\pi m_h B \int_{B_{max}^{-1}}^{B^{-1}} d\alpha \sqrt{1-\alpha B} \int_0^{\infty} dK \frac{K^{5/2} Q}{\omega - \bar{\omega}_{dh}} \bar{J} \begin{bmatrix} \alpha B/2(1-\alpha B) \\ 1 \end{bmatrix} \quad (8.175)$$

The internal kink instability is close to the marginal stability and has critical local poloidal beta value (β_{pc}) so that ideal fluid energy integral can be modeled as $W_{MHD} \sim (\beta_{pc}^2 - \beta_{ps}^2)$, where β_{ps} is the local poloidal beta at $q = 1$ singular surface. From the dispersion relation (8.134), L. Chen derived a model evolution equation for mode amplitude, $A = B_{r1}/B$ combined with fast particle beta evolution equation as, $dA/dt = \Gamma(\beta_h - \beta_{crit})A$, and $d\beta_h/dt = D - AZ\beta_{max}\theta(\beta_h - \beta_{min})$, where $\theta(\beta_h - \beta_{min})$ is a Heaviside function, $\Gamma \sim 1.1 \times 10^7 s^{-1}$, $\beta_{crit} \sim 0.0025, D \sim 0.5 s^{-1}$, $Z \sim 2.5 \times 10^6 s^{-1}$, $\beta_{max} = \beta_{crit}/(1 - f/2), f = 0.4$, $\beta_{min} = (1-f)\beta_{max}$.

This theory implies that introduction of the energetic particles produces two modes, one is discrete MHD mode and the other is energetic particle mode whose frequency is characteristic energetic particle frequency such as transit, bounce and precession.

8.7.5 Nonlinear AE Modes

Interaction of AE with energetic particles produces radial transport or loss of fast ions [180], damage of first wall [817] and reduction of fusion reaction [385, 687]. These processes may become strongly nonlinear. There are typically four nonlinear attractor behaviors, (1) saturation, (2) limit cycle oscillation, (3) chaotic nonlinear state, (4) explosive growth with mode frequency sweeping. Type (4) is observed by Shinohara [687–689] as shown in Fig. 8.37, called the "Fast Sweeping Mode", which could be explained by the Berk-Breizman model to be given below. The mode is associated with the "Abrupt Large Event (ALE)", which could be explained by the EPM [80].

So-called Berk-Breizman (BB) model provides a clear view of nonlinear behavior of AE, which naturally explains type (4) instability [57, 58].

Key idea is that the coherent structure with varying frequency represents nonlinear traveling waves in fast particle phase space. Like the beam modulation instabilities, the frequency modulation occurs spontaneously as a result of the resonant particle trapping by the excited wave. The initial modulation matches the frequency of a plasma eigenmode. frequency shifts from the initial frequency occurs as the coherent structure evolves similar to the nonlinear Bernstein-Greene-Kruskal (BGK) mode [61] rather than slow evolution of the linear eigenmode.

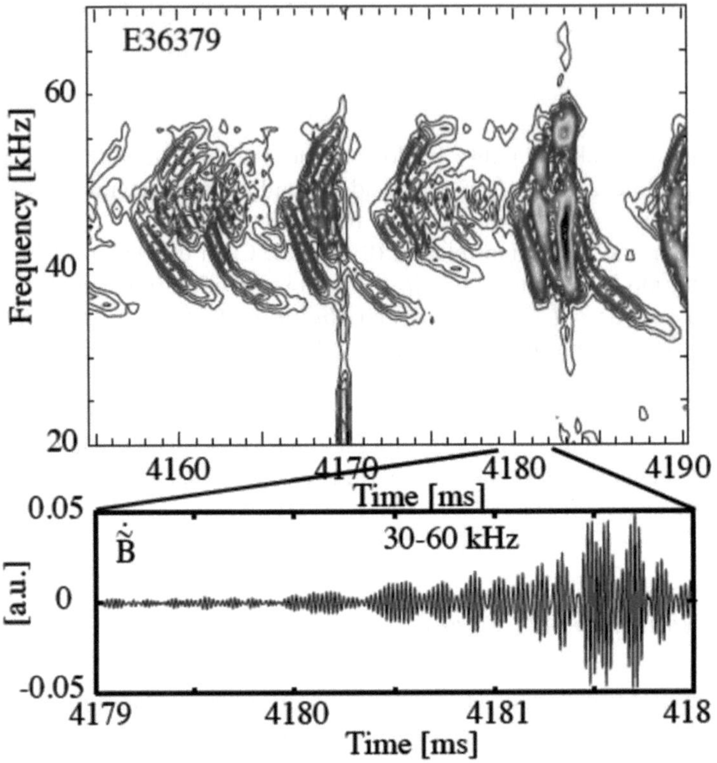

Fig. 8.37 Fast Frequency Sweeping (fast FS) mode observed in JT-60U. Reproduced with permission from Shinohara [687]. Copyright IAEA Vienna

Column 8-8: BGK Mode [473, 771]

The Bernstein-Greene-Kruskal (BGK) mode [61] is an exact static solution of Vlasov-Maxwell equations (Fig. 8.38). It can be found by taking $\partial f_a/\partial t = 0$ while keeping the finite electrostatic potential Φ. One dimensional Vlasov-Poisson equation can be expressed as,

$$v_x \frac{\partial f_a}{\partial x} - \frac{e_a}{m_a}\frac{\partial \Phi}{\partial x}\frac{\partial f_a}{\partial v_x} = 0 \tag{8.176}$$

$$\epsilon_0 \frac{\partial^2 \Phi}{\partial x^2} = e\int f_e(x, v_x)dv_x - eZ_i\int f_i(x, v_x)dv_x \tag{8.177}$$

The general solution of (8.180) is $f_a = f_a(m_a v_x^2/2 + e_a\Phi/m_a)$ $(a = e, i)$. This has a simple meaning that any function of the total energy $E = m_a v_x^2/2+$

(continued)

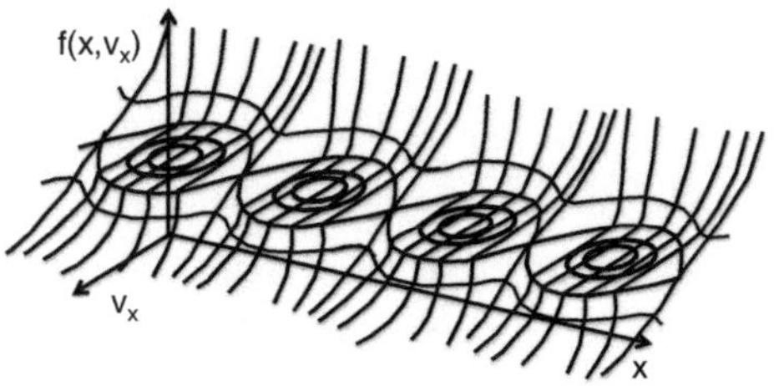

Fig. 8.38 Typical BGK solution showing the phase space vortex. If $V(\Phi)$ has a well in the region of interest, periodic wave solution with a discrete spectrum appears. Interaction of the large amplitude wave with the wave-trapped particles causes spatial modulation of the wave potential

$e_a\Phi$ is a solution of this Vlasov equation. Particle motion in the phase space (x, v_x) is along the constant E surface. If the phase space density (distribution function) is constant on constant E surface, it can be a steady state solution.

Converting the variable v_x to E, the velocity integral is given as follows,

$$\int_{-\infty}^{+\infty} f_a(v_x)dv_x = 2\int_{e_a\Phi}^{\infty} \frac{f_a(E)dE}{\sqrt{2m_a(E - e_a\Phi)}} \tag{8.178}$$

Substituting this expression into (8.177), right and side becomes a function of Φ denoted by $\epsilon_0 G(\Phi)$. The Poisson equation becomes $\partial^2\Phi/\partial x^2 = G(\Phi)$. Multiplying by $\partial\Phi/\partial x$ and integration by x, we have,

$$\frac{1}{2}\left(\frac{d\Phi}{dx}\right)^2 + V(\Phi) = const, \quad \text{where } V(\Phi) = -\int_{\Phi_0}^{\Phi} G(\Phi)d\Phi \tag{8.179}$$

We obtain following solution for $x(\Phi)$.

$$x - x_0 = \pm\int_{\Phi_0}^{\Phi} d\Phi/\sqrt{2(V(\Phi) - V(\Phi_0))} \tag{8.180}$$

Each Lagrangian is a summation of free-particle, field-particle, and field Lagrangian [443]. Simplified Lagrangian is given by assuming adiabatic response for the background plasma and L_w is approximated by the quadratic form of wave amplitude [57] and its contribution can be regarded as a part of electric field, $\tilde{E}$.

The nonlinear behavior of AE is determined by a competition among drive by resonant particles, the external damping, the particle relaxation to recover positive df/dv, and particle trapping to smooth it. Chirping solutions occurs in rare collision regime when hole and clump structure is formed in phase space and the clump propagates toward low energy (i.e. low frequency) and hole propagates toward

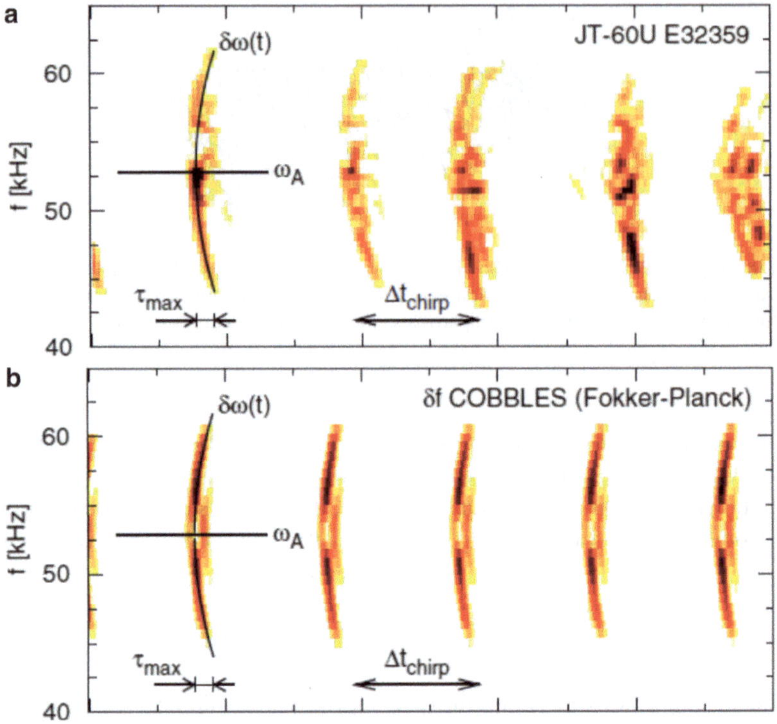

Fig. 8.39 (**a**) Comparison of fast FS mode in JT-60U and (**b**) simulation using Berk-Breizman model. Reproduced with permission from Lesur [500]. Copyright IAEA Vienna

high energy (i.e. high frequency). [58]. Lesur [499] developed a code to solve corresponding full f and delta-f Vlasov equation numerically and explained above frequency sweeping (FS) as shown in Fig. 8.39 showing the effectiveness of BB model and also importance of collisional drag and diffusion in velocity space [500, 509].

Salon: Phase Space Hole in Space Plasma Physics [603]
A quasi-static stationary structure is observed by an high time resolution electric field probe in 1977 in the auroral ionosphere, later called the electrostatic solitary wave (ESW). Tetreault [762] implied that the ESW is caused by the phase-space hole predicted as one of BGK mode. The BGK hole is associated with localized depletion of phase space densities and they propagate at thermal speed of the plasma. The observed slowly moving structure (a few tens km/s) corresponds to ion phase space hole and fast moving structure (up to 1000 km/s) is electron hole. These holes are also observed in the plasma sheet boundary layer, magnetosheath, and bow shock.

8.8 Resistive Wall Modes

8.8.1 RWM in the Cylindrical Tokamak

All advanced tokamak operating regimes (WS, NS, CH) for steady state tokamak operation require wall stabilization or are benefitted to increase operating plasma beta value by wall stabilization. As discussed in Sect. 1.5, maximum normalized beta attainable without wall stabilization is $\beta_N = 2.8$ according to Troyon [772].

The wall stabilization works for short time scale $t \ll \tau_w$ with the resistive wall time constant $\tau_s = \mu_0 \sigma b d$, where σ is the electrical conductivity, b is the radius of wall, d is the thickness of the wall. As discussed in Freidberg [211], the resistive wall do not stabilize the external kink modes but reduces the growth rate significantly, where we can assume $\omega \sim 0$. The mode is called the **resistive wall mode** (RWM). Since $\omega \sim 0$, the mode behavior in the cylindrical plasma is governed by the marginal stability (Newcomb) equation.

$$L \equiv \frac{d}{dr}\left(f(r)\frac{d\xi}{dr}\right) - g(r)\xi = 0 \tag{8.181}$$

Taking $\xi \times$ (8.181) and integrating over the plasma volume, Freidberg gives following form of energy integrals for the case of wall at infinity (δW_∞) and for the case of ideal wall at $r = b$ (δW_b).

$$\delta W_\infty = \frac{2\pi^2 R_p}{\mu_0}\left[\frac{F\hat{F}}{k_0^2} + \frac{r^2 \Lambda_\infty F^2}{|m|} + \frac{F^2}{k_0^2}\left(\frac{r\xi'}{\xi}\right)\right]_a \xi_a^2 \tag{8.182}$$

$$\delta W_b = \delta W_\infty + \frac{2\pi^2 R_p}{\mu_0}\left[\frac{r^2 F^2}{|m|}\right]_a (\Lambda_b - \Lambda_\infty)\xi_a^2 \tag{8.183}$$

$$\Lambda_b - \Lambda_\infty = \Lambda_\infty \frac{(I'_m(ka)/K'_m(ka)) - (I_m(ka)/K_m(ka))}{(I'_m(kb)/K'_m(kb)) - (I'_m(ka)/K'_m(ka)}, \tag{8.184}$$

where $\Lambda_\infty = -|m|K_m(ka)/kaK'(ka)$, $F = kB_z + mB_\theta/r$, $\hat{F} = kB_z - mB_\theta/r$, and $k_0^2 = k^2 + m^2/r^2$. The resistive wall mode occurs when $\delta W_\infty < 0 < \delta W_b$.

Solution of the perturbed magnetic field at two vacuum regions separated by the thin wall can be expressed by the scalar potential as $\boldsymbol{B}_1 = \nabla\phi$, where ϕ is given by the modified Bessel functions, $I_m(kr)$ and $K_m(kr)$. With appropriate boundary conditions at $r = b$ and $r = \infty$, Freidberg obtained following dispersion relation.

$$\gamma\tau_s = \frac{k^2b^2 + m^2}{k^2b^2K'_m(kb)I'_m(kb)[1 - (I'_m(ka)K'_m(kb)/I'_m(kb)K'_m(ka)]}\frac{\delta W_\infty}{\delta W_b} \tag{8.185}$$

This can be rewrite in a simple form by defining effective wall time constant τ_w.

$$\gamma\tau_w = -\frac{\delta W_\infty}{\delta W_b}, \tag{8.186}$$

where $\tau_w = \tau_s(1 - (a/b)^{2|m|})/2|m|$ in tokamak ordering ($kb \ll 1$).

8.8.2 RWM in General Tokamak Equilibrium

In this section, we sketch the formulation of the Haney-Freidberg [285] on the RWM stability in general tokamak equilibrium, which in the end gives same formula for RWM growth rate, Eq. (8.70). The procedure is surprisingly similar to that for the cylindrical plasma given in the previous section.

Due to the stabilizing effect of the eddy current in the resistive wall, the plasma inertial effect in (8.1) is negligible on the time scale of interest. So the relevant equation is $2D$ Newcomb equation.

$$\boldsymbol{F}(\boldsymbol{\xi}) = 0, \tag{8.187}$$

where $\boldsymbol{F}$ is the linear ideal MHD operator. Similar to the cylindrical tokamak case, we take $\boldsymbol{\xi}\cdot\boldsymbol{F}(\boldsymbol{\xi})$ and integrate over the plasma volume, we have following equation.

$$L \equiv \int_P \boldsymbol{\xi}\cdot\boldsymbol{F}(\boldsymbol{\xi})dV = 0, \tag{8.188}$$

which can be put in the following familiar form (see Column 8-1).

$$L = \delta W_P + \delta W_V \tag{8.189}$$

$$\delta W_p = \frac{1}{2}\int_P \left(\frac{|\delta\boldsymbol{B}|^2}{\mu_0} - \boldsymbol{\xi}\cdot(\boldsymbol{J}\times\boldsymbol{B}_1) - \Gamma p|\nabla\cdot\boldsymbol{\xi}|^2 + (\boldsymbol{\xi}\cdot\nabla p)\nabla\cdot\boldsymbol{\xi} \right) dV \tag{8.190}$$

Here we use Γ for adiabatic constant instead of γ to differentiate with the growth rate. The vacuum energy δW_V can be divided into the vacuum energies inside and outside of the resistive shell and the surface energy at thin resistive wall, by which L can be given in the following form.

$$L = \delta W_P + \delta W_{V_i} + \delta W_{V_o} + \frac{\sigma\gamma d}{2}\int_{S_b} |\boldsymbol{n}\times\boldsymbol{A}_1|^2 dS \tag{8.191}$$

$$\delta W_{V_i} = \frac{1}{2\mu_0}\int_{V_i} |\nabla\times\boldsymbol{A}_{1i}|^2 dV, \delta W_{V_o} = \frac{1}{2\mu_0}\int_{V_o} |\nabla\times\boldsymbol{A}_{1o}|^2 dV \tag{8.192}$$

Here, $\boldsymbol{n}$ is outward-facing unit normal vector to shell surface: S_b and σ is the electrical conductivity of the resistive wall. The energy integrals for the case of wall at infinity (δW_∞) and of ideal wall at $r = b$ (δW_b).

$$\delta W_\infty = \delta W_P + \delta W_V^\infty \tag{8.193}$$

$$\delta W_V^\infty = \frac{1}{2}\int_V \frac{|\boldsymbol{B}_1^\infty|^2}{\mu_0} dV = \frac{1}{2\mu_0}\int_{S_p} (\boldsymbol{n} \times \boldsymbol{A}_1^\infty) \cdot \boldsymbol{n} \times (\boldsymbol{n} \times \nabla \times \boldsymbol{A}_1^\infty) dS \tag{8.194}$$

$$\delta W_b = \delta W_P + \delta W_V^b \tag{8.195}$$

$$\delta W_V^b = \frac{1}{2}\int_V \frac{|\boldsymbol{B}_1^b|^2}{\mu_0} dV = \frac{1}{2\mu_0}\int_{S_p} (\boldsymbol{n} \times \boldsymbol{A}_1^b) \cdot \boldsymbol{n} \times (\boldsymbol{n} \times \nabla \times \boldsymbol{A}_1^b) dS \tag{8.196}$$

Here, no-wall eigenfunction is used commonly for both δW_b and δW_∞ and $\boldsymbol{n}$ is outward-facing unit normal vector to plasma surface: S_p. Expressing the perturbed vector potentials by a combination of $\boldsymbol{A}_1^\infty$ and $\boldsymbol{A}_1^b$, Haney-Freidberg obtained $\boldsymbol{A}_{1i} = (1 - c_2)\boldsymbol{A}_1^\infty + c_2\boldsymbol{A}_1^b$, and $\boldsymbol{A}_{1o} = (1 - c_2)\boldsymbol{A}_1^\infty$, where c_2 is a free parameter. They also give the energy integral L in the following form.

$$L(c_2) = \delta W_\infty + c_2^2(\delta W_b - \delta W_\infty) + \frac{\sigma\gamma d(1 - c_2)^2}{2}\int_{S_b} |\boldsymbol{n} \times \boldsymbol{A}_1^\infty|^2 dS \tag{8.197}$$

The free parameter c_2 can be determined from $\partial L/\partial c_2 = 0$, which is substituted into $L(c_2) = 0$. We obtain final form of the growth rate expression as:

$$\gamma\tau_w = -\delta W_\infty/\delta W_b, \tag{8.198}$$

where $\tau_w = \mu_0\sigma d\bar{b}$ and $\bar{b}$ is given by:

$$\bar{b} = \int_{S_b} |\boldsymbol{n} \times \boldsymbol{A}_1^\infty|^2 dS \Big/ \int_{S_b} (\boldsymbol{n} \times \boldsymbol{A}_1^\infty) \cdot \boldsymbol{n} \times (\boldsymbol{n} \times \nabla \times \boldsymbol{A}_1^b) dS \tag{8.199}$$

Hu-Betti [351] consider the mode resonance with particle drifts at the magnetic precession drift frequencies of trapped particles as a possible damping mechanism.

8.8.3 Ferromagnetic Wall Effect on RWM

Use of ferritic material in fusion DEMO and beyond is closely related to the choice of blanket structural material. Since reduced activation ferritite/Martensitic steel (RAF) is a primary candidate for the blanket structural material, effect of magnetization to RWM is an important subject. Ferromagnetism attracts perturbed magnetic field to wall and may destabilize RWM. Kurita [479] found that order of 8 % reduction in beta limit is expected for effective relative permeability $\mu_r = \mu/\mu_0 = 2$.

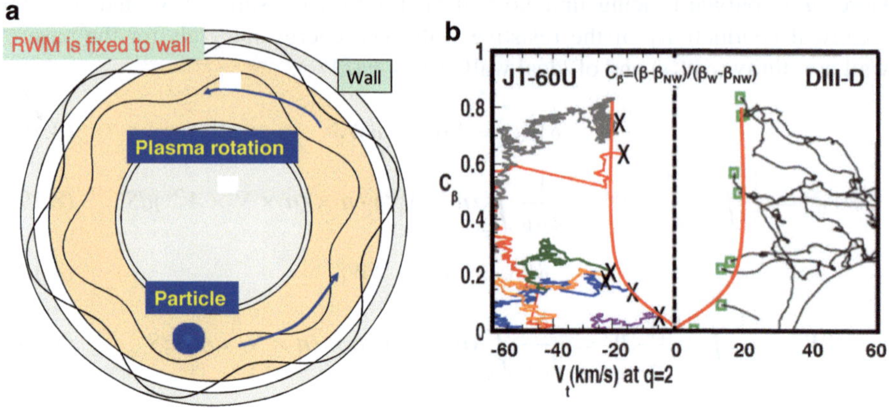

Fig. 8.40 (**a**) Schematics of rotating plasma, resistive wall and RWM fixed to the wall. (**b**) Discharge trajectories of DIII-D and JT-60U in $(V_t|_{q=2}, C_\beta)$ plane. *X* and *green open box* show onsets of RWM. Plasma is stable to RWM for absolute toroidal rotation above 20 km/s. Modified from Garofalo [245] and Takechi [737]

8.8.4 *Stabilization of RWM in Tokamak*

In a rotating tokamak, the RWM is fixed to wall and rotating plasma will slip with respect to the mode as shown in Fig. 8.40a. The plasma in the rotating frame sees RWM as a traveling wave and experiences resonant wave-particle interaction to damp this traveling wave by phase mixing of continuum modes (or by Landau damping of KAW produced by the mode conversion of the shear Alfven wave).

In 2007, both Reimerdes [623] in DIII-D and Takechi [737] in JT-60U showed that RWM is stabilized with small toroidal rotation as shown in Fig. 8.40b. Both machine showed that critical rotation speed is rather small at $\sim$20 km/s which is $\sim$0.3 % of Alfven velocity. This critical velocity to stabilize RWM is close to the expectation by the continuous damping of shear Alfven wave [623]. Such low toroidal rotation may be driven by so-called intrinsic rotation [167, 628]. While "small" toroidal rotation may stabilize RWM in ITER and DEMO, such rotation may be decelerated due to mode locking. In such case, rotational wall stabilization may be ineffective. To ensure wall stabilization, active feedback control of RWM using sector coils is important. Stabilization of RWM in high β, low toroidal rotation plasma is successfully demonstrated by Sabbagh [645].

8.8.5 *Energetic-Particle-Driven Wall Mode*

While RWM can be stabilized with toroidal rotation of >0.3 % of Alfven velocity, different bursting MHD instability called Energetic-particle-driven Wall Mode (EWM) was observed by Matsunaga [544] in beta regime between no-wall beta

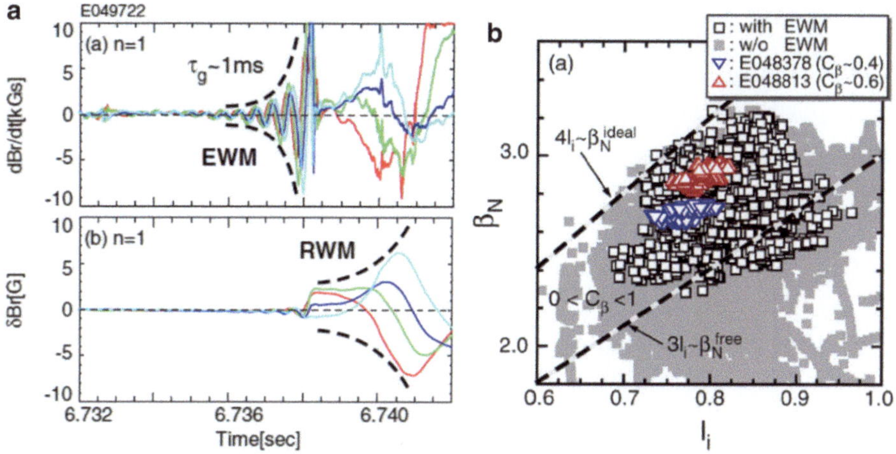

Fig. 8.41 (**a**) Observation of the Energetic-particle-driven Wall Mode by the perpendicular Neutral Beam Injection. (**b**) Observed range of EWM is between no-wall limit and with wall limit. Reproduced with permission from Matsunaga [544]. Copyright American Physical Society

limit and wall beta limits during the perpendicular neutral beam injection as shown in Fig. 8.41a. The mode is oscillatory with its real frequency close to the precession frequency of the trapped energetic beam ion. This mode has been excited by the interaction of trapped energetic particles with marginally stable mode in the wall-stabilized high β_N regime.

This observation is explained as a fishbone like bursting mode (FLM) with threshold energetic particle beta $\beta_c^* = 0.141(\beta^* = \beta_h/\beta)$ by Hao [287]. Hao solved the dispersion relation of the RWM $(\gamma - i\omega_r)\tau_w = -(\delta W_\infty + \delta W_{MHD,h} + \delta W_k)/(\delta W_b + \delta W_{MHD,h} + \delta W_k)$ using the approximate expressions for the variables, $\tau_w^* = \mu_0 \sigma b d(1 - a^{2m}/b^{2m})/2m$, $\boldsymbol{\xi}_\perp = am(r/a)^{m-1}(\boldsymbol{e}_r + i\boldsymbol{e}_\theta)e^{i(m\theta - n\phi)}/[(m - nq)a/Rq]$.

Quite recently, Shiraishi [693] formulated more exact kinetic energy integral for the rotation plasma based on the modern gyrokinetic formulation showing that the growth rate may be much reduced than the simple doppler shifted kinetic treatment.

Chapter 9
Technology Developments for Fusion Power

Abstract Introduction of key fusion technologies such as superconducting magnet, neutron damage and structural materials, tritium breeding and neutron multiplier materials, fusion neutronics, tritium and deuterium chemistry, negative ion source development, Gyrotron development for ECRF are described. Section 9.1 gives brief survey of superconductivity such as Ginzburg-Landau equation, type I and II superconductivities, superconducting materials, and the conductors. Section 9.2 gives brief introduction of neutron irradiation processes, and resulting important damages such as DBTT shift, creep swelling, and introduction of the reduced activation ferritic/martensitic steel including the optimization of C_r contents against DBTT shift and possible operation regimes, and the introduction of SiC/SiC composite and its issue, especially the reduction of thermal conductivity due to neutron irradiation. Section 9.3 introduces solid and liquid tritium breeding materials with their properties, L_i recovery from sea-water, and neutron multipliers such as B_e metal and inter metallic compounds of B_e. Section 9.4 introduces neutron slowing down process and its spectrum, and fusion neutronics calculation and facilities. Section 9.5 gives physical chemistry of hydrogen, chemical properties of hydrogen isotopes, isotope separation method based on chemical exchange reaction between liquid and gas. Section 9.6 and 9.7 are introduction of N-NBI and ECRF system including ion source and gyrotron operations.

Further reading :
Books: Tinkham (2004)[766] for introduction to superconductivity. Moore (1962) [558] for physical chemistry. Benedict (1981)[53] for Isotope separation. Souers (1986) [710] for hydrogen properties.

Reports: Both IAEA-TECDOC-855 and RS-G-1.7 are important to see derived values of concentration for exemption while giving quite different values.

M. Kikuchi, M. Azumi, *Frontiers in Fusion Research II*,
DOI 10.1007/978-3-319-18905-5_9

9.1 Superconducting Technology for Magnet System

In tokamak, magnetic field is produced by the coil current in the magnet and the power loss in the magnet becomes an important issue. Therefore, use of superconducting coils and its technical development becomes essential for fusion to be an efficient energy system.

9.1.1 Superconductivity

The discovery of superconductivity by K. Onnes (Fig. 9.1a) in 1911 [590] is just 3 years after the development of helium liquefaction technology. Meissner in 1933 [549] discovers the so-called Meissner effect, in which the magnetic field is expelled from the superconductor if the temperature is less than the critical temperature, T_c.

The superconductor which exhibits full Meissner effect is called type I superconductor and that exhibits the partial Meissner effect is called type II superconductor. J.E. Kunzler in 1961 [477] discovered a type II superconducting material Nb_3Sn having the upper critical magnetic field B_{c2} more than 10 T. This leads to the strong interests in the application of superconductivity to the magnet. Large superconducting coils such as those for ITER are developed along this line.

The type I superconductor becomes superconducting state only below the critical magnetic field B_c, which is relatively lows and not useful for the practical application. The type II superconductor has two critical magnetic field B_{c1} and B_{c2}. For $B < B_{c1}$, the magnetic field is completely expelled. For $B_{c1} < B < B_{c2}$, the superconductor is in the "vortex state" and the magnetic field is partially expelled.

Phenomenological equation to explain the Meissner effect is given by London in 1935 [525]. They assumed so-called London equation $\boldsymbol{J} = \boldsymbol{A}/\mu_0\lambda_L^2$, where the current is proportional to the vector potential instead of Ohm's law $\boldsymbol{J} = \sigma\boldsymbol{E}$. Taking the rotation of the Maxwell equation $\nabla \times \boldsymbol{B} = \mu_0\boldsymbol{J}$ and substituting London equation, we have diffusion type equation, $\nabla^2\boldsymbol{B} = \boldsymbol{B}/\lambda_\perp^2$. This equation does not have a solution of the uniform magnetic field ($B = B_0$) and also gives $\boldsymbol{J} = 0$ at $\boldsymbol{B} = 0$. The constant $\lambda_L = c/\omega_{pe} = (\epsilon_0 m_e c^2/n_e e^2)^{1/2}$, where m_e and n_e are mass and number density of superconducting electrons. This penetration length is called London's penetration length and is 1–10 μm for pure metals. This is nothing but a collision less skin depth in plasma physics. Ginzburg-Landau (GL) (Fig. 9.1b,a) theory in 1950 [247] is the first to explain that superconductivity is the state in which the quantum effect appears in the macroscopic scale. This theory naturally derives the London's penetration length and also leads to the concept of GL coherence length and explain vortex state in the type II superconductor.

While microscopic state is described by the wave function of quantum mechanics, it is the normal way to think that macroscopic state of the matter with large number of particles can be expressed by the deterministic parameters such as the density and the temperature of the matter. Ginzburg (Fig. 9.1b) and Landau (Fig. 9.1c) changed this view that the quantum mechanical wave function in addition to the fluid density and the temperature is necessary to describe the macroscopic

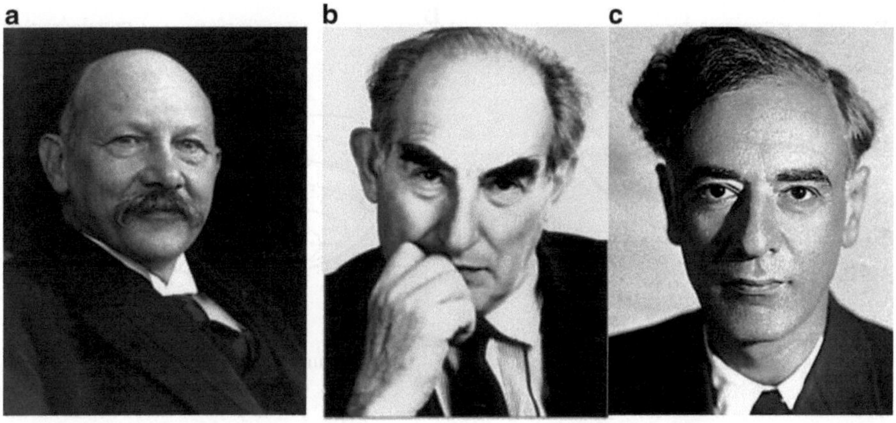

Fig. 9.1 (**a**) H.K. Onnes (1853–1926). (**b**) V. Ginzburg (1916–2009), (**c**) L.D. Landau (1908–1968)

state of the electron fluid under superconductivity, where the square of the wave function is assumed to be the electron density and not the probability density.

Let $n_s(\boldsymbol{r}) = |\psi(\boldsymbol{r})|^2$ is the local density of the superconducting electrons. The wave function $\psi(\boldsymbol{r})$ can be derived by the variational principle to minimize the action integral $S = \int F_s dV$, where $F_s = F_N + \alpha|\psi(\boldsymbol{r})|^2 + \beta|\psi(\boldsymbol{r})|^4/2 + m_c v^2/2$, where $m_c v^2/2 = (1/m_c)|(-i\hbar\nabla - e_c\boldsymbol{A})\psi|^2$ is a kinetic energy of electron and $m_c = 2m_e$ and $e_c = 2e$ are mass and charge of Cooper pair to be discussed later. The $F_{GL} = \alpha|\psi(\boldsymbol{r})|^2 + \beta|\psi(\boldsymbol{r})|^4/2$ is called Ginzburg-Landau potential, where $\alpha = \alpha'(T - T_c)/T_c$ and $\alpha' > 0$ and $\beta > 0$. GL potentials for $T > T_c$ and $T < T_c$ are shown in Fig. 9.2. The potential difference in case of $T < T_c$ is $\Delta F_{GL} = B_c^2/2\mu_0$. The resulting equation is the following Ginzburg-Landau (GL) equation.

$$\left[\frac{1}{2m_c}(-\hbar\nabla - e_c\boldsymbol{A})^2 + \alpha + \beta|\psi|^2\right]\psi = 0 \tag{9.1}$$

The one dimensional GL equation $-(\hbar/2m)d^2\psi/dx^2 = \alpha\psi$ obtained by neglecting $\boldsymbol{A}$ and the nonlinear term has an oscillatory solution with the characteristic length (GL coherence length) $\xi = (\hbar^2/2m\alpha)^{1/2}$.

The ratio of GL coherence length and London's penetration length $\kappa = \lambda_L/\xi$ is called the Ginzburg-Landau parameter. This GL parameter characterizes type I superconductor ($\kappa < 1/\sqrt{2}$) and type II superconductor ($\kappa > 1/\sqrt{2}$).

An important difference of the type I and type II superconductors is the difference in the mean free path at the normal conducting state. If the mean free path is short, the coherence length comes short and the penetration length is long, which leads to the larger κ and the type II superconductivity. The vortex state is a mixed state, in which some regions are in the normal state and are surrounded by the superconducting regions. Key concepts in type II superconductivity such as the quantization of the magnetic flux, vortex filament, upper and lower critical fields are established by Abrikosov [2].

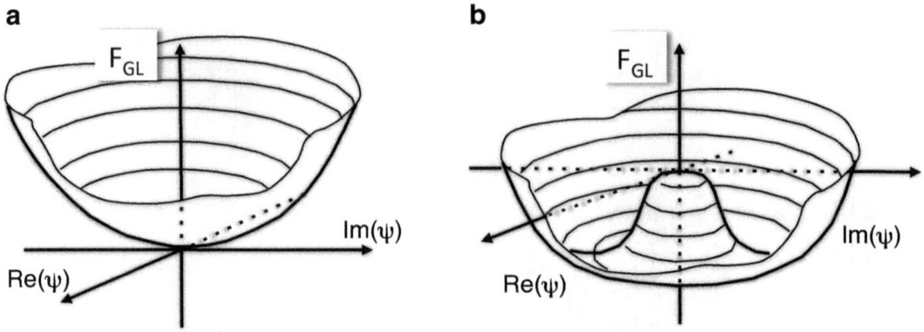

Fig. 9.2 (**a**) Ginzburg-Landau potential for $T > T_c$. (**b**) Ginzburg-Landau potential for $T < T_c$

Microscopic theory of superconductivity is established by the BCS (Bardeen, Cooper, and Schrieffer) theory in 1957 [44]. Absolute square of the quantum mechanical wave function gives the probability density but this gives the particle density and the gradient of the phase gives particle flux when the large number of Boson particles moves simultaneously. The superfluid state of the liquid ^4H_e is a typical example. In the GL theory, the electrons appear as if they are Boson particles.

Since electrons are Fermions, the Pauli exclusion principle prohibits that the two electrons take the same motion state. Why macro-scale number of electrons align the translational motion? The BCS theory gives the answer for this question. The key is a Cooper pair. The two electrons form a Cooper pair by the attraction between the electrons. Electron is a Fermi particle but the Cooper pair consisting of two electrons becomes a Boson.

Since Bosons can occupy in the same state, all Cooper pairs can enter the state of lowest energy. In this way, the macroscopic number of electron Cooper pairs fill inside the superconductor. Thus, macro-scale number of Cooper pairs has same phase leading to the superconducting current. Source of attractive force to create a Cooper pair is the electron-phonon interaction, where a phonon emitted from an electron is absorbed by another electron. Near the Fermi energy level, this becomes the attractive force. Viewing from the BCS theory, GL theory is the quantum mechanics of the phase-aligned Cooper pairs. Gor'kov in 1959 [257] derived the Ginzburg-Landau theory from the microscopic BCS theory.

The phase of the macroscopic wave function is the essential quantity to describe electron motion in a superconductor. Its phase velocity is $v_{ph} = \omega/k$ and the group velocity is $v_g = d\omega/dk$, where k is the wave number. Wave number $\boldsymbol{k}$ of the Cooper pair is given by the canonical momentum as $\boldsymbol{k} = (m_a \boldsymbol{v} + e_a \mathbf{A})/\hbar$.

The essence of the BCS theory is the formation of Cooper pairs via phonon attraction. Therefore, higher critical temperature may be expected for higher phonon attraction. But, the cooper pairs becomes smaller in size and becomes two-electron molecule if the phonon attraction is strong. When the number of two electron molecules increases in the conductor, this leads to a transition to the insulator.

Salon 9-1: Higgs Particle and Ginzburg-Landau Potential
In the quantum theory of fields, existence of materials is understood as "field" instead of "particle". The "particle" is expressed as the "oscillation" of the field.

The "photon" is a typical example that the "particle" of light is produced as the electromagnetic oscillation in the electromagnetic field. Known particles can be explained by some quantum fields.

Various particles and forces are explained by so-called the "spontaneous symmetry breaking". The spontaneous symmetry breaking produce "mass" to the massless gauge field by assuming the Higgs field immersed in the vacuum. The oscillation of the Higgs field is called "Higgs particle" [185, 320].

Concept originates from the Nambu's spontaneous symmetry breaking of the chiral symmetry to produce most of the hadron mass. To describe these processes, the Lagrangian having similar structure of Ginzburg-Landau potential shown in Fig. 9.2 is used. In case of BCS theory, this appears as sufficient condition to produce cooper-pair.

The potential difference in $|\psi|$ direction becomes an origin of mass and the massless freedom in azimuthal direction becomes an origin of photon. The Higgs boson with a mass of 125 GeV has been observed in 2012 at both CMS [131] and ATLAS [31]. Peter Higgs and Francois Englert received the Nobel prize in physics in 2013 after their theoretical discovery of Higgs particle.

9.1.2 Superconducting Materials

Superconducting materials for magnet are desired as much as possible to operate at high field, high current density, and high temperature. In addition, workability to form a superconducting wire and coiling is required. Furthermore, it must have sufficient strength to withstand the electromagnetic force (Lorentz force) and its effect on degradation of superconducting properties must be small. Important characteristics of superconducting materials are transition temperature T_c, upper critical field B_{c2}, and critical current density j_c. Table 9.1 shows the characteristics values of $NbTi$, Nb_3Sn, and Nb_3Al developed for large coil.

While $NbTi$ is alloy, both Nb_3Al and Nb_3Sn are inter-metallic compound. In the alloy material, it is processed to the form of conductor cable from the solid solution, and then improves its superconducting properties by the work hardening and age

Table 9.1 Critical temperature T_c and upper critical field B_{c2} for $NbTi$, Nb_3Sn and Nb_3Al

Superconducting material	$NbTi$	Nb_3Sn	Nb_3Al
Critical temperature T_c (K)	10.7	18.3	18.9
Upper critical field B_{c2} (T) at 4.2 K	17	22	32

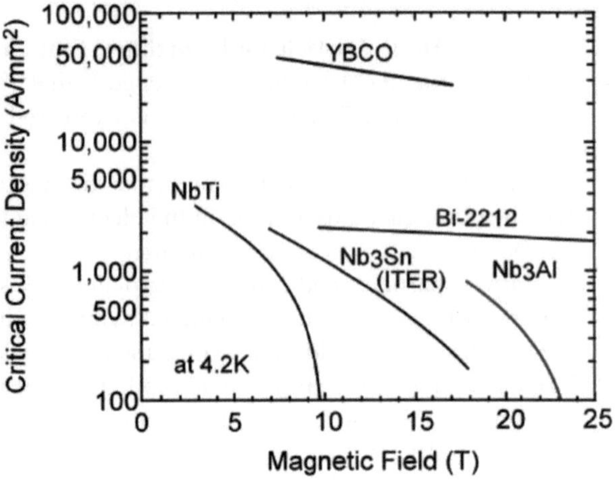

Fig. 9.3 Critical current density as a function of the magnetic field for various superconducting materials, *NbTi*, Nb_3Sn, Nb_3Al, *Bi*-2212, *YBCO*

precipitation hardening. Since inter-metallic compound has almost no workability, the raw material is processed into a cable and is transformed into the compound by the diffusion reaction.

For stability of the superconductor, the superconducting material must be combined with normal conductor with high electrical conductivity such as *Cu*. This is done by carrying out co-processed by hole die drawing. In this case, it is necessary that the mechanical hardness of a superconducting alloy is close to that of copper. NbTi meets this requirement and is successful industrialization. High-purity aluminum has a higher electrical conductivity than copper at low temperature as used in TRIAM-1M, but it is very soft. Formation of the multi-filament wire is processed using raw materials for the inter-metallic compound cable.

Strain tolerance characteristics of the superconducting wire varies greatly between intermetallic compounds and alloys. In the case of *NbTi* alloy, significant I_c degradation is not observed even at a strain of 0.5 %, and degradation of about 5 % for I_c at 1.5 %. On the other hand , in the case of *Nb3Sn*, I_c degradation starts from 0.1 % and causes the degradation in the level of 20 % at 0.3 %.

Mueller at IBM laboratory in Zurich predicts high critical temperature will be realized near the transition to the insulator. Bednorz-Muller in 1976 [52] discovered the oxide high-temperature superconductor made of mixed oxide of copper (Cu), lanthanum (La) and barium (Ba). They were awarded the Nobel Prize in Physics in 1989. Since then, various high-temperature superconductors are produced. Figure 9.3 shows the magnetic field dependence of the critical current density at 4.2 K of *YBCO* and *Bi*-2212 high-temperature superconductors as compared with that of *NbTi*, Nb_3Sn, and Nb_3Al.

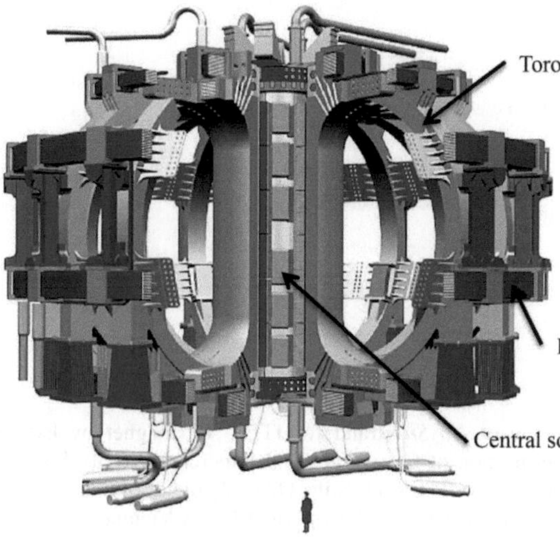

Fig. 9.4 Bird's eye view of ITER superconducting magnet system. With kind permission of ITER Organization (http://www.iter.org/)

9.1.3 Superconducting Magnet

Superconducting magnet for ITER consists of 18 Toroidal Field (TF) coils, a central solenoid (CS) coil, and 6 poloidal field (PF) coils and correction coils. The maximum field of TF coil is 11.8 T, the magnetic energy of TF coils is ~41 GJ, and the total weight is ~6000*t*. The maximum field of CS coil is 13 T and total weight is ~950*t*. The total weight of magnet system is ~10,500*t*. Figure 9.4 shows a Bird's eye view of ITER superconducting magnet system.

While Nb_3Al compound is a promising material for the magnet system, *NbTi* alloy and Nb_3Sn compound are used for ITER. *NbTi* magnet usually operate less than 8 T and Nb_3Sn magnet can operate up to 13 T. High T_c superconductor such as *BSCCO* and *YBCO* are used for the current leads.

For the magnet use, multi-filament wire (strand) is necessary to reduce hysteresis loss and improve stability. Major fabrication techniques of multi-filamentary Nb_3Sn strands are bronze process and internal tin diffusion process.

Figure 9.5a shows a cross section of Nb_3Sn strand for the ITER TF conductor. Main specifications of Nb_3Sn strand for ITER TF magnet are strand diameter of 0.82 ± 0.005 mm, twist pitch of 15 ± 2 mm, unit length of ≥ 1000 m, Copper ratio of 1.0 ± 0.1, Chrome plate thickness of $2.0 + 0.0/-1.0\,\mu$m critical current at 12 T, 4.2 K of >190 A, n-value of > 20, hysteresis loss ($\pm 3T$) of < 500 mJ/cm^3, residual resistivity ratio >100 [567]. Figure 9.5b shows a structure of ITER TF conductor made by bronze process in Japan [734]. The conductor for the ITER TF magnet composed of 900 Nb_3Sn strands, 522 *Cu* strands, *He* cooling channel formed

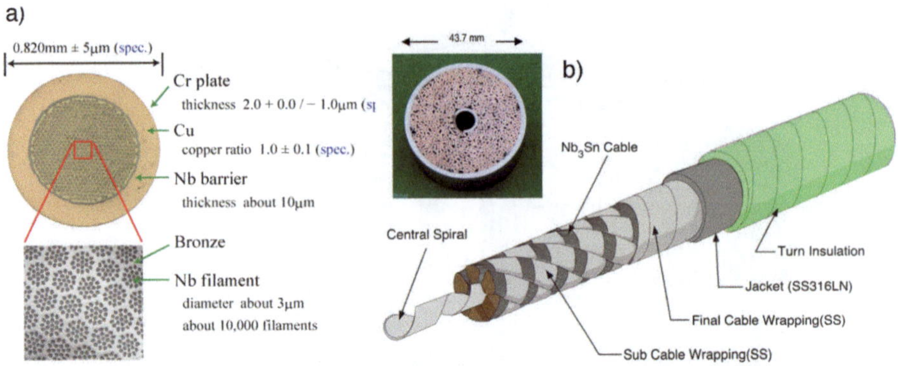

Fig. 9.5 (**a**) Cross sectional view of Nb_3Sn strand for ITER TF magnet by Bronze method. After Nabara [567] with the permission of Cryogenics and Superconductivity Society of Japan. (**b**) Nb_3Sn cable in conduit conductor for ITER TF coils. Outer diameter of conductor is 43.7 mm. Reproduced with permission from Takahashi [734]. Copyright IAEA Vienna

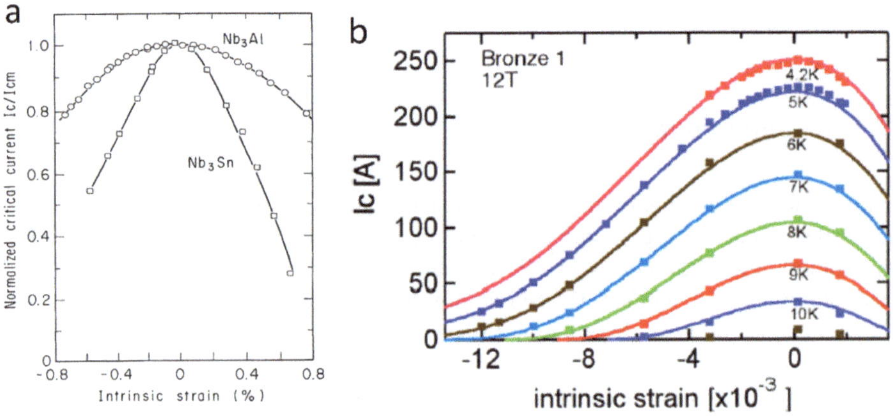

Fig. 9.6 (**a**) A comparison of J_c degradation curve of Nb_3Sn and Nb_3Al strand. After Ando [11]. (**b**) Degradation of the J_c with axial intrinsic strain for various temperatures in Nb_3Sn strand for ITER TF magnet. Reproduced with permission from Takahashi [734]. Copyright IAEA Vienna

by central spiral, and a jacket of 43.7 mm outer diameter made of ss 316LN. The strands are divided into 6 sub-cables wounded by the stainless-steel tapes.

Coupling loss is one of important dissipation mechanism in the superconductor. There are two kinds of coupling loss, coupling loss between filaments and between strands. To reduce the coupling loss between filaments, strand is twisted with a pitch of 15 mm for ITER TF strand. In order to reduce the coupling loss between strands, chromium plating on strand surface is made to increase contact resistance as well as stainless steel taping on sub-cables.

Nb_3Sn compound is more sensitive to the strain than Nb_3Al compound as well as $NbTi$ alloy. Figure 9.6a shows degradation of J_c as a function of axial intrinsic

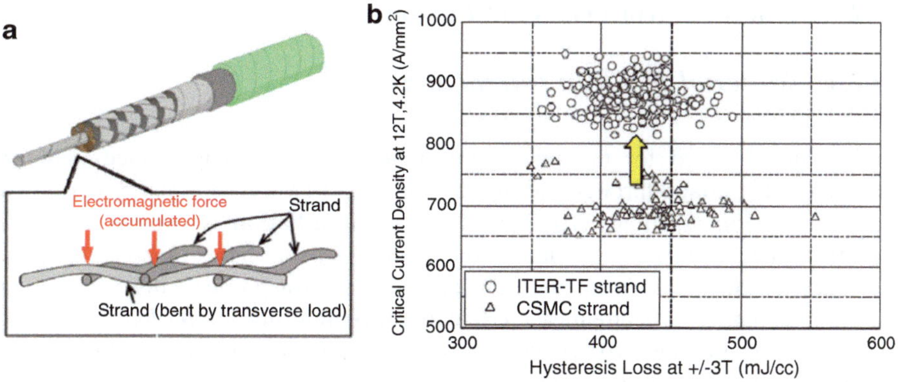

Fig. 9.7 (**a**) Schematics of bending in strands. (**b**) Critical current density (J_c at 12 T, 4.2 K) and hysteresis loss (Q_h) of bronze method strands for CS model coil (CSMC) and ITER TF magnet. Reproduced with permission from Takahashi [734]. Copyright IAEA Vienna

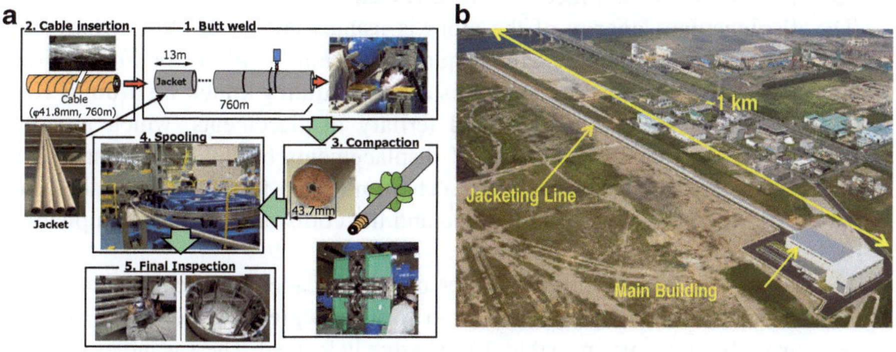

Fig. 9.8 (**a**) Schematics of jacketing processes of superconducting cable for ITER TF magnet. (**b**) Jacketing line for ITER TF conductor. Reproduced with permission from Takahashi [734]. Copyright IAEA Vienna

strain for Nb_3Sn and Nb_3Al. For example, 30 % reduction in J_c occurs by 0.4 % axial intrinsic strain. Figure 9.6b shows detailed measurement of J_c degradation as a function of axial intrinsic strain for ITER Nb_3Sn strand. As operating temperature increases, relative degradation becomes bigger. The degradation occurs by axial strain and may occur by periodic bending as schematically shown in Fig. 9.7a. Actual inspection of the model conductor after cyclic testing revealed a large deflection and bending strain on the lower loading side (not the higher loading side) by Hemmi [319]. Nb_3Sn strand for ITER TF magnet is manufactured to have higher J_c to compensate such a bending effect as shown in Fig. 9.7b.

The manufacturing process of superconductor is shown in Fig. 9.8a. After welding the 13 ms jacket pipes to form 760 m jacket, the cable is inserted into the jacket. Then the compaction is made to a specified size (43.7 mm) and spooled and final inspection is made. Figure 9.8b shows the jacketing line in Kyushu, Japan.

9.2 Structural Materials for Fusion

9.2.1 Material-Neutron Interaction

The materials used for the fusion are subject to the damage called **neutron irradiation damage** through the interaction with 14 MeV neutrons produced by the D-T fusion reaction [370, 765]. **Neutron irradiation** is expressed by the **neutron fluence** (MWa/m^2), which is the product of neutron flux (MW/m^2) and the irradiation time in year (a: annual). The energetic neutron produces displacement of atom in the lattice structure measured by **displacement per atom** (*dpa*). In case of ferrite/martensitic steel, 10 MWa/m^2 corresponds approximately to 100 dpa.

The dominated energy loss processes of the high energy neutrons are the **elastic collision** and the **inelastic collision** with the atoms of the materials. The atom displaced from its original position by the elastic scattering becomes an **interstitial atom** (IA) and a **vacancy** is produced at the original position. This pair of vacancy and the interstitial atom is called the **Frenkel Pair**.

The **displacement energy** of the atom is usually $E_b \sim 25$ eV for the metal, while higher energy is required for displacement $E_d \sim 40$ eV for *Fe* and steel. Atom directly displaced by the neutron is called **Primary knock-on atom** (PKA). This PKA can produce secondary and tertiary displacements, which is called the **collisional cascade**. The number of displacements by the PKA is called the **displacement damage function**, $\nu(E_p)$. The time scale for a collision takes only $\sim 10^{-14}$ s for 30 eV slow incident particle and the collisional cascade completes in $\sim 10^{-12}$ s.

IAs are formed away from the center of the cascade and vacancies are accumulated near the center of the cascade to form a **depleted zone**. When the free movement of IAs becomes possible, **IAs cluster** is formed. The vacancies form the **dislocation loop**.

For the neutron energy $E_n < 1$ MeV, neutron scattering is elastic scattering and is also isotropic since the **de Broglie wave length** is much longer than the nuclear radius. For $E_n > 1$ MeV, forward scattering increases.

As a result of neutron irradiation damage, material properties change. In fusion neutron damage, **nuclear transmutation** produces H and He atoms in the material. This He induces the **lattice defect**, the **swelling**, and the **ductile to brittle transition temperature (DBTT) shift**. Under the fusion 14 MeV neutron irradiation, large He production of 17–30 times that of fission occurs for SS (stainless steel) due to *Ni* reaction with the neutron. For the low neutron fluence in ITER (<5 dpa), *He* effect is negligible.

For the fusion demonstration reactor (DEMO), the **reduced activation ferrite/martensitic steel (RAF)** is the primary candidate as a blanket structural material. Due to the absence of *Ni*, *He* production is much smaller than SS but $\sim 10 appmHe/dpa$ occurs due to the nuclear transmutation of *Fe* and *Cr*. This *He* effect is expected to be important for RAF above 50 dpa. In this range of neutron fluence, material property can be changed due to the composition change.

Neutron irradiation changes the **mechanical strength** such as the **radiation hardening**, the **embrittlement**, and loss of the **creep strength**. In case of RAF, embrittlement occurs at $T < 450$ deg followed by the loss of **ductility**, the reduction of **rapture toughness**, and the increase in DBTT. At $T > 450$ deg, the **softening** of the material and reduction of the **creep strength** occurs.

Irradiation hardening is caused by the irradiation defects due to neutron irradiation. Irradiation defects are categorized into (a) **point defects** (vacancy and interstitial), (b) **vacancy cluster**, (c) **dislocation loop**, (d) **dislocation line**, (e) **impurity atoms** (nuclear transmutation product), (f) **cavity** (void, helium bubble), (g) **precipitates**, (h) **segregation**.

Mechanisms of embrittlement by the irradiation defect are, (1) the **yield stress** increases closer to the **rupture stress** and **total elongation** and **elongation at break** reduces. (2) DBTT increases by the increase in yield stress. At high temperatures, (1) the growth of the cavity near **grain boundary** leads to embrittlement, (2) helium bubbles at grain boundary induces embrittlement.

Since irradiation defect occurs in very short 10^{-12} s, the thermal recovery process contributes to the material property within μ s.

A semiconductor such as SiC/SiC composites as well as metal are also considered as a structural material of a fusion reactor. But the semiconductor shows the different response to the neutron irradiation compared with the metal. For SiC/SiC composites, which is expected as an ultimate structural material, the thermal conductivity is reduced by the neutron irradiation.

As is well known and given in the standard textbook of the solid state physics for example by Kittel [459], the electron conduction is responsible for the thermal conductivity in the case of the metal. But the heat conduction is mainly carried by the phonon conduction and the contribution of the electron conduction is small for the semi-conductor, especially for SiC, and this phonon conduction is more susceptible to the irradiation defects because the phonons are scattered by the defects.

Column 9-1: Glossary on Mechanical Property and Radiation Defects

Grain boundary: Material is an assembly of many crystalline grains. The boundary is called grain boundary, where He tends to concentrate.

Fracture toughness: Measure of the energy required to break a material, in contrast to strength (a measure of the stress required to break a material).

Creep: Time-dependent mechanical deformation under constant stresses below the ultimate tensile stress.

Dislocation: Line defect which allows propagation of deformation inside crystalline materials.

DBTT: DBTT is mainly observed for BCC materials (ferritic/martensitic and ferritic steels, their ODS variants, and the refractory metallic materials).

Embrittlement: Loss of elasticity or deformability of a material.

(continued)

Segregation: Concentration of some elements can be increased at grain boundary if the free energy can be reduced, called boundary segregation.
Swelling: Increase in the material volume due to irradiation-induced void.

9.2.2 Reduced Activation Ferrite/Martensitic Steel

Before the design of SSTR (Steady State Tokamak Reactor) in 1990, high-Z material (Mo and Nb alloy) and improved stainless steel such as (PCM) with FCC (Face Centered Cubic) structure are thought to be the structural material of a fusion reactor blanket. In 1988, Japan Atomic Energy Research Institute (JAERI) and Nippon Kokan (NKK) developed a reduced activation ferritic steel named F82H (may be better call it a martensitic steel) [746]. This RAF is adopted as a blanket structural material of SSTR due to its superior properties against neutron irradiation and the higher thermal conductivity ($32W/mK$) than that of austenitic steel ($16W/mK$). Recent progress is given by Baluk [42].

It is a difficult choice from the fact that at that time, the materials having ferromagnetism has been avoided in fusion devices as a cause of irregular magnetic field. The compatibility of high performance plasma and ferrite steel has been tested in JFT-2M by Kawashima [422], and Tsuzuki [777] and JT-60U by Shinohara [690] and the result is encouraging to show that high performance plasma can be maintained in an appropriate environment by suppressing the irregular magnetic field. Martensitic structure has a BCC (Body Centered Cubic) structure with carbide and high density of dislocations, which has a large number of point defects and trapping sites of H and He. This properties lead to an excellent tolerance to fast neutron irradiation. This martensitic steel is has a specific issue of DBTT at low temperature, which is a unique problem in martensitic steel with a BCC structure and is absent in the austenitic steel.

F82H is modified from HT-9 (12Cr-1Mo-V-W) by changing $1Mo + 0.5W$ to $2W$ in order to avoid molybdenum (Mo) producing long-lived radionuclide [746]. The F82H is developed as high performance martensitic steels with a target impact absorption energy above 265 J and a creep rapture strength above 98 MPa. The reduced activation ferritic steel F82H has lower DBTT and adequate weldability and high yield strength. F82H is adopted as a standard material under the IEA cooperation. The composition of F82H recommended by IEA is shown in Table 9.2 [331].

Metal structure varies significantly with the chromium concentration. Low Cr steel (example $2.25Cr$) shows a relatively high strength, but shows lower fracture toughness at low temperature. The strength decreases rapidly at a high temperature $>$ 773 K. It also show significant irradiation embrittlement by the generation of finer carbides. Conversely, the high chromium steel (example, $12Cr$), it becomes ferrite/martensite dual phase steel containing about 20 % of the δ ferrite,

Table 9.2 Composition of the major elements of the reduced activation ferritic steel F82H

Elements	Cr	W	V	C	Si	Mn	Ti	Ta
Typical wt %	8.0	2.0	0.2	0.1	0.1	0.1	0.005	0.04
wt % range	7.5–8.5	1.8–2.2	0.15–0.25	0.08–0.12	0.05–0.2	0.05–0.20	0.004–0.012	0.01–0.06

Other elements are $P < 0.01, B < 0.001, S < 0.01$ and $Nb, Mo, Ni, Cu, N, Co, Ag, Sn, O, As, Sb$ are LAP (low as possible). From Hishinuma [331]

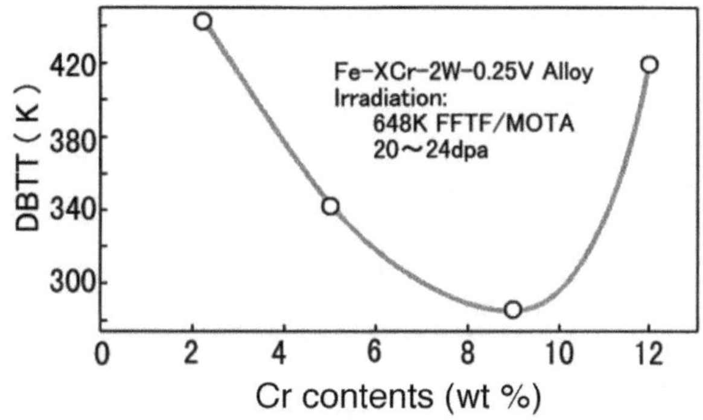

Fig. 9.9 DBTT as a function of Cr contents in weight %. Reproduced with permission from Hishinuma [331]. Copyright Elsevier

which induces degradation of toughness. The most important optimization for the composition is *Cr* contents to reduce DBTT after neutron irradiation as clarified by Kohyama [462]. Figure 9.9 shows the DBTT for $Fe - XCr - 2W - 0.25V$ alloys under 20–24 dpa neutron irradiation at the temperature of 648 K. This test clearly shows appropriate *Cr* content will reduce DBTT to less than the room temperature.

The 316 stainless steel exhibits a large swelling due to neutron irradiation, 1 %/dpa. But the Martensitic steel such as F82H shows excellent stability against swelling up to 200 dpa and swelling is not thought to be to be a major problem.

It is considered irradiation hardening almost saturated at 10 dps at ~648 K, while irradiation hardening occurs slowly until 70 dpa at ~703 K. Irradiation softening occurs at $T > 723$ K and the creep strain reduces with neutron fluence.

Conversion rate of *W* to *Os* by fusion neutron is calculated to be 10 %/100 dpa. In other words, $2W$ martensitic steel becomes $1.8W + 0.2(Os + Re)$ steel by 100 dpa irradiation. Deterioration of the irradiation resistance due to the transmutation is an important issue. For 100 dpa fusion neutron irradiation to martensitic steel, the generation of Mn is 0.5 %. The grain boundary segregation and precipitation of Mn produces the irradiation-induced grain boundary embrittlement cracking as well as the void swelling.

Helium and hydrogen produced by the neutron irradiation interact with point defects and change the damaged tissue. While hydrogen interaction with point defects is weak in the medium to high-temperature range, He interaction with point defects is strong. In low activation martensitic steels, void swelling is enhanced by the transmuted He. Helium embrittlement is remarkable in 300 ppmHe in a high temperature region for austenitic steels, but martensitic steels still shows the ductility. The test results of spallation neutron source implies the DBTT shift due

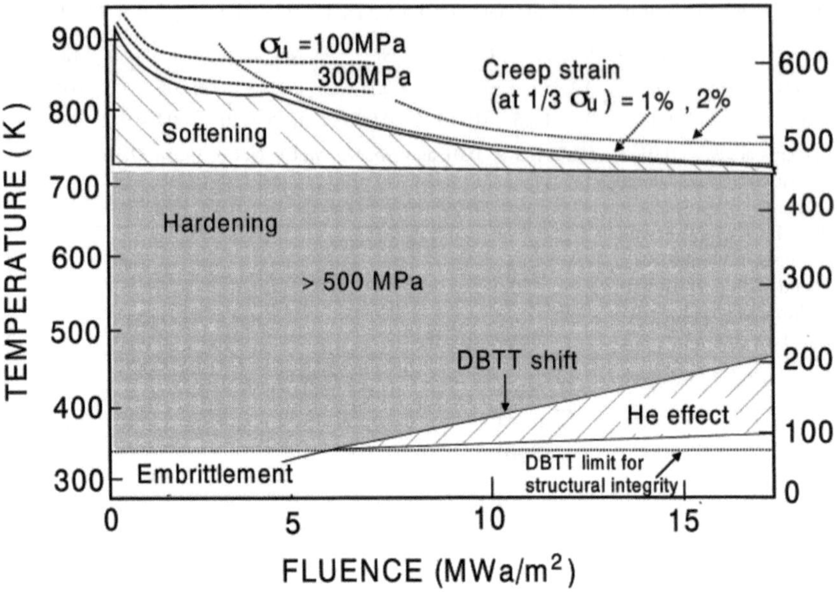

Fig. 9.10 Operation diagram of *F82H* in (fluence, temperature) space. Reproduced with permission from Hishinuma [331]. Copyright Elsevier

to He implantation may occur above 50 dpa. Considering these features, Hishinuma [331] gives an operating diagram shown in Fig. 9.10. F82H is an excellent structural material that for fusion reactor blanket, but the operating temperature is limited to 500–550 °C from the point of view of high-temperature strength. Oxide dispersion strengthened steel (ODS steel) is considered to improve high-temperature creep strength [454]. ODS steel can provide a lot of He trapping site as well as grain boundary. Hasegawa [292] reported that there is almost no DBTT shift due to He-implantation for both 9*Cr* and 14*Cr* ODS steels while F82H has large DBTT shift due to He-implantation. Effort to increase He trapping site in base structural material is important to mitigate He embrittlement [864].

Column 9-2: Steels

Iron is said to be found for the first time in the meteorite. In ancient years, iron is manufactured from Iron sand. Ion ore are oxidized and the reduction has been made by using charcoal, coal, coke. World steel production in 2013 is 1607 Mega Tonnes (*Mt*) of which 779 Mt in China, 111 Mt in Japan, 87 Mt in US, and 81 Mt in India according to World Steel Association. Pure iron has **Body-Centered Cubic (BCC)** crystal structure having magnetism, called the α-Fe. At around 900°, it changes to **Face-Centered Cubic (FCC)** crystal structure without magnetism, called the γ-Fe. British metallurgist Sir William

(continued)

Chandler Roberts-Austen (1843–1902) found the γ-Fe phase can be sustained for all temperature if alloying elements including *Ni* are included, named as **Austenitic steel** after him. In 1891, German metallurgist Adorf Martens created a very hard BCC-structured steel with high *C* solution by rapid cooling from Austenitic phase, named as **Martensite** after him. Stainless steels are categorized into austenitic steel and ferrite and martensitic steels.

Austenitic steel: Austenitic stainless steel has **FCC** structure and is metallic non-magnetic solid solution in γ-Fe phase. It contains 0.02–0.08 % *C*, 16–19 % *Cr* and 8–12 %*Ni* to retain γ-Fe phase.

Ferritic steel: Ferritic steels has **BCC** crystal structure and metallic magnetic solid solution in α-Fe phase. It contains less than 0.2 % *C*, 11.5–27 % *Cr* and free from *Ni*.

Ferrite/Martensitic steel: Ferrite/Martensitic steel has **BCC** crystal structure and is metallic non-magnetic solid solution in α-Fe phase. It contains 7–9 %*Cr* and is produced by the rapid cooling from austenite region to form a martensitic structure.

9.2.3 SiC/SiC Composite

Silicon carbide (*SiC*) is a refractory semiconductor. This material has an excellent thermal, mechanical and chemical stability and is suitable for use in extreme harsh environments. Pure *SiC* shows the lowest activation properties to 14 MeV neutron. Since SiC is a brittle material, it is thought difficult to be a structural material initially.

Yajima's invention of SiC fibre [840] leads to the fabrication of the SiC fiber-reinforced SiC based ceramic composite material (hereinafter, SiC/SiC composite material) proving possibility as a structural material. SiC/SiC composite material is composed of SiC fibers having a diameter of 10 μm, SiC matrix and the interphase. SiC fiber is responsible for loading and the matrix transfer load to the fiber.

For the blanket structural material, there are requirements for the strength, the thermal conductivity, the dimensional stability under irradiation, and the compatibility with coolant and neutron multiplier. Early SiC fibers showed dimensional instability under irradiation but the improvements were made subsequently. On the other hand, there is a concern on the air-tightness (hermeticity) of high pressure coolant such as He. The hermeticity could be improved by impregnation or coating. SiC has a low permeability against hydrogen and will function as a tritium permeability barrier.

Third Generation CVI Method

Since pure SiC shows an excellent irradiation resistance, the choice of fiber and interphase is important. The SiC fiber is made by crystalline SiC, which does

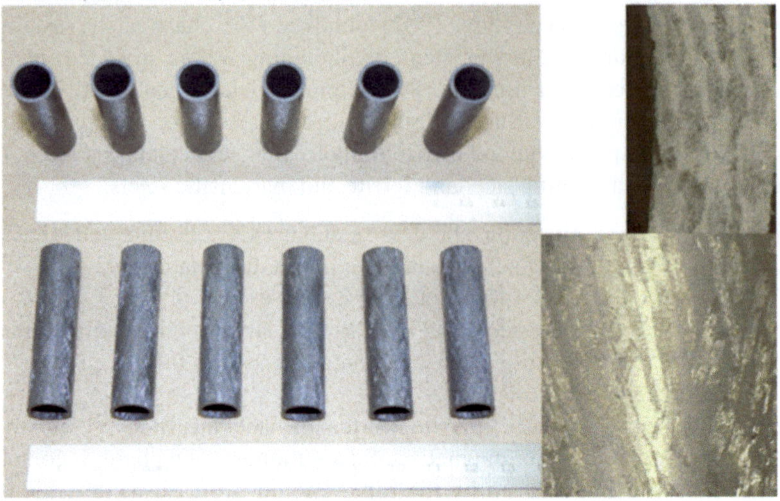

Fig. 9.11 SiC/SiC composite tubes made by third generation NITE method. Reproduced with permission from Katoh [418]. Copyright Elsevier

not exhibit degradation of strength by neutron irradiation and is chemically stable. Carbon is used for the interphase instead of BN subject to nuclear transmutation. Chemical Vapor Infiltration (CVI) method is the standard production methods. Third generation crystalline SiC fibers are available commercially. In the CVI method, silane-based organic gas flows in the gap between the fiber bundle and deposits by the chemical vapor deposition.

Thermal conductivity of the SiC/SiC composite increases with the improved process. In the early (second generation) CVI process, the thermal conductivity is $\kappa \sim 4\,\mathrm{W/mK}$ at 20 °C and 2 W/mK at 1000 °C, but the thermal conductivity in the third generation CVI is increased by higher crystallinity of the SiC fiber to $\kappa \sim 15\,\mathrm{W/mK}$ at 20 °C and 10 W/mK at 1000 °C.

NITE Method

Nano-infiltration and Transient Eutectic-phase (NITE) method is a method to form a matrix by the liquid phase sintering (LPS) using oxide additives. This process requires higher temperature to achieve a high density matrix while retaining an acceptable range for fiber damage operating at higher than temperature limit of highly crystalline SiC fiber. Kohyama group in Kyoto University developed NITE with low porosity and high tensile stress $\sim 400\,\mathrm{MPa}$, high proportional limit stress (corresponds to the yield stress of the metal) $\sim 220\,\mathrm{MPa}$, high thermal conductivity $\kappa \sim 30\,\mathrm{W/mK}$ at 20 °C and 20 W/mK at 1000 °C [417] (Fig. 9.11).

Neutron Irradiation Damages

Early SiC/SiC composite shows strong degradation of mechanical strength even at 1 dpa but the reduction of the difference in the swelling between fiber and matrix improved mechanical property of SiC/SiC composite such as Hi-Nicalon S and

Tyranno-SA up to 10 dpa [406]. Swelling tends to saturate in a few dpa except at high temperature. Near the Critical Amorphization Temperature (150 °C), distortion between interstitials and vacancies is the cause of the swelling. As the temperature increases, survived interstitial atoms of Si and C are reduced. The number of dislocation loops and voids increases at temperature $>$ 1000 °C and the mobility of vacancy is reduced leading to higher swelling.

Since the valence electrons are few in SiC, the heat transfer is based on the phonon transport. The phonon scattering as a heat resistance mechanism are categorized into the grain boundary scattering ($1/K_{gb}$), the phonon-phonon scattering (Umklapp process) ($1/K_u$), and the defect scattering ($1/K_d$). In other words, the inverse of thermal conductivity is given by $1/K = 1/K_{gb} + 1/K_u + 1/K_d$. Defects generated by neutron irradiation lower the thermal conductivity by scattering phonon effectively. Non-irradiated CVD SiC achieved the thermal conductivity up to 280 W/mK, which is reduced to 5–20 W/mK at $T < 100$ °C. The thermal resistance increases in proportion to the swelling suggesting an increase in the phonon scattering by irradiation defects. The thermal conduction in metal is electron conduction which is robust to radiation defects.

Thus the development of SiC/metal compound may be necessary for the SiC to be a promising structural material for fusion high heat flux environment. Here the metal component should have low nuclear heating rate to preserve superiority of the SiC in case of loss of coolant accident.

9.3 Blanket Materials

9.3.1 Tritium Breeding Material

Tritium breeding material is used to produce and recover tritium fuel in a fusion reactor. Following tritium generation reactions for *Li* are used.

$$^6Li + n \rightarrow ^4 He + T + 4.78\,MeV \tag{9.2}$$

$$^7Li + n \rightarrow ^4 He + T + n' - 2.47\,MeV \tag{9.3}$$

The Lithium is first discovered by Swedish chemist J.A. Arfwedson shown in Fig. 9.12a in 1817. Whereas ^6Li neutron cross section is 940 barn at $E_n = 0.025$ eV, ^7Li neutron cross section is three orders of magnitude smaller, 0.355 barn at $E_n = 14$ MeV (see Fig. 9.12b). Even taking into account the energy spectrum, the contribution of ^6Li for tritium production is two orders of magnitude greater. Furthermore, ^6Li reaction is an exothermic reaction and the energy multiplication occurs at the whole blanket. Natural lithium contains 7.5 % of ^6Li. If the breeder does not have a high tritium breeding ratio (TBR), it is necessary to concentrate ^6Li.

The tritium breeding materials are categorized by liquid breeders and ceramic-like solid breeders. Since the early 1980s, Li_2O, Li_2ZrO_3, $LiAlO_2$, Li_2TiO_3, and

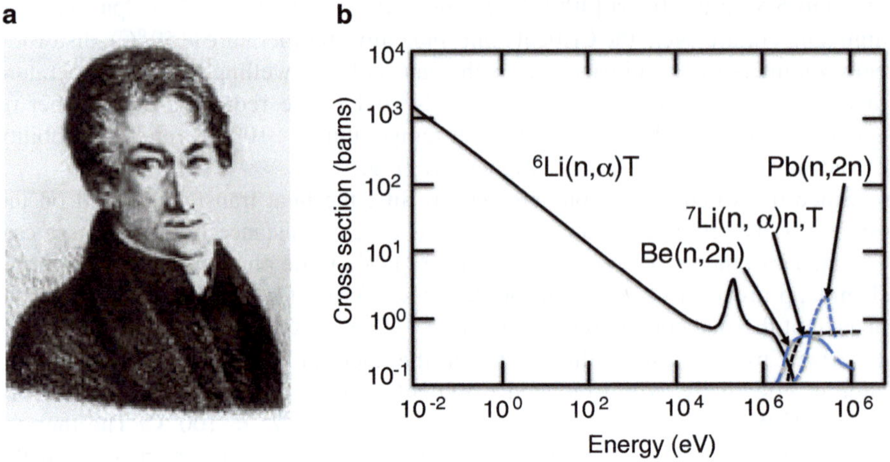

Fig. 9.12 (**a**) J.A. Arfwedson who discovered *Li*. (**b**) ^{6}Li, ^{7}Li, *Be* and *Pb* neutron reaction cross sections modified from Johnson [405]

Li_4SiO_4 are considered as candidate solid breeders reviewed by Johnson [405]. The liquid breeder are liquid *Li* metal, $Li_{17}Pb_{83}$, and *FLiBe* etc. The desired characteristics of the tritium breeding material are:

(1) High *Li* atom density and no elements with high neutron absorption.
(2) Short tritium residence time and easy to release tritium.
(3) Low chemical reactivity with structural material and coolant
(4) Excellent safety characteristics for accident.
(5) Small radioactive products.
(6) High thermal conductivity.
(7) Not excessive thermal expansion and swelling.

Table 9.3 shows characteristics of the solid and liquid breeders. The solid breeders approximately satisfy the above conditions (1) and (2) , but the liquid breeders have a problem in corrosion of the structural materials and are not easy to satisfy the condition (2).

Lithium Oxide (*Li_2O*) Li_2O has been proposed as a tritium breeding material for fusion reactor blanket first in 1973 by Sako [656]. Among many breeders, Li_2O shows high performance with its highest *Li* density and relatively smooth recovery of tritium (retention time: 8 h). Since the thermal conductivity is the highest among many solid breeders, it has a merit on the heat removal generated in the blanket. Furthermore, only ^{16}N is a radioactive material and decays by a half-life of 7 s. But, the reactivity with the water is high and the use of pressurized water as a coolant is less desirable. In such case, sophisticated safety measure is necessary (case for SSTR). The helium cooling has no reactivity with Li_2O and is excellent in safety while *He* pumping power is large and the maximum heat flux will be limited. The

Table 9.3 Characteristics of solid and liquid breeders

T breeder	Li_2O	Li_2TiO_3	Li_2ZrO_3	Li_4SiO_4	$LiAlO_2$	Li	$Li_{17}Pb_{83}$	Li_2BeF_4
Melting temperature (K)	1696	1808	1888	1523	1883	453	508	732
Theoretical density (g/m^3)	2.02	3.43	4.15	2.21	2.55	0.509	9.59	1.99
Li density (g/m^3)	0.92	0.43	0.38	0.51	0.27	0.509	0.065	0.279
Thermal cond. (W/mK)[a]	4.7	1.8	0.75	2.4	2.4	41.4	12.2	1.0
Water reactivity	Large	No	No	Small	Small	Violent	Modest	Small
T residence time (h)[a]	8	2	1.1	7	50	long	[b]	[b]
Thermal expansion (%)[a]	1.25	0.8	0.5	1.15	0.54	-	-	-
Swelling (vol.%)[a]	7	-	< 0.7	1.7	< 0.5	-	-	-

From Tanaka [748]
[a] Value at 500 °C
[b] Low solubility and problem in leakage

Li_2O has relatively large thermal expansion coefficient and also the swelling due to neutron irradiation up to 7 %. So the care must be made for the dimensional stability. In the case of Li_2O pellets with 70–85 % relative to the theoretical density, creep takes place for the swelling and may be internally adjusted [523].

Lithium Titanate (Li_2TiO_3) Li_2TiO_3 is inferior in thermal conductivity and Li atomic density compared to Li_2O, but there is no reaction with the water similar to Lithium Zirconate, Li_2ZrO_3. Furthermore, there are some merits than the Li_2ZrO_3 such as shorter half-life of the activation product and the high thermal conductivity. Tritium residence time is shorter than that of Li_2O, 2 h at 440 °C, which is advantageous in the tritium recovery in case of the high temperature pressurized water cooling. Li_2TiO_3 has been one of candidate breeders from early days [405] but actual manufacturing development has been made rather recently [775, 776].

DEMO (fusion demonstration reactor) require advanced tritium breeders that have high stability at high temperatures. Lithium titanate (Li_2TiO_3) is one of the most promising candidates among tritium breeders. However, a decrease in lithium mass of Li_2TiO_3 with time occurs in such environments as the DEMO blanket because of Li evaporation and Li burn-up. Therefore, an original material of Li_2TiO_3 with excess Li ($Li_{2+x}TiO_{3+y}$) as an advanced tritium breeder that can make up to the lithium loss has been proposed [346]. Furthermore, pebble fabrication using the emulsion method is one of the promising techniques for the mass production of the advanced tritium breeder pebbles. JAEA have been developing a technique of fabricating $Li_{2+x}TiO_{3+y}$ pebbles using the emulsion method [345, 348].

Lithium Silicate (Li_4SiO_4) and Lithium Aluminate ($LiAlO_2$) Li_4SiO_4 has similar properties on Li atom density, thermal conductivity, half-life of the activation

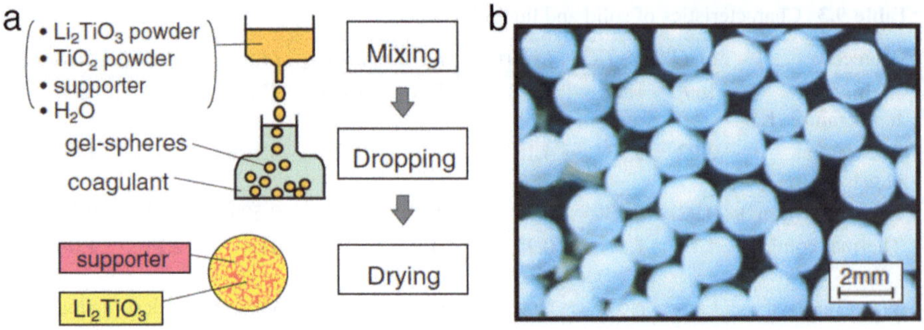

Fig. 9.13 (**a**) Schematics of wet process before calcination/sintering. (**b**) Photograph of TiO_2-doped Li_2TiO_3 pebbles by wet process. Reproduced with permission from Kawamura [420]. Copyright IAEA Vienna

product comparable to Li_2TiO_3, but has some reactivity with the water. $LiAlO_2$ has long T residence time of 50 h.

Liquid Breeders (Li, $Li_{17}Pb_{83}$, Li_2BeF_4) Merits of liquid breeders are negligible radiation damage and simultaneous cooling and T recovery. In addition, self-cooled blanket is possible with liquid metal (Li) or lithium-lead alloy ($Li_{17}Pb_{83}$). On the other hand, pipe corrosion and prevention of tritium permeation are the challenge for $Li_{17}Pb_{83}$ and Li_2BeF_4. For liquid Li, high chemical reactivity, challenging tritium recovery, and large MHD pressure loss are key issues.

Solid Breeder Microspheres Since tritium release is the rate-limiting by the surface, the surface area is increased by manufacturing the microspheres. Then the He gas is used as sweep gas to enhance the T diffusion and isotope exchange with hydrogen. Solid breeder microspheres of 0.3–3 mm diameter have been developed for this purpose. In order to increase tritium release, it is important to control the grain size to 1–10 μm.

There are a few methods to produce microspheres but we briefly introduce the wet process [775]. The wet process is the molding method with liquid and there are gel precipitation method and sol-gel method as shown in Fig. 9.13. Sol-gel method is especially effective to obtain high density microspheres. Li_2TiO_3 microspheres of 0.2–1.5 mm diameter can be produced at the rate of 150 kg/year.

The sintering process of the microspheres is divided to the pre-baking for de-binding and the main firing to form a dense ceramic body. For the solid breeders, we have to be careful on that Solid breeders (1) are hygroscopic and (2) tend to evaporate Li (especially Li_2O), and (3) are highly reactive. Sintering temperature is generally referred to $(2/3)T_m$ but is quite high ~1000–1400 °C. Here T_m is the melting temperature.

***Li* Recovery from Sea Water**

The Li is an important fuel resource for DT fusion. The Li reserve is just 3.7 M ton and the Li reserve base is 9.4 M ton, while gross mineral resource is estimated

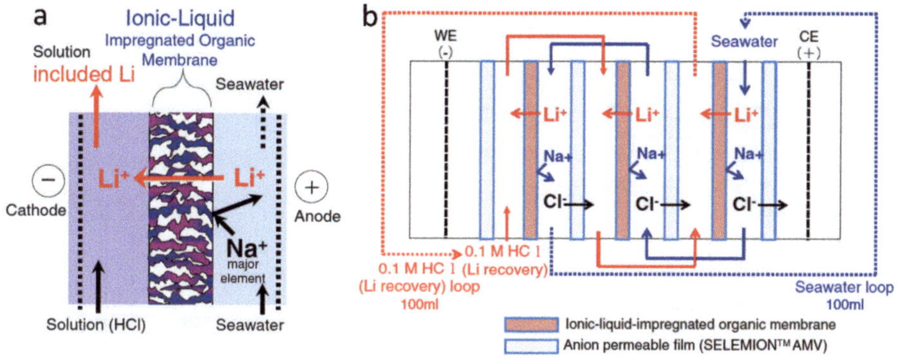

Fig. 9.14 (**a**) Schematics of selective *Li* permeation through membrane. (**b**) Experimental set up of electrodialysis. After Hoshino [347]

as 800 M ton. Salt lakes in South America is major source of *Li*. Especially *Li* production is made at Atakama Salt Lake in Chille and Hombre Muerto Salt Lake in Argentina.

The *Li* concentration in the sea water is 170 ppb and total resource in the sea is estimated as 233*G* ton. Since there are large future market needs for *Li* and the *Li* recovery from sea water is thought to be an important future technology. There is new development of *Li* recovery from sea water by Hoshino [347]. This method utilizes electrodialysis using ionic liquid membrane as shown in Fig. 9.14.

Figure 9.14a shows the principle of selective permeation of *Li* ion in the sea water through ionic liquid membrane. *Li* selectively permeates from anode side to the cathode side through an organic membrane impregnated with an ionic liquid (PP13-TFSI). In the experimental set up shown in Fig. 9.14b, the *Li* concentration of 5.94 % after 2 h with an applied voltage of 2 V, while other ions do not permeate the membrane. With improved membrane covered with a Nafion 324 overcoat to prevent outflow of ionic liquid, the *Li* concentration increased to 22.2 %.

This electrodialysis using ionic liquid membrane is also useful for the isotope separation of ^6Li and ^7Li as demonstrated by Hoshino-Terai [343, 344].

9.3.2 *Neutron Multiplier*

With the tritium production reaction $^6Li + n \rightarrow^4 He + T + 4.78\,\mathrm{MeV}$, the tritium breeding ratio (TBR) is intrinsically less than 1 and the tritium production reaction $^7Li + n \rightarrow^4 He + T + n' - 2.47\,\mathrm{MeV}$ does not contribute much since it is a threshold reaction. Multiplication of tritium fuel is crucially important for fusion and the neutron multiplier to induce $(n, 2n)$ reaction is an essential functional material. The typical neutron multiplication reactions are $^9Be + n \rightarrow 2n + 2He - 2.5\,\mathrm{MeV}$ and

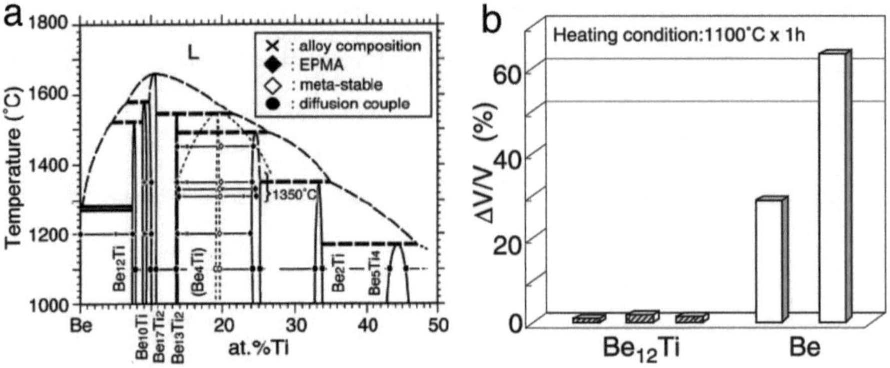

Fig. 9.15 (**a**) Phase diagram of *Be* − *Ti* inter-metallic compounds. From Kawamura [421]. (**b**) Comparison of swelling between *Be* and $Be_{12}Ti$. Reproduced with permission from Kawamura [420]. Copyright IAEA Vienna

$^{A}Pb + n \rightarrow 2n +^{A-1} Pb - 7\,\mathrm{MeV}$ ($A = 204, 206, 207, 208$). *Bi* and *Zr* are also candidates in addition to *Be* and *Pb*.

Beryllium has a hexagonal close-packed crystal structure at room temperature and pressure. Pure *Be* metal exhibits large swelling by the neutron irradiation and may produce large amount of hydrogen by the chemical reaction with water. Thus the inter-metallic compounds ($Be_{12}Ti$, etc.) attract strong attention with their high melting temperature and their chemical stability. The phase diagram of *Be* − *Ti* is shown in Fig. 9.15a.

Be − *Ti* inter-metallic compounds have high melting points as 1500–1700 °C and show the excellent creep strength and the high ductility at high temperature. While tritium breeding ratio is reduced by ∼10 % than that of Be metal, the reaction with RAF is significantly smaller (1/5 reaction layer thickness) and the swelling is 1/50 of the metal Be (see Fig. 9.15b). The desorption of deuterium in $Be_{12}Ti$ occurs at much lower temperature than *Be*. Reactivity to steam in $Be_{12}Ti$ is less than 1/1000 of Be metal at 1000 °C. While the Be metal thermal conductivity is 93 W/m/K at 700 °C, that of non-irradiated $Be_{12}Ti$ is 46 W/m/K and reduced to 28 W/m/K after irradiation. Properties of $Be_{12}Ti$ are reported by Kawamura [420, 421].

New granulation process has been established by combining a plasma sintering method for beryllide synthesis [572] and a rotating electrode method using a plasma-sintered electrode for granulation [573]. The beryllide granulation process and the granulation apparatus are optimized for mass production of beryllide pebbles by the improvement of ductility of plasma-sintered beryllide electrode [574] and the customization of apparatus. Prototypic beryllide pebbles were successfully fabricated with higher yield and larger volume.

9.4 Neutronics

9.4.1 Neutron-Material Interaction

There are scattering and absorption in the neutron interaction with matter. Scattering includes elastic and inelastic scatterings and absorption has capture and nuclear transmutation. Each process has partial cross section of the reaction and the sum is the total cross-section. In the fusion reactor core, 14 MeV neutrons are produced and the energy spectrum in the blanket surrounding the plasma has a sharp peak at 14 MeV. This is very different from the neutron spectrum in fission reactor having a broad peak near 2 MeV as seen in Fig. 9.16.

Fusion neutron energy of 14 MeV is much higher than the average binding energy of nuclei, 7 MeV. Thus, inelastic scattering and threshold reactions such as $(n, n'\alpha), (n, n'p), (n, 2n)$ occur and the secondary neutrons and the forward scattering are also important. Neutron decays into proton, electron and neutrino (β decay) with a half-life of 614 s but is absorbed into the nucleus in time from μ second to a few tens of milliseconds before reaching the lifetime.

Neutron elastic scattering with nucleus X is expressed as $X(n, n)X$ and it causes a change in the traveling direction and a reduction in the kinetic energy of the neutron. Assuming θ is the scattering angle in the center-of-mass system, the ratio of the neutron kinetic energy before (E_1) and after (E_2) the scattering is given by $E_2/E_1 = (A^2 + 2Acos\theta + 1)/(A + 1)^2$. Here, A is the mass number of the nucleus X. In case of head-on collision ($\theta = \pi$), it is given by $E_2/E_1 = (A - 1)^2/(A + 1)^2 \equiv \alpha$, which implies light elements are suitable for neutron slowing down. The natural logarithm of energy, $ln(E_0/E)$ is called "lethargy", where E_0 is a constant usually the largest energy of the system. The vertical axis of Fig. 9.16 is given by the neutron flux per unit lethargy. Average increase of lethargy due to the elastic scattering is called the "slowing down parameter" and can be expressed as $\xi = 1 + (\alpha/(\alpha + 1))ln\alpha \sim 1/(A + 2/3)$ for $A \gg 1$[490]. Average distance between scatterings is called the

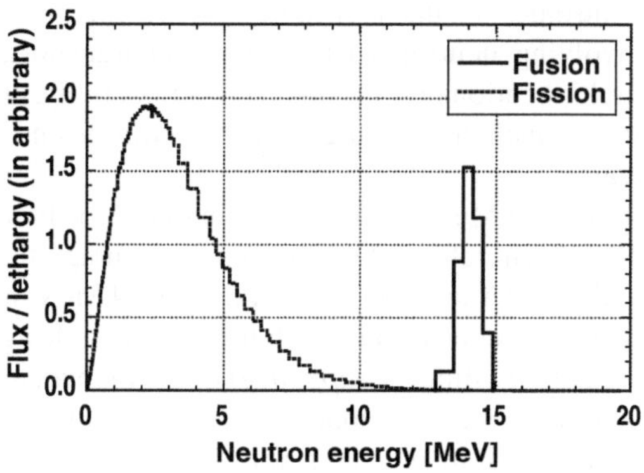

Fig. 9.16 Energy spectra of fission and fusion neutrons in flux/unit lethargy. Private communication from Dr. C. Konno

Table 9.4 Neutron scattering parameters for various materials [576]

Material	Slowing-down parameter ξ	Number of scattering (2 MeV $\rightarrow$ 0.025 eV)	Mean free path (14 MeV neutron) (cm)
H_2O	~ 0.8	23	11
D_2O	~ 0.5	36	9.4
Be	0.21	86	5.4
Fe	0.035	516	4.7
Pb	0.0096	1881	5.5
Li	0.27	67	14
Li_2O	0.175	103	5.2
$Li_{17}Pb_{83}$	0.015	1204	6.3

"mean free path ($\lambda = 1/\sigma n$)", where n is the atom density and σ is the scattering cross section. Table 9.4 shows slowing down parameter, number of scattering and mean free path for various materials.

In the neutron reaction with nuclei X, sometimes an excited state C^* is formed and is called "compound nucleus". The reaction is expressed as $X + n \rightarrow C^* \rightarrow Y + a$. Cross-section to form a compound nuclei becomes large when the neutron kinetic energy E is close to a resonance energy E_r. This cross section is given by a well-known Breit-Wigner formula. Most of the inelastic scattering occurs through the compound nucleus formation. In the Breit-Wigner formula, neutron emission width Γ_n is proportional to $E^{1/2}$. If the total width Γ is large enough, resonance term $(E - E_r)^2$ can be negligible and the cross section is proportional to $1/v$. In the tritium breeding reaction $^6Li(n, \alpha)T$, cross section has $1/v$ dependence.

9.4.2 Fusion Neutronics

Field of research dealing with the neutron behavior is called "Neutronics". Fusion neutronics is the neutronics for fusion research, which includes all processes involving neutron such as neutron transport, tritium breeding process, neutron multiplication process, induced radioactivity, nuclear heat generation, and shielding. It also include measurements of neutrons and γ rays and validation of the nuclear data.

The 14 MeV neutron, carrying 80 % of fusion energy as a mean free path of 5–10 cm. Therefore, there is some advantage that it is a volumetric nuclear heating and not the surface heat flux. The 14 MeV neutron generates secondary gamma rays and neutrons on which the shielding design must be carefully made taking the weldability limit ($<1 appmHe$) of the blanket and the vacuum vessel and the nuclear heating in the superconductors into account.

The tritium breeding ratio must be high enough consistent with heat transfer design of the first wall and the blanket. Material design must be consistent with the knowledge of radiation damage and care must be taken to short term induced activity for safety consideration and long term induced activities for disposal. Irradiation damage due to neutron irradiation may cause deterioration of the material characteristics and the transmutation into unstable nuclides may change the material composition itself. The calculation codes are developed such as one-dimensional transport code ANISN [184], two-dimensional transport code DOT3.5 [626], Monte Carlo code MCNP [547]. Monte Carlo method is a statistical method for simulating the transport of gamma rays and neutrons according to the probability distribution determined by the cross-section by using a pseudo-random number. The Monte Carlo method is started by von Neumann in LANL for application to the neutron diffusion. The transport equation of neutron and gamma ray is following Boltzmann equation.

$$\boldsymbol{\Omega} \cdot \nabla \Phi(\boldsymbol{r}, E, \boldsymbol{\Omega}) + \Sigma_t(\boldsymbol{r}, E, \boldsymbol{\Omega})\Phi(\boldsymbol{r}, E, \boldsymbol{\Omega}) =$$

$$\int d\boldsymbol{\Omega}' \int dE' \Sigma_s(\boldsymbol{r}, E' \to E, \boldsymbol{\Omega}' \to \boldsymbol{\Omega})\Phi(\boldsymbol{r}, E', \boldsymbol{\Omega}') + S(\boldsymbol{r}, E, \boldsymbol{\Omega}) \tag{9.4}$$

Here, $\boldsymbol{r}$: space coordinates, E: energy, $\boldsymbol{\Omega}$: solid angle, Φ: $\boldsymbol{\Omega}$ dependent flux, Σ_t: total cross section, Σ_s: scattering cross section, S: external source.

Multi-group method is the method to solve this Boltzmann equation by dividing energy into some groups, which leads the double integral to a single integral.

As for the cross-section data for neutron transport calculation, there are JENDL3 by Shibata [686], group datasets, Fusion-J3 (neutron 125 group, gamma 40 group) and Fusion40 (neutron 42 group, gamma 21 group) by Maki [533], continuous dataset for Monte Carlo calculation FSXLIB-J3 by Kosako [470].

Fusion nuclear data is provided as FENDL (Fusion Evaluation Nuclear Data Library) contributed from all over the world including contributions from Japan as coordinated by the IAEA Nuclear Data Center section [235].

The nuclear heating rate by gamma rays and neutron can be calculated using the KERMA factor by Maki [532] and local gamma and neutron fluxes obtained from transport calculations. Neutron radiation damage in dpa can be calculated by multiplying the neutron flux and displacement damage cross section by Maki [534]. Helium production rate can also be calculated by multiplying the neutron flux and He production cross section by S. Mori [560]. Induced activity in fusion is fairly large compared with fission since 14 MeV is larger than the average nuclear binding energy. Codes to calculated induced activity are developed in US, EU and Japan in 1980s. In Japan, THIDA-2 system is developed by Seki [674]. Substituting the neutron and gamma fluxes by the transport codes, activation cross sections, decay sequence data and delayed gamma data into ACT4 code, decay heat can be calculated with the induced activity and decay gamma data.

These neuronic calculations are sometimes different among various codes and also by geometry by duct streaming [12] and voids in addition to the uncertainties of cross sections. The 14 MeV fusion neutron sources such as FNS (Fusion Neutronics Source) [469], FNG (Frascati Neutron Generator) [48], ASP facility [601] are quite effective to check the validity of integrated consistency of neuronic calculations. Especially code variation against experiment is important for the physical quantities related to the tritium breeding ratio (TBR) since there are still some uncertainties of order of 5–10 % for important cross sections and 5 % difference in TBR has large impact for tritium self-sufficiency of fusion reactor. Tungsten (*W*) is thought to be an important plasma-facing component (PFC) and neutronics benchmarking has been done [49]. Sato [660] showed that *W* armor results in non-negligible reduction of tritium production in the blanket and the accurate measurement of the tritium production rate is important. The tritium production in the blanket mockup has been extensively studied using 14 MeV neutron sources and noble measuring techniques such as by Sato [661] and Tanigawa [753]. Neutronics and nuclear data issues in ITER and their validation are discussed by Batistoni [50] and validation of TBM neutronics has been done [51].

9.4.3 Neutronics Applications

Necessary shield thickness is usually determined by the nuclear heating in the superconducting TF coils in the inboard side. Using the group dataset (such as Fusion40), transport calculations are made by 1-D ANISN and 2-D DOT3.5 transport codes. These produce neutron and gamma flux distributions during the operation. Substituting the calculated neutron and gamma flux distributions into THIDA system, we can calculate induced activities. These calculations are used to modify shield design to meet design criterion for nuclear heating rate and absorption dose limit of insulators. Figure 9.17a shows attenuation of neutron flux for advanced shielding materials such as TiH_2, ZrH_2 and $Mn(BH_4)_2$ [306]. Neutronics calculations are also used to design the neutron shielding materials as shown in Fig. 9.17b, c.

Induced activity is also important for waste management. It is critical for fusion to minimize radioactive waste disposal and maximize clearance level materials. The clearance is discussed for solid wastes initially in IAEA-TECDOC-855 [360] and the exclusion, exemption and clearance in RS-G-1.7 [358]. Table 9.5 shows the derived values of concentration for exemption given in RS-G-1.7 which is more conservative than IAEA-TECDOC-855.

Tobita [768] made a classification of fusion waste into clearance level using similar values defined in IAEA-TECDOC-855 to show that choice of shielding material and thickness drastically changes clearance level waste (CLW).

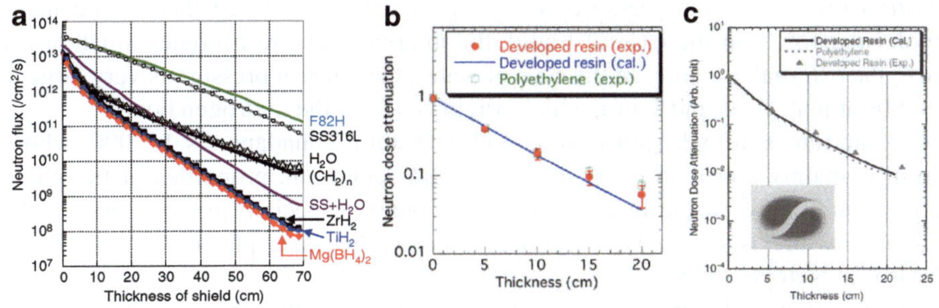

Fig. 9.17 (**a**) Attenuation of neutron flux for various shielding materials. Reproduced with permission from Hayashi [306]. Copyright Elsevier. (**b**) Characteristics of heat resistant neutron shielding material. Reproduced with permission from Morioka [561]. Copyright Elsevier. (**c**) Characteristics of flexible resign for neutron shielding. Reproduced with permission from Sukegawa [723]. Copyright Taylor and Francis

Table 9.5 Derived values of concentration for exemption

Isotopes	H-3	Be-7	C-14	Na-22	Cl-36	Ca-45	Sc-46	Mn-54	Fe-55	Fe-59	Co-56	Co-57	Co-58
Bq/g	100	10	1	0.1	1	100	0.1	0.1	1000	1	0.1	1	1

From RS-G-1.7 [358]

9.5 Tritium and Deuterium Chemistry

9.5.1 Physical Chemistry of Hydrogen

Hydrogen molecule (H_2) consists of two electrons and two protons and the two hydrogen atoms superpose their 1*s* orbitals each other, in which not only one's own electron but also other's electron are in the orbital. Two electrons are shared by two hydrogen atoms. This is called shared electron pair. The potential energy is reduced by sharing the electrons, which produces binding of two atoms. This form of binding is called "covalent bond". Actual measurement of the dissociation energy of hydrogen is 109.5 kcal/mol and the inter-nuclei distance is 0.74 Å, whose explanation needs the wave function of about 40 rather than the simple sum of the wave function of the 1*s* orbital.

Water (H_2O) is an molecule with two hydrogen atoms bonded to an oxygen atom. Oxygen has six electrons in the L shell and two in the K shell. L shell consists of 2*s* and three 2*p*, while electrons occupy from a lower energy level, i.e. enter the *s*-orbital, then enters the *p*-orbitals. Six electrons are distributed as two in 2*s* and two,one and one in 2*p*. Single electron in the 2*p* orbital is called "unpaired electron", to which 1*s* orbital of hydrogen atom couples to form H_2O.

Below the boiling temperature, H_2O becomes a liquid state by the "hydrogen bonding". Above the boiling temperature, H_2O become gaseous state close to the ideal gas. The strength of binding of the liquid is represented by the "cohesive force", whose index is the thermodynamical internal pressure ($P_i = (\partial E/\partial V)_T =$

$T(\partial P/\partial T)_V P)$. Internal pressure is zero in the ideal gas since there is no intermolecular force. In the case of liquid, the internal pressure becomes much larger than the external pressure. In the case of water, the internal pressure is very large at 20,000 atm at 25 °C and 1 atm, which originates from the hydrogen bond.

Deuterium is a hydrogen isotope and its natural abundance is 1/7000, whose form is mainly *HDO*. In order to concentrate deuterium, properties of the phase equilibrium between the liquid and the gas phase, especially the property of the vapor pressure is used. We introduce some of the basic physical chemistry of hydrogen isotopes [558].

In the liquid (*L*) and gas (*G*) phase equilibrium, the chemical potential μ_i introduced by Gibbs takes same value ($\mu_i^G = \mu_i^L$: This condition is obtained from the change in the Gibbs's free energy is zero: $dG = \mu_i^L(-dn_i) + \mu_i^G dn_i = (\mu_i^L - \mu_i^G)dn_i = 0$, where dn_i is the molecule transfer from liquid to gas).

In 1986, French chemist Francois Marie Raoult (1830–1901) gives a law for the partial pressure of component *A* as $P_A = X_A P_A^0$ from the experimental data of vapor pressure. Here, X_A is the molar fraction of the component *A*, P_A is the vapor pressure of component A at X_A, P_A^0 is the vapor pressure of a solution of 100 % A. This Raoult law is important in the isotope separation. If the interaction between components in the solution is the same, it is called the "ideal solution". In the two component (A, B) ideal solution, cohesive force between $A - A, A - B$, and $B - B$ are the same. Raoult's law holds in the ideal solution. Two component solution with same molecular structure with different isotope such as *HDO* and H_2O is a good example of an ideal solution.

The chemical potential of component *A*, μ_A^L in the ideal solution in phase equilibrium with the gas phase is given by $\mu_A^L = \mu_A^G = \mu_A^{G0} + RT_A lnP_A$ (this can be obtained by the integration of the thermodynamic relation $\partial\mu_A/\partial P_A = V_A = RT_a/P_A$ under constant T_A). Here, μ_A^{G0} is the chemical potential at $P_A = 1$ atm.

For the chemical reaction $A + B \rightleftharpoons C + D$, the reaction rate to the right is proportional to the density of A and B, while the reaction rate to the left is proportional to the density of C and D, which is known as the "law of mass action". The chemical equilibrium is a state of left and right reactions are the same $(k_R[A][B] = k_L[C][D])$, where $[A]$ is the density of component *A*. The equilibrium constant is defined as $K = k_R/k_L = [C][D]/[A][B]$. If the reaction is given by $aA + bB \rightleftharpoons cC + dD$, the equilibrium constant is given by $K = [C]^c[D]^d/[A]^a[B]^b$.

If each component is close to the ideal gas for a chemical reaction between the gas phases, the change in Gibbs's free energy is given by $dG = VdP = RTdlnP$ for isothermal process. Integrating from $P = 1$ atm to *P*, we obtain $G - G^0 = RTlnP$. Therefore change due to chemical reaction is $\Delta G - \Delta G^0 = RT\sum \nu_i lnP_i$. Here ν_i is the number of mole. We obtain $\Delta G^0 = -RT\sum \nu_i lnP_i = RTln(P_C)^c(P_D)^d/(P_A)^a(P_B)^b = RTlnK$ using $\Delta G = 0$. Gibbs's free energy at 25 °C for various materials are given in Table 9.6.

Table 9.6 Gibbs's free energy at 25 °C for various materials

Material	H_2O	D_2O	HDO	H_2S	H_2O	D_2O	HDO
Phase	Gas	Gas	Gas	Gas	Liquid	Liquid	Liquid
ΔG_{298} (kcal/mole)	−54.635	−24.73	−55.828	−7.892	−56.69	−58.206	−57.925

After Moore [558]

Fig. 9.18 (**a**) British Physicist H. Cavendish (1731–1810). (**b**) French Physicist J. Charles (1746–1823)

Salon: Hydrogen
The hydrogen is discovered in 1766 by Henry Cavendish (1731–1810) (Fig. 9.18a) through chemical reaction between metal (such as Fe) and strong acid (such as sulfuric acid). In 1783, French physicist J. Charles (1746–1823) (Fig. 9.18b) considered hydrogen gas can be used for a balloon vehicle filled with hydrogen. Manned balloon (Charles and co-pilot) flies ∼ 36 km at 550 m height (Fig. 9.18).

9.5.2 *Hydrogen Isotopes*

In 1931, American chemist H. Urey (Fig. 9.19a) showed that the hydrogen isotope of mass 2 exists by carefully investigating the spectrum of the residue in evaporated hundreds of liters of liquid hydrogen. In 1932, Washburn and Urey found that heavy water (D_2O) is concentrated in the residual liquid in the electrolysis of water. Tritium hardly exists in nature but can be produced by the nuclear reaction of deuteriums first produced by Australian physicist M. Olifant (Fig. 9.19b). The physico-chemical properties of Hydrogen, Deuterium and Tritium are slightly different each other. In addition, physical properties also differ among light water,

Fig. 9.19 (**a**) US Chemist Harold Urey (1893–1981) who discovered Deuterium. (**b**) Australian physicist Mark Oliphant (1901–2000) who discovered Tritium

Table 9.7 Physico-chemical properties of hydrogen isotopes

Item	H_2	D_2	T_2
Boiling temperature (K)	20.41	23.67	25.04
Vaporization heat (cal/mol)	216	293	333
Critical temperature (K)	33.19	38.34	40.44
Critical pressure (atm)	12.77	16.28	18.26
Critical density (mole/m^3)	15,200	16,700	17,700
Triple point temperature (K)	13.96	18.73	20.62
Triple point pressure (mmHg)	54.0	128.6	162.0

After Souers [710]

Table 9.8 Physico-chemical properties of isotopic waters

Item	Boiling temp. (K)	Vaporization heat (cal/mol)	Triple point temp. (K)
H_2O	100	9.72	0.01
D_2O	101.42	9.90	3.82
T_2O	101.51	10.1	4.49

After Souers [710]

heavy water, and the tritiated water. The physical properties of the hydrogen isotopes and isotopic waters are shown in Tables 9.7 and 9.8, respectively.

9.5.3 Hydrogen Isotope Exchange Reaction

Using the difference in physic-chemical properties of isotopic waters, hydrogen-isotope exchange reaction is an efficient method to produce deuterium water and to separate hydrogen, deuterium and tritium. This method is based on the principle that the chemical equilibrium between the gas and liquid phases is different by isotope [53].

H. Urey considered various isotope exchange reactions at Columbia University between 1940–1943 years in order to produce a heavy water, which includes the

Table 9.9 Separation factors in the deuterium exchange reaction in liquid-gas system After Benedict [53]

Chemical reaction $L \rightleftharpoons G$	Separation factor (25 °C)	Separation factor (125 °C)	Ratio
$H_2O + HDS \rightleftharpoons HDO + H_2S$	2.37	1.84	1.29
$H_2O + HD \rightleftharpoons HDO + H_2$	3.81	2.43	1.57

isotope exchange reaction of hydrogen sulfide and water. Spevack invented dual temperature water-hydrogen sulfide exchange method in this period. In 1949, a large amount of heavy water was required at the Savannah River reactor and Du Pont assessed that this method is most efficient for the heavy water production. To improve further this manufacturing method, Spevack takes measures for the corrosion by hydrogen sulfide. Girdler Inc. manufactures a heavy water production plant under the guidance of Du Pont and this method was called Girdler-Salfide (GS) method.

Separation Factor α is an important factor in the isotope separation. For example, the separation factor α of HD is defined by $\alpha = [x(1-x)]/[y(1-y)]$ for H_2, HD, and D_2 system, where x and y are concentration of HD in liquid and gas phase, respectively.

Table 9.9 shows the separation factors at 125 °C and 25 °C in the isotope exchange reaction typical of purifying the deuterium. If the ratio of the separation factors in two temperatures is high, dual temperature isotope separation method is efficient. The merit of hydrogen sulfide process that it does not require catalyst for the reaction.

The reaction $H_2O(liquid) + HDS(gas) \rightleftharpoons HDO(liquid) + H_2S(gas)$ is practically most important reaction. Let K is equilibrium constant of liquid-gas phases and α is the separation factor. Separation factor is defined as the ratio of the deuterium in the liquid phase and that in the gas phase.

$$\alpha = \frac{(x_{HDP} + 2x_{D_2O})/(2x_{H_2O} + x_{HDO})}{(y_{HDS} + 2y_{D_2S})/(2y_{H_2S} + y_{HDS})} \tag{9.5}$$

If D_2O concentration is much smaller than HDO concentration, the separation factor is equal to the equilibrium constant K, i.e. $\alpha = [x_{HDO}/x_{H_2O}]/[y_{HDS}/y_{H_2S}] = K$.

Girdler-Sulfide method utilizes the fact that the separation factor for $H_2O(liquid) + HDS(gas) \rightleftharpoons HDO(liquid) + H_2S(gas)$ changes depending on the temperature. Its operation principle is schematically shown in Fig. 9.20a. Deuterium is transferred to the liquid phase (water) in the low temperature tower and to the gas phase (hydrogen sulfide) in the high temperature tower because of the difference in the separation factor.

System consists of cold and hot towers. Cold tower is assumed to have n_c stages and hot tower to have n_h stages. Let F (Feed), P (Product), W (Waste), G (Gas) as the flow rate, and x_F, x_p, x_w, y_F, and y_p as the deuterium concentration.

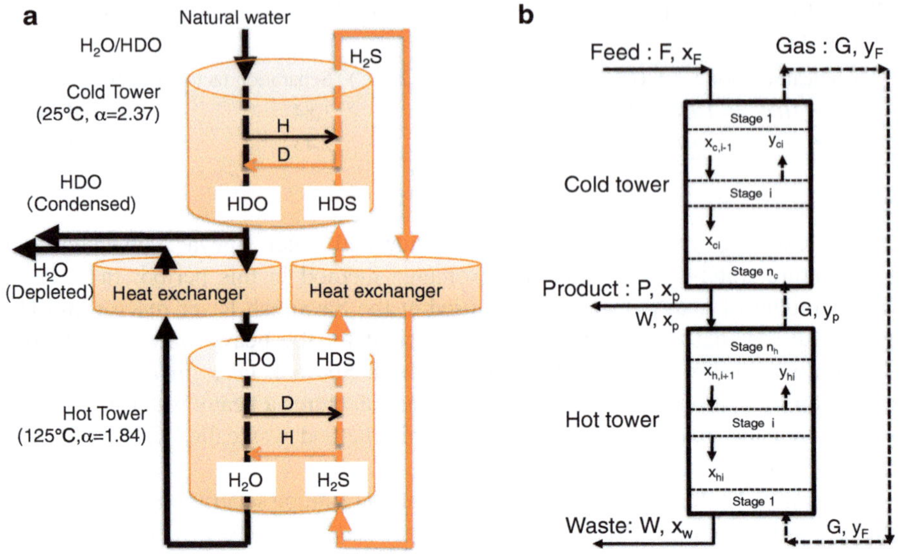

Fig. 9.20 (**a**) Schematics of GS method dual-temperature water-hydrogen sulfide exchange system to enrich deuterium. *Black line* for liquid and *orange line* for gas. Modified from Benedict [53]. (**b**) Mass flow diagram of GS process. Modified from Benedict [53]

Natural water containing deuterium of 0.014 % is supplied from the top of the cold tower. Since the separation factor (equilibrium constant) of deuterium transportation to liquid is large at low temperature, deuterium is transported actively to water from the hydrogen sulfide in the cold tower. Some of the deuterium-rich water enriched at low temperature column is heated via a heat exchanger and fed to a high temperature tower. Since the separation factor is smaller in high temperature, the deuterium is transferred from deuterium-rich water to hydrogen sulfide in hot tower. Hydrogen sulfide with high deuterium concentration required at cold tower is purified in this process.

Ignoring the evaporation of water and the solubility of hydrogen sulfide in water, mass flow diagram of GS process is given in Fig. 9.20b. Assuming deuterium concentration is small, we have $x_{ci} = \alpha_c y_{ci}$ and $x_{hi} = \alpha_h y_{hi}$.

The mass balance of deuterium in cold tower is given by $F(x_p - x_F) = G(y_p - y_F)$ and that above stage i is given by $F(x_{c,i-1} - x_F) = G(y_{ci} - y_F)$. The operation line ($y_{ci}$ as a function of x_{ci}) is expressed as $y_{ci} = y_F + (F/G)(x_{c,i-1} - x_F)$ from these relations. This gives the relation $F/G = (y_p - y_F)/(x_p - x_w)$. By selecting the ratio of water (F) and hydrogen sulfide (G) flow rates, it is possible to transport deuterium from water to hydrogen sulfide at hot tower. The operation lines are expressed in (x, y) diagram, called McCale-Thiele diagram. For detail, see M. Benedict [53].

9.5.4 Fuel Circulation System in the Fusion Reactor

Fuel cycle of a fusion reactor consists of "fuel injection system", "vacuum exhaust system", "fuel purification and impurity processing system", "hydrogen isotope separation system", and "fuel storage system".

The fuel injection system consists of the pellet and gas injection systems. Pellet injection system supplies deuterium and tritium in the plasma periphery. The gas injection system is used mainly to control the heat and particle fluxes to the diverter.

The vacuum system exhausts neutralized fuel, He ash and the impurities. He ash and impurities are removed by Pd/Ag alloy membrane diffuser etc. in the fuel purification and impurity processing system.

The Isotope Separation System (ISS) purifies deuterium and tritium by utilizing the difference in the boiling temperatures of $H_2, HD, HT, D_2, DT, T_2$. The purified fuel are stored in the hydride bed metal such as U and ZrCo.

Tritium is generated by the (Li, n) reaction inside the blanket and recovered by He sweep gas flowing around the breeder.

Tritium is permeated into the coolant from the plasma side and from tritium breeding zone in the blanket. Thus, the tritium recovery system from coolant is also required.

The system also includes confinement system of radio active material and de-tritiation system and stack release system.

The T availability for the first DEMO is an important issue since we have relatively small production of T from CANDU reactors. The T self sufficiency is also an important issue as discussed in Sect. 3.6.

9.6 Negative Ion Source and N-NBI System

9.6.1 Negative-Ion-Based Neutral Beam Injection

In the ITER and fusion power reactor, negative-ion-based neutral beam injector (N-NBI) is expected for the heating the plasma high enough to fuse deuterium and tritium to produce fusion energy. The reason to use negative ion is the neutralization efficiency of positive ion decreases significantly with the beam energy $\geq$ 100 keV, while negative ion produced by attaching the electron to the neutral atom has high stripping efficiency to produce neutral beam at high beam energy as seen in Fig. 9.21. In case of negative ion, neutralizing efficiency of $\sim$60 % is expected and achieved for gas neutralizing target. In case of plasma neutralizing target, neutralizing efficiency close to $\sim$90 % is expected.

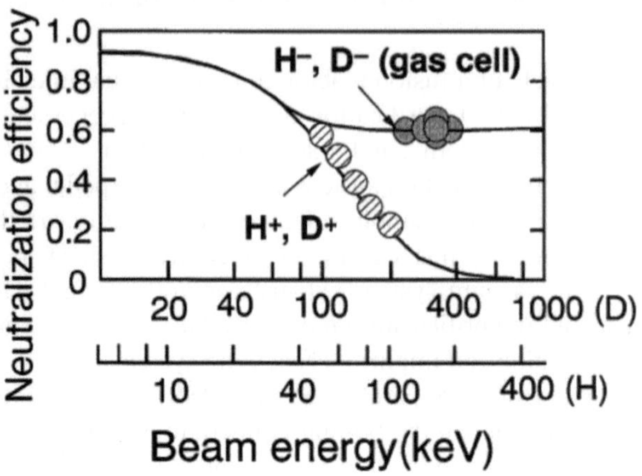

Fig. 9.21 Neutralizing efficiency as a function of beam energy. Reproduced with permission from Kuriyama [480]. Copyright Elsevier

9.6.2 Negative Ion Source

Negative hydrogen ion has been used as the ion beam species in the nuclear experiments since the 1930s but its production is based on the double charge exchange and the surface production. Bacal in 1979 [34] showed that the negative ion density inside the ion source can be as high as 30 % of the positive ion density by adjusting the arc discharge condition and gas pressure as measured by the laser-induced photo-detachment of the electron [35]. It is called the 'Volume Production' of negative ion source. Later, it is found theoretically that the mechanism for the volume production is two-step reaction: $H_2 + e_{fast} \rightarrow H_2^*(v) + e_{slow}$ and $H_2^*(v) + e_{slow} \rightarrow H^- + H^0$.

In the first step reaction, electrons of several eV collide with the hydrogen molecules to produce vibrational excited states (v), called the Driver region.

In the second step, a negative ion is formed when the low energy electrons with <1 eV are absorbed by these excited molecules. This process called dissociative electron attachment (DA). The cross section of this DA increases by five orders of magnitude from vibrationally cold $v = 0$ molecule state to vibrationally excited $v = 5$ state [792]. This reaction should be located near the Extraction region and should have low electron temperature since the formed negative ions are easy to be broken by the fast electrons.

Since the bonding energy of e^- is low (~ 0.75 eV), the electron in the negative ion is easily separated by collisions. There are three important destruction processes.

$$H^- + H^+(H_2^+, H_3^+) \rightarrow H + - : \text{mutual neutralization (MN) by collision} \quad (9.6)$$

$$H^- + e \rightarrow H + 2e : \text{electron detachment (ED) by electron collision} \quad (9.7)$$

$$H^- + H \rightarrow H_2(v'') + e : \text{associative detachment (AD) in atomic collision} \quad (9.8)$$

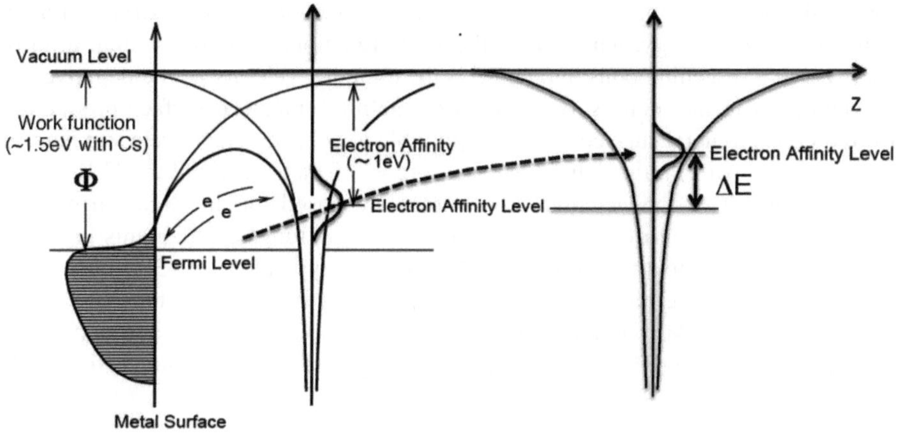

Fig. 9.22 Energy diagram of H^0 at a distance z from the metal surface. Modified from Bacal [36]

It is a contradictory requirement to have high electron temperature to excite hydrogen molecule, while high electron temperature is dangerous to sustain negative ions. Due to the difference in the optimum electron temperatures, It is necessary to separate the Driver region and the Extraction region in the negative ion source.

Magnetic Filter is conceived to separate the high and low electron temperature regimes. The magnetic field of the magnetic filter is ~50–100 Gauss and it prevents to destroy the negative ions by the high energy electron in the driver region. The external magnetic filter [373] is effective to produce high negative ion current density $\sim$ 220 A/m^2. The $\boldsymbol{B} \times \nabla B$ drift due to magnetic filter produces non-uniformity of current density distribution in a large ion source and the tent filter magnetic concept [767] successfully produces uniform negative ion current density distribution since $\boldsymbol{B} \times \nabla B$ drift round the ion source periphery in this configuration.

In addition to the volume production mechanism, *Cs* coating on the metal surface enhances the negative ion production and is called the "surface production". Negative ions are generated through the jumping of the electron by the tunnel effect for H^0 or H^+, which are produced by the dissociation of molecular hydrogen in low pressure arc discharge ($H_2 + e_{fast} \rightarrow H^0 + H^+ + e$). This jumping occurs easily if the surface work function is low. *Cs* has the minimum work function is as low as 1.52 eV (see Fig. 9.22) and the electrons easily jump out from the surface by *Cs* deposition. Combination of surface production produces 3 times larger negative ion current density.

In front of the metal surface, hydrogens are neutralized with high probability. Most important mechanism is resonant neutralization into an excited state followed by Auger de-excitation to the ground state. Proton incident onto *Cs* coated metal may penetrates the cesium layer and reflects from metal lattice as neutral. After reflection, the hydrogen neutral picks up an additional electron. A negative ion induces an image charge in the metal and the associated attractive interaction results in a shift ΔE of the atomic affinity level. The probability of negative ion production on a metal surface, $P_i = (2/\pi)exp[-\pi(\Phi - E_a)/2av]$ increases with the atomic energy, where Φ is the work function, E_a is the atomic affinity energy, a is a screening constant, and v is the escaping velocity of the negative ion perpendicular to the metal surface. With the help of enhanced surface production from *Cs* coated metal surface, it is now possible to achieve reactor relevant current density $> 200\,A/m^2$ at the neutral gas pressure $< 0.3\,Pa$ as required in ITER.

Development of the negative ion sources are based on the high current arc discharges in the multi cusp field. Cathodes such as tungsten filaments are used which are subject to the lifetime of the filament typically 100–300 h. A radio frequency ion-source has no filament and has long life. An RF ion source has been developed at Max Planck institute for plasma physics (IPP) and applied to the negative ion source [790]. IPP further developed RF ion sources for ITER [196, 197, 210].

9.6.3 *Negative Ion Accelerator*

Extraction and acceleration of the negative ions is associated with the acceleration of the electrons. To suppress electron acceleration, a weak magnetic field is applied so that electrons are bended but ions can be extracted. Care must be taken on the beam optics for a small bending of negative ion beam, which may cause excessive heat load to the apertures.

Neutral beam injection heating device with a large negative ion source was developed for the first time in JT-60U with the energy of 500 keV as shown in Fig. 9.23 [467]. Deuterium negative ion beam is accelerated to 500 kV in the three-stage acceleration with a beam current of 22 A. While break down voltage scaling in proportion to $1/2$ power of vacuum insulation gap by Cranberg [151] (Clump hypothesis) was used for the accelerator design, the locally intensified electric field at the aperture of negative ion extraction and the corners of electrodes reduces the break down voltage to about half of the scaling. Stable 500 kV acceleration is achieved by performing optimization of the gap in each stage [467]. Beam acceleration up to 1 MeV is planned in ITER based on such progresses.

9.7 Gyrotron and ECRF System

Electron cyclotron range of frequency (ECRF) wave heating made a great progress by the development of Gyrotron as a high-frequency power source.

Gyrotron is a coherent microwave radiation source [38]. In the former Soviet Union during the 1970s, the tapered open-ended waveguide cavity and the magnetron injection gun are developed, by which electron beam formation method having most of the energy in the form of cyclotron motion is established. This is the technical breakthrough produced by Zaytsev [856] and Baird [37].

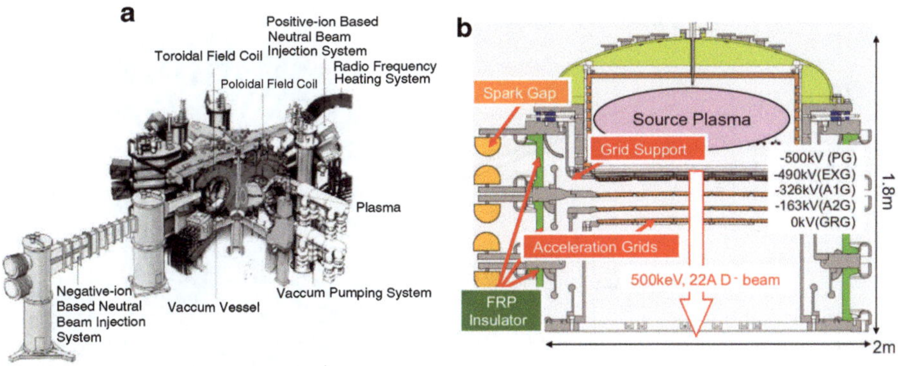

Fig. 9.23 (**a**) JT-60U Negative Ion based Neutral Beam Injector [445]. (**b**) JT-60U Negative Ion Source with 3 stage acceleration grids. Reproduced with permission from Kikuchi [445] and Kojima [467]. Copyright IAEA Vienna

The key principle of excitation is the Cyclotron Resonance Maser (CRM) effect to convert the rotational kinetic energy of weakly relativistic electron beam to the electromagnetic energy. In the CRM effect, electron beams make the spacial bunching due to the interaction with RF waves, and also the phase bunching occurs to transfer kinetic energy to RF wave. These two processes occurs simultaneously.

In order to keep CRM effect as the dominant process, the Weibel instability must be suppressed by having the parallel phase velocity $\omega/k_{\parallel} > c$.

The parameters characterizing the magnetron injection gun (MIG) are (1) beam energy (V_0), (2) Beam current (I_0), (3) cyclotron frequency (B_0), (4) radius of the guiding center (r_g), (5) ratio of parallel and perpendicular velocity ($\alpha = v_{\perp}/v_{\parallel}$).

Important three technical innovations are carried out in the gyrotron development during 2000s in Japan, EU and Russia. These are significant improvement in energy efficiency by the energy recovery [648], significant increase in transmission power density by the development of diamond window [416], robust high power operation using the hard self-excitation shown in Fig. 9.24a [653]. While Sakamoto uses magnetic field to control hard self-excitation, it is also possible to use anode voltage to enter hard self-excitation regime as shown by Kobayashi [460].

With these technical break through, the gyrotrons capable of MW steady output are realized. Figure 9.25 shows the 170 GHz gyrotron for ITER and a schematics showing the basic structure [654]. There is a electron gun (magnetron gun) incident at the bottom, and the electrons are extracted from the cathode potential of −47 kV. Due to the CRM effect, the extracted fast electrons immersed in the strong magnetic field by the superconductor excite the fast wave near the resonator with a mode structure shown in Fig. 9.25b. The remaining energy of the decelerated electrons are recovered at the anode with anode potential of 0 V.

The fast wave amplified near the resonator is transmitted to the transmission line through the diamond window having high thermal conductivity after passing the quasi-optical launcher above the resonator and the multiple reflection by the mirrors. ECRF power is transmitted through the transmission lines and injected into the plasma via ECRF launcher shown in Fig. 9.26.

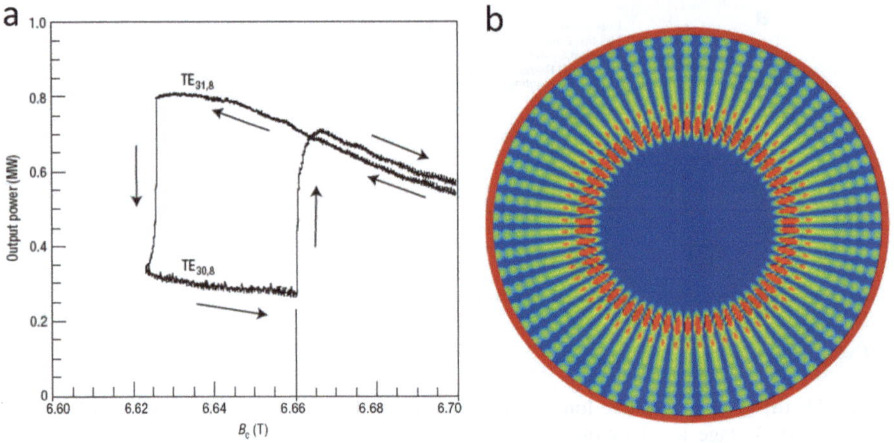

Fig. 9.24 (**a**) Hysteresis curve of hard-self excitation regime in Gyrotron. Reproduced with permission from Sakamoto [653]. Copyright Nature. (**b**) Gyrotron mode structure near the resonator ($TE_{31,8}$). Private communication from K. Sakamoto

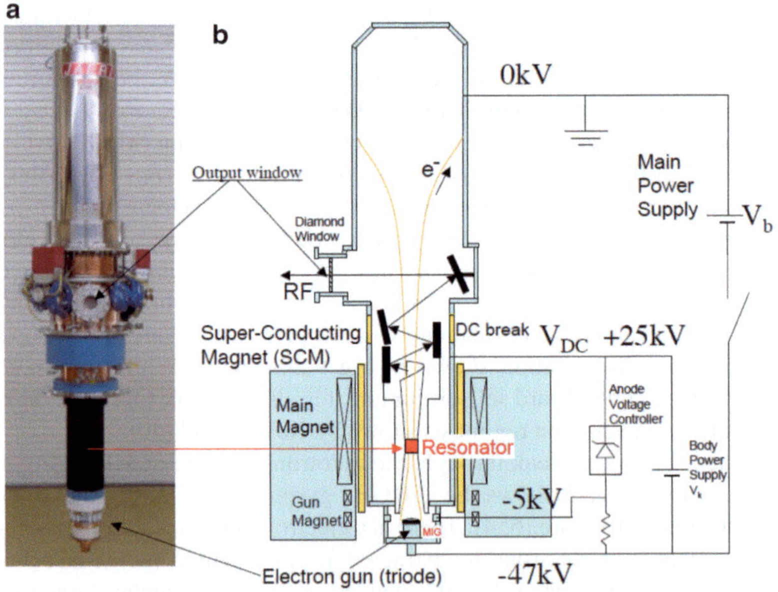

Fig. 9.25 (**a**) 170 GHz Gyrotron for ITER and (**b**) its schematic structure. Reproduced with permission from Sakamoto [655]. Copyright IAEA Vienna

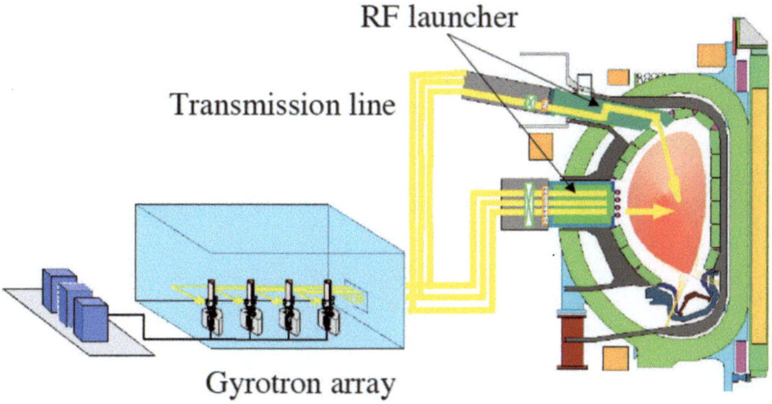

Fig. 9.26 Schematic arrangement of Fusion reactor (ITER), ECRF launcher, and the gyrotron array

Appendix A
Mathematical Basics

Useful formulas are given in NRL plasma formulary by J.D. Huba [353], and also by J.D. Callen [92].

Flanders [206] is a good textbook for the elementary introduction of the differential form. Arnolds [26] gives geometrical introduction of differential forms. Lovelock [526] is also useful to see difference between differential form and tensor algebra. Frankel [209] gives modern framework of the differential geometry.

Cary-Littlejohn [99] is best introduction of Lie transform techniques. Cary-Brizard [100] and Brizard-Hahm [82] gives some fundamentals. Ince [372] Chap. IV gives introduction to continuous transformation-group. Frankel [209] Chap. 4 gives introduction to Lie derivatives.

The characteristics of L_2 space is well described by Kolmogolov-Fomin [468].

For ordinary differential equation, see, for example, Tenenbaum [761] and Ince [372].

A.1 Vector Identities and Differential Operators

Vector Formula: We list here well-known vector formulas. Here, $\boldsymbol{a}$, $\boldsymbol{b}$, $\boldsymbol{c}$, $\boldsymbol{d}$ are arbitrary vectors, f, g are scalar.

$$\boldsymbol{a} \cdot (\boldsymbol{b} \times \boldsymbol{c}) = \boldsymbol{b} \cdot (\boldsymbol{c} \times \boldsymbol{a}) = \boldsymbol{c} \cdot (\boldsymbol{a} \times \boldsymbol{b}) \tag{A.1}$$

$$\boldsymbol{a} \times (\boldsymbol{b} \times \boldsymbol{c}) = \boldsymbol{b}(\boldsymbol{a} \cdot \boldsymbol{c}) - \boldsymbol{c}(\boldsymbol{a} \cdot \boldsymbol{b}) \tag{A.2}$$

$$(\boldsymbol{a} \times \boldsymbol{b}) \cdot (\boldsymbol{c} \times \boldsymbol{d}) = (\boldsymbol{a} \cdot \boldsymbol{c})(\boldsymbol{b} \cdot \boldsymbol{d}) - (\boldsymbol{a} \cdot \boldsymbol{d})(\boldsymbol{b} \cdot \boldsymbol{c}) \tag{A.3}$$

$$\nabla \times \nabla f = 0 \tag{A.4}$$

$$\nabla \cdot (\nabla \times \boldsymbol{a}) = 0 \tag{A.5}$$

$$\nabla \cdot (\nabla f \times \nabla g) = 0 \tag{A.6}$$

M. Kikuchi, M. Azumi, *Frontiers in Fusion Research II*,
DOI 10.1007/978-3-319-18905-5

$$\nabla \cdot (f\boldsymbol{a}) = f\nabla \cdot \boldsymbol{a} + \boldsymbol{a} \cdot \nabla f \tag{A.7}$$

$$\nabla \times (f\boldsymbol{a}) = \nabla f \times \boldsymbol{a} + f\nabla \times \boldsymbol{a} \tag{A.8}$$

$$\nabla \times (\nabla \times \boldsymbol{a}) = \nabla(\nabla \cdot \boldsymbol{a}) - \nabla^2 \boldsymbol{a} \tag{A.9}$$

$$\nabla \times (\boldsymbol{a} \times \boldsymbol{b}) = \boldsymbol{a}(\nabla \cdot \boldsymbol{b}) - \boldsymbol{b}(\nabla \cdot \boldsymbol{a}) + (\boldsymbol{b} \cdot \nabla)\boldsymbol{a} - (\boldsymbol{a} \cdot \nabla)\boldsymbol{b} \tag{A.10}$$

$$\nabla \cdot (\boldsymbol{a} \times \boldsymbol{b}) = \boldsymbol{b} \cdot (\nabla \times \boldsymbol{a}) - \boldsymbol{a} \cdot (\nabla \times \boldsymbol{b}) \tag{A.11}$$

$$\nabla(\boldsymbol{a} \cdot \boldsymbol{b}) = (\boldsymbol{b} \cdot \nabla)\boldsymbol{a} + \boldsymbol{b} \times (\nabla \times \boldsymbol{a}) + (\boldsymbol{a} \cdot \nabla)\boldsymbol{b} + \boldsymbol{a} \times (\nabla \times \boldsymbol{b}) \tag{A.12}$$

$$\nabla^2 \boldsymbol{a} \equiv (\nabla \cdot \nabla)\boldsymbol{a} = \nabla(\nabla \cdot \boldsymbol{a}) - \nabla \times (\nabla \times \boldsymbol{a}) \tag{A.13}$$

We have following integral formula for vectors.

$$\int_V dV \nabla \cdot \boldsymbol{A} = \oint d\boldsymbol{S} \cdot \boldsymbol{A} : \text{Gauss theorem} \tag{A.14}$$

$$\int_V dV \nabla f = \oint d\boldsymbol{S} f \tag{A.15}$$

$$\int_V dV \nabla \times \boldsymbol{A} = \oint d\boldsymbol{S} \times \boldsymbol{A} \tag{A.16}$$

$$\int_V dV (f\nabla^2 g - g\nabla^2 f) = \oint d\boldsymbol{S} \cdot (f\nabla g - g\nabla f) \tag{A.17}$$

$$\int_S d\boldsymbol{S} \times \nabla f = \oint_C d\boldsymbol{l} f \tag{A.18}$$

$$\int_S d\boldsymbol{S} \cdot \nabla \times \boldsymbol{A} = \oint_C d\boldsymbol{l} \cdot \boldsymbol{A} : \text{Stokes theorem} \tag{A.19}$$

$$\int_S (d\boldsymbol{S} \times \nabla) \times \boldsymbol{A} = \oint_C d\boldsymbol{l} \times \boldsymbol{A} \tag{A.20}$$

For integral formulas, V is a volume enclosed by a surface S and the surface element is given by $d\boldsymbol{S} = \boldsymbol{n} dS$, where $\boldsymbol{n}$ is the unit normal vector outward from V. For an open surface S, an line element is given by $d\boldsymbol{l}$.

Tensor Formula: We list here tensor formulas. Here, $\boldsymbol{T}$ is a 2nd order tensor $\boldsymbol{T} = T^{ij}\boldsymbol{e}_i\boldsymbol{e}_j = T_{ij}\boldsymbol{e}^i\boldsymbol{e}^j$, $\boldsymbol{ab} = a^i b^j \boldsymbol{e}_i\boldsymbol{e}_j = a_i b_j \boldsymbol{e}^i\boldsymbol{e}^j$ is a dyadic tensor constructed from two vectors $\boldsymbol{a} = a^i\boldsymbol{e}_i = a_i\boldsymbol{e}^i$ and $\boldsymbol{b} = b_j\boldsymbol{e}^j = b^j\boldsymbol{e}_j$, $\boldsymbol{I}$ is unit tensor. Here $\boldsymbol{e}^i = \nabla u^i$ and $\boldsymbol{e}_i = \partial \boldsymbol{x}/\partial u^i$ in general curvilinear coordinates [443]. Dot product between tensors is a tensor defined by dyadic tensor such as $(\boldsymbol{e}_i\boldsymbol{e}_j) \cdot (\boldsymbol{e}^k\boldsymbol{e}^l) = (\boldsymbol{e}_j \cdot \boldsymbol{e}^k)\boldsymbol{e}_i\boldsymbol{e}^l = \delta_{jk}\boldsymbol{e}_i\boldsymbol{e}^l$ using the orthogonal relation. Double dot product between two tensor is a scalar defined such as $(\boldsymbol{e}_i\boldsymbol{e}_j) : (\boldsymbol{e}^k\boldsymbol{e}^l) = \boldsymbol{e}_i(\boldsymbol{e}_j \cdot \boldsymbol{e}^k)\boldsymbol{e}^l = \delta_{jk}\delta_{il}$ to see $\boldsymbol{F} : \boldsymbol{G} = f^{ij}g_{ji}$. Hints of proof of following formulas are shown in the Descartes coordinates $\{x_i\}$.

$$\boldsymbol{I} : \boldsymbol{ab} \equiv (\boldsymbol{I} \cdot \boldsymbol{a}) \cdot \boldsymbol{b} = \boldsymbol{a} \cdot \boldsymbol{b} \qquad (\delta_{ij} a_j b_i = a_i b_i) \tag{A.21}$$

$$\boldsymbol{ab} : \boldsymbol{cd} \equiv \boldsymbol{a} \cdot (\boldsymbol{b} \cdot \boldsymbol{c}) \cdot \boldsymbol{d} = (\boldsymbol{b} \cdot \boldsymbol{c})(\boldsymbol{a} \cdot \boldsymbol{d}) \qquad (a_i b_j c_j d_i = b_i c_i a_j d_j) \tag{A.22}$$

$$\boldsymbol{I} : \boldsymbol{T} = tr(\boldsymbol{T}) \qquad (\delta_{ij} T_{ji} = T_{ii}) \tag{A.23}$$

$$\boldsymbol{a} \cdot (\boldsymbol{cb} - \boldsymbol{bc}) = (\boldsymbol{a} \cdot \boldsymbol{c})\boldsymbol{b} - (\boldsymbol{a} \cdot \boldsymbol{b})\boldsymbol{c} = \boldsymbol{a} \times (\boldsymbol{b} \times \boldsymbol{c}) \tag{A.24}$$

Cross product of tensor:

$$\boldsymbol{a} \cdot (\boldsymbol{b} \times \boldsymbol{T}) = (\boldsymbol{a} \times \boldsymbol{b}) \cdot \boldsymbol{T} \qquad (a_i \epsilon_{ijk} b_j T_{kl} = (\epsilon_{kij} a_i b_j) T_{kl}) \tag{A.25}$$

$$\boldsymbol{a} \cdot (\boldsymbol{T} \times \boldsymbol{b}) = (\boldsymbol{a} \cdot \boldsymbol{T}) \times \boldsymbol{b} \tag{A.26}$$

$$(\boldsymbol{a} \times \boldsymbol{T}) \cdot \boldsymbol{b} = \boldsymbol{a} \times (\boldsymbol{T} \cdot \boldsymbol{b}) \tag{A.27}$$

$$\nabla \cdot (\boldsymbol{ab}) = (\nabla \cdot \boldsymbol{a})\boldsymbol{b} + (\boldsymbol{a} \cdot \nabla)\boldsymbol{b} \qquad (\partial_i (a_i b_j) = (\partial_i a_i) b_j + (a_i \partial_i) b_j) \tag{A.28}$$

$$\boldsymbol{ab} : \nabla \boldsymbol{c} + \boldsymbol{cb} : \nabla \boldsymbol{a} = \boldsymbol{a} \cdot (\boldsymbol{b} \cdot \nabla)\boldsymbol{c} + \boldsymbol{c} \cdot (\boldsymbol{b} \cdot \nabla)\boldsymbol{a} = (\boldsymbol{b} \cdot \nabla)(\boldsymbol{a} \cdot \boldsymbol{c}) \tag{A.29}$$

$$(\boldsymbol{ab} - \boldsymbol{ba}) : \nabla \boldsymbol{c} = \boldsymbol{b} \cdot \nabla \boldsymbol{c} \cdot \boldsymbol{a} - \boldsymbol{a} \cdot \nabla \boldsymbol{c} \cdot \boldsymbol{b} = (\boldsymbol{b} \times \boldsymbol{a}) \cdot \nabla \times \boldsymbol{c} \tag{A.30}$$

$$\boldsymbol{I} : \nabla \boldsymbol{a} = \nabla \cdot \boldsymbol{a} \qquad (\delta_{ij} \partial_j a_i = \partial_i a_i) \tag{A.31}$$

$$\boldsymbol{a} \times \boldsymbol{I} : \nabla \boldsymbol{b} = \boldsymbol{a} \cdot \nabla \times \boldsymbol{b} \qquad (\epsilon_{ijk} a_j \delta_{kl} \partial_l b_i = a_j \epsilon_{jki} \partial_k b_i = a_j (\nabla \times \boldsymbol{b})_j) \tag{A.32}$$

$$\nabla \boldsymbol{a} : \boldsymbol{T} = \boldsymbol{T} : \nabla \boldsymbol{a} = \boldsymbol{a} \cdot \nabla \cdot \boldsymbol{T} - \nabla \cdot (\boldsymbol{a} \cdot \boldsymbol{T}) \tag{A.33}$$

$$(\nabla \boldsymbol{a}) \cdot \boldsymbol{b} + (\nabla \boldsymbol{b}) \cdot \boldsymbol{a} = \nabla(\boldsymbol{a} \cdot \boldsymbol{b}) \qquad ((\partial_i a_j) b_j + (\partial_i b_j) a_j = \partial_i (a_j b_j)) \tag{A.34}$$

$$\nabla \cdot (\boldsymbol{ab}) = (\nabla \cdot \boldsymbol{a})\boldsymbol{b} + (\boldsymbol{a} \cdot \nabla)\boldsymbol{b} \qquad (\partial_i (a_i b_j) = (\partial_i a_i) b_j + a_i \partial_i b_j) \tag{A.35}$$

$$\nabla \cdot (\boldsymbol{I} \times \boldsymbol{a}) = \nabla \times \boldsymbol{a} \tag{A.36}$$

$$\nabla \cdot (\boldsymbol{ab} - \boldsymbol{ba}) = \nabla \times (\boldsymbol{b} \times \boldsymbol{a}) \qquad (\text{use (A.35) and (A.10)}) \tag{A.37}$$

$$\boldsymbol{a} \times (\boldsymbol{b} \times (\nabla \times \boldsymbol{c})) = \boldsymbol{a} \times \nabla \boldsymbol{c} \cdot \boldsymbol{b} + \boldsymbol{b} \cdot \nabla \boldsymbol{c} \times \boldsymbol{a} \tag{A.38}$$

$$\boldsymbol{I} \cdot \nabla \boldsymbol{a} = \nabla \boldsymbol{a} \cdot \boldsymbol{I} = \nabla \boldsymbol{a} \tag{A.39}$$

$$\nabla \times \boldsymbol{ab} = (\nabla \times \boldsymbol{a})\boldsymbol{b} - \boldsymbol{a} \times \nabla \boldsymbol{b} \tag{A.40}$$

$$\nabla \cdot (f\boldsymbol{T}) = \nabla f \cdot \boldsymbol{T} + f \nabla \cdot \boldsymbol{T} \tag{A.41}$$

$$\nabla \times (f\boldsymbol{T}) = \nabla f \times \boldsymbol{T} + f \nabla \times \boldsymbol{T} \tag{A.42}$$

We illustrate one integral relation for tensor.

$$\int_V dV \nabla \cdot \boldsymbol{T} = \oint d\boldsymbol{S} \cdot \boldsymbol{T} \tag{A.43}$$

Differential formula in orthogonal coordinates: Let (x^1, x^2, x^3) are orthogonal coordinate system, we have following formula for differential operators.

$$\nabla f = \frac{1}{h_1} \frac{\partial f}{\partial x^1} \boldsymbol{e}_1 + \frac{1}{h_2} \frac{\partial f}{\partial x^2} \boldsymbol{e}_2 + \frac{1}{h_3} \frac{\partial f}{\partial x^3} \boldsymbol{e}_3 \tag{A.44}$$

$$\nabla^2 f = \frac{1}{h_1 h_2 h_3}\left[\frac{\partial}{\partial x^1}\left(\frac{h_2 h_3 \partial f}{h_1 \partial x^1}\right) + \frac{\partial}{\partial x^2}\left(\frac{h_3 h_1 \partial f}{h_2 \partial x^2}\right) + \frac{\partial}{\partial x^3}\left(\frac{h_1 h_2 \partial f}{h_3 \partial x^3}\right)\right] \tag{A.45}$$

$$\nabla \cdot \boldsymbol{F} = \frac{1}{h_1 h_2 h_3}\left[\frac{\partial}{\partial x^1}(h_2 h_3 F_1) + \frac{\partial}{\partial x^2}(h_3 h_1 F_2) + \frac{\partial}{\partial x^3}(h_1 h_2 F_3)\right] \tag{A.46}$$

$$\begin{aligned}\nabla \times \boldsymbol{F} =& \frac{1}{h_2 h_3}\left[\frac{\partial}{\partial x^2}(h_3 F_3) - \frac{\partial}{\partial x^3}(h_2 F_2)\right]\boldsymbol{e}_1 \\ &+ \frac{1}{h_3 h_1}\left[\frac{\partial}{\partial x^3}(h_1 F_1) - \frac{\partial}{\partial x^1}(h_3 F_3)\right]\boldsymbol{e}_2 \\ &+ \frac{1}{h_1 h_2}\left[\frac{\partial}{\partial x^1}(h_2 F_2) - \frac{\partial}{\partial x^2}(h_1 F_1)\right]\boldsymbol{e}_3,\end{aligned} \tag{A.47}$$

where the line element ds is given by $ds^2 = h_1^2(dx^1)^2 + h_2^2(dx^2)^2 + h_3^2(dx^3)^2$.

As an important coordinate system, we show differential formula in the cylindrical coordinates (R, ζ, Z), where a differential length is given by $ds^2 = dR^2 + R^2 d\zeta^2 + dZ^2$.

$$\nabla f = \frac{\partial f}{\partial R}\boldsymbol{e}_R + \frac{1}{R}\frac{\partial f}{\partial \zeta}\boldsymbol{e}_\zeta + \frac{\partial f}{\partial Z}\boldsymbol{e}_Z \tag{A.48}$$

$$\nabla^2 f = \frac{1}{R}\frac{\partial}{\partial R}\left(R\frac{\partial f}{\partial R}\right) + \frac{1}{R^2}\frac{\partial^2 f}{\partial \zeta^2} + \frac{\partial^2 f}{\partial Z^2} \tag{A.49}$$

$$\nabla \cdot \boldsymbol{F} = \frac{1}{R}\frac{\partial}{\partial R}(RF_R) + \frac{1}{R}\frac{\partial F_\zeta}{\partial \zeta} + \frac{\partial F_Z}{\partial Z} \tag{A.50}$$

$$\nabla \times \boldsymbol{F} = \left(\frac{\partial F_Z}{R\partial \zeta} - \frac{\partial F_\zeta}{\partial Z}\right)\boldsymbol{e}_R + \left(\frac{\partial F_R}{\partial Z} - \frac{\partial F_Z}{\partial R}\right)\boldsymbol{e}_\zeta + \left(\frac{\partial RF_\zeta}{R\partial R} - \frac{\partial F_R}{R\partial \zeta}\right)\boldsymbol{e}_Z \tag{A.51}$$

$$\boldsymbol{A} \cdot \nabla \boldsymbol{B} = \left(\boldsymbol{A} \cdot \nabla B_R - \frac{A_\zeta B_\zeta}{R}\right)\boldsymbol{e}_R + \left(\boldsymbol{A} \cdot \nabla B_\zeta - \frac{A_\zeta B_R}{R}\right)\boldsymbol{e}_\zeta + (\boldsymbol{A} \cdot \nabla B_Z)\boldsymbol{e}_Z \tag{A.52}$$

Components of the Laplacian of the vector $\boldsymbol{F}$ in the cylindrical coordinates are given as follows,

$$(\nabla^2 \boldsymbol{F})_R = \nabla^2 F_R - \frac{2}{R^2}\frac{\partial F_\zeta}{\partial \zeta} - \frac{F_R}{R^2} \tag{A.53}$$

$$(\nabla^2 \boldsymbol{F})_\zeta = \nabla^2 F_\zeta + \frac{2}{R^2}\frac{\partial F_R}{\partial \zeta} - \frac{F_\zeta}{R^2} \tag{A.54}$$

$$(\nabla^2 \boldsymbol{F})_Z = \nabla^2 F_Z, \tag{A.55}$$

where Laplacian operator to a scalar is given by (A.49).

The divergence of a Tensor $\boldsymbol{T}$ in the cylindrical coordinates is given as follows,

$$(\nabla \cdot \boldsymbol{T})_R = \frac{1}{R}\frac{\partial}{\partial R}(RT_{RR}) + \frac{1}{R}\frac{\partial T_{\zeta R}}{\partial \zeta} + \frac{\partial T_{ZR}}{\partial Z} - \frac{T_{\zeta\zeta}}{R} \tag{A.56}$$

$$(\nabla \cdot \boldsymbol{T})_\zeta = \frac{1}{R}\frac{\partial}{\partial R}(RT_{R\zeta}) + \frac{1}{R}\frac{\partial T_{\zeta\zeta}}{\partial \zeta} + \frac{\partial T_{Z\zeta}}{\partial Z} + \frac{T_{\zeta R}}{R} \tag{A.57}$$

$$(\nabla \cdot \boldsymbol{T})_Z = \frac{1}{R}\frac{\partial}{\partial R}(RT_{RZ}) + \frac{1}{R}\frac{\partial T_{\zeta Z}}{\partial \zeta} + \frac{\partial T_{ZZ}}{\partial Z} \tag{A.58}$$

In the spherical coordinates (r, θ, ϕ), we have

$$ds^2 = dr^2 + r^2 d\theta^2 + r^2 sin^2\theta d\phi^2 \tag{A.59}$$

$$\nabla f = \frac{\partial f}{\partial r}\boldsymbol{e}_r + \frac{1}{r}\frac{\partial f}{\partial \theta}\boldsymbol{e}_\theta + \frac{1}{rsin\theta}\frac{\partial f}{\partial \phi}\boldsymbol{e}_\phi \tag{A.60}$$

$$\nabla \cdot \boldsymbol{F} = \frac{1}{r^2}\frac{\partial}{\partial r}(r^2 F_1) + \frac{1}{rsin\theta}\frac{\partial}{\partial \theta}(sin\theta F_2) + \frac{1}{rsin\theta}\frac{\partial F_3}{\partial \phi} \tag{A.61}$$

$$\nabla \times \boldsymbol{F} = \frac{1}{rsin\theta}\left[\frac{\partial}{\partial \theta}(sin\theta F_3) - \frac{\partial F_2}{\partial \phi}\right]\boldsymbol{e}_r$$

$$+\frac{1}{r}\left[\frac{1}{sin\theta}\frac{\partial F_1}{\partial \phi} - \frac{\partial}{\partial r}(rF_3)\right]\boldsymbol{e}_\theta + \frac{1}{r}\left[\frac{\partial}{\partial r}(rF_2) - \frac{\partial F_1}{\partial \theta}\right]\boldsymbol{e}_\phi \tag{A.62}$$

$$\nabla^2 f = \frac{1}{r^2}\frac{\partial}{\partial r}(r^2\frac{\partial f}{\partial r}) + \frac{1}{r^2 sin\theta}\frac{\partial}{\partial \theta}(sin\theta\frac{\partial f}{\partial \theta}) + \frac{1}{r^2 sin^2\theta}\frac{\partial^2 f}{\partial \phi^2} \tag{A.63}$$

A.2 Curvilinear Coordinates

Descartes coordinates (x, y, z) is given by, $\boldsymbol{x} = x\hat{\boldsymbol{e}}_x + y\hat{\boldsymbol{e}}_y + z\hat{\boldsymbol{e}}_z$, where $\hat{\boldsymbol{e}}_x, \hat{\boldsymbol{e}}_y, \hat{\boldsymbol{e}}_z$ are unit vectors satisfying $\hat{\boldsymbol{e}}_i \cdot \hat{\boldsymbol{e}}_j = \delta_{ij}$, where δ_{ij} is Kronecker delta function. Let (u^1, u^2, u^3) as general curvilinear coordinates, "gradient" and "tangent" vectors are defined as,

$$\nabla u^i = \partial_x u^i \hat{\boldsymbol{e}}_x + \partial_y u^i \hat{\boldsymbol{e}}_y + \partial_z u^i \hat{\boldsymbol{e}}_z \tag{A.64}$$

$$\partial_i \boldsymbol{x} = \partial_i x \hat{\boldsymbol{e}}_x + \partial_i y \hat{\boldsymbol{e}}_y + \partial_i z \hat{\boldsymbol{e}}_z \tag{A.65}$$

$\nabla u^i (= \partial u^i/\partial \boldsymbol{x})$ is called the "co-variant basis vector" and $\partial_i \boldsymbol{x} (= \partial \boldsymbol{x}/\partial u^i)$ is called the "contravariant basis vector". Neither (∇u^i) nor $(\partial_i \boldsymbol{x})$ are orthogonal set. But, gradient and tangent vectors are orthogonal called "orthogonal relation".

$$\nabla u^i \cdot \partial_j \boldsymbol{x} = \delta_{ij} \tag{A.66}$$

Let $(i,j,k) = (1,2,3), (2,3,1), (3,1,2)$, we have following **dual relations** between gradient and tangent vectors.

$$\nabla u^i = J^{-1}\partial_j \boldsymbol{x} \times \partial_k \boldsymbol{x} \; , \; \partial_i \boldsymbol{x} = J\nabla u^j \times \nabla u^k, \tag{A.67}$$

where $J \equiv \partial_1 \boldsymbol{x} \cdot (\partial_2 \boldsymbol{x} \times \partial_3 \boldsymbol{x})$ is called the Jacobian. Any vector $\boldsymbol{a}$ can be expanded using the covariant basis vector as $\boldsymbol{a} = a^i \nabla u^i$ called the "covariant representation", where $a^i = \boldsymbol{a} \cdot \partial_i \boldsymbol{x}$, or using the contravariant basis vector as $\boldsymbol{a} = a_i \partial_i \boldsymbol{x}$ called the "contravariant representation", where $a_i = \boldsymbol{a} \cdot \nabla u^i$. Metric tensor $[g_{ij}]$ and $[g^{ij}]$ and g are defined as follows.

$$g_{ij} = \partial_i \boldsymbol{x} \cdot \partial_j \boldsymbol{x} \; , \; g \equiv |g_{ij}| = J^2 \tag{A.68}$$

$$g^{ij} = \nabla u^i \cdot \nabla u^j \tag{A.69}$$

Each metric tensor is inverse of other metric tensor, e.g. $[g_{ij}][g^{ij}] = [g^{ij}][g_{ij}] = \boldsymbol{I}$, where $\boldsymbol{I}$ is unit tensor. Differential length ds^2 is given by using metric g_{ij} as,

$$ds^2 = dx^2 + dy^2 + dz^2 = g_{ij} du^i du^j \tag{A.70}$$

Differential volume $dV = dxdydz$ is given by, $dV = \partial_1 \boldsymbol{x} \cdot (\partial_2 \boldsymbol{x} \times \partial_3 \boldsymbol{x}) du^1 du^2 du^3 = J du^1 du^2 du^3$. Rotation and divergence of vector $\boldsymbol{a}$ are given by,

$$\nabla \times \boldsymbol{a} = J^{-1} \left[\partial_i a_j - \partial_j a_i \right] \partial_k \boldsymbol{x} \tag{A.71}$$

$$\nabla \cdot \boldsymbol{a} = J^{-1} \partial_i (J a^i) \tag{A.72}$$

The line, surface and volume integrals are given by,

$$\int \boldsymbol{a} \cdot d\boldsymbol{x} = \int \boldsymbol{a} \cdot \partial_i \boldsymbol{x} du^i \tag{A.73}$$

$$\int \boldsymbol{a} \cdot d\boldsymbol{S} = \int \boldsymbol{a} \cdot \nabla u^k J du^i du^j \tag{A.74}$$

$$\int f dV = \int f J du^1 du^2 du^3 \tag{A.75}$$

A.3 Vectors and Tensors on the Surface

In the plasma theory, we use tensor analysis in differential geometry. We gives elementary introduction of tensor. Lovelock [526] is useful for tensor algebra on manifolds. Consider a coordinates (u^1, u^2) and $(\bar{u}^1, \bar{u}^2)$ on the surface S which are related by the transformation, $\bar{u}^j = \bar{u}^j(u^1, u^2) \; : \; i = 1, 2$. The difference du^i is given by,

$$du^i = \frac{\partial u^i}{\partial \bar{u}^j} d\bar{u}^j \tag{A.76}$$

A set of number (a^1, a^2) is called a "contravariant vector" or "contravariant tensor of first order" if the following "contravariant transformation" relation is satisfied under the coordinate transformation from (u^1, u^2) to $(\bar{u}^1, \bar{u}^2)$.

$$a^i = \frac{\partial u^i}{\partial \bar{u}^j} \bar{a}^j \tag{A.77}$$

Since $du^i = (\partial u^i / \partial \bar{u}^j) d\bar{u}^j$, (du^1, du^2) is a contravariant vector. A set of number (a_1, a_2) is called a "covariant vector" or "covariant tensor of first order" if the following "covariant transformation" relation is satisfied.

$$a_i = \frac{\partial \bar{u}^j}{\partial u^i} \bar{a}_j \tag{A.78}$$

The "gradient" of a scalar function $(\partial \phi / \partial u^1, \partial \phi / \partial u^2)$ is a covariant vector since,

$$\frac{\partial \phi}{\partial u^i} = \frac{\partial \bar{u}^j}{\partial u^i} \frac{\partial \phi}{\partial \bar{u}^j} \tag{A.79}$$

The differential length ds is an invariant quantity to coordinate transformation.

$$ds^2 = g_{ij} du^i du^j = \bar{g}_{kl} d\bar{u}^k d\bar{u}^l \tag{A.80}$$

Therefore, we obtain following transformation relation for metric tensor g_{ij}.

$$g_{ij} = \frac{\partial \bar{u}^k}{\partial u^i} \frac{\partial \bar{u}^l}{\partial u^j} \bar{g}_{kl} \tag{A.81}$$

A set of number (T_{ij}) is called a "covariant tensor of second order" if the following "covariant transformation" relation is satisfied under the coordinate transformation from (u^1, u^2) to $(\bar{u}^1, \bar{u}^2)$.

$$T_{ij} = \frac{\partial \bar{u}^k}{\partial u^i} \frac{\partial \bar{u}^l}{\partial u^j} \bar{T}_{kl} \tag{A.82}$$

Thus, metric tensor g_{ij} $(i, j = 1, 2)$ is a covariant tensor of second order. Definition of contravariant and mixed tensor of second order is similarly defined. Here, it is stressed that co- and contra-variant vectors and tensors are defined in this section by using the transformation relation in S and without using the basis vectors in R^3.

A.4 Differential Form

In the modern gyro kinetic theory, differential form is used as powerful tool. Here we give an elementary introduction of the differential form.

Consider region $U \subset \boldsymbol{R}^3$, differential 0 form is a function in U. Differential 1 form ω in U is following form of summation.

$$\omega = a_1(\boldsymbol{x})dx^1 + a_2(\boldsymbol{x})dx^2 + a_3(\boldsymbol{x})dx^3 \tag{A.83}$$

Differential 2 form in U is following form of summation.

$$a_1(\boldsymbol{x})dx^2 \wedge dx^3 + a_2(\boldsymbol{x})dx^3 \wedge dx^1 + a_3(\boldsymbol{x})dx^1 \wedge dx^2 \tag{A.84}$$

Here, symbol $\wedge$ is called "wedge" operator to define "external product", which satisfies following rules.

$$dx \wedge dx = dy \wedge dy = dz \wedge dz = 0 \tag{A.85}$$

$$dx \wedge dy = -dy \wedge dx, dy \wedge dz = -dz \wedge dy, dx \wedge dz = -dz \wedge dx \tag{A.86}$$

External product between p-form u and q form v becomes $(p+q)$ form written as $u \wedge v$. Let $= u_i dx^i$ and $v = v_i dx^i$ are one-forms with components u_i and v_i, we have

$$(u \wedge v)_{ij} = u_i v_j - u_j v_i \tag{A.87}$$

Rule is simple except multiplication of same differential form is 0 and change in order change sign. Hereafter, "wedge operator" $\wedge$ is dropped for simplicity unless otherwise specified.

External derivative d for scalar (f) is given by $df = (\partial f/\partial x^i)dx^i = \nabla f \cdot d\boldsymbol{x}$. Consider one form $u = u_i dx^i$ its external derivative du is given by,

$$du = d(u_i dx^i) = \frac{\partial u_i}{\partial x^j} dx^j dx^i = \sum_{i<j} \left[\frac{\partial u_i}{\partial x^j} - \frac{\partial u_j}{\partial x^i} \right] dx^i dx^j, \tag{A.88}$$

i.e. $d(\boldsymbol{a} \cdot d\boldsymbol{x}) = \nabla \times \boldsymbol{a} \cdot d\boldsymbol{S}$. External derivative for differential two form $(\boldsymbol{a} \cdot d\boldsymbol{S})$ is given as $d(\boldsymbol{a} \cdot d\boldsymbol{S}) = (\nabla \cdot \boldsymbol{a})dxdydz$.

We introduce the "interior product" or "inner product" as a product of vector field and the differential one form. Here we call a differential operator, $X = X^i \partial/\partial x^i$ as a vector field. Let $u = u_i dx^i$ is a differential 1 form and $X = X^i \partial/\partial x^i$ is a differential operator, the scalar $u_i X^i$ is a scalar which is invariant to the selection of the coordinates and we call this quantity an "interior product" of a vector field and a differential 1 form. The calculation rule is $X \cdot u = (X^i \partial/\partial x^i) \cdot (u_i dx^i) = u_i X^i$. It is clear that the differential form unifies differential operators such as grad, rot, div in vector analysis. This has natural extension to any dimension.

A.5 Spectrum and Initial Value Problem of Linear Operator

There are many cases in plasma physics where we see linear operator. This section introduces categories of spectrum (point, continuous and residual spectrums) and the initial value problem. See Kolmogolov-Fomin [468] and Friedman [214].

The eigenvalue problem is to obtain the eigenvalue $\lambda \in C$ and eigenvector u to satisfy $Au = \lambda u$. This can be rewritten as $(\lambda I - A)u = 0$ and the problem is to find a set of null points of the linear operator $(\lambda I - A)$. In the operator in infinite dimensional linear space, spectrum analysis is used to investigate singularity of $(\lambda I - A)^{-1}$. The operator $R_\lambda = (\lambda I - A)^{-1}$ if exists, is called the resolvent operator of A. For complex values of λ, the following three classes of the spectrum arise.

1. **Point spectrum**: If $(\lambda I - A)u = 0$ has a non-trivial u, the resolvent operator R_λ does not exists. The corresponding set of λ is called a "point spectrum". Example: $A = -\partial_x^2$ defined at $x \in [0, 1]$ satisfying $u(0) = u(1) = 0$. Eigenvalues of $(\lambda I - A)u = 0$ are $\lambda = \{(n\pi)^2; n = 1, 2, --\}$.
2. **Continuous spectrum**: In this case, the unbounded inverse R_λ exists. The corresponding set is called a "continuous spectrum". Example: $A = x$. The solution for $(\lambda - x)u = 0$ is $u = \delta(x)$. This Dirac delta function is not square integrable and does not belong to Hilbert space.
3. **Residual spectrum**: In the case where inverse $(\lambda I - A)^{-1}$ exists and is bounded, the corresponding set called a "residual spectrum". Example of the residual spectrum is shown in p126 of Friedman [214]. Here, a linear operator A is said to be "bounded" if there exists a constant N for all $u \in H$ such that $||Au|| \leq N||u||$. It is important to note that if λ is in the residual spectrum of A, λ is in the point spectrum of the adjoint operator A^* in p127 of Friedman [214]. Since Hermitian (self adjoint) operator satisfy $A^* = A$, there is no residual spectrum in the Hermitian operator.

Dunford Integral:

The Dunford integral is a natural extension of following Cauchy integral formula for a regular function $F(\lambda)$ and complex constant a.

$$F(a) = \frac{1}{2\pi i} \oint_\Gamma \frac{F(\lambda) d\lambda}{\lambda - a} \tag{A.89}$$

The Dunford integral is defined for a regular function $F(\lambda)$ by,

$$F(A) = \frac{1}{2\pi i} \oint_\Gamma F(\lambda)(\lambda I - A)^{-1} d\lambda \tag{A.90}$$

Consider the following initial value problem for the linear operator A.

$$du/dt = Au \ , \ u(0) = u_0 \tag{A.91}$$

The solution may be given as $u = e^{tA}u_0$, where the exponential operator e^{tA} is given by the Dunford integral as follows,

$$e^{tA} = \frac{1}{2\pi i}\oint_{\Gamma} e^{\lambda t}(\lambda I - A)^{-1}d\lambda, \tag{A.92}$$

where the Γ is a closed line to see all singular points of the integrant left.

A.6 Important Ordinary Differential Equations

Hypergeometric Equation:
Let α, β and γ are constants and γ is not integer.

$$(x^2 - x)y'' + [(1 + \alpha + \beta)x - \gamma]y' + \alpha\beta y = 0 \tag{A.93}$$

This equation has regular singular points at $x = 0, 1, \infty$. The indicial equation is $\kappa(\kappa - 1 + \gamma) = 0$ and $\kappa = 0$ gives the hypergeometric function.

Legendre Equation:
Legendre equation appears in collisional transport, when we expand the velocity distribution function in velocity coordinates (v, θ, φ).

$$(1 - x^2)y'' - 2xy' + n(n + 1)y = 0, \tag{A.94}$$

where $x = cos\theta$. This equation has regular singular points at $x = \pm 1$. The indicial equation is $\kappa(\kappa - 1) = 0$. This Legendre equation is a special case of hypergeometric equation. For positive integer n, this gives Legendre polynomials, $P_n(x)$.

$$P_n(x) = \frac{1}{2^n n!}\frac{d^n}{dx^n}(x^2 - 1)^n \tag{A.95}$$

Lower order Legendre polynomials are $P_0(x) = 1, P_1(x) = x, P_2(x) = (3x^2 - 1)/3$. The Legendre polynomials satisfy following orthogonal relation.

$$\int_{-1}^{1} P_n(x)P_m(x)dx = \frac{2}{2n + 1}\delta_{mn} \tag{A.96}$$

Sonnine Equation:
Sonnine equation appears in collisional transport, when we expand the velocity distribution function in velocity coordinates (v, θ, φ).

$$xy'' + (\alpha + 1 - x)y' + ny = 0, \tag{A.97}$$

where $x = v/v_{Ta}$ and v_{Ta} is thermal speed of the species a. The solution is Sonnine polynomials $L_n^\alpha(x)$.

$$L_n^\alpha(x) = \frac{e^x x^{-\alpha}}{n!}\frac{d^n}{dx^n}(e^{-x}x^{n+\alpha}) \tag{A.98}$$

Lower order Sonnine polynomials are $L_0^\alpha(x) = 1, L_1^\alpha(x) = 1 + \alpha - x, L_2^\alpha(x) = (\alpha+1)(\alpha+2)/2 - (\alpha+2)x + x^2/2$. The Sonnine polynomials satisfy following orthogonal relation.

$$\int_0^\infty e^{-x}x^\alpha L_n^\alpha(x)L_m^\alpha(x)dx = \frac{\Gamma(\alpha+n+1)}{n!}\delta_{mn} \tag{A.99}$$

This property is important to solve the drift kinetic equation to derive analytical parallel viscosity for the Maxwellian distribution function (e^{-x} term in the integral) in the collisional transport (see Sect. 4.3).

Sturm-Liouville Equation:

Sturm-Liouville equation is a following type of ordinary differential equation.

$$\frac{d}{dx}\left(f(x)\frac{dy}{dx}\right) + [\lambda\rho(x) - g(x)]y(x) = 0 \tag{A.100}$$

Here λ is the eigenvalue and $\rho > 0$. This type of ODE is first studied by Sturm (J. de Math, 1 (1836)106), which leads to Comparison Theorems and Oscillation Theorems (see Ince [372]). The Sturm-Liouville type equation appears in Sect. 8.2 (8.45). This equation is much simpler than the Hain-Lust equation (8.22). The singular point in the Hain-Lust equation (8.22) changes for different values of ω. But the singular point is fixed in the Sturm-Liouville type equation (8.58). This property is useful to have fast convergence of eigenvalue.

A.7 Square Integrable Functions and L_2 Space

In MHD and Vlasov spectrum theories, we sometimes see square integrable and non square integrable functions. We gives elementary introduction on such functions.

Hilbert space is one of important linear normed space in the Functional Analysis. Hilbert space is a natural infinite-dimensional analogue of Euclidean n-space. One of important Hilbert space is the space L_2 defined below. We say that $f(x)$ is a square integrable function on a set R (where a measure μ is defined on R) if the following integral exists and finite.

$$\int f^2(x)d\mu \tag{A.101}$$

Set of all square integrable functions is called L_2. The inner product in L_2 is:

$$(f, g) = \int f(x)g(x)d\mu \tag{A.102}$$

The space L_2 is the Euclidean space (a linear space with an inner product) whose elements are square integrable functions. This L_2 has following important property.

This L_2 has following important properties.

$L_2(1)$; The product of two square integrable functions is an integrable function.
$L_2(2)$; A square integrable function is integrable.
$L_2(3)$; Let $f(x), g(x) \in L_2$, then $f(x) + g(x) \in L_2$.
$L_2(4)$; Let α is arbitrary constant and $f(x) \in L_2$, then $\alpha f(x) \in L_2$.

A.8 Poisson Summation Formula

Let $g(x)$ is an arbitrary function of x, we define its Fourier transform $G(k)$ by,

$$G(k) = \int_{-\infty}^{+\infty} g(x)e^{-ikx}dx \tag{A.103}$$

And, $g(x)$ can be expressed as:

$$g(x) = \int_{-\infty}^{+\infty} \frac{dk}{2\pi} G(k)e^{ikx} \tag{A.104}$$

Noting the following δ-function formula,

$$\sum_{m=-\infty}^{+\infty} e^{imx} = 2\pi \sum_{j=-\infty}^{+\infty} \delta(x - 2\pi j) \tag{A.105}$$

We have:

$$\sum_{m=-\infty}^{+\infty} g(mT) = \sum_{m=-\infty}^{+\infty} \int \frac{dk}{2\pi} G(k)e^{ikmT} = \int dk \sum_{j=-\infty}^{+\infty} \delta(kT - 2\pi j)G(k) \tag{A.106}$$

Namely, we have following Poisson Summation formula.

$$\sum_{m=-\infty}^{+\infty} g(mT) = \frac{1}{T} \sum_{j=-\infty}^{+\infty} G(\frac{2\pi j}{T}) \tag{A.107}$$

A.9 Algebraic Solutions

Some of the dispersion relation in plasma physics become the quadratic, cubic or quartic equations. The instability threshold is given by $D = 0$, where D is the discriminant of the algebraic equation. For the quadratic equation $x^2 + ax + b = 0$, the instability threshold is given by $D = a^2 - 4b = 0$.

For the cubic equation $x^3 + ax^2 + bx + c = 0$ $(a \neq 0)$, we first convert to a reduced cubic equation $y^3 + py + q = 0$, where $y = x + a/3$, $p = b - a^2/3$, $q = c - ab/3 - 2a^3/27$. The instability threshold is given by Gerolamo Cardano (1501–1576):

$$D = 4p^3 + 27q^2 = 4(b - a^2/3)^3 + 27(c - ab/3 - 2a^3/27)^2 = 0 \tag{A.108}$$

Appendix B
Elementary Physics

B.1 Physical Constants

Physical constant	Symbol	Value
Electron charge	e	1.602×10^{-19} C
Electron mass	m_e	9.110×10^{-31} kg
Proton mass	m_p	1.673×10^{-27} kg
mass ratio	m_p/m_e	$1836(= 42.9^2)$
Speed of light	c	2.998×10^8 m/s
Boltzmann const.	k_B	1.38×10^{-23} J/K
Vac. permeability	μ_0	1.257×10^{-6} H/m
Vac. dielectric c.	ε_0	8.854×10^{-12} F/m

B.2 Electro Dynamics

B.2.1 Maxwell Equation

Maxwell equation in vector form is given by,

$$\nabla \cdot \boldsymbol{B} = 0 \tag{B.1}$$

$$\nabla \cdot \boldsymbol{E} = \frac{\rho}{\varepsilon_0} \tag{B.2}$$

M. Kikuchi, M. Azumi, *Frontiers in Fusion Research II*,
DOI 10.1007/978-3-319-18905-5

$$\nabla \times \boldsymbol{B} = \mu_0 \left[\boldsymbol{J} + \varepsilon_0 \frac{\partial \boldsymbol{E}}{\partial t} \right] \tag{B.3}$$

$$\nabla \times \boldsymbol{E} = -\frac{\partial \boldsymbol{B}}{\partial t}, \tag{B.4}$$

where $\boldsymbol{B}$ is the magnetic field, $\boldsymbol{E}$ is the electric field, $\boldsymbol{J}$ is the current density, ρ is the charge density, respectively. Here, $c^2 = 1/\mu_0\varepsilon_0$ and c is speed of light.

Taking $\nabla\cdot$ to (B.3), we have following charge continuity equation.

$$\nabla \cdot \boldsymbol{J} + \frac{\partial \rho}{\partial t} = 0 \tag{B.5}$$

B.2.2 Vector and Scalar Potentials

Magnetic field has vector potential since $\boldsymbol{B}$ satisfies (B.1).

$$\boldsymbol{B} = \nabla \times \boldsymbol{A} \tag{B.6}$$

Substitution of this equation into (B.4) gives,

$$\nabla \times \left(\boldsymbol{E} + \frac{\partial \boldsymbol{A}}{\partial t} \right) = 0, \tag{B.7}$$

which leads to the following form of the electric field $\boldsymbol{E}$.

$$\boldsymbol{E} = -\nabla \Phi - \frac{\partial \boldsymbol{A}}{\partial t} \tag{B.8}$$

B.2.3 Gauge

The vector and the scalar potentials have following gauge arbitrariness.

$$\boldsymbol{A}' = \boldsymbol{A} - \nabla G\,,\ \Phi' = \Phi + \frac{\partial G}{\partial t}, \tag{B.9}$$

which means $\boldsymbol{A}', \Phi'$ can also produce same Maxwell's equation. Substitutions of (B.6) into (B.3) and of (B.8) into (B.2) gives,

$$\nabla \times \nabla \times \boldsymbol{A} + \mu_0\varepsilon_0 \left[\nabla \frac{\partial \Phi}{\partial t} + \frac{\partial^2 \boldsymbol{A}}{\partial t^2} \right] = \mu_0 \boldsymbol{J} \tag{B.10}$$

$$\nabla^2 \Phi + \nabla \cdot \frac{\partial \boldsymbol{A}}{\partial t} = -\frac{\rho}{\varepsilon_0} \tag{B.11}$$

We use the gauge arbitrariness to impose following gauge selection.

$$\nabla \cdot \boldsymbol{A} + \mu_0 \varepsilon_0 \frac{\partial \Phi}{\partial t} = 0 \tag{B.12}$$

Then Eqs. (B.10) and (B.11) becomes as follows,

$$\nabla^2 \boldsymbol{A} - \mu_0 \varepsilon_0 \frac{\partial^2 \boldsymbol{A}}{\partial t^2} = -\mu_0 \boldsymbol{J}\,,\ \nabla^2 \Phi - \mu_0 \varepsilon_0 \frac{\partial^2 \Phi}{\partial t^2} = -\frac{\rho}{\varepsilon_0} \tag{B.13}$$

B.2.4 Poynting Vector

Poynting vector $\boldsymbol{S}$ is electromagnetic energy flux density and is given by,

$$\boldsymbol{S} = \boldsymbol{E} \times \boldsymbol{B}/\mu_0 \tag{B.14}$$

From Maxwell's equation, we have following "Poynting Energy Theorem".

$$\frac{\partial}{\partial t}\left(\frac{\boldsymbol{B}^2}{2\mu_0} + \frac{\varepsilon_0 \boldsymbol{E}^2}{2}\right) = -\boldsymbol{J} \cdot \boldsymbol{E} - \nabla \cdot \boldsymbol{S} \tag{B.15}$$

Left hand side is rate of change of the electromagnetic field energy. $-\boldsymbol{J} \cdot \boldsymbol{E}$ is the joule energy loss. $-\nabla \cdot \boldsymbol{S}$ is sink term of electromagnetic energy. Integral form of the Poynting Energy Theorem can be obtained using the Gauss's theorem.

$$\int \left[\frac{\partial}{\partial t}\left(\frac{\boldsymbol{B}^2}{2\mu_0} + \frac{\varepsilon_0 \boldsymbol{E}^2}{2}\right) + \boldsymbol{J} \cdot \boldsymbol{E}\right] dV = -\oint \boldsymbol{S} \cdot d\boldsymbol{\sigma}, \tag{B.16}$$

where $d\boldsymbol{\sigma}$ is a surface element. This means that an increase of the electromagnetic energy and the Joule dissipation in a volume is supplied by the Poynting energy influx from the surface.

B.2.5 Maxwell Stress

The force acting on continuous electromagnetic medium $\boldsymbol{F}$ is given by,

$$\boldsymbol{F} = \rho \boldsymbol{E} + \boldsymbol{J} \times \boldsymbol{B} = \nabla \cdot \boldsymbol{T} - \mu_0 \varepsilon_0 \frac{\partial \boldsymbol{S}}{\partial t}, \tag{B.17}$$

where $\boldsymbol{T}$ is the electromagnetic stress tensor whose i,j component is given by,

$$T_{ij} = \varepsilon_0 \left(E_i E_j - \frac{1}{2} E^2 \delta_{ij} \right) + \frac{1}{\mu_0} \left(B_i B_j - \frac{1}{2} B^2 \delta_{ij} \right) \tag{B.18}$$

2nd term in the RHS is much larger than the 1st term in the magnetized plasma. Then,

$$T_{ij} = \frac{1}{\mu_0} \left(B_i B_j - \frac{1}{2} B^2 \delta_{ij} \right) \tag{B.19}$$

This electromagnetic stress tensor is sometimes called Maxwell stress tensor.

B.3 Lagrange and Hamilton Mechanics

Newton equation of motion in potential field can be converted to following Euler-Lagrange equation,

$$\frac{d}{dt}\left(\frac{\partial L}{\partial \dot{q}_i} \right) - \frac{\partial L}{\partial q_i} = 0 \tag{B.20}$$

Here $L = T - V$ is Lagrangian (T is kinetic energy and V is potential energy), $\dot{q}_i = dq_i/dt$, q_i is ith coordinate. Variational principle to give this Euler-Lagrange equation is given by $\delta S = 0$ with following action integral.

$$S = \int_{t_1}^{t_2} L dt \tag{B.21}$$

The Lagrangian L to give Newton equation of motion is not unique. Assuming G is an arbitrary function of $(\boldsymbol{q}, t)$, $L + dG/dt$ also gives same Euler-Lagrange equation. The Lagrangian for the charged particle motion is,

$$L(\boldsymbol{x}, \dot{\boldsymbol{x}}, t) = \frac{1}{2} m_a |\dot{\boldsymbol{x}}|^2 + e_a \dot{\boldsymbol{x}} \cdot \boldsymbol{A}(\boldsymbol{x}, t) - e_a \Phi(\boldsymbol{x}, t), \tag{B.22}$$

where $\Phi(\boldsymbol{x}, t)$ and $\boldsymbol{A}(\boldsymbol{x}, t)$ are scalar and the vector potentials, respectively.

We define canonical momentum $\boldsymbol{p}$ by using L as

$$\boldsymbol{p} = \frac{\partial L}{\partial \dot{\boldsymbol{q}}} \tag{B.23}$$

Then, Euler-Lagrange equation can be converted to following Hamilton equation if we double the independent variables from $\boldsymbol{q}$ to $\boldsymbol{q}$ and $\boldsymbol{p}$ as follows,

$$\frac{d\boldsymbol{q}}{dt} = \frac{\partial H}{\partial \boldsymbol{p}}, \frac{d\boldsymbol{p}}{dt} = -\frac{\partial H}{\partial \boldsymbol{q}}, \tag{B.24}$$

where the Hamiltonian H is defined as follows,

$$H(\boldsymbol{q}, \boldsymbol{p}, t) = \boldsymbol{p} \cdot \dot{\boldsymbol{q}} - L \tag{B.25}$$

In case of charged particle motion, $\boldsymbol{p}$ and H are given by,

$$\boldsymbol{p} = m\dot{\boldsymbol{x}} + e\boldsymbol{A} \tag{B.26}$$

$$H = \frac{1}{2m}|\boldsymbol{p} - e\boldsymbol{A}|^2 + e\Phi \tag{B.27}$$

Variational principle to give this Hamilton equation is given by $\delta S = 0$ with following action integral.

$$S = \int_{t_1}^{t_2} [\boldsymbol{p} \cdot \dot{\boldsymbol{q}} - H(\boldsymbol{q}, \boldsymbol{p}, t)]\, dt \tag{B.28}$$

B.4 Special Relativity

Variational principle of special relativity for free particle is given by $\delta S = 0$ with following action integral similar to Fermat's principle in Optics.

$$S = -m_a c^2 \int_{\tau_1}^{\tau_2} d\tau \tag{B.29}$$

where m_a is the rest mass. Proper time τ is time in rest frame and is given by,

$$\tau = t[1 - v^2]^{1/2} \tag{B.30}$$

where $v = d\boldsymbol{x}/dt$ is velocity. Using (B.30),

$$S = -m_a c^2 \int_{t_1}^{t_2} \sqrt{1 - \left(\frac{v}{c}\right)^2}\, dt \tag{B.31}$$

This gives Lagrangian of free particle in relativistic motion L as,

$$L = -m_a c^2 \sqrt{1 - \left(\frac{v}{c}\right)^2} \tag{B.32}$$

Under the electromagnetic fields, the charged particle feels a potential field called generalized potential V given by,

$$V = e_a(\Phi - \boldsymbol{v} \cdot \boldsymbol{A}) \tag{B.33}$$

Then, the relativistic Lagrangian L is given by,

$$L = -m_a c^2 \sqrt{1 - \left(\frac{v}{c}\right)^2} - e_a(\Phi - \boldsymbol{v} \cdot \boldsymbol{A}) \tag{B.34}$$

Relativistic canonical momentum $\boldsymbol{p} = \partial L / \partial \boldsymbol{v}$ is,

$$\boldsymbol{p} = \frac{m_a \boldsymbol{v}}{\sqrt{1 - \left(\frac{v}{c}\right)^2}} + e_a \boldsymbol{A} \tag{B.35}$$

Hamiltonian (Energy) of the particle $H = \boldsymbol{p} \cdot \dot{\boldsymbol{x}} - L$ is given by,

$$H = \frac{m_a c^2}{\sqrt{1 - \left(\frac{v}{c}\right)^2}} + \varphi(\boldsymbol{x}) \tag{B.36}$$

Hamilton equation $d\boldsymbol{p}/dt = -\partial H / \partial \boldsymbol{x}$ gives,

$$\frac{d}{dt} \frac{m_a \boldsymbol{v}}{\sqrt{1 - \left(\frac{v}{c}\right)^2}} = e_a(\boldsymbol{E} + \boldsymbol{v} \times \boldsymbol{B}) \tag{B.37}$$

Appendix C
Appendix to Plasmas Physics

In this appendix, banana and fast ion parallel viscosities are derived from the drift kinetic equation.

C.1 DKE Solution in Banana Regime

We shows the derivation of the parallel viscous in banana regime. It is noted that the normalization is slightly different from main text.

The collisionless plasma in tokamak is characterized by the collisionality (4.97) $\nu_a^* \ll 1$. The relevant DKE (4.103) may be expressed as,

$$(\mathbf{v}_\parallel + \mathbf{v_d}) \cdot \nabla f_a = C_a(f_a) \text{ or, } \mathbf{v}_\parallel \cdot \nabla\big[f_a + \frac{m_a F}{e_a B} v_\parallel \frac{\partial f_a}{\partial \psi}\big] = C_a(f_a) \tag{C.1}$$

Hereafter, we omit "$_a$" for simplicity. Then, we have following solvability condition.

$$\oint \frac{dl}{v_\parallel} C(f) = 0 \tag{C.2}$$

We expand f in powers of ρ and ν as $f = f^0 + f^1 + f^2 \ldots\ldots \;;\; f^n = 0(\rho^n)$ and $f^n = f_0^n + f_1^n + f_2^n \ldots\ldots \;;\; f_m^n = 0(\nu^m)$. The lowest order solution is Maxwellian.

$$f^0 = f_M(\psi, \mathbf{v}) \tag{C.3}$$

And the 1st order solution is given as,

$$f_0^1 = -\frac{mF}{eB} v_\parallel \frac{\partial f^0}{\partial \psi} + g(\psi, v, \lambda), \text{ where } \lambda \equiv v_\perp^2 / B v^2 \tag{C.4}$$

M. Kikuchi, M. Azumi, *Frontiers in Fusion Research II*,
DOI 10.1007/978-3-319-18905-5

Defining G as $g \equiv (mF/e)v(df^0/d\psi)G(\psi, v, \lambda)$, we have:

$$f_0^1 = \frac{mF}{eB} v \frac{\partial f^0}{\partial \psi} (BG - \zeta), \text{ where } \zeta \equiv v_\parallel / v \tag{C.5}$$

Then, the solvability condition (C.2) is given by:

$$\oint \frac{dl}{v_\parallel} C\Big(\frac{mF}{eB} v \frac{\partial f^0}{\partial \psi} (BG - \zeta) \Big) = 0 \tag{C.6}$$

Hereafter, considering the collisionless limit, the collision term is approximated as:

$$C(f^1) = \nu_D \mathscr{L}(f^1) + P[f^1]f_M \approx \nu_D \mathscr{L}(f^1), \tag{C.7}$$

$$\text{where } \mathscr{L}(f^1) = \frac{2\zeta}{B} \frac{\partial}{\partial \lambda} \lambda \zeta \frac{\partial f^1}{\partial \lambda}$$

Then the explicit form of solvability condition Eq. (C.6) may be given by:

$$\oint_p \frac{dl}{v_\parallel} \Big[\frac{\zeta}{B} \frac{\partial}{\partial \lambda} \lambda \zeta \frac{\partial}{\partial \lambda} (BG - \zeta) \Big] = 0, \tag{C.8}$$

where $\oint_p$ is the integral along particle trajectory: = 0 for trapped particle.

Calculating the l.h.s as

$$\oint_p \frac{dl}{v_\parallel} \Big[\frac{\zeta}{B} \frac{\partial}{\partial \lambda} \lambda \zeta \frac{\partial}{\partial \lambda} (BG - \zeta) \Big] = \oint_p \frac{dl}{B} \Big[\frac{\partial}{\partial \lambda} \lambda \zeta \frac{\partial}{\partial \lambda} (BG - \zeta) \Big]$$

$$= \oint_p dl \Big[\frac{\partial}{\partial \lambda} \lambda \zeta \frac{\partial G}{\partial \lambda} \Big] + \frac{1}{2} \oint_p dl$$

and introducing the average $< X >= (\oint X dl/B)/(\oint dl/B)$ along particle trajectory, the solvability condition C.8 is expressed as

$$\frac{\partial}{\partial \lambda} \lambda < \zeta > \frac{\partial G}{\partial \lambda} + \frac{1}{2} H(\lambda_c - \lambda) = 0, \text{ where } \lambda_c = 1/B \tag{C.9}$$

From this equation, the function G is given by:

$$G(\psi, v, \lambda) = \frac{1}{2} \int_\lambda^{\lambda_c} \frac{d\lambda}{< \zeta >} \tag{C.10}$$

This solution G is the function of (ψ, ζ) and does not have v-dependence. In order to allow v-dependence of G, we introduce $\hat{G}(v)$ as

$$G(\psi, v, \lambda) = \frac{1}{2}\hat{G}(\psi, v)\int_{\lambda}^{\lambda_c} \frac{d\lambda}{<\zeta>} \tag{C.11}$$

Using G function and noting $d^3v = \pi B v^2 dv d\lambda/\zeta$, the parallel particle and heat fluxes are calculated as,

$$\begin{aligned}
&< nBu_{\|} >=< B\int d^3v v_{\|} f_0^1 > \\
&=< B\int d^3v v_{\|}\frac{mF}{e}v\frac{df^0}{d\psi}(G-\frac{\zeta}{B}) > \\
&= -\frac{mF}{e} < \int d^3v v^2\zeta^2\frac{df^0}{d\psi} > +\frac{mF}{e} < B\int d^3v v^2\frac{df^0}{d\psi}\zeta G > \\
&= nBV_1 + \frac{mF}{e}\pi\int dv v^4\frac{df^0}{d\psi}\langle\oint d\lambda\sigma B^2 G\rangle \ , \ \sigma = \pm 1 \\
&= nBV_1 + < B^2 > \frac{mF}{e}\pi\int dv v^4\hat{G}\frac{df^0}{d\psi}\oint d\lambda\int_{\lambda}^{\lambda_c}\frac{d\lambda'}{<|\zeta'|>} \ , \\
&= nBV_1 + \frac{mF}{e}\frac{4\pi}{3}f_c\int dv v^4\hat{G}\frac{df^0}{d\psi} \ ,
\end{aligned} \tag{C.12}$$

$$\begin{aligned}
&< Bq_{\|} >=< B\int d^3v\frac{m}{2}v^2 v_{\|} f_0^1 > -\frac{5}{2}p < Bu_{\|} > \\
&=< B\int d^3v\frac{m}{2}v^2 v_{\|}\frac{mF}{e}v\frac{df^0}{d\psi}(G-\frac{\zeta}{B}) > -\frac{5}{2}p < Bu_{\|} > \\
&= -\frac{mF}{e} < \int d^3v\frac{m}{2}v^4\zeta^2\frac{df^0}{d\psi} > +\frac{mF}{e} < B\int d^3v\frac{m}{2}v^4\frac{df^0}{d\psi}\zeta G > -\frac{5}{2}p < Bu_{\|} > \\
&= \frac{5}{2}pBV_2 + \frac{mF}{e}\pi\int dv(\frac{m}{2}v^2 - \frac{5}{2}\frac{p}{n})v^4\frac{df^0}{d\psi}\langle\oint d\lambda\sigma B^2 G\rangle \\
&= \frac{5}{2}pBV_2 + < B^2 > \frac{mF}{e}\pi\int dv(\frac{m}{2}v^2 - \frac{5}{2}\frac{p}{n})v^4\hat{G}\frac{df^0}{d\psi}\oint d\lambda\int_{\lambda}^{\lambda_c}\frac{d\lambda'}{<|\zeta'|>} \\
&= \frac{5}{2}pBV_2 + \frac{mF}{e}\frac{4\pi}{3}f_c\int dv(\frac{m}{2}v^2 - \frac{5}{2}\frac{p}{n})v^4\hat{G}\frac{df^0}{d\psi}
\end{aligned} \tag{C.13}$$

where,

$$V_1 \equiv -\frac{F}{eB}\left[\frac{1}{n}\frac{dp}{d\psi} + e\frac{d\Phi}{d\psi}\right], \quad V_2 \equiv -\frac{F}{eB}\frac{dT}{d\psi} \tag{C.14}$$

$$f_c = \frac{3}{4} < B^2 > \int_0^{\lambda_c} \frac{\lambda d\lambda}{< |\zeta| >} \tag{C.15}$$

On the other hand, moment equations give the following relations

$$< nBu_\parallel >= nBV_1 + nu_\theta < B^2 >, \tag{C.16}$$

$$< Bq_\parallel >= \frac{5}{2}pBV_2 + q_\theta < B^2 >, \tag{C.17}$$

Then,

$$nu_\theta < B^2 >= \frac{mF}{e}\frac{4\pi}{3}f_c \int dv v^4 \hat{G}\frac{df^0}{d\psi} \tag{C.18}$$

$$q_\theta < B^2 >= \frac{mF}{e}\frac{4\pi}{3}f_c \int dv \left(\frac{m}{2}v^2 - \frac{5}{2}\frac{p}{n}\right) v^4 \hat{G}\frac{df^0}{d\psi} \tag{C.19}$$

From these relations, $\hat{G}$ can be expressed, in series of Sonine polynomials of order (3/2), as

$$\frac{mF}{e}\hat{G}\frac{df^0}{d\psi} = \frac{2}{v_T^2}\frac{< B^2 >}{f_c}\left[u_\theta - \frac{2}{5}\frac{q_\theta}{p}L_1^{(3/2)}\right] f_M \tag{C.20}$$

Next, the parallel viscous force $< \boldsymbol{B}\cdot\nabla\cdot\boldsymbol{\Pi} >$ and heat viscous force $< \boldsymbol{B}\cdot\nabla\cdot\boldsymbol{\Theta} >$ are given by:

$$< \boldsymbol{B}\cdot\nabla\cdot\boldsymbol{\Pi} >=< B \int dv^3 m v_\parallel^2 \nabla_\parallel f^1 >=< mB \int dv^3 v_\parallel C(f^1) > \tag{C.21}$$

$$< \boldsymbol{B}\cdot\nabla\cdot\boldsymbol{\Theta} >= \frac{m}{2T} < \boldsymbol{B}\cdot\nabla\cdot\boldsymbol{r} > - \frac{5}{2} < \boldsymbol{B}\cdot\nabla\cdot\boldsymbol{\Pi} > \tag{C.22}$$

$$< \boldsymbol{B}\cdot\nabla\cdot\boldsymbol{r} >=< mB \int dv^3 v^2 v_\parallel C(f^1) > \tag{C.23}$$

To calculate these parallel viscous forces, instead of Eqs. (C.21), (C.23), we employ the following expressions,

$$\begin{aligned} < \boldsymbol{B}\cdot\nabla\cdot\boldsymbol{\Pi} > &=< mB \int dv^3 v_\parallel C(f^1) > \\ &- < mB \int dv^3 v_\parallel \frac{V_\parallel}{f_c}\frac{\sqrt{< B^2 >}}{B^2} < \frac{B}{v_\parallel}C(f^1) >>, \end{aligned} \tag{C.24}$$

$$< \boldsymbol{B} \cdot \nabla \cdot \boldsymbol{r} > = < mB \int dv^3 v^2 v_\parallel C(f^1) >$$
$$- < mB \int dv^3 v^2 v_\parallel \frac{V_\parallel}{f_c} \frac{\sqrt{< B^2 >}}{B^2} < \frac{B}{v_\parallel} C(f^1) >>, \tag{C.25}$$

where

$$V_\parallel \equiv \frac{\sigma v}{2} \sqrt{< B^2 >} \int_\lambda^{\lambda_c} \frac{d\lambda}{< \zeta >}$$

Here note that $< (B/v_\parallel) C(f^1) > = 0$.

$$< mB \int dv^3 v_\parallel C(f^1) > = < mB \int 2\pi B \nu_D \frac{v^2 dv d\lambda}{2|\zeta|} v\zeta \frac{2\zeta}{B} \frac{\partial}{\partial \lambda} \lambda \zeta \frac{\partial}{\partial \lambda} \frac{mF}{eB} v \frac{\partial f_0}{\psi} (BG - \zeta) >$$
$$= < 2m\pi \frac{mF}{e} \int \nu_D v^4 dv \frac{\partial f_0}{\partial \psi} \int d\lambda \zeta \frac{\partial}{\partial \lambda} \lambda \zeta \frac{\partial}{\partial \lambda} (BG - \zeta) >$$
$$= \frac{2m^2 \pi F}{e} < \int \nu_D v^4 dv \frac{\partial f_0}{\partial \psi} \int d\lambda \frac{B}{2\zeta} \lambda \zeta \frac{\partial}{\partial \lambda} (BG - \zeta) >$$
$$; \frac{\partial \zeta}{\partial \lambda} = -\frac{B}{2\zeta}$$
$$= \frac{m^2 \pi F}{e} < B \int \nu_D v^4 dv \frac{\partial f_0}{\partial \psi} \int d\lambda \lambda \frac{\partial}{\partial \lambda} (BG - \zeta) >$$
$$= -\frac{m^2 \pi F}{e} < B \int \nu_D v^4 dv \frac{\partial f_0}{\partial \psi} \int d\lambda (BG - \zeta) >$$
$$= -\frac{m^2 \pi F}{e} \int \nu_D v^4 dv \frac{\partial f_0}{\partial \psi} (< B^2 > \int d\lambda G - \int d\lambda < B\zeta >)$$
$$= -\frac{m^2 \pi F}{e} \int \nu_D v^4 dv \frac{\partial f_0}{\partial \psi} (< B^2 > \int d\lambda G - \frac{4}{3})$$
$$; B \sum_{\sigma = \pm} \int_0^{\lambda_c} \zeta (-2\zeta d\zeta / B) = 4/3$$
$$= \frac{m^2 \pi F}{e} \int \nu_D v^4 dv \frac{\partial f_0}{\partial \psi} (\frac{4}{3} - < B^2 > \int d\lambda G)$$

$$< mB \int dv^3 v_\parallel \frac{V_\parallel}{f_c} \frac{\sqrt{< B^2 >}}{B^2} < \frac{B}{v_\parallel} C(\frac{mF}{eB} v \frac{\partial f_0}{\partial \psi} BG) >>$$
$$= < mB \int 2\pi B \nu_D \frac{v^2 dv d\lambda}{2|\zeta|} v\zeta \frac{V_\parallel}{f_c} \frac{\sqrt{< B^2 >}}{B^2} < \frac{B}{v_\parallel} \frac{2\zeta}{B} \frac{\partial}{\partial \lambda} \lambda \zeta \frac{\partial}{\partial \lambda} \frac{mF}{eB} v \frac{\partial f_0}{\partial \psi} BG >>$$

$$
\begin{aligned}
&= \frac{m^2\pi F}{ef_c} < B^2 >< \int \nu_D v^2 dv \frac{\partial f_0}{\partial \psi} \int d\lambda \Big(\int_\lambda^{\lambda_c} \frac{d\lambda'}{<\zeta>} \Big) < \frac{1}{B} \frac{\partial}{\partial \lambda} \lambda \zeta \frac{\partial}{\partial \lambda} BG >> \\
&= \frac{m^2\pi F}{ef_c} < B^2 > \int \nu_D v^4 dv \frac{\partial f_0}{\partial \psi} < \int \frac{d\lambda}{<\zeta>} \lambda \zeta \frac{\partial}{\partial \lambda} G > \\
&= -\frac{m^2\pi F}{ef_c} < B^2 > \int \nu_D v^4 dv \frac{\partial f_0}{\partial \psi} \int d\lambda G
\end{aligned}
\tag{C.26}
$$

$$
\begin{aligned}
&< mB \int dv^3 v_\parallel \frac{V_\parallel}{f_c} \frac{\sqrt{<B^2>}}{B^2} < \frac{B}{v_\parallel} C\Big(\frac{mF}{eB} v \frac{\partial f_0}{\partial \psi} \zeta\Big) >> \\
&= < mB \int 2\pi B \nu_D \frac{v^2 dv d\lambda}{2|\zeta|} v\zeta \frac{V_\parallel}{f_c} \frac{\sqrt{<B^2>}}{B^2} < \frac{B}{v_\parallel} \frac{2\zeta}{B} \frac{\partial}{\partial \lambda} \lambda \zeta \frac{\partial}{\partial \lambda} \frac{mF}{eB} v \frac{\partial f_0}{\partial \psi} \zeta >> \\
&= \frac{m^2\pi F}{ef_c} < B^2 >< \int \nu_D v^2 dv \frac{\partial f_0}{\partial \psi} \int d\lambda \Big(\int_\lambda^{\lambda_c} \frac{d\lambda'}{<\zeta>} \Big) < \frac{1}{B} \frac{\partial}{\partial \lambda} \lambda \zeta \frac{\partial \zeta}{\partial \lambda} >> \\
&= \frac{m^2\pi F}{ef_c} < B^2 >< \frac{1}{B} \int \nu_D v^2 dv \frac{\partial f_0}{\partial \psi} \Big(-\frac{1}{2}\Big) \int d\lambda \Big(\int_\lambda^{\lambda_c} \frac{d\lambda'}{<\zeta>} \Big) \\
&= \frac{m^2\pi F}{ef_c} < B^2 > \int \nu_D v^4 dv \frac{\partial f_0}{\partial \psi} \int_0^{\lambda_c} \frac{\lambda d\lambda}{<\zeta>} \\
&= -\frac{m^2\pi F}{e} \frac{4}{3} \int \nu_D v^4 dv \frac{\partial f_0}{\partial \psi}
\end{aligned}
\tag{C.27}
$$

Then

$$
\begin{aligned}
< \boldsymbol{B} \cdot \nabla \cdot \boldsymbol{\Pi} > &= \frac{m^2\pi F}{e} \frac{f_t}{f_c} < B^2 > \int \nu_D v^4 dv \frac{\partial f_0}{\partial \psi} \int d\lambda G \ ; \ f_t = 1 - f_c \\
&= \frac{4\pi}{3} f_t m \int \nu_D v^4 dv \Big(\frac{mF}{e} \hat{G}(v, \psi) \frac{\partial f_0}{\partial \psi} \Big) \\
&= \frac{4\pi}{3} \frac{f_t}{f_c} \frac{2m}{v_T^2} \int \nu_D v^4 dv \Big[u_\theta - \frac{2}{5} \frac{q_\theta}{p} L_1^{(3/2)} \Big] f_M
\end{aligned}
\tag{C.28}
$$

Defining $\{A(x)\} = (8/3\sqrt{\pi}) \int A(x) x^4 e^{-x^2} dx$,

$$
< \boldsymbol{B} \cdot \nabla \cdot \boldsymbol{\Pi} > = \frac{f_t}{f_c} \frac{nm}{\tau} \Big[\{\nu_D \tau\} u_\theta + \big(\{x^2 \nu_D \tau\} - \frac{5}{2} \{\nu_D \tau\} \big) \frac{2}{5} \frac{q_\theta}{p} \Big]
\tag{C.29}
$$

In the same way, the parallel heat viscous force is

$$
\begin{aligned}
< mB\int dv^3v^2v_{\|}C(f^1) > &= < mB\int 2\pi B\nu_D\frac{v^4dvd\lambda}{2|\zeta|}v\zeta\frac{2\zeta}{B}\frac{\partial}{\partial\lambda}\lambda\zeta\frac{\partial}{\partial\lambda}\frac{mF}{eB}v\frac{\partial f_0}{\psi}(BG-\zeta) > \\
&= < 2m\pi\frac{mF}{e}\int \nu_D v^6 dv\frac{\partial f_0}{\partial\psi}\int d\lambda\zeta\frac{\partial}{\partial\lambda}\lambda\zeta\frac{\partial}{\partial\lambda}(BG-\zeta) > \\
&= \frac{2m^2\pi F}{e} < \int \nu_D v^6 dv\frac{\partial f_0}{\partial\psi}\int d\lambda\frac{B}{2\zeta}\lambda\zeta\frac{\partial}{\partial\lambda}(BG-\zeta) > \\
&= \frac{m^2\pi F}{e} < B\int \nu_D v^6 dv\frac{\partial f_0}{\partial\psi}\int d\lambda\lambda\frac{\partial}{\partial\lambda}(BG-\zeta) > \\
&= -\frac{m^2\pi F}{e} < B\int \nu_D v^6 dv\frac{\partial f_0}{\partial\psi}\int d\lambda(BG-\zeta) > \\
&= -\frac{m^2\pi F}{e}\int \nu_D v^6 dv\frac{\partial f_0}{\partial\psi}(< B^2 > \int d\lambda G - \int d\lambda < B\zeta >) \\
&= -\frac{m^2\pi F}{e}\int \nu_D v^6 dv\frac{\partial f_0}{\partial\psi}(< B^2 > \int d\lambda G - \frac{4}{3}) \\
&= \frac{m^2\pi F}{e}\int \nu_D v^6 dv\frac{\partial f_0}{\partial\psi}(\frac{4}{3} - < B^2 > \int d\lambda G)
\end{aligned}
$$

$$
\begin{aligned}
&< mB\int dv^3v^2v_{\|}\frac{V_{\|}}{f_c}\frac{\sqrt{< B^2 >}}{B^2} < \frac{B}{v_{\|}}C(\frac{mF}{eB}v\frac{\partial f_0}{\partial\psi}BG) >> \\
&=< mB\int 2\pi B\nu_D\frac{v^4dvd\lambda}{2|\zeta|}v\zeta\frac{V_{\|}}{f_c}\frac{\sqrt{< B^2 >}}{B^2} < \frac{B}{v_{\|}}\frac{2\zeta}{B}\frac{\partial}{\partial\lambda}\lambda\zeta\frac{\partial}{\partial\lambda}\frac{mF}{eB}v\frac{\partial f_0}{\partial\psi}BG >> \\
&= \frac{m^2\pi F}{ef_c} < B^2 >< \int \nu_D v^4 dv\frac{\partial f_0}{\partial\psi}\int d\lambda(\int_{\lambda}^{\lambda_c}\frac{d\lambda'}{< \zeta >}) < \frac{1}{B}\frac{\partial}{\partial\lambda}\lambda\zeta\frac{\partial}{\partial\lambda}BG >> \\
&= \frac{m^2\pi F}{ef_c} < B^2 > \int \nu_D v^6 dv\frac{\partial f_0}{\partial\psi} < \int\frac{d\lambda}{< \zeta >}\lambda\zeta\frac{\partial}{\partial\lambda}G > \\
&= -\frac{m^2\pi F}{ef_c} < B^2 > \int \nu_D v^6 dv\frac{\partial f_0}{\partial\psi}\int d\lambda G
\end{aligned}
\tag{C.30}
$$

$$
\begin{aligned}
&< mB \int dv^3 v^2 v_\parallel \frac{V_\parallel}{f_c} \frac{\sqrt{< B^2 >}}{B^2} < \frac{B}{v_\parallel} C(\frac{mF}{eB} v \frac{\partial f_0}{\partial \psi} \zeta) >> \\
&= < mB \int 2\pi B \nu_D \frac{v^4 dv d\lambda}{2|\zeta|} v\zeta \frac{V_\parallel}{f_c} \frac{\sqrt{< B^2 >}}{B^2} < \frac{B}{v_\parallel} \frac{2\zeta}{B} \frac{\partial}{\partial \lambda} \lambda \zeta \frac{\partial}{\partial \lambda} \frac{mF}{eB} v \frac{\partial f_0}{\partial \psi} \zeta >> \\
&= \frac{m^2 \pi F}{e f_c} < B^2 >< \int \nu_D v^4 dv \frac{\partial f_0}{\partial \psi} \int d\lambda (\int_\lambda^{\lambda_c} \frac{d\lambda'}{< \zeta >}) < \frac{1}{B} \frac{\partial}{\partial \lambda} \lambda \zeta \frac{\partial \zeta}{\partial \lambda} >> \\
&= \frac{m^2 \pi F}{e f_c} < B^2 >< \frac{1}{B} \int \nu_D v^4 dv \frac{\partial f_0}{\partial \psi} (-\frac{1}{2}) \int d\lambda (\int_\lambda^{\lambda_c} \frac{d\lambda'}{< \zeta >}) \\
&= \frac{m^2 \pi F}{e f_c} < B^2 > \int \nu_D v^6 dv \frac{\partial f_0}{\partial \psi} \int_0^{\lambda_c} \frac{\lambda d\lambda}{< \zeta >} \\
&= -\frac{m^2 \pi F}{e} \frac{4}{3} \int \nu_D v^6 dv \frac{\partial f_0}{\partial \psi} \qquad \text{(C.31)}
\end{aligned}
$$

$$
\begin{aligned}
< \boldsymbol{B} \cdot \nabla \cdot \boldsymbol{r} > &= \frac{m^2 \pi F}{e} \frac{f_t}{f_c} < B^2 > \int \nu_D v^6 dv \frac{\partial f_0}{\partial \psi} \int d\lambda G \\
&= \frac{4\pi}{3} f_t m \int \nu_D v^6 dv (\frac{mF}{e} \hat{G}(v, \psi) \frac{\partial f_0}{\partial \psi}) \\
&= \frac{4\pi}{3} \frac{f_t}{f_c} \frac{2m}{v_T^2} \int \nu_D v^6 dv \Big[u_\theta - \frac{2}{5} \frac{q_\theta}{p} L_1^{(3/2)} \Big] f_M \\
&= nm v_T^2 \frac{f_t}{f_c} [\{\nu_D x^2\} u_\theta - (\frac{5}{2} \{\nu_D x^2\} - \{\nu_D x^4\}) \frac{2}{5} \frac{q_\theta}{p}] \qquad \text{(C.32)}
\end{aligned}
$$

$$
< \boldsymbol{B} \cdot \nabla \cdot \boldsymbol{\Theta} >= \frac{f_t}{f_c} \frac{nm}{\tau} \Big[(\{x^2 \nu_D \tau\} - \frac{5}{2} \{\nu_D \tau\}) u_\theta + (\{x^4 \nu_D \tau\} - 5\{x^2 \nu_D \tau\} + \frac{25}{4} \{\nu_D \tau\}) \frac{2}{5} \frac{q_\theta}{p} \Big] \qquad \text{(C.33)}
$$

C.2 DKE Solution for Fast Ion

This section gives fast ion parallel viscosity [803]. An approximate Coulomn collision operator for the fast ion (and α particle) distribution function are given under the condition of $v_{Ti} \ll v \ll v_{Te}$. Since $x_e \ll 1$ and $G(x_e) \sim 2x_e/3\sqrt{\pi}$, we find the fast ion-electron slowing down frequency is given by $\nu_s^{fe} \sim 1/\tau_{se}$, where the slowing down time τ_{se} is given by,

$$\tau_{se} = \frac{3(2\pi)^{3/2}\varepsilon_0^2 m_f T_e^{3/2}}{e^4 Z_f^2 n_e m_e^{1/2} ln\Lambda} \tag{C.34}$$

For the fast-ion slowing down, $x_i \gg 1$ and $G(x_i) \sim 1/2x_i^2$ to obtain,

$$\nu_s^{fi} = \left(1 + \frac{m_i}{m_f}\right)\left(\frac{v_c}{v}\right)^3 \frac{1}{\tau_{se}} \tag{C.35}$$

$$v_c = \left(\frac{3\sqrt{\pi} m_e n_i Z_i^2}{4 m_i n_e}\right)^{1/3} v_{Te}, \tag{C.36}$$

where $E_c = m_f v_c^2/2$ is called the critical energy. The deflection frequency ν_D^{fe} and the parallel velocity diffusion frequency $\nu_{\parallel}^{fe}$ are smaller than ν_s^{fe} and can be neglected since the slowing down process is dominant for the fast ion-electron collision since $m_f \gg m_e$. But fast ion-ion collision is dominated by the deflection. since $\nu_D^{fi} = v_c/v)^3(m_i/m_f)/\tau_s$, we have following kinetic equation for the fast particle in the uniform plasma:

$$\tau_{se}\frac{\partial f_f}{\partial t} = \frac{1}{v^2}\frac{\partial}{\partial v}[(v^3 + v_c^3)f_f(v)] + \frac{v_c^3 m_i}{v^3 m_f}\mathscr{L}(f_f) + \tau_{se}S(v,\theta,\phi) \tag{C.37}$$

In case of isotropic fast ion distribution (such as in case of α particle) with a birth speed of v_0, we have $S = S_0\delta(v - v_0)/4\pi v_0^2$ and the solution of the kinetic equation is the Stix solution given as:

$$f_f(v) = \frac{S\tau_{se}}{4\pi(v^3 + v_c^3)} \tag{C.38}$$

The drift kinetic equation in a rare collision regime is given by,

$$(v_{\parallel}\boldsymbol{b} + \boldsymbol{v}_d)\cdot\nabla f_a = C(f_a) + S \tag{C.39}$$

$$S = S(\psi)\delta(v - v_b) \tag{C.40}$$

$$\tau_{se}C(f_a) = \frac{1}{v^2}\frac{\partial}{\partial v}[(v^3 + v_c^3)f] + \frac{\hat{Z}}{2}\frac{v_c^3}{v^3}\frac{\partial}{\partial\zeta}\left[(1-\zeta^2)\frac{\partial f_a}{\partial\zeta}\right], \tag{C.41}$$

where $\zeta = v_{\parallel}/v$. Rewriting the collision term from $\zeta = \sqrt{1 - \lambda B}$ to the pitch angle variable $\lambda = (1 - \zeta^2)/B$, the collision term can be given by,

$$\tau_{se}C(f_a) = \frac{1}{v^2}\frac{\partial}{\partial v}[(v^3 + v_c^3)f] + \frac{\hat{Z}}{2}\frac{v_c^3}{v^3}\frac{4\zeta}{B}\frac{\partial}{\partial\lambda}\left[\lambda\zeta\frac{\partial f_a}{\partial\lambda}\right] \tag{C.42}$$

We use two smallness parameters ν/ν_b and ρ/a satisfying $\nu/\nu_b \ll \rho/a \ll 1$, where ν_b is a bounce time of the trapped particle. We expand f_a in powers as $f = f^0 + f^1 + --, f^1 = f_0^1 + f_1^1 + --$ and $f_\beta^\alpha = o((\rho/a)^\alpha, (\nu/\nu_b)^\beta)$. The lowest order equation and it solution are,

$$v_\parallel \boldsymbol{b} \cdot \nabla f^0 = C(f^0) + S(\psi)\delta(v - v_b) \tag{C.43}$$

$$f^0(\psi, v) = \frac{S\tau_{se}}{v^3 + v_c^3} H(v_b - v) = \frac{n_f}{4\pi} \frac{\tau_{se}}{\tau_{th}} \frac{1}{v^3 + v_c^3} H(v_b - v) \tag{C.44}$$

The conservation of toroidal angular momentum conservation give rise to the relation,

$$\boldsymbol{v}_d \cdot \nabla f^0 = v_\parallel \boldsymbol{b} \cdot \nabla \left[\frac{F v_\parallel}{\Omega} \frac{\partial f^0}{\partial \psi} \right], \tag{C.45}$$

the first order equation is given by,

$$v_\parallel \boldsymbol{b} \cdot \nabla \left[f^1 + \frac{mF}{eB} v_\parallel \frac{\partial f^0}{\partial \psi} \right] = C(f^1) \tag{C.46}$$

The f^1 is expanded in orders of ν/ν_b as $f^1 = f_0^1 + f_1^1$,

$$v_\parallel \boldsymbol{b} \cdot \nabla \left[f_0^1 + \frac{mF}{eB} v_\parallel \frac{d f^0}{d\psi} \right] = 0 \tag{C.47}$$

$$v_\parallel \boldsymbol{b} \cdot \nabla f_1^1 = C(f_0^1) \tag{C.48}$$

The general solution of (4.123) is given by,

$$f_0^1 = -\frac{mF}{eB} v_\parallel \frac{\partial f^0}{\partial \psi} + g(\psi, v, \lambda) \tag{C.49}$$

Equation (4.124) gives solvability condition for f_0^1 or g:

$$\oint_p \frac{dl}{v_\parallel} C(f_0^1) = \oint_p C(-\frac{mF}{eB} v_\parallel \frac{\partial f^0}{\partial \psi} + g) = 0 \tag{C.50}$$

By using explicit form of the Coulomb collision operator (4.118) and rewriting $g = (mF/e)v(\partial f^0/\partial\psi)G$, this condition can be rewritten as,

$$\frac{1}{v^2} \frac{\partial}{\partial v} \left[(v^3 + v_c^3) \left(\langle \frac{B}{\zeta} \rangle_p G - H(\lambda_c - \lambda) \right) v \frac{\partial f^0}{\partial \psi} \right]$$

$$+\frac{\hat{Z}v_c^3}{v^3}v\frac{\partial f^0}{\partial \psi}\frac{\partial}{\partial \lambda}\left[\lambda\left(2\langle\zeta\rangle_p\frac{\partial G}{\partial \lambda}+H(\lambda_c-\lambda)\right)\right]=0 \tag{C.51}$$

By using G, the first order flows in the poloidal direction and parallel to the magnetic field are expressed by,

$$nu_\theta = \frac{mF\pi}{e}\int dvv^4\frac{\partial f^0}{\partial \psi}\int d\lambda\sum\sigma G \tag{C.52}$$

$$q_\theta = \frac{mF\pi}{e}\int dv(\frac{m}{2}v^2-\frac{5}{2}\frac{p}{n})v^4\frac{\partial f^0}{\partial \psi}\int d\lambda\sum\sigma G \tag{C.53}$$

$$\langle Bnv_\parallel\rangle = nBV_1+\langle B^2\rangle nu_\theta \tag{C.54}$$

$$\langle Bq_\parallel\rangle = \frac{5}{2}pBV_2+\langle B^2\rangle q_\theta, \tag{C.55}$$

where

$$nBV_1 = -\frac{F}{e}\left[\frac{\partial p}{\partial \psi}+en\frac{\partial \Phi}{\partial \psi}\right] \tag{C.56}$$

$$pBV_2 = -\frac{2F}{5e}\left[m\frac{\partial r}{\partial \psi}-\frac{5}{2}\frac{p}{n}\frac{\partial p}{\partial \psi}\right] \tag{C.57}$$

$$p = \frac{1}{3}\int d^3vmv^2f^0 \tag{C.58}$$

$$r = \frac{1}{6}\int d^3vmv^4f^0 \tag{C.59}$$

For the Maxwellian velocity distribution,

$$p = nT \tag{C.60}$$

$$r = \frac{5}{2}\frac{T}{m}p \tag{C.61}$$

$$BV_2 = -\frac{F}{e}\frac{\partial T}{\partial \psi} \tag{C.62}$$

Using the first order flows, the first order Legendre component of the perturbed distribution function can be expressed as,

$$f^1 = -\left[u_\parallel+\frac{q_\parallel}{\hat{r}}\left(\frac{1}{5}\frac{mn}{p}v^2-1\right)+--\right]\frac{\partial f^0}{\partial v_\parallel}, \tag{C.63}$$

where

$$\hat{r} = \frac{7}{5}\frac{nmr}{p} - \frac{5}{2}p \quad \text{(C.64)}$$

Here $\hat{r} = p$ for the Maxwell distribution.

We define the parallel viscous forces by,

$$\langle \boldsymbol{B} \cdot \nabla \cdot \boldsymbol{\pi} \rangle = \langle B \int d^3 v m v_{\|}^2 \boldsymbol{b} \cdot \nabla f \rangle \quad \text{(C.65)}$$

$$\langle \boldsymbol{B} \cdot \nabla \cdot \boldsymbol{r} \rangle = \langle B \int d^3 v m v^2 v_{\|}^2 \boldsymbol{b} \cdot \nabla f \rangle \quad \text{(C.66)}$$

By using the relation,

$$v_{\|} \boldsymbol{b} \cdot \nabla f_1^1 = C(f_0^1) \quad \text{(C.67)}$$

the parallel viscous forces caused by the 1st order deformation of the velocity distribution function are calculated from the following equations:

$$\langle \boldsymbol{B} \cdot \nabla \cdot \boldsymbol{\pi} \rangle = \langle mB \int d^3 v v_{\|} C(f^1) \rangle \quad \text{(C.68)}$$

$$\langle \boldsymbol{B} \cdot \nabla \cdot \boldsymbol{r} \rangle = \langle mB \int d^3 v v^2 v_{\|} C(f^1) \rangle \quad \text{(C.69)}$$

Using Hirshman-Sigmar technique [330] to use the relation $\langle (B/v_{\|} C(f^1)) \rangle = 0$ (Eq. (4.56)), these are rewritten as,

$$\langle \boldsymbol{B} \cdot \nabla \cdot \boldsymbol{\pi} \rangle = \langle mB \int d^3 v v_{\|} \left[C(f^1) - \frac{V_{\|}}{f_c} \frac{\sqrt{\langle B^2 \rangle}}{B^2} \langle \frac{B}{v_{\|}} C(f^1) \rangle \right] \rangle \quad \text{(C.70)}$$

$$\langle \boldsymbol{B} \cdot \nabla \cdot \boldsymbol{r} \rangle = \langle mB \int d^3 v v^2 v_{\|} \left[C(f^1) - \frac{V_{\|}}{f_c} \frac{\sqrt{\langle B^2 \rangle}}{B^2} \langle \frac{B}{v_{\|}} C(f^1) \rangle \right] \rangle \quad \text{(C.71)}$$

Here, $V_{\|}$ and f_c are given by,

$$V_{\|} = \frac{\sigma v}{2} \sqrt{\langle B^2 \rangle} \int_{\lambda}^{\lambda_c} \frac{d\lambda}{\langle \zeta \rangle} \quad \text{(C.72)}$$

$$f_c = \frac{3}{4} \langle B^2 \rangle \int_0^{\lambda_c} \frac{\lambda d\lambda}{\langle |\zeta| \rangle} \quad \text{(C.73)}$$

After some manipulations, we obtain following forms for the parallel viscous forces.

$$\langle \boldsymbol{B} \cdot \nabla \cdot \boldsymbol{\pi} \rangle = \frac{2\pi}{\tau_{se}} \frac{mF}{e} m \frac{< B^2 >}{2f_c} \int v dv \frac{\partial f^0}{\partial \psi} \int d\lambda \sum_\sigma \sigma G$$

$$\times \left[f_t \hat{Z} v_c^3 - (v^3 + v_c^3) \left(\frac{\partial < |\zeta| >}{\partial \lambda} \int_\lambda^{\lambda_c} \frac{d\lambda}{< |\zeta| >} + f_c \right) \right] \tag{C.74}$$

$$\langle \boldsymbol{B} \cdot \nabla \cdot \boldsymbol{r} \rangle = \frac{2\pi}{\tau_{se}} \frac{mF}{e} m \frac{< B^2 >}{2f_c} \int v dv \frac{\partial f^0}{\partial \psi} \int d\lambda \sum_\sigma \sigma G$$

$$\times \left[f_t \hat{Z} v_c^3 - 3(v^3 + v_c^3) \left(\frac{\partial < |\zeta| >}{\partial \lambda} \int_\lambda^{\lambda_c} \frac{d\lambda}{< |\zeta| >} + f_c \right) \right] \tag{C.75}$$

In the pitch angle scattering dominant regime $\hat{Z} v_c^3 / v^3 \gg 1$, G is given by,

$$G = -\frac{1}{2} \int_\lambda^{\lambda_c} \frac{d\lambda}{< \zeta >_p} \tag{C.76}$$

Allowing the v dependence of G in the following form,

$$G = -\frac{1}{2} \hat{G}(v, \psi) \int_\lambda^{\lambda_c} \frac{d\lambda}{< \zeta >_p} \tag{C.77}$$

The parallel viscous forces are given by,

$$\langle \boldsymbol{B} \cdot \nabla \cdot \boldsymbol{\pi} \rangle = \frac{2\pi}{\tau_{se}} \frac{mF}{e} m \frac{2}{3} f_t \hat{Z} v_c^3 \int dv v \hat{G} \frac{\partial f^0}{\partial \psi} \tag{C.78}$$

$$\langle \boldsymbol{B} \cdot \nabla \cdot \boldsymbol{r} \rangle = \frac{2\pi}{\tau_{se}} \frac{mF}{e} m \frac{2}{3} f_t \hat{Z} v_c^3 \int dv v^3 \hat{G} \frac{\partial f^0}{\partial \psi} \tag{C.79}$$

The corresponding flows in the poloidal direction are,

$$n u_\theta = \frac{mF}{e} \pi \frac{4}{3} f_c \int dv v^4 \hat{G} \frac{\partial f^0}{\partial \psi} \tag{C.80}$$

$$q_\theta = \frac{mF}{e} \pi \frac{4}{3} f_c \int dv \left(\frac{1}{2} m v^2 - \frac{5}{2} \frac{p}{n} \right) v^4 \hat{G} \frac{\partial f^0}{\partial \psi} \tag{C.81}$$

References

1. M.A. Abdou, C.C. Baker et al., Plasma physics and controlled nuclear fusion research 1980, vol. 2 (IAEA, Vienna, 1981), p. 119
2. A.A. Abrikosov, Sov. Phys. JETP **5**, 1174 (1957)
3. J.W. Ahn, H.S. Kim, Y.S. Park et al., Nucl. Fusion **52**, 114001 (2012)
4. N. Aiba, S. Tokuda, T. Ishizawa, M. Okamoto, Comp. Phys. Comm. **175**, 269 (2006)
5. N. Aiba, S. Tokuda, M. Furukawa, N. Oyama, T. Ozeki, Nucl. Fusion **49**, 065015 (2009)
6. N. Aiba, M. Furukawa, M. Hirota, S. Tokuda, Nucl. Fusion **50**, 045002 (2010)
7. N. Aiba, M. Furukawa, M. Hirota, N. Oyama et al., Nucl. Fusion **51**, 073012 (2011)
8. N. Aiba, N. Oyama, Nucl. Fusion **52**, 114002 (2012)
9. H. Alfven, Ark. Mat. Astron. Fys. **27A**, 22 (1940)
10. H. Alfven, Nature **150**, 405 (1942)
11. T. Ando, M. Sugimoto, H. Tsuji et al., Cryogenics (ICEC Suppl.) **34**, 509 (1994)
12. M. Angelone, P. Batistone et al., Fusion Eng. Des. **51–52**, 653 (2000)
13. C. Angioni, A.G. Peeters, G.V. Pereverzev, F. Ryter et al., Phys. Rev. Lett. **90**, 205003 (2003)
14. C. Angioni, A.G. Peeters, G.V. Pereverzev, F. Ryter et al., Plasma Phys. Controlled Fusion **10**, 3225 (2003)
15. C. Angioni, A.G. Peeters, X. Garbet et al., Nucl. Fusion **44**, 827 (2004)
16. C. Angioni, A.G. Peeters, F. Jenko, T. Dannert, Phys. Plasmas **12**, 112310 (2005)
17. C. Angioni, H. Weisen et al., Nucl. Fusion **47**, 1326 (2007)
18. C. Angioni, H. Weisen et al., Phys. Plasmas **14**, 055905 (2007)
19. C. Angioni, E. Fable, M. Greenwald et al., Plasma Phys. Controlled Fusion **51**, 124017 (2009)
20. C. Angioni, R.M. McDermott, F.J. Casso et al., Phys. Rev. Lett. **107**, 215003 (2011)
21. C. Angioni, Y. Camenen, F.J. Casson et al., Nucl. Fusion **52**, 114003 (2012)
22. T.M. Antonsen Jr., B. Lane, Phys. Fluids **23**, 1205 (1980)
23. T. Antonsen, K.R. Chu, Phys. Fluids **25**, 1295 (1982)
24. K. Appert, R. Gruber, J. Vaclavik, Phys. Fluids **17**, 1471 (1974)
25. K. Appert, R. Gruber, F. Troyon, J. Vaclavik, Plasma Phys. **24**, 1147 (1982)
26. V.I. Arnolds, *Mathematical methods of Classical Mechanics* (Springer, Berlin, 1978)
27. L.A. Artsimovich, Research on controlled thermonuclear reactions in the USSR, in *Proceedings of 2nd UN International Conference on Peaceful Uses of Atomic Energy*, vol. 31, Paper P/2298, 1958
28. L. Artsimovich, Nucl. Fusion **12**, 215 (1972) (Review Article)
29. N. Asakura, Y. Koide, K. Itami et al., J. Nucl. Mater. **241–243**, 559 (1997)
30. I. Asimov, *Atom* (Truman Talley Books, New York, 1991)
31. ATLAS Collaboration, Phys. Lett. B **716**, 1 (2012)

M. Kikuchi, M. Azumi, *Frontiers in Fusion Research II*,
DOI 10.1007/978-3-319-18905-5

32. R. Aymar, Equipe Tore Supra, Plasma physics and controlled nuclear fusion research 1988, vol. 1 (IAEA, Vienna, 1989), p. 9
33. M. Azumi et al., Plasma physics and controlled nuclear fusion research 1980, vol. 1 (IAEA, Vienna, 1981), p. 293
34. M. Bacal, G.W. Hamilton, Phys. Rev. Lett. **42**, 1538 (1979)
35. M. Bacal, G.W. Hamilton et al., Rev. Sci. Instrum. **50**, 719 (1979)
36. M. Bacal, Chem. Phys. **398**, 3 (2012)
37. J.M. Baird, Int. Electron. Dev. Meet., Washington DC, 1979, p. 156
38. J.M. Baird, Gyrotron theory, in *High Power Microwave Source*, ed. by V.L. Granatstein, I. Alexeff, Chap. 4, (Artech House, Boston/London, 1987)
39. P. Bak, C. Tang, K. Wiesenfeld, Phys. Rev. Lett. **59**, 381 (1987)
40. C.C. Baker, M.A. Abdou et al., STARFIRE - A Commercial Tokamak Fusion Power Plant Study, Technical Report for Argonne National Laboratory, ANL/EPP-80-1, (1980
41. R. Balescu, *Aspects of Anomalous Transport in Plasmas* (IOP, Bristol, 2005)
42. N. Baluk, K. Abe, J.L. Boutard, V.M. Chernov et al., Nucl. Fusion **49**, S696 (2007)
43. P. Barabaschi, Y. Kamada, S. Ishida, JT-60SA Team, in *Proceedings of IAEA FEC OV/3-2*, 2014
44. J. Bardeen, L.N. Cooper, J.R. Schrieffer, Phys. Rev. **108**, 1175 (1957)
45. G. Bateman, *MHD iInstabilities.*, (MIT Press, Cambridge, 1978)
46. E. Barston, Ann. Phys. **29**, 282 (1964)
47. S. Batha, F.M. Levinton, A.T. Ramsey et al., Phys. Plasmas **4**, 3614 (1997)
48. P. Batistoni, M. Angelonem et al., Fusion Eng. Des. **28**, 504 (1995)
49. P. Batistoni, M. Angelone et al., J. Nucl. Mater. **329–333**, 1610 (2004)
50. P. Batistoni, U. Fischer, K. Ochiai et al., Fusion Eng. Des. **83**, 834 (2008)
51. P. Batistoni, M. Angelone, U. Fischer et al., Nucl. Fusion **52**, 083014 (2012)
52. J.G. Bednorz, K.A. Muller, Z. Phys. **B64**, 189 (1986)
53. M. Benedict, T.H. Pigford, H.W. Levi, *Nuclear Chemical Engineering*, Chap. 13 (McGraw Hill College, Blacklick, 1981)
54. S. Bernabei, N. Sauthoff, E. Sindoni, *ISPP-16 Tokamak Concept Improvement* (Varenna, 29 August – 3 September 1994 Pag. XIV + 410, Prezzo EUR 51, 65, ISBN 88-7794-067-0)
55. H. Berk, J.W. Van Dam, Z. Guo, D.M. Lindberg, Phys. Fluids **B4**, 1806 (1992)
56. H. Berk, J.W. Van Dam, D. Borba, J. Candy et. al., Phys. Plasmas **2**, 3401 (1995)
57. H. Berk, B.N. Breizman, M. Pekker, Phys. Plasmas **2**, 3007 (1995)
58. H.L. Berk, B.N. Breizman, J. Candy, M. Pekker et al., Phys. Plasmas **6**, 3102 (1999)
59. H. Berk, D.N. Borba, B.N. Breizman et al., Phys. Rev. Lett. **87**, 185002 (2001)
60. H. Berk, C.J. Boswell, D. Borba, A.C.A. Figueiredo et al., Nucl. Fusion **46**, S888 (2006)
61. I.B. Bernstein, J.M. Greene, M.D. Kruskal, Phys. Rev. **108**, 546 (1957)
62. I. Bernstein, E.A. Frieman, M.D. Kruskal, R.M. Kulsrud, Proc. Roy. Soc. **A244**, 17 (1958)
63. R. Betti, J.P. Freidberg, Phys. Fluids **B3**, 1865 (1991)
64. R. Betti, J.P. Freidberg, Phys. Fluids **B4**, 1465 (1992)
65. R. Bickerton, J.W. Connor, J.B. Taylor, Nat. Phys. **229**, 110 (1971)
66. H. Biglari, P.H. Diamond, M.N. Rosenbluth, Phys. Fluids **B1**, 109 (1989)
67. H. Biglari, P.H. Diamond, P.W. Terry, Phys. Fluids **B2**, 1 (1990)
68. D. Biskamp, *Nonlinear Magnetohydrodynamics* (Cambridge University Press, Cambridge, 1993)
69. D. Biskamp, *Magnetic Reconnection in Plasmas* (Cambridge University Press, Cambridge, 2000)
70. A. Boozer, Phys. Fluids **24**, 1999 (1981)
71. A. Boozer, Phys. Fluids **26**, 1288 (1983)
72. A. Boozer, Phys. Rev. Lett. **86**, 5059 (2001)
73. A. Boozer, Rev. Mod. Phys. **76**, 1071 (2004) (Review Article)
74. A. Bortolon, B.P. Duval, A. Pochelon, A. Scarabosia, Phys. Rev. Lett. **97**, 235003 (2006)
75. T. Boutros-Ghali, T. H. Dupree, Phys. Fluids **24**, 1839 (1981)
76. S.I. Braginskii, Rev. Plasma Phys. **1**, 205 (1965)

77. B.N. Breizman, H.L. Berk, M.S. Pekker et al., Phys. Plasmas **10**, 3649 (2003)
78. B.N. Breizman, S.E. Sharapov, Plasma Phys. Controlled Fusion **53**, 054001 (2011)
79. B.N. Breizman, M.S. Pekker, S.A. Sharapov et al., Phys. Plasmas **12**, 112506 (2005)
80. S. Briguglio, G. Fogaccia, G. Vlad, F. Zonca, K. Shinohara, M. Ishikawa, M. Takechi, Phys. Plasmas **14**, 055904 (2007)
81. L. Brillouin, Comptes Rendus **183**, 24 (1926)
82. A.J. Brizard, T.S. Hahm, Rev. Mod. Phys. **79**, 421 (2007) (Review Article)
83. S. Brunner, M. Fivaz, T.M. Tran, J. Vaclavik, Phys. Plasmas **5**, 3929 (1998)
84. J. Bucalossi et al., Fusion Eng. Des. **86**, 684 (2011)
85. E. Buckingham, Phys. Rev. **4**, 345 (1914)
86. K.H. Burrell et al., Phys. Plasmas **8**, 2153 (2001)
87. K.H. Burrell et al., Plasma Phys. Controlled Fusion **44**, A253 (2002)
88. K.H. Burrell et al., Phys. Rev. Lett. **102**, 155003 (2009)
89. K.H. Burrell et al., Nucl. Fusion **49**, 085024 (2009)
90. R.J. Buttery, R.J. La Haye, P. Gohil et al., Phys. Plasmas **15**, 056115 (2008)
91. J.D. Callen, R.A. Dory, Phys. Fluids **15**, 1523 (1972)
92. J.D. Callen, Fundamentals of plasma physics (2006), http://homepages.cae.wisc.edu/~callen/book.html (Appendix D)
93. Y. Camenen, A.G. Peeters, C. Angioni et al., Phys. Rev. Lett. **102**, 125001 (2009)
94. Y. Camenen, A. Bortolon, B.P. Duval et al., Phys. Rev. Lett. **105**, 135003 (2010)
95. J. Candy, M.N. Rosenbluth, Phys. Plasmas **1**, 356 (1994)
96. J. Candy, D. Borba, H.L. Berk et al., Phys. Plasmas **4**, 2597 (1997)
97. J. Candy, R.E. Waltz, M.N. Rosenbluth, Phys. Plasmas **11**, 1879 (2004)
98. R. Carrera, R.D. Hazeltine, M. Kotschenreuther, Phys. Fluids **29**, 899 (1986)
99. J.H. Cary, R.G. Littlejohn, Ann. Phys. **151**, 1 (1983)
100. J.R. Cary, A.J. Brizard, Rev. Mod. Phys. **81**, 693 (2009) (Review Article)
101. P.J. Catto, Plasma Phys. **20**, 719 (1978)
102. S. Chandrasekhar, *Principles of Stelar Dynamics* (University of Chicago Press, Chicago, 1942)
103. S. Chandrasekar, *Hydrodynamic and Hydromagnetic Stability* (Oxford University Press, Oxford, 1962), p. 158
104. Z. Chang, J.D. Callen, E.D. Fredrickson et al., Phys. Rev. Lett. **74**, 4663 (1995)
105. S. Chapman, T.G. Cowling, *The Mathematical Theory of Non-Uniform Gases* (Cambridge University Press, New York, 1953)
106. I. Chapman, N.R. Walkden et al., Plasma Phys. Controlled Fusion **53**, 125002 (2011)
107. M. Chatelier, Equipe Tore Supra, Nucl. Fusion **47**, S579 (2007)
108. F.F. Chen, Phys. Fluids **8**, 912 (1965)
109. L. Chen, R.B. White, M.N. Rosenbluth, Phys. Rev. Lett. **53**, 454 (1984)
110. L. Chen, *Waves and Instabilities in Plasmas* (World Scientific, Singapore, 1987)
111. L. Chen, Phys. Plasmas **1**, 1519 (1994)
112. L. Chen, Z. Lin, R.B. White, Phys. Plasmas **7**, 3129 (2000)
113. L. Chen, Z. Lin, R.B. White, F. Zonca, Nucl. Fusion **41**, 747 (2001)
114. L. Chen, F. Zonca, Nucl. Fusion **47**, S727 (2007)
115. C.Z. Cheng, L. Chen, M.S. Chance, Ann. Phys. **161**, 21 (1985)
116. C.Z. Cheng, M.S. Chance, Phys. Fluids **29**, 3695 (1986)
117. C.Z. Cheng, M.S. Chance, J. Comp. Phys. **71**, 124 (1987)
118. C.Z. Cheng, H.P. Furth, A.H. Boozer, Plasma Phys. Controlled Fusion **29**, 351 (1988)
119. C.Z. Cheng, Phys. Rep. **211**, 1 (1992)
120. J. Cheng, J.Q. Dong, K. Itoh, L.W. Yan et al., Phys. Rev. Lett. **110**, 265002 (2013)
121. G.F. Chew, M.L. Goldberger, F.E. Low, Proc. R. Soc. Lond. Ser. A **236**, 112 (1956)
122. M.S. Chu, J.M. Green, L.L. Lao, A.D. Turnbull, M.S. Chance, Phys. Fluids **B4**, 3713 (1992)
123. M. Chu, J.M. Green, L.L. Lao, R.L. Miller et al., Phys. Rev. Lett. **77**, 2710 (1996)
124. M.S. Chu, L. Chen, L.-J. Zheng, C. Ren, A. Bonderson, Nucl. Fusion **39**, 2107 (1999)
125. M. Chu, P.B. Parks, Phys. Plasmas **9**, 5036 (2002)

126. M. Chu, A. Bonderson, M.S. Chance, Y.Q. Liu et al., Phys. Plasmas **11**, 4859 (2004)
127. M. Chu, M. Okabayashi, Plasma Phys. Controlled Fusion **52**, 123001 (2010)
128. J. Citrin, F. Jenko, P. Mantica et al., Phys. Rev. Lett. **111**, 155001 (2013)
129. W.H.M. Clark, J.G. Cordey, M. Cox, R.D. Gill, J. Hugill, Phys. Rev. Lett. **45**, 1101 (1980)
130. J.F. Clarke, D.J. Sigmar, Phys. Rev. Lett. **38**, 70 (1977)
131. CMS Collaboration, Phys. Lett. B **716**, 30 (2012)
132. S. Coda, O. Sauter et al., in *Proceedings of 22nd IAEA Fusion Energy Conference*, EX/2-3, Geneva, 2008
133. S. Coda, M. Porkolab, K.H. Burrel, Phys. Rev. Lett. **86**, 4835 (2001)
134. R. Cohen, Phys. Fluids **30**, 2442 (1987)
135. L. Colas, V. Basiuk, B. Beaumont et al., Nucl. Fusion **46**, S500 (2006)
136. A.J. Cole, J.D. Callen, W.M. Solomon et al., Phys. Rev. Lett. **106**, 225002 (2011)
137. R.W. Conn et al., Plasma physics and controlled nuclear fusion research 1976, vol. 3 (IAEA, Vienna, 1977), p. 203
138. G. Conway, C. Angioni, F. Ryter, P. Sauter et al., Phys. Rev. Lett. **106**, 065001 (2011)
139. R. Conn et al., Plasma physics and controlled nuclear fusion research 1990, vol. 3 (IAEA, Vienna, 1991), p. 659
140. R.W. Conn, F. Najmabadi, M.S. Tilack, Fusion Eng. Des. **41**, 337 (1998)
141. J.W. Connor, R.J. Hastie, J.B. Taylor, Phys. Rev. Lett **40**, 396 (1978)
142. J.W. Connor, R.J. Hastie, J.B. Taylor, Proc. Roy. Soc. Lond. A **42**, 1 (1979)
143. J. Connor, Plasma Phys. Controlled Fusion **30**, 619 (1988) (Review Article)
144. J.W. Connor, S.C. Cowley, R.J. Hastie, Plasma Phys. Controlled Fusion **32**, 799 (1990)
145. J.W. Connor, J.B. Taylor, H.R. Wilson, Phys. Rev. Lett. **71**, 1803 (1993)
146. J. Connor, H. Wilson, Plasma Phys. Controlled Fusion **36**, 719 (1994) (Review Article)
147. B. Coppi, C. Spicht, Phys. Rev. Lett **41**, 551 (1978)
148. J.G. Cordey, Nucl. Fusion **16**, 499 (1976)
149. J.G. Cordey, C.D. Challis, P.M. Stubberfield, Plasma Phys. Controlled Fusion **30**, 1625 (1988)
150. R. Courant, D. Hilbert, *Methods of Mathematical Physics* (Interscience, NY, 1953) vol. 1
151. L. Cranberg, J. Appl. Phys. **23**, 18 (1952)
152. B. Davies, J. Comput. Phys. **66**, 36 (1986)
153. A. de Chambrier, Plasma Phys. **24**, 893 (1982)
154. L. Degtyarev, A. Martynov, S. Medvedev et al., Comp. Phys. Comm. **103**, 10 (1997)
155. R.L. Dewar, R.C. Grimm, J.L. Johnson et al., Phys. Fluids **17**, 930 (1974)
156. R.L. Dewar et al., PPPL-1587 (1979)
157. R. Dewar, J. Manickam, R.C. Grimm, M.S. Chance, Nucl. Fusion **21**, 493 (1981)
158. R.L. Dewar, A.H. Glasser, Phys. Fluids **26**, 3038 (1983)
159. R.L. Dewar, B. Davies, J. Plasma Physics **32**, 443 (1984)
160. R. Dewar, in *Proceedings Workshop on Theory of Fusion Plasmas*, Varenna, 1987, ed. by A. Bonderson, E. Sindoni, F. Troyon, 1988, p. 107
161. R. Dewar, Plasma Phys. Controlled Fusion **39**, 453 (1997)
162. P. Diamond, Y.M. Liang, B.A. Carreras, P.W. Terry, Phys. Rev. Lett. **72**, 2565 (1994)
163. P. Diamond, T.S. Hahm, Phys. Plasmas **2**, 3640 (1995)
164. P. Diamond, M. Rosenbluth, F. Hinton, in *18th IAEA Fus. Energy Conf.*, CN-69/TH3/1, 1998
165. P.H. Diamond, M.N. Rosenbluth, E. Sanchez et al., Phys. Rev. Lett. **84**, 4842 (2000)
166. P.H. Diamond, S.-I. Itoh, K. Itoh, T.S. Hahm, Plasma Phys. Controlled Fusion **47**, R35 (2005) (Review Article)
167. P.H. Diamond, C.J. McDevitt, O.D. Gurcan et al., Nucl. Fusion **49**, 045002 (2009)
168. P.D. Diamond, S.I. Itoh, K. Itoh, *Modern Plasma Physics Volume 1: Physical Kinetics of Turbulent Plasmas* (Cambridge University Press, New York, 2010)
169. A. Dimits, B. I. Cohen, N. Mattor, W.M. Nevins et al., Nucl. Fusion **40**, 661 (2000)
170. Y. Dnestrovskij, D.P. Kostomarov, A.A. Lukyanitsa et al., Nucl. Fusion **28**, 267 (1988)
171. H. Doerk, F. Jenko, M. J. Pueschel, and D. R. Hatch, Phys. Rev. Lett. **106**, 155003 (2011)
172. R. Dominguez, G. M. Staebler, Phys. Fluids **B5**, 3876 (1993)
173. W. Dorland, F. Jenko, M. Kotschenreuther, B.N. Rogers, Phys. Rev. Lett. **85**, 5579 (2000)

174. R.A. Dory, Y.K.M. Peng, Nucl. Fusion **17**, 21 (1977)
175. E. Doyle, W.A. Houlberg, Y. Kamada et al., Nucl. Fusion **47**, S18 (2007)
176. J.F. Drake, Y.C. Lee, Phys. Fluids **20**, 1341 (1977)
177. J.F. Drake, N.T. Gladd, C.S. Liu, C.L. Chang, Phys. Rev. Lett. **44**, 994 (1980)
178. D.H.E. Dubin, J.A. Krommes, C. Oberman, W.W. Lee, Phys. Fluids **26**, 3524 (1983)
179. R.J. Dumont, M. Goniche, A. Ekedahl, B. Sautic et al., Plasma Phys. Controlled Fusion **56**, 075020 (2014)
180. H.H. Duong, W.W. Heidbrink, E.J. Strait, T.W. Petrie et al., Nucl. Fusion **33**, 749 (1993)
181. T.H. Dupree, Phys. Fluids **15**, 334 (1972)
182. P. Efthimion, D.K. Mansfield, B.C. Stratton et al., Phys. Rev. Lett. **66**, 421 (1991)
183. A. Ekedahl, J. Bucalossi, V. Basiuk et al., Nucl. Fusion **49**, 095010 (2009)
184. W.W. Engle, *A User's Manual for ANISN*, K-1693, Union Carbide Cooperation, Computing Technology Center (1967)
185. F. Englert, R. Brout, Phys. Rev. Lett. **3** 321 (1964)
186. EFDA, *Roadmap to the Realisation of Fusion Energy* (2012), https://www.euro-fusion.org/
187. T. Evans, P.M. Valanju, J.F. Benesch et al., Phys. Rev. Lett. **53**, 1743 (1984)
188. T.E. Evans, C.J. Lasnier, D.N. Hill et al., Phys. Plasmas **2**, 2408 (1995)
189. T.E. Evans, R.K.W. Roeder, J.A. Carter B.I. Rapoport, Contrib. Plasma Phys. **44**, 235 (2004)
190. T.E. Evans, R.A. Moyer, J.G. Watkins, T.H. Osborne et al., Nucl. Fusion **45**, 595 (2005)
191. T. E. Evans, R.K.W. Roeder, J.A. Carter et al., J. Phys. Conf. Ser. **7**, 174 (2005)
192. T.E. Evans, Implications of topological complexity and hamiltonian chaos in the edge magnetic field of toroidal fusion plasmas, in *Chaos, Complexity and Transport: Theory and Applications*, ed. by C. Chandre, X. Leoncini, G. Zaslavsky (World Scientific Publishing, Singapore, 2008), pp. 147–176
193. T. Evans, M.E. Fenstermacher, R.A. Moyer et al., Nucl. Fusion **48**, 024002 (2008)
194. E. Fable, C. Angioni, O. Sauter, Plasma Phys. Controlled Fusion **50**, 115005 (2008)
195. E. Fable, C. Angioni, O. Sauter, Plasma Phys. Controlled Fusion **52**, 015007 (2010)
196. U. Fantz, P. Franzen, W. Kraus et al., Nucl. Fusion **49**, 125007 (2009)
197. U. Fantz, P. Franzen, D. Wunderlich, Chem. Phys. **398**, 7 (2012)
198. A. Fasoli, D. Borba, G. Bosia, D.J. Campbell et al., Phys. Rev. Lett. **75**, 645 (1995)
199. A. Fasoli, J.B. Lister, S.E. Sharapov, S. Ali-Arshad et al., Nucl. Fusion **35**, 1485 (1995)
200. A. Fasoli, J.B. Lister, S. Sharapov, D. Borba et al., Phys. Rev. Lett. **76**, 1067 (1996)
201. J. Ferron, M.S. Chu, F.J. Helton, W. Howl et al., Phys. Fluids **B2**, 1280 (1990)
202. N. Fisch, A.H. Boozer, Phys. Rev. Lett. **45**, 720 (1980)
203. N. Fisch, Nucl. Fusion **24**, 378 (1984)
204. N.J. Fisch, Rev. Mod. Phys. **59**, 175 (1987) (Review Article)
205. R. Fitzpatrick, Phys. Plasmas **2**, 825 (1995)
206. H. Flanders, *Differential Forms with Applications to the Physical Sciences* (Dover, New York, 1989)
207. C. Forest, K. Kupfer, T.C. Luce, P.A. Politzer et al., Phys. Rev. Lett. **73**, 2444 (1994)
208. R. Fowler et al., ORNL/TM-5845 (1979)
209. T. Frankel, *The Geometry of Physics* (Cambridge University Press, Cambridge, 2004)
210. P. Franzen, H.D. Falter, U. Fantz et al., Nucl. Fusion **47**, 264 (2007)
211. J. Freidberg, *Ideal Magneto-Hydrodynamics* (Plenum, New York, 1987)
212. E. Frieman, M. Rotenberg, Rev. Mod. Phys. **34**, 898 (1960)
213. E.A. Frieman, Liu Chen, Phys. Fluids **25**, 502 (1982)
214. B. Friedman, *Principles and Techniques of Applied Mathematics* (Dover, New York, 1990)
215. U. Frisch, *Turbulence - The Legacy of A.N. Kolmogolov* (Cambridge University Press, Cambridge, 1995)
216. G.Y. Fu, J.W. Van Dam, Phys. Fluids **B1**, 1949 (1989)
217. G.Y. Fu, C.Z. Cheng, Phys. Fluids **B2**, 985 (1990)
218. G.Y. Fu, C.Z. Cheng, Phys. Fluids **B4**, 3722 (1992)
219. G.Y. Fu, Phys. Plasmas **2**, 1029 (1995)

220. G.Y. Fu, Phys. Rev. Lett. **101**, 185002 (2008)
221. A. Fujisawa, K. Itoh, H. Iguchi et al., Phys. Rev. Lett. **93**, 165002 (2004)
222. A. Fujisawa, Nucl. Fusion **49**, 013001 (2009) (Review Article)
223. T. Fujita, S. Ide, H. Kimura, Y. Koide et al., Fusion energy 1996, vol. 1, in *Sixteenth Conference Proceedings*, Montreal, 7–11 October 1996 (IAEA, Vienna, 1997), p. 227
224. T. Fujita, S. Ide, H. Shirai, M. Kikuchi et al., Phys. Rev. Lett. **78**, 2377 (1997)
225. T. Fujita, T. Hatae, T. Oikawa et al., Nucl. Fusion **38**, 207 (1998)
226. T. Fujita, Y. Kamada, S. Ishida et al., Nucl. Fusion **39**, 1627 (1999)
227. T. Fujita, T. Oikawa, T. Suzuki, S. Ide et al., Phys. Rev. Lett. **87**, 245001 (2001)
228. T. Fujita, T. Suzuki, T. Oikawa, A. Isayama et al., Phys. Rev. Lett. **95**, 075001 (2005)
229. T. Fujita, Nucl. Fusion **50**, 113001 (2010) (Review Article)
230. A. Fukuyama et al., in *Proceedings of 5th IAEA TCM on α Particles in Fusion Res.*, 1997
231. A. Fukuyama, E. Yokota, in *Proceedings of 6th IAEA TCM on Energetic particles in Magnetic Confinement Systems* (JAERI, Naka, 1999) p. 61, http://jolissrch-inter.tokai-sc.jaea.go.jp/pdfdata/JAERI-Conf-2000-004.pdf
232. H.P. Furth, J. Killeen, M.N. Rosenbluth, Phys. Fluids **6**, 459 (1963)
233. H.P. Furth, J. Kileen, M.N. Rosenbluth, B. Coppi, Plasma physics and controlled nuclear fusion research 1965, vol. 1 (IAEA, Vienna, 1966), p. 103
234. H.P. Furth, P.H. Rutherford, H. Selberg, Phys. Fluids **16**, 1054 (1973)
235. *Fusion Evaluated Nuclear Data Library FENDL 3.0*, https://www-nds.iaea.org/fendl30/ (2012)
236. Fusion Reactor System Laboratory, Concept study of the steady state Tokamak reactor (SSTR), JAERI-M 91-081 (1991), http://jolissrch-inter.tokai-sc.jaea.go.jp/pdfdata/JAERIM-91-081.pdf
237. A. Galeev, Zh. Eksp.Teor. Fiz. **59**, 1378 (1970) [Sov. Phys. JETP **32**, 752 (1971)]
238. G. Gantenbein, H. Zohm, G. Giruzzi, S. Gunter et al., Phys. Rev. Lett. **85**, 1242 (2000)
239. X. Garbet, R.E. Waltz, Phys. Plasmas **5**, 2836 (1998)
240. X. Garbet, C. Bourdelle, G.T. Hoang, P. Maget et al., Phys. Plasmas **8**, 2793 (2001)
241. X. Garbet, Y. Sarazin, P. Ghendrih, S. Benkadda et al., Phys. Plasmas **9**, 3893 (2002)
242. X. Garbet, L. Garzotti, P. Mantica, H. Nordman et al., Phys. Rev. Lett. **91**, 035001 (2003)
243. X. Garbet, P. Mantica et al., Plasma Phys. Controlled Fusion **46**, B557 (2004) (Review Article)
244. X. Garbet, D. Esteve, Y. Sarazin et al., Phys. Plasmas **20**, 072502 (2013)
245. A.M. Garofalo, G.L. Jackson, R.J. LaHaye et al., Nucl. Fusion **47**, 1121 (2007)
246. A. Garofalo, K.H. Burrell, J.C. DeBoo et al., Phys. Rev. Lett. **101**, 195005 (2008)
247. V.L. Ginzburg, L.D. Landau, Zh. Eksperim. i Theor. Fiz. **20**, 1064 (1950)
248. G. Giruzzi et al., Nucl. Fusion **49**, 104010 (2009)
249. H.A. Glasser, J.M. Green, J.L. Johnson, Phys. Fluids **18**, 875 (1975)
250. J.P. Goedbloed, Physica **12D**, 107 (1984)
251. H. Goedbloed, S. Poedts, *Principles of Magnetohydrodynamics* (Cambridge University Press, Cambridge, 2004)
252. R. Goldston, S.H. Batha et al., Plasma Phys. Controlled Fusion **36**, B213 (1994)
253. N. Gorelenkov, H.L. Berk, R. Budny et al., Nucl. Fusion **43**, 594 (2003)
254. N. Gorelenkov, H.L. Berk, E. Fredrickson, S.E. Sharapov, Phys. Lett. **A 370**, 70 (2007)
255. N.N. Gorelenkov, M.A. Van Zeeland, H.L. Berk et al., Phys. Plasmas **16**, 056107 (2009)
256. N.N. Gorelenkov, S.D. Pinches, K. Toi, Nucl. Fusion **54**, 125001 (2014) (Review Paper)
257. L.P. Gor'kov, Sov. Phys. JETP **9**, 1364 (1959)
258. C. Gormezano, JET Team, Nucl. Fusion **39**, 1875 (1999)
259. C. Gormezano, A.C.C. Sips, T.C. Luce, S. Ide et al., Nucl. Fusion **47**, S285 (2007)
260. H. Grad, Phys. Fluids **10**, 137 (1967)
261. H. Grad, P.N. Hu, D.C. Stevens, Proc. Natl. Acad. Sci. **72**, 9789 (1975)
262. A. Grosman, J. Bucalossi et al., Fusion Eng. Des. **88**, 497 (2013)
263. J.M. Green, M.S. Chance, Nucl. Fusion **21**, 453 (1981)
264. M. Greenwald et al., Nucl. Fusion **28**, 2199 (1988)

265. M. Greenwald, Plasma Phys. Controlled Fusion **44**, R27 (2002)
266. R. Grimm, R.L. Dewar, J. Manickam, J. Comp. Phys. **49**, 94 (1983)
267. R.J. Groebner, W. Pfeiffer, F.P. Blau et al., Nucl. Fusion **26**, 543 (1986)
268. R.A. Gross, *Fusion Energy* (Wiley, New York, 1984)
269. R. Gruber, S. Semenzato, F. Troyon et al., Comput. Phys. Comm. **24**, 363 (1981)
270. S. Guo, F. Romanelli, Phys. Fluids **B5**, 520 (1993)
271. O. Gurcan, P.H. Diamond, T.S. Hahm, R. Singh, Phys. Plasmas **14**, 042306 (2007)
272. O.D. Gurcan, P.H. Diamond, Phys. Rev. Lett. **100**, 135001 (2008)
273. O. Gurcan, P.H. Diamond, P. Hennequin et al., Phys. Plasmas **17**, 112309 (2010)
274. W. Guttenfelder, J. Candy, S. Kaye, W. Nevins et al., Phys. Rev. Lett. **106**, 155004 (2011)
275. T.S. Hahm, Phys. Fluids **31**, 2670 (1988)
276. T.S. Hahm, W.W. Lee, A. Brizard, Phys. Fluids **31**, 1940 (1988)
277. T.S. Hahm, W.M. Tang, Phys. Fluids **B1**, 1185 (1989)
278. T.S. Hahm, Phys. Plasmas **1**, 2940 (1994)
279. T. Hahm, K.H. Burrel, Phys. Plasmas **2**, 1648 (1995)
280. T. Hahm, P.D. Diamond, O.D. Gurcan, G. Rewoldt, Phys. Plasmas **14**, 072302 (2007)
281. K. Hallatschek, D. Biskamp, Phys. Rev. Lett. **86**, 1223 (2001)
282. S. Hamada, Nucl. Fusion **2**, 23 (1962)
283. Y. Hamada, A. Nishizawa, T. Ido et al., Nucl. Fusion **45**, 81 (2005)
284. K. Hamamatsu, A. Fukuyama, Fusion Eng. Des. **53**, 53 (2001)
285. S. Haney, J. Freidberg, Phys. Fluids **B1**, 1637 (1989)
286. E. Hameiri, Phys. Fluids **26**, 230 (1983)
287. G.Z. Hao, A.K. Wang, Y.Q. Liu, X.M. Qiu, Phys. Rev. Lett. **107**, 015001 (2011)
288. A. Hasegawa, K. Mima, Phys. Rev. Lett. **39**, 205 (1977)
289. A. Hasegawa, K. Mima, Phys. Fluids **21**, 87 (1978)
290. A. Hasegawa, C. Maclennan, Y. Kodama, Phys. Fluids **22**, 2122 (1979)
291. A. Hasegawa, Adv. Phy. **34**, 1 (1985)
292. A. Hasegawa, M. Ejiri, S. Nogami et al., J. Nucl. Mat. **386–388**, 241 (2009)
293. A. Hasegawa, L. Chen, Phys. Rev. Lett. **35**, 370 (1975)
294. A. Hasegawa, L. Chen, Phys. Fluids **19**, 1924 (1976)
295. R. Hastie, J.B. Taylor, Nucl. Fusion **21**, 187 (1981)
296. R.J. Hastie, J.J. Ramos, F. Porcell, Phys. Plasmas **10**, 4405 (2003)
297. T. Hatae, Y. Kamada, S. Ishida et al., Plasma Phys. Controlled Fusion **40**, 1073 (1998)
298. T. Hatae et al., Plasma Phys. Controlled Fusion **42**, A283 (2000)
299. D.R. Hatch, M.J. Pueschel, F. Jenko et al., Phys. Rev. Lett. **108**, 235002 (2012)
300. C. Hawkes, B.C. Stratton, T. Tala et al., Phys. Rev. Lett. **87**, 115001 (2001)
301. R. Hawryluk, Rev. Mod. Phys. **70**, 537 (1998) (Review Article)
302. R.J. Hawryluk, D.J. Campbell, G. Janeschitz et al., Nucl. Fusion **49**, 065012 (2009)
303. N. Hayashi, A. Isayama, K. Nagasaki, T. Ozeki, J. Plasma Fusion Res. **80**, 605 (2004)
304. N. Hayashi, T. Ozeki, K. Hamamatsu, T. Takizuka, Nucl. Fusion **44**, 477 (2004)
305. N. Hayashi, T. Takizuka, T. Ozeki, Nucl. Fusion **45**, 933 (2005)
306. T. Hayashi, K. Tobita, S. Nishio et al., Fusion Eng. Des. **81**, 1285 (2006)
307. N. Hayashi, T. Takizuka, N. Aiba, N. Oyama et al., Nucl. Fusion **49**, 095015 (2009)
308. R.D. Hazeltine, J.D. Meiss, *Plasma Confinement* (Dover, New York, 2003)
309. R.D. Hazeltine, F.L. Waelbloeck, *The Framework of Plasma Physics* (Westview, Boulder, 2004)
310. R.D. Hazeltine, D. Dobrott, T.S. Wang, Phys. Fluids **18**, 1778 (1975)
311. J. Heading, *An Introduction to Phase-Integral Methods* (Dover, New York, 2013)
312. W.W. Heidbrink, G.J. Sadler, Nucl. Fusion **34**, 535 (1994)
313. W.W. Heidbrink, E.J. Strait, M.S. Chu, A.D. Turnbull, Phys. Rev. Lett. **71**, 855 (1993)
314. W. Heidbrink, Phys. Plasmas **15**, 055501 (2008)
315. D.B. Heifetz, in *Physics of Plasma-Wall Interactions in Controlled Fusion* ed. by D.E. Post, R. Behrisch (Plenum Press, New York, 1986), p. 695. For DEGAS-2, D.P. Stotler, C.F.F. Karney, Contrib. Plasma Phys. **34**, 392 (1994)

316. P. Helander, P.J. Catto, Phys. Plasmas **8**, 1988 (2001)
317. P. Helander, D.J. Sigmar, *Collisional Transport in Magnetized Plasmas* (Cambridge University Press, Cambridge, 2005)
318. P. Helander, T. Fulop, M. Lisak, Phys. Plasmas **13**, 102506 (2006)
319. T. Hemmi, S. Harjo, Y. Nunoya et al., Supercond. Sci. Technol. **26**, 084002 (2013)
320. P. Higgs, Phys. Lett. **12** 132 (1964)
321. F.L. Hinton, R.D. Hazeltine, Rev. Modern Phys. **48**, 239 (1976) (Review Article)
322. F. Hinton, Phys. Fluids **B3**, 696 (1991)
323. F. Hinton, G.M. Staebler, Phys. Fluids **B5**, 1281 (1993)
324. M. Hirota, Y. Fukumoto, Phys. Plasmas **15**, 122101 (2008)
325. S. Hirshman et al., Phys. Fluids **19**, 1532 (1976)
326. S. Hirshman, Phys. Fluids **20**, 589 (1977)
327. S. Hirshman et al., Phys. Fluids **20**, 418 (1977)
328. S. Hirshman, Nucl. Fusion **18**, 917 (1978)
329. P. Hirshman, Phys. Fluids **21**, 224 (1978)
330. S. Hirshman, D.J. Sigmar, Nucl. Fusion **21**, 1079 (1981) (Review Article)
331. A. Hishinuma, A. Kohyama, R.L. Klueh et al., J. Nucl. Mater. **258–263**, 193 (1998)
332. G.T. Hoang, P. Monier-Garbet, T. Aniel et al., Nucl. Fusion **40**, 913 (2000)
333. G. Hoang, C. Bourdelle, X. Garbet, G. Giruzzi et al., Phys. Rev. Lett. **87**, 4096 (2001)
334. G. Hoang, C. Bourdelle, B. Pegourie et al., Phys. Rev. Lett. **90**, 155002 (2003)
335. G.T. Hoang et al., in *Proceedings of 16th Topical Conference on RF Power, AIP Conference Proc.*, 787, 2005
336. M. Honda, M. Kikuchi, M. Azumi, Nucl. Fusion **52**, 023021 (2012)
337. M. Honda, Jpn. J. Appl. Phys. **52**, 108002 (2013)
338. M. Honda, Phys. Plasmas **21**, 092508 (2014)
339. B.G. Hong, W. Horton, D-I. Choi, Phys. Fluids **B1**, 1589 (1989)
340. W. Horton, R.K. Varma, Phys. Fluids **15**, 620 (1972)
341. W. Horton, Rev. Mod. Phys. **71**, 735 (1999) (Review Article)
342. K. Hoshino, M. Mori, T. Yamamoto, H. Tamai et al., Phys. Rev. Lett. **69**, 2208 (1992)
343. T. Hoshino, T. Terai, J. Nucl. Mater. **417**, 696 (2011)
344. T. Hoshino, T. Terai, Fusion Eng. Des. **86**, 2168 (2011)
345. T. Hoshino, M. Nakamichi, Fusion Eng. Des. **87**, 486 (2012)
346. T. Hoshino, Fusion Eng. Des. **88**, 2264 (2013)
347. T. Hoshino, Desalination **317**, 11 (2013)
348. T. Hoshino, Fusion Eng. Des. **89**, 1431 (2014)
349. W. Houlberg, K.C. Shaing, S.P. Hirshman, M.C. Zarnstorff, Phys. Plasmas **4**, 3230 (1997)
350. W. Howl, A.D. Turnbull, T.S. Taylor, L.L. Lao et al., Phys. Fluids **B4**, 1734 (1992)
351. B. Hu, R. Betti, Phys. Rev. Lett. **93**, 105002 (2004)
352. Q. Hu, G. Zhuang, Q. Yu et al., Nucl. Fusion **54**, 064013 (2014)
353. J.D. Huba, in *2007 NRL Plasma Formulary*, Naval Research Laboratory, 2007
354. M. Hugon, B.Ph. van Milligen, P. Smeulders et al., Nucl. Fusion **32**, 33 (1992)
355. I.H. Hutchinson, A.H. Morton, Nucl. Fusion **16**, 447 (1976)
356. G. Huysmans, T.C. Hender, N.C. Hawkes, X. Litaudon, Phys. Rev. Lett. **87**, 245002 (2001)
357. T. Hwa, M. Kardar, Phys. Rev. A **45**, 7002 (1992)
358. IAEA Safety Standards Series RS-G-1.7, Application of the concepts of exclusion, exemption and clearance (2004), http://www-pub.iaea.org/MTCD/publications/PDF/Pub1202_web.pdf
359. IAEA (ed. by G. Mank), Fifty Years of Magnetic Confinement Fusion Research - A Retrospective, IAEA Booklet, Vienna, 2008
360. IAEA-TECDOC-855, Clearance Levels for radio nulcides in solid materials: application of exemption principles (1996),http://www-pub.iaea.org/MTCD/publications/PDF/te_855_web.pdf
361. K. Ida, Y. Miura, T. Matsuda, K. Itoh et al., Phys. Rev. Lett. **74**, 1990 (1995)
362. K. Ida, Y. Sakamoto, H. Takenaga, N. Oyama et al., Phys. Rev. Lett. **101**, 055003 (2008)
363. T. Ido, Y. Miura, K. Kamiya, Y. Hamada et al., Plasma Phys. Controlled Fusion **48**, S41 (2006)

364. T. Ido, Y. Miura et al., Nucl. Fusion **46**, 512 (2006)
365. Y. Idomura, M. Wakatani, S. Tokuda, Phys. Plasmas **7**, 3551 (2000)
366. Y. Idomura, S. Tokuda, Y. Kishimoto, Nucl. Fusion **43**, 234 (2003)
367. Y. Idomura, S. Tokuda, Y. Kishimoto, Nucl. Fusion **45**, 1571 (2005)
368. Y. Idomura, Phys. Plasmas **13**, 080701 (2006)
369. Y. Idomura, H. Urano, N. Aiba, S. Tokuda, Nucl. Fusion **49**, 065029 (2009)
370. N. Ikata (Ed.), *Fusion Reactor Materials (In Japanese)* (Baifukan, Tokyo, 1986)
371. K. Ikeda, Nucl. Fusion **50**, 014002 (2010)
372. E.L. Ince, *Ordinary Differential Equations* (Dover, New York, 1956)
373. T. Inoue, M. Araki et al., Nucl. Instrum. Methods Phys. Res. **B37/38**, 111 (1989)
374. A. Isayama, Y. Kamada, S. Ide et al., Plasma Phys. Controlled Fusion **42**, L37 (2000)
375. A. Isayama, N. Oyama, H. Urano, T. Suzuki et al., Nucl. Fusion **47**, 773 (2007)
376. S. Ishida, Y. Koide, T. Ozeki, M. Kikuchi, S. Tsuji et al., Phys. Rev. Lett. **68**, 1531 (1992)
377. S. Ishida, Y. Neyatani, Y. Kamada, A. Isayama et al., Fusion energy 1996, vol. 1, in *Sixteenth Conference Proceedings*, Montreal, 7–11 October 1996 (IAEA, Vienna, 1997), p. 315
378. S. Ishida, T. Fujita et al., Phys. Rev. Lett. **79**, 3917 (1997)
379. S. Ishida, P. Barabaschi, Y. Kamada et al., Nucl. Fusion **51**, 094018 (2011)
380. Y. Ishii, T. Ozeki, S. Tokuda, T. Fujita et al., Plasma Phys. Controlled Fusion **40**, 1607 (1998)
381. Y. Ishii, M. Azumi, G. Kurita, T. Tuda, Phys. Plasmas **7**, 4477 (2000)
382. Y. Ishii, M. Azumi, Y. Kishimoto, Phys. Rev. Lett. **89**, 205002 (2002)
383. Y. Ishii, M. Azumi, Y. Kishimoto, J.N. Leboeuf, Nucl. Fusion **43**, 539 (2003)
384. Y. Ishii, A.I. Smolyakov, M. Takechi, Nucl. Fusion **49**, 085006 (2009)
385. M. Ishikawa, M. Takechi, K. Shinohara, C.Z. Cheng et al., Nucl. Fusion **49**, 849 (2007)
386. S. Itoh, K. Nakamura, M. Sakamoto, K. Makino et al., Fusion energy 1996, vol. 3, in *Sixteenth Conference Proceedings*, Montreal, 7–11 October 1996 (IAEA, Vienna, 1997), p. 351
387. ITER Physics Expert Groups, Nucl. Fusion **39**, 2175 (1999) (Review Article)
388. S. Itoh, K.N. Sato, K. Nakamura, H. Zushi et al., Nucl. Fusion **39**, 1257 (1999)
389. J. Jacquinot, G.T. Hoang, Plasma Sci. Technol. **6**, 2101 (2004)
390. J. Jacquinot, Nucl. Fusion **50**, 014001 (2010)
391. R. Janev et al., *Elementary Processes in Hydrogen - Helium Plasmas, Cross Sections and Reaction Rate Coefficients* (Springer, Berlin, 1987)
392. R. Janev, C.D. Boley, D.E. Post, Nucl. Fusion **29**, 2125 (1989)
393. R.K. Janev, J.J. Smith, Atomic and plasma-material interaction data for fusion. Nucl. Fusion Suppl. **4** (1993)
394. S. Jardin, *Computational Methods in Plasma Physics* (CRC Press, Boca Raton, 2010)
395. S.C. Jardin, C.G. Bathke, D.A. Ehst et al., Fusion Eng. Des. **48**, 281 (2000)
396. H. Jeffreys, Proc. Lond. Math. Soc. 2 **23**, 428 (1923)
397. H. Jeffreys, *Asymptotic Approximations* (Oxford University Press, Oxford, 1962)
398. F. Jenko, Phys. Plasmas **7**, 514 (2000)
399. F. Jenko, W. Dorland, M. Kotschenreuther, B.N. Rogers, Phys. Plasmas **7**, 1904 (2000)
400. F. Jenko, W. Dorland, Phys. Rev. Lett. **89**, 225001 (2002)
401. Y.M. Jeon, J.K. Park, S.W. Yoon et al., Phys. Rev. Lett. **109**, 035004 (2012)
402. JET Team, Plasma physics and controlled nuclear fusion research 1988, vol. 1 (IAEA, Vienna, 1989), p. 41
403. E. Joffrin et al., in *Proceedings of 32nd EPS Conf. Control. Fusion Plasma Phys*, vol. 29C, P-5.073, 2005
404. E. Joffrin, O. Barana, D. Manzon et al., Nucl. Fusion **47**, 1664 (2007)
405. C.E. Johnson, R.G. Clemmer, G.W. Hollenberg, J. Nucl. Mater. **103–104**, 547 (1981)
406. R.H. Jones, L. Giancarli, A. Hasegawa et al., J. Nucl. Mater. **303–311**, 1057 (2002)
407. B.B. Kadomtsev, *Plasma Turbulence* (Academic, New York, 1965)
408. B.B. Kadomtsev, O.P. Pogutse, Nucl. Fusion **11**, 67 (1971)
409. B.B. Kadomtzev, *Collective Phenomena in Plasmas* (Nauka, Moscow, 1976)
410. S. Kajita, W. Sakaguchi, N. Ohno et al., Nucl. Fusion **49**, 095005 (2009)
411. Y. Kamada, T. Ozeki, M. Azumi, Phys. Fluids **B4**, 124 (1992)

412. Y. Kamada, T. Oikawa, L. Lao et al., Plasma Phys. Controlled Fusion **42A**, 247 (2000)
413. Y. Kamada, T. Fujita, S. Ishida, M. Kikuchi et al., Fusion Sci. Tech. **42**, 185 (2002)
414. Y. Kamada, P. Barabaschi, S. Ishida et al., Nucl. Fusion **53**, 104010 (2013)
415. K. Kamiya, K. Ida, Y. Sakamoto et al., Phys. Rev. Lett. **105**, 045004 (2010)
416. A. Kasugai et al., Rev. Sci. Instrum. **70**, 208 (1999)
417. Y. Katoh, A. Kohyama, T. Hinoki, L.L. Snead, Fusion Sci. Technol. **44**, 155 (2003)
418. Y. Katoh, L.L. Snead, C.H. Henager Jr. et al., J. Nucl. Mater. **367–370**, 659 (2007)
419. A.N. Kaufman, Phys. Fluids **3**, 610 (1960)
420. H. Kawamura, S. Ishitsuka, K. Tsuchiya et al., Nucl. Fusion **43**, 675 (2003)
421. H. Kawamura, H. Takahashi, N. Yoshida et al., J. Nucl. Mater. **329–333**, 112 (2004)
422. H. Kawashima, M. Sato, K. Tsuzuki et al., Nucl. Fusion **41**, 257 (2001)
423. M. Keilhacker, A. Gibson, C. Gormezano, P.J. Lomas et al., Nucl. Fusion **39**, 209 (1999)
424. M. Keilhacker, A. Gibson, C. Gormezano, P.H. Rebut, Nucl. Fusion **41**, 1925 (2001) (Topical Review)
425. D. Kelliher, N.C. Hawkes, P.J. McCarthy, Plasma Phys. Controlled Fusion **47**, 1459 (2005)
426. C. Kennel, F. Engelmann, Phys. Fluids **9**, 2377 (1966)
427. C. Kessel, J. Manckam, G. Rewoldt, W.M. Tang, Phys. Rev. Lett. **72**, 1212 (1994)
428. S. Kida, S. Yanase, *Dynamics of Turbulence (In Japanese)* (Asakura, Tokyo, 1999)
429. M. Kikuchi, Nucl. Fusion **26**, 223 (1986)
430. M. Kikuchi, Nucl. Fusion **30**, 265 (1990)
431. M. Kikuchi, M. Azumi, S. Tsuji, K. Tani, H. Kubo, Nucl. Fusion **30**, 343 (1990)
432. M. Kikuchi, R.W. Conn, F. Najmabadi, Y. Seki, Fusion Eng. Des. **16**, 253 (1991) (Review Article)
433. M. Kikuchi, Y. Seki, A. Oikawa, T. Ando, Y. Ohara et al., Fusion Eng. Des. **18**, 195 (1991)
434. M. Kikuchi, Physics of plasma rotation and its application to high Ti mode in JT-60, in *3rd IAEA TM on H-mode Physics*, JET, Culham, 10–12 June 1991
435. M. Kikuchi, K. Kikuchi, K. Tani, Kinetic database of the JT-60 Tokamak during 1985–1987 experiments, JAERI-M 91-057 (1991), http://jolissrch-inter.tokai-sc.jaea.go.jp/pdfdata/JAERI-M-91-057.pdf
436. M. Kikuchi et al., Plasma physics and controlled nuclear fusion research 1992, vol. 1 (IAEA, Vienna, 1993), p. 189
437. M. Kikuchi, Plasma Phys. Controlled Fusion **35**, 39 (1993) (Review Article)
438. M. Kikuchi, M. Sato, Y. Koide et al., in *Proceedings of 20th EPS Conf. on Plasma Physics and Controlled fusion*, Lisboa, vol I7C, part I, p179, 1993
439. M. Kikuchi et al., Plasma physics and controlled nuclear fusion research 1994, vol. 1 (IAEA, Vienna, 1995), p. 31
440. M. Kikuchi, M. Azumi, Plasma Phys. Controlled Fusion **37**, 1215 (1995) (Review Article)
441. M. Kikuchi, S. Nishio, G. Kurita et al., Fusion Eng. Des. **81**, 1285 (2006)
442. M. Kikuchi, Energies **3**, 1741 (2010) (Review Article)
443. M. Kikuchi, *Frontier in Fusion Research - Physics and Fusion* (Springer, Berlin, 2011)
444. M. Kikuchi, M. Azumi, Rev. Mod. Phys. **84**, 1807 (2012) (Review Article)
445. M. Kikuchi, K. Lackner, M. Tran (ed.), *Fusion Physics*, IAEA, Vienna, 2012, Free download available at: http://www-pub.iaea.org/books/IAEABooks/8879/Fusion-Physics
446. Y.B. Kim, P. Diamond, R.J. Groebner, Phys. Fluids **B3**, 2050 (1991)
447. J. Kim, K.H. Burrell, P. Gohil et al., Phys. Rev. Lett. **72**, 2199 (1994)
448. J.Y. Kim, Y. Kishimoto, M. Wakatani, T. Tajima, Phys. Plasmas **3**, 3689 (1996)
449. Y.B. Kim, Moment approach to neoclassical flows, currents and transport in auxiliary heated Tokamaks, Ph.D. thesis, Physics Department, University of Wisconsin (1988)
450. E.J. Kim, P.H. Diamond, Phys. Rev. Lett. **90**, 185006 (2003)
451. K. Kim, J.K. Park, A. Boozer, Phys. Rev. Lett. **110**, 185004 (2013)
452. H. Kimura, S. Moriyama, M. Saigusa, Y. Kusama, T. Ozeki et al., Fusion energy 1996, vol. 3, in *Sixteenth Conference Proceedings*, Montreal, 7–11 October 1996 (IAEA, Vienna, 1997), p. 295
453. H. Kimura, Y. Kusama, M. Saigusa, G.J. Kramer et al., Nucl. Fusion **38**, 1303 (1998)

454. A. Kimura, T. Sawai, K. Shiba et al., Nucl. Fusion **43**, 1246 (2003)
455. A. Kirk, B. Koch, R. Scannell, H. Wilson et al., Phys. Rev. Lett. **96**, 185001 (2006)
456. A. Kirk, J. Harrison, Y. Liu et al., Phys. Rev. Lett. **108**, 255003 (2012)
457. Y. Kishimoto, J.-Y. Kim et al., Plasma Phys. Controlled Fusion **40**, A664 (1998)
458. H. Kishimoto, S. Ishida, M. Kikuchi, H. Ninomiya, Nucl. Fusion **45**, 986 (2005) (Topical Review)
459. C. Kittel, *Introduction of Solid State Physics* (Wiley, London, 1971)
460. T. Kobayashi, A. Isayama, K. Yokokura et al., Nucl. Fusion **51**, 103037 (2011)
461. T. Kobayashi, K. Itoh, T. Ido et al., Phys. Rev. Lett. **111**, 035002 (2013)
462. A. Kohyama, A. Hishinuma, D.S. Gelles et al., J. Nucl. Mater. **233–237**, 138 (1996)
463. Y. Koide, M. Kikuchi, M. Mori, S. Tsuji et al., Phys. Rev. Lett. **72**, 3662 (1994)
464. Y. Koide, T. Takizuka, S. Takeji et al., Plasma Phys. Controlled Fusion **38**, 1011 (1996)
465. Y. Koide, M. Mori, T. Fujita et al., Plasma Phys. Controlled Fusion **40**, 641 (1998)
466. Y. Koide, K.H. Burrell et al., Plasma Phys. Controlled Fusion **40**, 97 (1998)
467. A. Kojima, M. Hanada, Y. Tanaka, M. Kawai et al., Nucl. Fusion **51**, 083049 (2011)
468. A.N. Kolmogolov, S.V. Fomin, *Elements of the Theory of Functions and Functional Analysis* (Dover, New York, 1999)
469. C. Konno, F. Maekawa, Y. Kasugai et al., Nucl. Fusion **41**, 333 (2001)
470. K. Kosako, Y. Oyama, H. Maekawa, FSXLIB-J3: MCNP continuous energy cross section library based on JENDL-3, JAERI-M 91-187 (1991), http://jolissrch-inter.tokai-sc.jaea.go.jp/pdfdata/JAERIM-91-187.pdf
471. M. Kotschenreuther, G. Rewoldt, W.M. Tang, Comp. Phys. Commun. **88**, 128 (1995)
472. R.H. Kraichnan, Phys. Fluids **10**, 1417 (1967)
473. N.A. Krall, A.W. Trivelpiece, *Principles of Plasma Physics* (McGraw-Hill, New York, 1973)
474. G. Kramer, M. Saigusa, T. Ozeki, Y. Kusama et al., Phys. Rev. Lett. **80**, 594 (1998)
475. G.J. Kramer, N.N. Gorelenkov et al., Plasma Phys. Controlled Fusion **46**, L23 (2004)
476. H.A. Kramers, Zeits. f. Physik. **39**, 828 (1926)
477. J.E. Kunzler, Rev. Mod. Phys. **33**, 501 (1961)
478. G. Kurita, T. Tuda, M. Azumi, T. Takizuka, T. Takeda, Nucl. Fusion **34**, 1497 (1994)
479. G. Kurita, J. Bialek, T. Tuda, M. Azumi, S. Ishida et al., Nucl. Fusion **46**, 383 (2006)
480. M. Kuriyama, N. Akino, T. Aoyagi et al., Fusion Eng. Des. **39–40**, 115 (1998)
481. Y. Kusama, H. Kimura, T. Ozeki, M. Saigusa et al., Nucl. Fusion **38**, 1215 (1998)
482. J.G. Kwak, Y.D. Bae, D.H. Chang et al., Nucl. Fusion **47**, 463 (2007)
483. J.G. Kwak, Y.K. Oh, H.L. Yang et al., Nucl. Fusion **53**, 104005 (2013)
484. M. Kwon, Y.S. Na, J.H. Han et al., Fusion Eng. Des. **83**, 883 (2008)
485. M. Kwon, Y.K. Oh, H.L. Yang, H.K. Na, S.Y. Kim et al., Nucl. Fusion **51**, 094006 (2011)
486. B. LaBombard, J.E. Rice, A.E. Hubbard et al., Nucl. Fusion **44**, 1047 (2004)
487. R. La Haye, Phys. Plasmas **13**, 055501 (2006)
488. R. LaHaye, R.J. Buttery, Phys. Plasmas **16**, 022107 (2009)
489. R. LaHaye, D.P. Brennan, R.J. Buttery, S.P. Gerhardt, Phys. Plasmas **17**, 056110 (2010)
490. J.R. Lamarsh, *Introduction to Nuclear Reactor Theory*, Chap. 6 (Addison-Wesley, Reading, 1965)
491. H. Lamb, *Hydrodynamics* (Dover, New York, 1979), p. 682
492. L.D. Landau, Phys. Z. Sov. **10**, 154 (1936)
493. L.D. Landau, J. Phys. USSR, **10**, 25 (1946)
494. L.D. Landau, E.M. Lifshitz, *Quantum Mechanics* (Elsevier, Amsterdam, 1977)
495. L.D. Landau, E.M. Lifshitz, *Fluid Mechanics* (Elsevier, Amsterdam, 1987)
496. Y.C. Lee, J.W. Van Dam, in *Proceedings of Finite Beta Theory Workshop*, Varenna, 1977, p. 93
497. G.S. Lee, J. Kim, S.M. Hwang, C.S. Chang et al., Nucl. Fusion **40**, 575 (2000)
498. G.S. Lee, M. Kwon, C.J. Doh, B.G. Hong et al., Nucl. Fusion **41**, 1515 (2001)
499. M. Lesur, Y. Idomura, X. Garbet, Phys. Plasmas **16**, 092305 (2009)
500. M. Lesur, Y. Idomura, K. Shinohara, X. Garbet et al., Phys. Plasmas **17**, 122311 (2010)
501. V. Levedev, P.H. Diamond, Phys. Plasmas **4**, 1087 (1997)

502. F. Levinton, M.C. Zarnstroff, S.H. Batha et al., Phys. Rev. Lett. **75**, 4417 (1995)
503. Z.X. Li, M.Q. Wie, B. Liu, Nucl. Fusion **48**, 125003 (2008)
504. J. Li, B. Wan, EAST Team, Nucl. Fusion **51**, 094007 (2011)
505. J. Li, H.Y. Guo, B.N. Wan, X.Z. Gong et al., Nat. Phys. **9**, 817 (2013)
506. Q. Li et al., The Construction Progress of HL-2M till Jan 2013, in *25th IEEE Symposium on Fusion Engineering (SOFE)*, San Francisco, CA, US (No. TPO-113), 10–14 June 2013
507. Y. Liang, H.R. Koslowski, P.R. Thomas et al., Phys. Rev. Lett. **98**, 265004 (2007)
508. A. Lichtenberg, M. Lieberman, *Regular and Chaotic Dynamics* (Springer, Berlin, 2010)
509. M.K. Lilley, B.N. Nreizman, S.E. Sharapov, Phys. Rev. Lett. **102**, 195003 (2009)
510. Z. Lin, T.S. Hahm, W.W. Lee, W.M. Tang, R.B. White, Science **281**, 1835 (1998)
511. J. Lingertat et al., J. Nucl. Mater. **266–269**, 124 (1999)
512. A.D. Liu, T. Lan, C.X. Yu, H.L. Zhao et al., Phys. Rev. Lett. **103**, 095002 (2009)
513. Y.R. Lin-Liu, F. Hinton, Phys. Plasmas **4**, 4179 (1997)
514. Y.R. Lin-Liu, A.D. Turnbull, M.S. Chu et al., Phys. Plasmas **6**, 3934 (1999)
515. Y. Lin-Liu, V.S. Chan, R. Prater, Phys. Plasmas **10**, 4064 (2003)
516. J. Liouville, J. Math. **2**, 16, 418 (1837)
517. X. Litaudon, R. Arslanbekov, G.T. Hoang et al., Plasma Phys. Controlled Fusion **38**, 1603 (1996)
518. R.G. Littlejohn, J. Math. Phys. **20**, 2445 (1979)
519. R.G. Littlejohn, Phys. Fluids **24**, 1730 (1981)
520. R.G. Littlejohn, J. Math. Phys. **23**, 742 (1982)
521. R.G. Littlejohn, Physica Scripta **T2/1**, 119 (1982)
522. R.G. Littlejohn, J. Plasma Phys. **29**, 111 (1983)
523. Y.Y. Liu, B. Arthur, J. Nuc. Mater. **133–134**, 216 (1985)
524. A. Loarte, G. Saibene, R. Satori et al., Plasma Phys. Controlled Fusion **45**, 1549 (2003)
525. F. London, H. London, Proc. Roy. Soc. Lond. **A149**, 71 (1935)
526. D. Lovelock, H. Rund, *Tensors, Differential Forms, and Variational Principles* (Dover, New York, 1989)
527. T. Luce, M.R. Wade, J.R. Ferron, A.W. Hyatt et al., Nucl. Fusion **43**, 321 (2003)
528. S.M. Mahajan, D. Ross, G.L. Chen, Phys. Fluids **26**, 2195 (1983)
529. S.M. Mahajan, Phys. Fluids **27**, 2238 (1984)
530. S.M. Mahajan, Physica Scripta **T60**, 160 (1995) (Review Article)
531. D. Maisonnier, I. Cook, S. Pierre et al., Fusion Eng. Des. **81**, 1123 (2006)
532. K. Maki, H. Kawasaki, K. Kosako, Y. Seki, Nuclear heating constant KERMA library for fusion nuclear group constant set fusion-J3, JAERI-M 91-073 (1991) http://jolissrch-inter.tokai-sc.jaea.go.jp/pdfdata/JAERIM-91-073.pdf
533. K. Maki, K. Kosako, Y. Seki, H. Kawasaki, Nuclear group constant set fusion-J3 for fusion reactor nuclear calculations based on JENDL-3, JAERI-M 91-072 (1991) http://jolissrch-inter.tokai-sc.jaea.go.jp/pdfdata/JAERIM-91-072.pdf
534. K. Maki, S. Satoh, H. Kawasaki, *Development of Displacement Cross Section Set for Evaluating Radiation Damage by Neutron Irradiation in Materials used for Fusion Reactors*, JAERI-Data-Code-97-002 (1997), http://jolissrch-inter.tokai-sc.jaea.go.jp/pdfdata/JAERI-Data-Code-97-002.pdf
535. J. Manickam, N. Pomphrey, A.M.M. Todd, Nucl. Fusion **27**, 1461 (1987)
536. J. Manickam, Phys. Fluids **B4**, 1901 (1992)
537. J. Manickam, M.S. Chance, S.C. Jardin, C. Kessel et al., Phys. Plasmas **1**, 1601 (1994)
538. J. Manickam, T. Fujita, N. Gorelenkov, A. Isayama et al., Nucl. Fusion **39**, 1819 (1999)
539. P. Mantica, D. Strintzi, T. Tala, C. Giroud et al., Phys. Rev. Lett. **102**, 175002 (2009)
540. P. Mantica, C. Angioni, C. Challis et al., Phys. Rev. Lett. **107**, 135004 (2011)
541. J. Marshall, R.A. Plumb, *Atmosphere, Ocean, and Climaite Dynamics* (Elsevier, Amsterdam, 2008)
542. M. Maslov, C. Angioni, H. Weisen, JET-EFDA Team, Nucl. Fusion **49**, 075037 (2009)
543. T. Matsumoto, S. Tokuda, J. Plasma Fusion Res. Ser. **9**, 568 (2010)
544. G. Matsunaga, N. Aiba, K. Shinohara et al., Phys. Rev. Lett. **103**, 045001 (2009)

545. K. McGuire et al., Phys. Rev. Lett. **50**, 891 (1983)
546. K. McGuire et al., Plasma Phys. Controlled Fusion **30**, 1391 (1988)
547. MCNP-A general Monte Carlo N-Particle Transport Code, Version 5, LA-UR-03-1987, Los Alamos National Laboratory, 2003
548. D. Meade, Nucl. Fusion **50**, 014004 (2010)
549. W. Meissner, R. Ochsenfeld, Naturwissenschaften **21**, 787 (1933)
550. R. Mett, S.M. Mahajan, Phys. Fluids **B4**, 2885 (1992)
551. K. Miki, Y. Kishimoto, N. Miyato, J.Q. Dong, Phys. Rev. Lett. **99**, 145003 (2007)
552. K. Miyamoto *Plasma Physics for Nuclear Fusion* (MIT Press, Cambridge, 1980)
553. K. Miyamoto, *Plasma Physics and Controlled Nuclear Fusion* (Springer, Berlin, 2005)
554. N. Miyato, Y. Kishimoto, Phys. Plasmas **11**, 5557 (2004)
555. N. Miyato, J.Q. Li, Y. Kishimoto, Nucl. Fusion **45**, 425 (2005)
556. N. Miyato, Y. Kishimoto, J.Q. Li, Nucl. Fusion **47**, 929 (2007)
557. H.K. Moffatt, The energy spectrum of knots and links. Nature **347**, 367 (1990)
558. W.J. Moore, *Physical Chemistry* (Prentice-Hall, Englewood Cliffs, 1962)
559. M. Mori et al., Plasma physics and controlled nuclear fusion research 1992, vol. 2 (IAEA, Vienna, 1993), p. 567
560. S. Mori, S. Zimin, H. Takatsu, Multi-group helium and hydrogen production cross section libraries for fusion neutronics design, JAERI-M 93-175 (1993), http://jolissrch-inter.tokai-sc.jaea.go.jp/pdfdata/JAERIM-93-175.pdf
561. A. Morioka et al., J. Nucl. Mater. **367–370**, 1085 (2007)
562. A.I. Morozov, L.S. Solovev, Rev. Plasma Phys. **2**, 2 (1966)
563. N.F. Mott, H.S.W. Massey, *The Theory of Atomic Collision* (Clarendon Press, London, 1961)
564. R.A. Moyer, G.R. Tynan, C. Holland, M.J. Burin et al., Phys. Rev. Lett. **87**, 135001 (2001)
565. M. Murakami, G.H. Neilson, H.C. Howe et al., Phys. Rev. Lett. **42**, 655 (1979)
566. T. Mutoh, T. Seki, R. Kumazawa, K. Saito et al., Nucl. Fusion **53**, 063017 (2013)
567. Y. Nabara, Y. Nunoya, T. Isono et al., Teion Kogaku **47**, 140 (2012)
568. K. Nagashima, Y. Koide, H. Shirai, Nucl. Fusion **34**, 449 (1994)
569. K. Nagashima, K. Hoshino, H. Ogawa, K. Hamamatsu, Nucl. Fusion **35**, 994 (1995)
570. Y. Nagashima, K. Hoshino, A. Ejiri et al., Phys. Rev. Lett. **95**, 095002 (2005)
571. F. Najmabadi, The ARIES Team, Fusion Eng. Des. **41**, 365 (1998)
572. M. Nakamichi, K. Yonehara, J. Nucl. Mater. **417**, 765 (2011)
573. M. Nakamichi, J.H. Kim, Fusion Eng. Des. **88**, 611 (2013)
574. M. Nakamichi, J.H. Kim, Fusion Eng. Des. **89**, 7 (2014)
575. T. Nakano, N. Asakura, H. Takenaga et al. et al., Nucl. Fusion **46**, 626 (2006)
576. M. Nakazawa, Kakuyugokenkyu **56**, 196 (1986)
577. R. Nazikian, G.Y. Fu, S.H. Batha, M.G. Bell et al., Phys. Rev. Lett. **78**, 2976 (1997)
578. R. Nazikian, G.J. Kramer, C.Z. Cheng et al., Phys. Rev. Lett. **91**, 125003 (2003)
579. R. Nazikian, K. Shinohara, G.J. Kramer, E. Valeo et al., Phys. Rev. Lett. **94**, 135002 (2005)
580. W. Newcomb, Ann. Phys. **10**, 232 (1960)
581. W.A. Newcomb, Nucl. Fusion Supplement **Part 2**, 451(1962).
582. G. Nicolis, I. Prigogine, *Self-Organization in Nonequilibrium Systems - From Dissipative Structures to Order through Fluctuations* (Wiley, London, 1977)
583. K. Oasa et al., Plasma physics and controlled nuclear fusion research 1994, vol. 2 (IAEA, Vienna, 1995), p. 279
584. T. Ohkawa, Steady-state operation of tokamaks by r-f heating, GA-A 13847 (1976)
585. T. Oikawa, Y. Kamada, A. Isayama, T. Fujita et al., Nucl. Fusion **40**, 435 (2000)
586. T. Oikawa, Y. Kamada, A. Isayama, T. Fujita, T. Suzuki et al., Nucl. Fusion **41**, 1575 (2001)
587. T. Oikawa, A. Isayama, T. Fujita, T. Suzuki et al., Phys. Rev. Lett. **95**, 075001 (2005)
588. T. Oikawa, T. Suzuki, A. Isayama, N. Hayashi et al., Nucl. Fusion **45**, 1101 (2005)
589. K. Okano, Y. Asaoka, R. Hiwatari et al., Fusion Eng. Des. **41**, 511 (1998)
590. H.K. Onnes, Leiden Comm. 120b, 122b, 124c (1911)
591. S.A. Orszag, L.D. Kells, J. Fluid Mech. **96**, 159 (1980)
592. T.H. Osbone et al., Plasma Phys. Controlled Fusion **40**, 845 (1998)

593. N. Oyama, Y. Kamada, A. Isayama et al., Plasma Phys. Controlled Fusion **49**, 249 (2007)
594. N. Oyama, JT-60Team, Nucl. Fusion **49**, 104007 (2009)
595. N. Oyama, A. Kojima, N. Aiba et al., Nucl. Fusion **50**, 064014 (2010)
596. T. Ozeki et al., Plasma physics and controlled nuclear fusion research 1992, vol. 2 (IAEA, Vienna, 1993), p. 187
597. T. Ozeki, M. Azumi, S. Tokuda, S. Ishida, Nucl. Fusion **33**, 1025 (1993)
598. T. Ozeki, M. Azumi, Y. Kamada, S. Ishida et al., Nucl. Fusion **35**, 861 (1995)
599. T. Ozeki, JT-60 Team, Plasma Phys. Controlled Fusion **45**, 645 (2003)
600. T. Ozeki et al., in *Proceedings of 30th EPS Conf. Control. Fusion Plasma Phys.*, vol. 27A, P2-111, 2003
601. L.W. Packer, M. Gilbert et al., Fusion Eng. Des. **87**, 662 (2012)
602. A. Pankin et al., Comp. Phys. Comm. **159**, 157 (2004)
603. G.K. Park, *Physics of Space Plasmas* (Westview, Boulder, 2004)
604. H.K. Park, N.C. Luhmann, A.J.H. Donne et al., Phys. Rev. Lett. **96**, 195003 (2006)
605. H.K. Park, A.J.H. Donne, N.C. Luhmann et al., Phys. Rev. Lett. **96**, 195004 (2006)
606. J.K. Park, Y.M. Jeon, J.E. Menard et al., Phys. Rev. Lett. **111**, 095002 (2013)
607. S.E. Parker, W.W. Lee, R.A. Santoro, Phys. Rev. Lett. **71**, 2042 (1993)
608. A.G. Peeters, C. Angioni, D. Strintzi, Phys. Rev. Lett. **98**, 265003 (2007)
609. G. Pereverzev et al., IPP 5/42 (1991)
610. C.C. Petty, R. Prater, J. Lohr, T.C. Luce et al., Nucl. Fusion **42**, 1366 (2002)
611. C. Petty, R. Prater, T.C. Luce, R.A. Ellis et al., Nucl. Fusion **43**, 700 (2003)
612. A. Pletzer et al., J. Plasma Phys. **45**, 427 (1991)
613. H.J. Poincare, *Les mettes nouvelles de la de mécanique cleste (New Methods of Celestial Mechanics)* (Gauthier-Vilars, Paris, 1892)
614. A.R. Polevoi, S.Y. Medvedev, T. Casper, V. Gribov et al., 37th EPS **2**, 187 (2010)
615. R. Prater, Phys. Plasmas **11**, 2349 (2004)
616. P.L. Pritchett, Y.C. Lee, J.F. Drake, Phys. Fluids **23**, 1368 (1980)
617. M. Pueschel, M. Kammerer, F. Jenko, Phys. Plasmas **15**, 102310 (2008)
618. M. Pueschel, F. Jenko, Phys. Plasmas **17**, 062307 (2010)
619. M.J. Pueschel, P.W. Terry, F. Jenko et al., Phys. Rev. Lett. **111**, 265001 (2013)
620. J. Raeder, Fusion Eng. Des. **29**, 121 (1995)
621. J. Ramos, Phys. Fluids **B3**, 2247 (1991)
622. L. Rayleigh, Proc. Roy. Soc. **A86**, 207 (1912)
623. H. Reimerdes, A.M. Garofalo, G.L. Jackson et al., Phys. Rev. Lett. **98**, 055001 (2007)
624. C. Ren, M.S. Chu, J.D. Callen, Phys. Plasmas **6**, 1203 (1999)
625. P. Rhines, J. Fluid Mech. **69**, 417 (1975)
626. W.A. Rhoades, F.R. Mynatt, The DOT-III Two dimensional discrete ordinates transport code, ORNL-TN-4280 (1973)
627. J. Rice, E.S. Marmar, F. Bombarda, L. Qu, Nucl. Fusion **37**, 421 (1997)
628. J.E. Rice, A. Ince-Cushman, J.S deGlassie et al., Nucl. Fusion **47**, 1618 (2007)
629. J. Rice, I. Cziegler, P.H. Diamond, B.P. Duval et al., Phys. Rev. Lett. **106**, 215001 (2011)
630. A.B. Richester, M.N. Rosenbluth, Phys. Rev. Lett. **40**, 38 (1978)
631. A.C. Riviere, Nucl. Fusion **11**, 363 (1971)
632. B.B. Robinson, I.B. Bernstein, Ann. Phys. **18**, 110 (1962)
633. D. Robinson, K. McGuire, Nucl. Fusion **19**, 115 (1979)
634. P. Rodrigues, J.P.S. Bizarro, Phys. Rev. Lett. **99**, 125001 (2007)
635. F. Romanelli, Phys. Fluids **B1**, 1018 (1989)
636. F. Romanelli, F. Zonca, Phys. Fluids **B5**, 4081 (1993)
637. D.J. Rose, M. Clarke Jr., *Plasmas and Controlled Fusion* (MIT Press, Cambridge, 1961)
638. M.N. Rosenbluth, W.M. MacDonald, D.L. Judd, Phys. Rev. **107**, 1 (1957)
639. M.N. Rosenbluth, H.L. Berk, J.W. Van Dam, D.M. Lindberg, Phys. Fluids **B4**, 2189 (1992)
640. M. Rosenbluth, F. L. Hinton, Phys. Rev. Lett. **59**, 1581 (1987)
641. D.W. Ross, G.L. Chen, S. Mahajan, Phys. Fluids **25**, 662 (1982)
642. P.H. Rutherford, E.A. Frieman, Phys. Fluids **11**, 569 (1968)

643. P. Rutherford, Phys. Fluids **16**, 1903 (1973)
644. F. Ryter, C. Angioni, A.G. Peeters, F. Leuterer et al., Phys. Rev. Lett. **95**, 085001 (2005)
645. S. Sabbagh, R.E. Bell, J.E. Menard, D.A. Gates et al., Phys. Rev. Lett. **97**, 045004 (2006)
646. R. Sagdeev, A. Galeev, Sov. Phys. -Docl. **14**, 1198 (1970)
647. F. Saint-Laurent, Equipe Tore Supra, Nucl. Fusion **40**, 1047 (2000)
648. K. Sakamoto, M. Tsuneoka, A. Kasugai et al., Phys. Rev. Lett. **73**, 3532 (1994)
649. Y. Sakamoto, Y. Kamada, S. Ide, T. Fujita, H. Shirai et al., Nucl. Fusion **41**, 865 (2001)
650. M. Sakamoto, S. Itoh, K. Nakamura et al., Nucl. Fusion **42**, 165 (2002)
651. Y. Sakamoto, T. Suzuki, S. Ide, T. Koide et al., Nucl. Fusion **44**, 876 (2004)
652. Y. Sakamoto, T. Fujita, S. Ide, A. Isayama et al., Nucl. Fusion **45**, 574 (2005)
653. K. Sakamoto, A. Kasugai, K. Takahashi, R. Minami et al., Nat. Phys. **3**, 411 (2007)
654. Y. Sakamoto, G. Matsunaga, N. Oyama, T. Suzuki et al., Nucl. Fusion **49**, 095017 (2009)
655. K. Sakamoto, A. Kasugai, K. Kajiwara et al., Nucl. Fusion **49**, 095019 (2009)
656. K. Sako, M. Ohta, Y. Seki, H. Yamato, T. Hiraoka, K. Tanaka, N. Asami, S. Mori, Conceptual Design of a gas cooled tokamak reactor, JAERI-M 5502 (1973), http://jolissrch-inter.tokai-sc.jaea.go.jp/pdfdata/JAERIM-73-5502.pdf
657. B. Saoutic et al., Nucl. Fusion **51**, 094014 (2011)
658. M. Sasaki, K. Itoh, A. Ejiri, Y. Takase, Contrib. Plasma Phys. **48**, 1 (2008)
659. S. Satake, J.K. Park, H. Sugama, R. Kanno, Phys. Rev. Lett. **107**, 055001 (2011)
660. S. Sato, M. Nakao, Y. Verzilov et al., Nucl. Fusion **45**, 656 (2005)
661. S. Sato, K. Ochiai, Y. Verzilov et al., Nucl. Fusion **47**, 517 (2007)
662. O. Sauter, R.J. La Haye, Z. Chang, D.A. Gates et al., Phys. Plasmas **4**, 1654 (1997)
663. O. Sauter, C. Angioni, Y.R. Lin-Liu, Phys. Plasmas **6**, 2834 (1999)
664. O. Sauter, M.A. Henderson, F. Hofmann et al., Phys. Rev. Lett. **84**, 3322 (2000)
665. O. Sauter, R.J. Buttery, R. Felton et al., Plasma Phys. Controlled Fusion **44**, 1999 (2002)
666. N. Sauthoff, S. Von Goeler, W. Stodiek, Nucl. Fusion **18**, 1445 (1978)
667. Y.C. Saxena, SST-1 Team, Nucl. Fusion **40**, 1069 (2000)
668. A. Scarabosio et al., Plasma Phys. Controlled Fusion **48**, 663 (2006)
669. L.I. Schiff, *Quantum Mechanics*, 3rd edn. (McGraw-Hill, New York, 1968)
670. L. Schmitz, L. Zeng, T.L. Rhodes et al., Phys. Rev. Lett. **108**, 155002 (2012)
671. B. Scott, Phys. Lett. A **320**, 53 (2003)
672. B. Scott, J. Smirnov, Phys. Plasmas **17**, 112302 (2010)
673. Z. Sedlacek, J. Plasma Phys. **5**, 230 (1971)
674. Y. Seki, H. Iida, H. Kawasaki, K. Yamada, THIDA-2: an advanced code system for calculation of transmutation, activation, decay heat and dose rate, JAERI-1301 (1986), http://jolissrch-inter.tokai-sc.jaea.go.jp/pdfdata/JAERI-1301.pdf
675. Y. Seki, M. Kikuchi et al., Plasma physics and controlled nuclear fusion research 1990, vol. 3 (IAEA, Vienna, 1991), p. 473
676. J. Seol, S.G. Lee, B.H. Park et al., Phys. Rev. Lett. **109**, 195003 (2012)
677. V. Shafranov, Rev. Plasma Phys. **2**, 103 (1966)
678. K.C. Shaing, J.D. Callen, Phys. Fluids **26**, 3315 (1983)
679. K. Shaing, M. Yokoyama, M. Wakatani, C.T. Hsu, Phys. Plasmas **3**, 965 (1996)
680. K. Shaing, A.Y. Aidemir, Y.R. Lin-Liu, R.L. Miller, Phys. Rev. Lett. **79**, 3652 (1997)
681. K. Shaing, Phys. Plasmas **10**, 1443 (2003)
682. K.C. Shaing, P. Cahyna, M. Becoulet, J.-K. Park et al., Phys. Plasmas **15**, 082506 (2008)
683. K.C. Shaing, M.S. Chu, S.A. Sabbagh, Nucl. Fusion **50**, 125012 (2010)
684. S. Sharapov, D. Testa, B. Alper, D.N. Borba, A. Fasoli et al., Phys. Lett. A **289**, 127 (2001)
685. Y. Shi, G.S. Xu, F. Wang, M. Wang et al., Phys. Rev. Lett. **106**, 235001 (2011)
686. K. Shibata et al., Japanese Evaluated Nuclear Data Library, Version-3 -JENDL-3, JAERI-1319 (1990), http://jolissrch-inter.tokai-sc.jaea.go.jp/pdfdata/JAERI-1319.pdf.
687. K. Shinohara, Y. Kusama, M. Takechi, A. Morioka et al., Nucl. Fusion **41**, 603 (2001)
688. K. Shinohara, M. Takechi, M. Ishikawa et al., Nucl. Fusion **42**, 942 (2002)
689. K. Shinohara, M. Takechi, M. Ishikawa et al., Plasma Phys. Controlled Fusion **46**, S31 (2004)
690. K. Shinohara, S. Sakurai, M. Ishikawa et al., Nucl. Fusion **47**, 997 (2007)

691. H. Shirai, M. Kikuchi, T. Takizuka, T. Fujita et al., Nucl. Fusion **39**, 1713 (1999)
692. H. Shirai, M. Kikuchi, T. Takizuka et al., Plasma Phys. Controlled Fusion **42**, A109 (2000)
693. J. Shiraishi, N. Miyato, G. Matsunaga, Plasma Fusion Res. **9**, 3403027 (2014)
694. W.M. Shu, E. Wakai, T. Yamanishi, Nucl. Fusion **47**, 201 (2007)
695. W.M. Shu, Appl. Phys. Lett. **92**, 211904 (2008)
696. A. Sips et al., Plasma Phys. Controlled Fusion **44**, B69 (2002)
697. D.V. Sivukhin, Rev. Plasma Phys. **4**, 93 (1966)
698. V.P. Smirnov, Nucl. Fusion **50**, 014003 (2010)
699. A.I. Smolyakov, A. Hirose, E. Lazzaro, G.B. Re, J.D. Callen, Phys. Plasmas **2**, 1581 (1995)
700. A. Smolyakov, E. Lazzaro et al., Plasma Phys. Controlled Fusion **43**, 1661 (2001)
701. P.B. Snyder, Gyrofluid theory and simulation of electromagnetic turbulence and transport in tokamak plasmas, Ph.D. thesis, Princeton University, 1999
702. P. Snyder, G.W. Hammett, Phys. Plasmas **8**, 744 (2001)
703. P.B. Snyder, H.R. Wilson, J.R. Ferron et al., Nucl. Fusion **44**, 320 (2004)
704. R. Snyder, K.H. Burrel, H.R. Wilson, M.S. Chu et al., Nucl. Fusion **47**, 961 (2007)
705. P.B. Snyder, R.J. Groebner et al., Phys. Plasmas **16**, 056118 (2009)
706. P.B. Snyder, N. Aiba, M. Beurskens et al., Nucl. Fusion **49**, 085035 (2009)
707. P.B. Snyder, R.J. Groebner et al., Nucl. Fusion **51**, 103016 (2011)
708. Y. Sokolov, Fusion Eng. Des. **25**, 159 (1994)
709. W. Solomon, K.H. Burrell, J.S deGlassie et al., Plasma Phys. Controlled Fusion **49**, B313 (2007)
710. P. C. Souers, *Hydrogen Properties for Fusion Energy* (University of California Press, Berkeley, 1986)
711. G.O. Spies, D.B. Nelson, Phys. Fluids **17**, 1879 (1974)
712. L. Spitzer, *Physics of Fully Ionized Gases* (Interscience Publishers, New York, 1962)
713. G.M. Staebler, F.L. Hinton, J.C. Wiley et al., Phys. Plasmas **1**, 909 (1994)
714. R. Stambaugh et al., Plasma physics and controlled nuclear fusion research 1984, vol. 1 (IAEA, Vienna, 1985), p. 217
715. R. Stambaugh et al., Plasma physics and controlled nuclear fusion research 1994, vol. 1 (IAEA, Vienna, 1995), p. 83
716. D.F.H. Start, J.G. Cordey, Phys. Fluids **23**, 1477 (1980)
717. T. Stix, *The Theory of Plasma Waves* (McGraw Hill, New York, 1962)
718. T.H. Stix, Phys. Rev. Lett. **36**, 521 (1976)
719. J. Strachan et al., Phys. Rev. Lett. **58**, 1004 (1987)
720. E. Strait, L.L. Lao, M.E. Mauel, B.W. Rice et al., Phys. Rev. Lett. **75**, 4421 (1995)
721. T.E. Stringer, in *Plasma Physics - An Introductory Course*, ed. by R. Dendy, Chap. 14 (Cambridge University Press, Cambridge, 1993)
722. H. Sugama, T.H. Watanabe, M. Nunami, Phys. Plasmas **16**, 112503 (2009)
723. A. Sukegawa, Y. Anayama, S. Ohnishi, S. Sakurai et al., J. Nucl. Sci. Technol. **48**, 585 (2011)
724. Y. Sun, Y. Liang, K.C. Shaing et al., Phys. Rev. Lett. **105**, 145002 (2010)
725. Y. Sun, Y. Liang, K.C. Shaing et al., Nucl. Fusion **52**, 083007 (2012)
726. Y. Sun, K.C. Shaing, Y. Liang, B. Shen et al., Nucl. Fusion **53**, 073026 (2013)
727. W. Suttrop, T. Eich, J. Fuchs, S. Guenter et al., Phys. Rev. Lett. **106**, 225004 (2011)
728. S. Suzuki, T. Shirai, M. Nemoto et al., Plasma Phys. Controlled Fusion **40**, 2097 (1998)
729. T. Suzuki, S. Ide, K. Hamamatsu, A. Isayama et al., Nucl. Fusion **44**, 699 (2004)
730. T. Suzuki, S. Ide, K. Hamamatsu, C.C. Petty et al., J. Plasma Fusion Res. **80**, 511 (2004)
731. T. Suzuki, R.J. Akers, D.A. Gates, S. Guenter et al., Nucl. Fusion **51**, 083020 (2011)
732. M. Taguchi, J. Phys. Soc. Jpn. **52**, 2035 (1983)
733. M. Taguchi, Plasma Phys. Controlled Fusion **31**, 241 (1989)
734. Y. Takahashi, T. Isono, K. Hamada, Y. Nunoya et al., Nucl. Fusion **51**, 113015 (2011)
735. Y. Takase et al., in *Proceedings of 21st IAEA Fusion Energy Conference*, EX/1-4, Chengdu, 2006
736. M. Takechi, A. Fukuyama, M. Ishikawa et al., Phys. Plasmas **12**, 0825009 (2005)
737. M. Takechi, G. Matsunaga, N. Aiba, T. Fujita et al., Phys. Rev. Lett. **98**, 055001 (2007)

738. T. Takeda, S. Tokuda, J. Comp. Phys. **93**, 1 (1991) (Review Article)
739. S. Takeji, Y. Kamada, T. Ozeki, S. Ishida et al., Phys. Plasmas **4**, 4283 (1997)
740. S. Takeji, S. Tokuda, T. Fujita, T. Suzuki, A. Isayama et al., Nucl. Fusion **42**, 5 (2002)
741. H. Takenaga, T. Nakano, N. Asakura et al. et al., Nucl. Fusion **46**, S39 (2006)
742. T. Takizuka, J. Plasma Fusion Res. **78**, 1282 (2002)
743. T. Takizuka, Y. Sakamoto, T. Fukuda et al., Plasma Phys. Controlled Fusion **44**, A423 (2002)
744. T. Takizuka, K. Shimizu, N. Hayashi et al., Nucl. Fusion **49**, 075038 (2009)
745. T. Takizuka, K. Hoshino, K. Shimizu, M. Yagi, Contrib. Plasma Phys. **50**, 267 (2010)
746. M. Tamura, H. Hayakawa, A. Yoshitake et al., J. Nucl. Mater. **155–157**, 620 (1988)
747. T. Tanabe, J. Nucl. Mater. **438**, S19 (2013)
748. S. Tanaka (ed.), *Fusion Technology of Blanket, Material and Safety (In Japanese)*, Special Committee of Atomic Energy Society of Japan (2004)
749. W. Tang, J.W. Connor, R.J. Hastie, Nucl. Fusion **20**, 1439 (1980)
750. W. Tang, R.L. Dewar, J. Manickam, Nucl. Fusion **22**, 1079 (1982)
751. I. Tani, *Nagare Gaku (Fluid Mechanics) in Japanese* (Iwanami, Tokyo, 1967)
752. K. Tani, M. Azumi, R.S. Devoto, J. Comp. Physics **98**, 332 (1992)
753. H. Tanigawa, T. Hoshino, Y. Kawamura et al., Nucl. Fusion **49**, 055021 (2009)
754. J.B. Taylor, Phys. Fluids **6**, 1529 (1963)
755. J.B. Taylor, R.J. Hastie, Plasma Phys. **10**, 479 (1968)
756. J.B. Taylor, Phys. Rev. Lett. **33**, 1139 (1974)
757. J.B. Taylor, J. Connor, H.R. Wilson, Plasma Phys. Controlled Fusion **35**, 1063 (1993)
758. J.B. Taylor, H.R. Wilson, J.W. Connor, Plasma Phys. Controlled Fusion **38**, 243 (1996)
759. T. Taylor, Plasma Phys. Controlled Fusion **39**, B47 (1997)
760. E. Teller, Peaceful use of fusion, in *Proceedings of 2nd UN International Conference on Peaceful Uses of Atomic Energy*, vol. 31, Paper P/2410, 1958
761. M. Tenenbaum, H. Pollard, *Ordinary Differential Equations* (Dover, New York, 1985)
762. D. Tetreault, Geophys. Res. Lett. **15**, 164 (1988)
763. P.W. Terry, P.H. Diamond, Phys. Fluids **28**, 11419 (1985)
764. F. Terry, Rev. Mod. Phys. **72**, 109 (2000) (Review Article)
765. M.W. Thompson, *Defects and Radiation Damage in Metals* (Cambridge University Press, Cambridge, 1969)
766. M. Tinkham, *Introduction to Superconductivity* (Dover, New York, 2004)
767. H. Tobari, M. Hanada, M. Kashiwagi et al., Rev. Sci. Instrum. **79**, 02C111 (2008)
768. K. Tobita, S. Nishio, S. Konishi, S. Jitsukawa, J. Nucl. Mater. **329–333**, 683 (2004)
769. S. Tokuda et al., J. Plasma Fusion Res. **73**, 1141 (1997)
770. S. Tokuda, T. Watanabe, Phys. Plasmas **6**, 3012 (1999)
771. R.A. Treumann, W. Baumjohann, *Advanced Space Plasma Physics* (Imperial College, London, 1997)
772. F. Troyon, R. Gruber, H. Saurenmann et al., Plasma Phys. Controlled Fusion **26**, 209 (1984)
773. B.A. Trubnikov, Rev. Plasma Phys. **1**, 105 (1965)
774. K. Tsang, J. Callen, Phys. Fluids **19**, 667 (1976)
775. K. Tsuchiya, H. Kawamura, K. Fuchinoue et al., J. Nucl. Mater. **258–263**, 1985 (1998)
776. K. Tsuchiya, H. Kawamura, J. Nucl. Mater. **283–287**, 1380 (2000)
777. K. Tsuzuki, K. Kamiya, K. Shinohara et al., Nucl. Fusion **46**, 966 (2006)
778. A.D. Turnbull, E.J. Strait, W.W. Heidbrinck et al., Phys. Fluids **B5**, 2546 (1993)
779. A. Turnbull, T.S. Taylor, M.S. Chu et al., Nucl. Fusion **38**, 1467 (1998)
780. C. Uberoi, Phys. Fluids **15**, 1673 (1972)
781. H. Urano, T. Takizuka, Y. Kamada et al., Phys. Rev. Lett. **95**, 035003 (2005)
782. H. Urano, T. Takizuka, H. Takenaga, N. Oyama et al., Nucl. Fusion **46**, 781 (2006)
783. H. Urano, T. Takizuka, Y. Kamada, N. Oyama et al., Nucl. Fusion **48**, 045008 (2008)
784. H. Urano, T. Takizuka, M. Kikuchi et al., Phys. Rev. Lett. **109**, 125001 (2012)
785. B. van der Holst, A.J.C. Belien, J.P. Goedbloed et al., Phys. Plasmas **6**, 1554 (1999)
786. D.van Houtte, Equipe Tore Supra, Nucl. Fusion **33**, 137 (1993)
787. D. van Houtte, G. Martin, A. Becoule et al., Nucl. Fusion **44**, L11 (2004)

788. L. Vermale, F. Ryter, C. Angioni, A.G. Peeters et al., Nucl. Fusion **47**, 490 (2007)
789. S. von Goeler, W. Stodiek, N. Sauthoff, Phys. Rev. Lett. **33**, 1201 (1974)
790. O. Vollmer, H. Falter et al., Rev. Sci. Instru. **71**, 939 (2000)
791. M. Wade, M. Murakami, P.A. Politzer, Phys. Rev. Lett. **92**, 235005 (2004)
792. J.M. Wadehra, J.N. Bardsley, Phys. Rev. Lett. **41**, 1795 (1978)
793. F. Waelbloeck, J.W. Connor, H.R. Wilson, Phys. Rev. Lett. **87**, 215003 (2001)
794. F. Wagner, G. Becker et al., Phys. Rev. Lett. **49**, 1408 (1982)
795. F. Wagner, G. Fussmann, T. Grave et al., Phys. Rev. Lett. **53**, 1453 (1984)
796. R.E. Waltz, Phys. Rev. Lett. **55**, 1098 (1985)
797. R. Waltz, G.D. Kerbel, J. Milovich, Phys. Plasmas **1**, 2229 (1994)
798. R. Waltz, G.D. Kerbel, J. Milovich, G.W. Hammett, Phys. Plasmas **2**, 2408 (1995)
799. R.E. Waltz, Phys. Plasmas **17**, 072501 (2010)
800. Y. Wan, HT-7 Team, HT-7U Team, Nucl. Fusion **40**, 1057 (2000)
801. Y.X. Wan, Mission of CFETR, in *ITER Training Forum and Second Workshop on MFE Development Strategy*, Hefei, 30 May 2012
802. B.N. Wan, J. Li, H. Guo, Y. Liang et al., Nucl. Fusion **53**, 104006 (2013)
803. J. Wang, M. Azumi, K. Tani, Nucl. Fusion **34**, 231 (1994)
804. S. Wang, Phys. Rev. Lett. **93**, 155007 (2004)
805. W. Wang, P.D. Diamond, T.S. Hahm, S. Ethier et al., Phys. Plasmas **17**, 072511 (2010)
806. W. Wang, T.S. Hahm, S. Ethier, P.H. Diamond et al., Phys. Plasmas **18**, 042502 (2011)
807. Z.X. Wang, L. Wei, X.G. Wang, Y. Liu, Phys. Plasmas **18**, 050701 (2011)
808. L. Wang, P.H. Diamond, Phys. Rev. Lett. **110**, 265006 (2013)
809. H.Q. Wang, G.S. Xu, B.N. Wan, S.Y. Ding et al., Phys. Rev. Lett. **112**, 185004 (2014)
810. A. Ware, Phys. Rev. Lett. **25**, 916 (1970)
811. T.-H. Watanabe, H. Sugama, Nucl. Fusion **46**, 24 (2006)
812. G. Wentzel, Zeits. f. Physik. **38**, 518 (1926)
813. L. Wei, Z. Wang, Nucl. Fusion **51**, 123005 (2011)
814. J. Weiland, A.B. Jarmen, H. Nordman, Nucl. Fusion **29**, 1810 (1989)
815. J. Weiland, *Collective Modes in Inhomogeneous Plasma* (IOP, Bristol, 2000)
816. H. Weisen, K. Appert, G.G. Borg, B. Joye et al., Phys. Rev. Lett. **63**, 63 (1989)
817. R. White, E. Fredrickson, D. Darrow, M. Zarnstorff et al., Phys. Plasmas **2**, 2871 (1995)
818. R. White, *The Theory of Toroidally Confined Plasmas* (Imperial College, London, 2006)
819. R.B. White, *Asymptotic Analysis of Differential Equations* (Imperial College, London, 2010)
820. R. White, Phys. Plasmas **20**, 0422105 (2013)
821. R. White, Phys. Plasmas **20**, 042116 (2013)
822. D.G. Whyte, A.E. Hubbard et al., Nucl. Fusion **50**, 10505 (2010)
823. H. Wilson, J. Connor, R.J. Hastie, C.C. Hegna, Phys. Plasmas **3**, 248 (1996)
824. H. Wilson, P.B. Snyder, G.T.A. Huysmans, R.L. Miller, Phys. Plasmas **9**, 1277 (2002)
825. A. Wingen, T.E. Evans, K.H. Spatschek, Nucl. Fusion **49**, 055027 (2009)
826. N. Winsor, J.L. Johnson, J.M. Dawson, Phys. Fluids **11**, 2448 (1968)
827. E.T. Wittaker, G.N. Watson, *A Course of Modern Analysis* (Cambridge University Press, Cambridge, 1962)
828. L. Woltjer, A theorem of force-free magnetic fields. Proc. Natl. Acad. Sci. **44**, 489 (1958)
829. K. Wong, R.J. Fonck, S.F. Paul, D.R. Roberts et al., Phys. Rev. Lett. **66**, 1874 (1991)
830. K. Wong, Plasma Phys. Controlled Fusion **41**, R1 (1999)
831. C. Wong, J. Wesley, R. Stambaugh, E.T. Cheng, Nucl. Fusion **42**, 547 (2002)
832. K.L. Wong, S. Kaye, D.R. Mikkelsen et al., Phys. Rev. Lett. **99**, 135003 (2007)
833. K.L. Wong, S. Kaye, D.R. Mikkelsen et al., Phys. Plasmas **15**, 056108 (2008)
834. Y. Xiao, Neoclassical Polarization, Ph.D. thesis, Department of Nuclear Science and Engineering, MIT, 2006
835. W.W. Xiao, X.L. Zou, X.T. Ding et al., Phys. Rev. Lett. **104**, 215001 (2010)
836. W.W. Xiao, P.H. Diamond, W.C. Kim, L.H. Yao et al., Nucl. Fusion **54**, 023003 (2014)
837. G.S. Xu, B.N. Wan, M. Song, J. Li, Phys. Rev. Lett. **91**, 125001 (2003)
838. G.S. Xu, B.N. Wan, H.Q. Wang, H.Y. Guo et al., Phys. Rev. Lett. **107**, 125001 (2011)

839. M. Xu, G.R. Tynan, P.H. Diamond et al., Phys. Rev. Lett. **108**, 245001 (2012)
840. S. Yajima, J. Hayashi, M. Omori, K. Okamura, Nature **261**, 683 (1976)
841. T. Yamamoto, T. Imai, M. Shimada, N. Suzuki et al., Phys. Rev. Lett. **45**, 716 (1980)
842. V. Yankov, JETP Lett. **60**, 171 (1994)
843. S.W. Yoon, J.W. Ahn, Y.M. Jeon et al., Nucl. Fusion **51**, 113009 (2011)
844. Y. Yoshioka, S. Kinoshita, T. Kobayashi, Nucl. Fusion **24**, 565 (1984)
845. M. Yoshida, Y. Kamada, H. Takenaga et al., Phys. Rev. Lett. **100**, 105002 (2008)
846. M. Yoshida, Y. Sakamoto, H. Takenaga et al., Phys. Rev. Lett. **103**, 065003 (2009)
847. M. Yoshida, T. Tanabe, A. Adachi et al., J. Nucl. Mater. **438**, S1261 (2013)
848. S. Yoshikawa, Phys. Fluids **7**, 278 (1964)
849. G.S. Yun, W. Lee, M.J. Choi et al., Phys. Rev. Lett. **107**, 045004 (2011)
850. G.S. Yun, H.K. Park, W. Lee, M.J. Choi et al., Phys. Rev. Lett. **109**, 145003 (2012)
851. L.E. Zakharov et al., Plasma physics and controlled nuclear fusion research 1978, vol. 1 (IAEA, Vienna, 1979), p. 689
852. M. Zarnstorff, S.C. Prager, Phys. Rev. Lett. **53**, 454 (1984)
853. M. Zarnstorff, M.G. Bell, M. Bitter et al., Phys. Rev. Lett. **60**, 1306 (1988)
854. M. Zarnstorff et al., Plasma physics and controlled nuclear fusion research 1988, vol. 1 (IAEA, Vienna, 1989), p. 183
855. M. Zarnstorff, K. McGuire, M.G. Bell, B. Grek et al., Phys. Fluids **B2**, 1852 (1990)
856. N.I. Zaytsev et al., Radio Eng. Electron. Phys. **32**, 454 (1974)
857. Y.Z. Zhang, S.M. Mahajan, Phys. Lett. A **157**, 133 (1991)
858. K.J. Zhao, T. Lan, J.Q. Dong et al., Phys. Rev. Lett. **96**, 255004 (2006)
859. W.L. Zhong, X.L. Zou et al., Phys. Rev. Lett. **111**, 265001 (2013)
860. W. Zhu, S.A. Sabbagh, R.E. Bell, J.M. Bialek et al., Phys. Rev. Lett. **96**, 225002 (2006)
861. P. Zhu, C.C. Hegna, Phys. Plasmas **15**, 092306 (2008)
862. P. Zhu, C.C. Hegna, C.R. Sovinec, Phys. Plasmas **19**, 032503 (2012)
863. G. Zhuang, K.W. Gentle, B. Rao et al., Nucl. Fusion **53**, 104014 (2013)
864. S.J. Zinkle, A. Moslang, T. Muroga, H. Tanigawa, Nucl. Fusion **53**, 104024 (2013)
865. F. Zonca, L. Chen, Phys. Rev. Lett. **68**, 592 (1992)
866. F. Zonca, L. Chen, Phys. Fluids **B5**, 3668 (1993)
867. F. Zonca, Continuum damping of Toroidal Alfven Eigenmodes in finite β Tokamak Equilibria, Ph.D. thesis, Princeton University, 1993
868. F. Zonca, S. Briguglio, L. Chen et al., Plasma Phys. Controlled Fusion **48**, B15 (2006) (Review Article)
869. H. Zushi, K. Nakamura, S. Itoh, K.N. Sato et al., Nucl. Fusion **39**, 1955 (1999)
870. H. Zushi, S. Itoh, K.N. Sato et al., Nucl. Fusion **40**, 1183 (2000)
871. H. Zushi, S. Itoh, K. Nakamura et al., Nucl. Fusion **41**, 1483 (2001)
872. H. Zushi, S. Itoh, K. Hanada et al., Nucl. Fusion **43**, 1600 (2003)
873. H. Zushi, K. Nakamura, K. Hanada et al., Nucl. Fusion **45**, 656 (2005)

Index

M. Kikuchi, M. Azumi, *Frontiers in Fusion Research II*,
DOI 10.1007/978-3-319-18905-5

C

D

P

Q

R

S

W

Z

MIX
Papier aus verantwortungsvollen Quellen
Paper from responsible sources
FSC® C105338

If you have any concerns about our products,
you can contact us on
ProductSafety@springernature.com

In case Publisher is established outside the EU,
the EU authorized representative is:
Springer Nature Customer Service Center GmbH
Europaplatz 3, 69115 Heidelberg, Germany

Printed by Libri Plureos GmbH
in Hamburg, Germany